# FOREST RESOURCE ECONOMICS

G. ROBINSON GREGORY
UNIVERSITY OF MICHIGAN

JOHN WILEY & SONS
New York Santa Barbara London Sydney Toronto

ISBN 0 471 06833-0

Library of Congress Catalog Card Number: 75–176158

PRINTED IN THE UNITED STATES OF AMERICA

10 9 8 7 6 5 4 3

*To*

*Ann, Bonnie, Rob and Shari*

# Preface

Use of forest land and of its many products is changing so rapidly that the forester who proudly received his degree a few years ago now questions the adequacy of his professional education. The forester, thoroughly grounded in the physical and natural sciences, finds that many of his most difficult problems stem from economic causes. To traditional strengths in biology, therefore, must now be added a knowledge of economic principles and of our economic society.

This text has been written to help forestry students and practicing professionals apply the tools of economic analysis to today's and tomorrow's forestry problems. It is a study in *applied* economics, aimed at integrating forestry with our economic system, and should serve as the basis of a course in forest resource economics. It is assumed that the course will follow completion of the now usually required "Economic Principles" sequence.

Part I provides a basic review of economic principles within a timber production context. Mathematical treatment is neither emphasized nor avoided. Graphs are generously used in exposition, and the mathematic and graphic developments are usually made directly comparable. Timber products are used as examples almost exclusively in this review to avoid the complications (introduced later) caused by non-market products such as recreation, water, and wildlife. Classes already possessing a solid foundation in basic theory are advised to begin directly with the timber products section.

Major markets for timber products are analyzed in Part II in terms of demand, production, supply, and price formation, using historical data throughout. The primary products are covered in detail so that the text alone will provide the required background, thus freeing class time for discussion and for the introduction of new developments.

Part III develops economic guides to timber production, purposely ignoring traditional lines between forest management and forest economics, with the result that a firm economic base is provided upon which a modern timber management can be built. The problem of rotation determination is discussed as a topic in capital theory; public and private objectives of timber management are placed in an economic framework. Topics such as interest, taxes, and credit are discussed within a national rather than a strictly forestry setting. A "theory of stumpage price" is developed from derived demand theory, making possible a much-needed integration of product demand with stumpage supply and virtually forcing recognition of the two-way interaction of (for example) stumpage and lumber prices.

Part IV of the text provides an economic approach to products other than timber—products such as recreation, water, and wildlife. An introduction to benefit/cost analysis, to welfare economics, and to public production will, it is hoped, introduce these elements into the more traditional framework of forest economics.

This book, like almost every text, draws on the work of hundreds who have brought the contributing disciplines to their present stage of development, and few of these are given specific recognition. I do, however, want to acknowledge the leadership of Henry Vaux, who had the unenviable task of guiding me through my formal training in economics, and who undoubtedly initiated many of the ideas developed in these pages. I want also to thank Kenneth Davis; as both colleague and friend, he encouraged me in this initial textbook effort and, over many hours of give-and-take debate, did much to clarify my thinking.

I have received much help from my graduate students here at the University of Michigan. To John Gray and Frank Shirley I owe a special debt for their careful review of the chapters on capital theory and taxation. Hans Gregersen was particularly helpful with the final chapter concerning economic development, and, with Beverly Driver, also contributed to the recreation chapter. Per Endsjo and my daughter, Shari, drafted many of the illustrations. My friend of many years, Irving Holland, of the University of Illinois, reviewed the entire manuscript and suggested many improvements.

I also owe a very large debt to Dean John Zivnuska of the University of California. His careful review and numerous comments resulted in many revisions and additions, some of a very substantial nature.

G. Robinson Gregory

Ann Arbor, Michigan
February, 1972

# Contents

## PART IV • NON-TIMBER PRODUCTS ECONOMICS

# FOREST RESOURCE ECONOMICS

# 1

# Introduction: Forestry in an Economic Environment

## STRAWS IN THE WIND

Not long ago I was introduced to a building contractor. Since he knew I was a forester it was only natural that he soon turned the conversation to construction, and particularly to the changing way in which wood is used in building family homes. I commented on how the natural beauty of wood was being used to advantage for exteriors as well as for interiors of houses, mentioning redwood siding, the use of wany-edged hardwood for rustic homes, the possibility of pre-painted panels, and so on. But the contractor took a different approach. "Speaking of siding," he said, "what do you think about the growing use of aluminum? You know, it's now being pre-finished with an enamel guaranteed to last twenty years. You simply wash it off with a hose and it looks like new. Will aluminum take over the siding market?"

I muttered that one could never be sure about changing tastes in a dynamic world, then thought of something more specific. "When aluminum window sash first came out, many people thought wooden windows would become a thing of the past. Then housewives discovered that aluminum sash became coated with ice during cold weather, and the windows rattled in the winter winds. Manufacturers of wooden windows got busy and designed windows treated against stain and decay, which could be slipped out for washing, and still held all the natural advantages of appearance, low heat transfer, and stability to temperature change. Now wood has recaptured virtually all the window sash market. Storm windows are no longer made of wood, because here aluminum has genuine advantages. But if the wood products people get on the ball there is little reason to believe the siding market will disappear."

The contractor wasn't convinced. "You may be right," he said, "but I heard that the largest builder of prefabricated homes in the country is going to feature the 'All Aluminum Home' this year! Anyway—what's your opinion of the new stuff—aluminum foam, or whipped aluminum? You can plane it, saw it, or nail it up—all with regular carpenter's tools. It's light, never rusts, takes paint well, and its heat conductance is relatively low. And what about plastics, and ceramic tile? How about the trend to glass walls? And on another front: are we going to become a nation of suburban dwellers, commuting to work from green belts? Or will we again follow our European ancestors and decide that country living is too costly in time—go instead to high-rise apartments and learn to live without the worry of mowing lawns, spraying elms, etc? If it's to be suburbia I'll bet on good lumber markets for a little longer, but if we shift to apartment living—look out. High rise apartments don't use much lumber!"

We were interrupted at this point so I didn't have to answer the contractor's questions. But that night I slept badly. I dreamed I was struggling through oceans of aluminum foam, climbing mountains of colored plastic, then wandering through miles of green belts into a forest—of high-rise apartments. I woke up when I heard a voice asking "Where does the forester fit in *this* world?"

I wonder! Do foresters know the role of wood in our modern society and the importance of wood in world markets? Are foresters aware that these markets are vital to the profession they have studied? Will the future forester be concerned with timber production, or will mankind's need for wood parallel his need for buggy-whips—and disappear as technology finds new and better materials?

Let's see what the wind brings from a different quarter. During the late 1950's and early 1960's the newspapers and other periodicals carried stories about the relative significance of "tight money" and "easy money" policies. In 1969 interest rates hit an all time high. Most people, reading headlines of "Federal Reserve Raises Interest Rate" probably thought "That's high finance and doesn't concern me." Those who were planning to buy a house or a new car may have wondered if the local finance company would increase monthly payment requirements. But how many foresters realized that a one per cent increase in interest rates could cancel out all the financial gain produced by a 25 per cent increase in yield from a plantation on a forty-year rotation? Did lumbermen realize that when interest rates jumped from six to eight and one-half per cent, monthly payments on a $20,000, twenty-year mortgage would rise by 21 per cent, which might so reduce building (and lumber sales) that their profits would drop more than they would under the impact of a ten per cent wage increase?

One more straw: the next time you visit a State or National forest—or for that matter, a private one—ask the forester what products he is expecting to take from the plantations he has established during the past five years. The most likely, and eminently sensible, answer is "Sawtimber, and maybe piling, with the thinnings going into pulpwood." Then put the question in broader terms and ask "What kind of forest do you think this nation will need in the year 2000?" Many foresters will reply, very bluntly, "Anyone trying to look that far into the future is crazy!" Yet the plantation established in 1970 or 1975 could just be coming up for a harvest cut in the year 2000. Will it then provide the kinds of products needed by the nation? Will the country require sawtimber and pulpwood—the products toward which most of our forest efforts are pointed today? Or will lumber and paper follow kerosine lamps down the road called "obsolete"?

## A FORESTRY PARADOX

Trees and forests change slowly. Even today most principles and methods of forestry are studied by observing the results of past treatments initiated by prior generations of foresters. Perhaps as a consequence we tend to be a conservative group: the forest based industries are tradition-oriented, and as a profession we are similarly inclined.

A paradox exists because forestry, perhaps more than any other profession, demands a willingness to act for the *future*. Every time a seedling is planted it is evidence of an implicit belief that the nation will need that tree when it matures. This means looking ahead a good many years—a paradoxical requirement indeed for a tradition-oriented profession. The silvicultural action taken today has little effect on this year's wood quality or quantity, but it largely determines the forest of thirty, fifty, or seventy years from now. What kind of forest should this be? Should it be primarily a source of wood or of recreation? Or will forests then be managed primarily for watershed protection, or for "open space," or even for use as a natural air purifier?

These questions, just as those raised by the contractor, the news stories on interest rates, or the forester asked to look ahead to the year 2000, illustrate the significance of *change* to forestry. New products, new services, and new ways of doing things are developing so rapidly that one can only be certain that many products and methods deemed valuable today will be worthless by the next decade—or next year.

Not long ago employment officials looked for their particular brand of "organization man." Today the progressive counterpart searches for the "change seeker"—an individual who will consciously seek and find, then conscientiously introduce, new concepts and new ideas into his organization. Some degree of conformity may still be required in areas of personal conduct and social behavior, but conformity in production, sales, or research—or even in administration—is no longer an asset.

For many years the forestry profession has been accused of being overly concerned with growing trees. Foresters have been urged to become "product oriented," and a great deal of professional effort is being spent today to increase the efficiency of timber production. As a matter of fact, much of this text is directed toward this same objective of efficient timber production, for it would be misleading—even foolish—to ignore the importance of timber today or in the coming decade.

But it is absolutely essential for forestry students to recognize that within the time context of forestry an overemphasis on product orientation could spell our professional demise, just as it has for many other industries of the recent past. In the early 1900's the continued success of railroads was unquestioned. Train schedules were improving, dining car and pullman service had been inaugurated, freight transfers were becoming more efficient, and train speeds were constantly increasing. But consumers proved to be less interested in trains than in transportation, and the railroads have come upon troubled times. Automobile manufacturers show little evidence of having learned the obvious lesson and may someday find themselves in a similar situation. For example, how will the "Big Three" respond to increasing pollution controls?

Forestry can no more afford undue emphasis on products than could the railroads. Few consumers really care whether the water, the board, or the paper they use comes from the forest or elsewhere. As Joseph Fisher observed (Fisher, 1969)[1]:

> This is a chastening thought for the forestry profession and a protection against the arrogance that there is something absolute, permanent, and ultimate about the work they do and the trees they love. It is just barely possible that trees for lumber will go the way of whalebone for corset stays or whale oil for lamps—not likely, perhaps, but possible.

Biologists know that organisms which fail to adapt to a changing environment soon become extinct. In this respect professions are no different from biological entities. Today's forestry needs to emphasize its forward-looking aspects—to look beyond the biological facts of timber production and seek an understanding of forests as they relate to people. This in no way minimizes the biological foundations of our profession: it only means that biology is not enough for a modern professional forester.

## WHY FOREST ECONOMICS?

The biological aspects of professional forestry education are a necessary basis for understanding how trees grow, and in addition lay a foundation for discovering means of increasing tree growth and quality. But if one is to understand *why* trees should be grown—if wise decisions are to be made about how and why forest resources are allocated and used—then economics must be brought into the picture.

Economics is the science which deals with allocating scarce resources among competing means to satisfy the wants of mankind. When forests could be regarded as the natural use of any and all land that was not needed for agriculture or grazing, the forester could remain relatively aloof from the every-day world and mind his

[1] References to the bibliography on pages 533–39 will be cited in this manner throughout the text.

own business of growing trees. Now the important forces shaping forestry are coming from outside the profession, and among the most important of these are economic forces.

If forestry is to contribute its full share to a more abundant life for the world's increasing population, foresters must become familiar with the economic and political forces that determine much of our way of life. We must understand, and be able to evaluate, the significance of the inevitable changes through the coming years. This means we must develop competence in applying the basic principles of economics to the production of all the goods and services forests can yield—a competence at least equal to that possessed by other production managers.

When increasing incomes meant bigger and better wooden houses the lumberman needed to worry only about whether he would have enough timber to supply the hungry saw. But when larger incomes mean houses made of steel, glass, and brick instead of wood, the lumberman should give serious thought to markets. When multiple use meant permitting other land uses whenever and wherever they didn't interfere with timber production the timber-trained forester could continue down the path established by his European counterpart of the 1800's. But when forest camping becomes financially feasible to 80 per cent of the nation's families, when demands for municipal water take precedence over wood production, and when the per acre income from hunting rights begins to rival that from timber sales, then multiple use requires more than lip service: it requires a careful appraisal of economic alternatives.

There is no special brand of economics that one must learn as a forester. The same principles of supply and demand, of production and distribution, that apply to manufacturing automobiles, Coca-Cola, or Scotch whiskey, also apply to recreation, water, forage, wildlife, and timber. Sometimes the application calls for a little ingenuity (and sometimes direct application becomes downright difficult) because of peculiarities of the forest production process or of forest products, but the basic economic *principles* remain the same. And the principles of production and distribution, of supply and demand apply just as surely in non-capitalistic countries as in those for which the profit motive supplies the driving force (though application of the results may be quite different).

This should not be interpreted as meaning that there is no need for "*forest* economics" or for "*forest* economists." It simply means that forest economics is an applied field, and that the major problems in the field arise from application difficulties rather than from theoretical ones. There remain, of course, problems of theory, usually the kind one expects to encounter whenever a general theory is extended to a very specific area. Forest economics deals with those economic problems involved in owning, buying, selling, taxing, and managing forest land, whether it is used for producing water, wildlife, wood, or some other product. It is also concerned with the economic problems of growing, protecting, harvesting, and marketing the products of that land, and with examining the interplay of forces that determine the product or group of products a particular forest should produce. There is no more reason for the forestry profession to have specialists in forest economics than to have specialists in forest mensuration, forest genetics, forest recreation, forest ecology, forest entomology, forest zoology, watershed protection, or forest management—and there is no less. The need of the modern forest manager for economic knowledge is certainly no less than his need for a knowledge of ecology, mensuration, recreation, or photogrammetry.

The justification of forest economics as a specialization distinct from the general field of economics lies primarily in the detailed knowledge that the peculiarities of forestry require the general economist to acquire. The approach to, and use of, economic concepts and principles in the field of forestry must always be tempered by the biological nature of the resource itself and by the time span involved in forest production. Few would argue that it would be impossible for an economist to obtain such knowledge, but the number willing to spend the time required to gain familiarity with the nature of forest production is small. Economists specializing in industry tend toward banking, manufacturing, or marketing. The necessity to economize our use of natural resources has not yet been universally accepted, while the non-economic approach of early conservationists (and of some groups today) could scarcely serve to attract economists to forestry in a role other than that of critic.

The fact that time must be brought very specifically into the forest production picture has both good and bad features; time is one of those variables that is more often ignored in economic analysis than it is given explicit recognition. In forestry the importance of time simply cannot be ignored, so here we may be able to bring up some problems not ordinarily encountered in economics. From the standpoint of intellectual stimulation this should be an advantage. But time has a way of complicating economic analysis. One of the most famous of all classical economists, Alfred Marshall, wrote of time as ". . . the center of the chief difficulty of almost every economic problem" (Marshall, 1947).

## PURPOSE OF ECONOMIC ANALYSIS

Is there any single, central purpose or objective toward which we might say our economic analysis will be pointed? In a very general way the answer can be "Yes," for we will be striving to determine the optimum allocation of the scarce forest resources, and of all those other resources which must be used in the utilization of forests, among the competing means by which mankind's wants can be met.

But optimum from whose standpoint? An owner might want to know the optimum harvest pattern for his particular forest. To answer this requires that we first know the goal of the owner. Even after learning that his primary interest is timber production (and it could as well be deer, or upland game, or any of a variety of products) it would be necessary to spell out his objective more carefully. If he wants to maximize his cash income from timber over a three-year time span, then a particular harvest schedule could be established; to maximize cash income over a twenty-year period would probably call for a quite different cutting schedule, while maximization of the property's present net worth could require a still different program. And clearly, the objective of a private owner might be very different from that of a public agency such as the U. S. Forest Service or a State Conservation Department, and these again different from the objectives of a large paper corporation or the National Forest Service of Liberia or Mexico.

The purpose of economic analysis will be interpreted here as one of analyzing the various economic quantities involved and giving the owner or the manager—the decision maker—the results of following alternative courses of action. The manager will then be expected to make his decision—quite frequently as much, or even more, on non-economic grounds than on economic. He may well, for example, base his decision primarily on political reasoning, but the economist still has the responsibility to point out the results of his selected alternative from an economic point of view.

The decision to adopt or not adopt water pollution controls is usually made through the political process. Yet the results of economic analysis should be available to the decision makers whether they happen to be congressmen, state legislators, or the voting public. Substantial costs in terms of employment may be involved, and large economic benefits as well. The determining and weighing of these costs and benefits is an important analytical task for the resource economist.

In an almost parallel fashion one might argue that it is not the forest economist who should decide whether a particular area should be devoted to forest camping, to use as a wilderness, or to the production of sawtimber. The task of the forest economist is to analyze the costs involved in the various alternatives and to determine the returns that will follow a decision to take a particular course of action. The results of his analysis should be made available to the decision maker, perhaps in a simple presentation such as a series of benefit/cost ratios, or (probably more often) in detailed analyses that spell out not only the dollar costs and returns and to whom they accrue, but also the costs and returns to individuals and to society that cannot be valued in present-day markets, and that describe the opportunities for future action that may be lost if certain of the alternatives are elected. An economic analysis should present those facts of an economic nature that will aid in making an intelligent choice between alternative courses of action.

## ECONOMIC MODELS

Economic analysis does not pretend to consider all the variables that are important. In fact, in most cases it is impossible to consider even all of the *economic* variables that are involved. Economic analysis aims at producing a map of the realistic situation, not a photograph, for only by abstracting from reality can the interactions of the key variables be worked out. This is probably best seen in the process of "model building," an exercise we will be attempting at several stages in the very near future.

An economic model is constructed by making certain basic assumptions, then fitting into this framework the particular situation to be analyzed. Economics, and particularly economic theory, is frequently criticized as being "abstract," and many people consider economic models overly theoretical. Yet one should recognize that it is precisely this abstraction which makes the model a powerful analytical device. The ability to reason abstractly is man's unique intellectual achievement, as well as his most powerful tool for advancing mankind. This fact is well recognized in all of science, and the process of model building is certainly not confined to economics. An example from physical science may assist in establishing several points.

### A Model from Astronomy

Suppose we wanted to construct a model to "explain" in some way the motion of the planets as they rotate about the sun. In order to keep the model simple, we will (a) represent each of these planets by a single point, and (b) assume that each point has a mass equal to that of the body it represents. Next we will say that each pair of points experiences a "mutual attraction"—an attraction that can be approximated by multiplying the mass of one point by that of another, then dividing that product by the square of the distance between the two points. In other words, we postulate a model of the following form:

$$F = \frac{M^1 \cdot M^2}{d^2}$$

where $M^1$ and $M^2$ = mass of planets 1 and 2
where $F$ = force of attraction
where $d$ = distance between planets 1 and 2

Finally, we will assume that the effect of any other variables is negligible.

Here is a simplified model of an extremely complex phenomenon. Yet this simple model was adequate for predicting the motions of the planets for some two and a half centuries. For the sake of scientific accuracy it must be noted that, after a lot of checking, some one discovered that after one hundred years the planet Mercury had drifted one fifth of a mil (a mil represents an angle such that it would intercept one foot at a distance of one mile) out of its predicted orbit. This, I suppose, only proves that the theory—the model—is pretty rough. Mr. Einstein explained this error.

It is also interesting to note that Sir Isaac Newton, the "inventor" of this model, first tested it by applying data others had gathered from observing the moon. Thereby he wasted several years of his life, because the data that astronomers had collected on the movements of the moon were inaccurate!

### Applicability to Forest Economics

There are a few lessons we should learn from this model, lessons that are directly applicable to the models we will be constructing and applying as we study forest economics. First: theories may be very simple even though the phenomena that they model are not. Anyone who has studied astronomy or navigation will know that the motion of planets is far from simple. Second: a model (or theory) can be very general without being sterile. The model given above applies to *all* bodies, not just the heavenly ones.[2] Third: although extremely accurate the model covers only *one* of the factors that may affect the motion of bodies. For example, the gravitational attraction between two airplanes flying in tight formation is roughly equivalent to the weight of a cigarette ash 1/16 of an inch long. Obviously there are a good many other things that the pilots of these airplanes must bear in mind when they are judging the motion of the two bodies for which they bear responsibility! Fourth: (as Newton so painfully discovered) there is a need for data if the model is to be tested, and that data must be reliable. We will discover as we proceed that one of the major problems of forest economics, and in fact of economics in general, is a lack of reliable quantitative data.

We will be doing quite a bit of this "model building" as we study the subject before us. Quite frequently we will begin with a set of assumptions that seem to be only distantly related to the real world. But through this procedure we may be able to discover what happens as we relax these assumptions—as we alter our model to something more closely approximating the world in which we live. We may even find that under certain conditions some of our most cherished assumptions must be displaced if we are to achieve a particular goal!

## PLAN OF APPROACH

Being an applied field it will be helpful to divide forest economics into several areas, each of which calls for a particular set of economic principles or body of theory, or which makes a logical unit purely from an application standpoint. These areas, three in number, are as follows:

[2] As a matter of fact, current research indicates that the same model even applies to a good many social phenomena.

1. Timber products
2. Timber production
3. Multiple use and non-timber products

Along the way a good many guides to management decisions will be developed, and occasionally insights into forest policy may be gathered.

The initial concentration on timber could be misleading, for it might convey to some students an impression that forestry and timber production are synonymous (at least in the mind of the author). There are, however, several reasons for accepting this risk and beginning with timber product study. The basic economic principles are more easily illustrated by reference to timber than to other forest products; application of economic analysis to non-timber products of the forest—particularly to recreation, water, and wildlife—is much more difficult than to timber. To imply, for example, that demand for forest campsites and for fuelwood are equally simple to derive, or to quantify, or even to understand, would not only mislead but would be grossly incorrect. Moreover, quantitative illustrations are far less easy to find.

One could also argue that the first two sections should be reversed in their order of treatment, since timber production must chronologically precede its marketing. Timber, however, is the objective of management on many forest areas, and a knowledge of timber markets is essential to an understanding of the role of forestry in the world of today. Forests are managed, and products and services produced, to satisfy human wants. Throughout much of the world these wants are expressed in the market place, and the professional forester must appreciate how these markets work. The fact that many important forest products are not marketed (at least in the usual sense) only increases this need for understanding, for unless the function and operation of markets is properly understood one can scarcely hope to comprehend the difficulties inherent to non-marketed products such as forest recreation, esthetics, wildlife, or even water.

One further introductory note may be helpful in explaining the organization of this volume. The text is predicated on the assumption that the student—whether he is presently a junior or senior in college, or has been working professionally for the last thirty years (and some of the best students of forestry are included in this latter category)—has completed an introductory course in "Principles of Economics." We will begin, therefore, with a review of the basic tools of economic analysis. From time to time as this review proceeds various more advanced analytical concepts, not ordinarily encountered in the usual introductory course, will be introduced. Immediately following this section we will go directly to the many problems of application that make up the fascinating field called "Forest Economics."

# PART I

# REVIEW OF ECONOMIC PRINCIPLES

# 2

# Market Structure and Demand

## "MICRO" AND "MACRO" ECONOMICS

Economics is usually divided into two major parts, *micro*economics and *macro*economics. *Microeconomics* deals with individual firms and industries, and with individual products, prices, and production systems. It is here that concepts such as demand, supply, and marginal analysis are of greatest usefulness. *Macroeconomics* deals with aggregates; with the output of the whole economic system rather than with that of a single industry; with the general price level instead of with the price of an individual product; with national employment levels rather than employment by an industry. Macroeconomics is therefore particularly significant from the standpoint of economic policy.

Forest economics is an applied field, primarily concerned with applying the tools of economic analysis to problems of production, demand, supply, and pricing of recreation, water, stumpage, lumber, plywood, fuelwood, wildlife, and similar products, and with problems of the industries that produce them. Necessarily, then, the major focus of this text and of the review will be within the area of microeconomics.

But there are important aspects of macroeconomics that must also be understood. Many factors that influence economic decisions in forestry are generated outside the forestry sector itself—factors such as interest rates, construction activity, wage levels, credit sources, and tax levels. Moreover, policy decisions arising in public forestry often involve major problems of aggregation. And clearly, in some nations (as well as in some states or regions of this country) the level of forestry activity can have an important bearing on national employment and even on national income levels. Specific aspects of macroeconomic theory, then, will be introduced whenever the need arises.

## ORGANIZATION OF MICROECONOMIC THEORY

Supply, demand, price, profit, competitive structure—all are so closely related that regardless of the selected starting point there will be occasions when a different one would seem preferable. Micro theory might be regarded as a wheel whose rim has sections blending into one another, labeled with the terms "competitive structure," "supply," "demand," "production," "price," and "distribution." The spokes of the wheel should be regarded as interconnecting pathways extending from a hub called "micro theory."

If we are to see how this wheel is assembled and how it functions we obviously have to start someplace. And regardless of the place we choose to begin it is important to hold in mind the unity of the system of thought referred to as "micro theory," or the "theory of the firm." We will begin with a brief recapitulation of industrial competitive structure, then review the major concepts of demand, production, and supply, and finally tie these together with a review of pricing and distribution. To avoid complications introduced by nonmarket products such as wildlife and recreation, only timber products will be used as examples in this review section. Later, after applying these tools to the analysis of major timber products and to timber production, the more difficult (but in many ways more interesting) problems encountered in analyzing other kinds of products will be tackled.

### Market Structure

We will begin with a look at the particular market structure called "pure competition." Often derided by so-called "practical" men as being completely divorced from reality, pure competition provides a very simple model which can be made more complicated through step-by-step alterations, additions, and modifications. It is an absolutely necessary starting point for anyone hoping to achieve an understanding of economic reality.

Economic theorists usually attach many conditions to "pure competition," but for nearly all work in forest economics three major conditions will be sufficient to identify a purely competitive situation:

1. *Large number of producers and buyers:* The number of firms and the number of buyers within the market area concerned must be large enough so that no one firm or group of firms, and no one buyer or group of buyers, can significantly influence the product price by changing the quantity produced or purchased.[1]
2. *Product is undifferentiated:* That is, the product of one firm must not be recognized by buyers as being significantly different from that produced by other firms. A southern pine 2 × 4, for example, produced by any one of several thousand sawmills, is equally desirable to the average buyer.
3. *No barriers to entry or exit:* There can be no real or artificial barriers to firms wishing to enter or leave the market. In effect this means no controlling patents, no secret processes that significantly affect costs, and a relatively low required capitalization.

Market structure is actually a continuum, with pure competition on one end of the scale and pure monopoly on the other. In "monopoly" a single firm produces the product—a situation difficult to

Pure Competition — Monopoly

maintain in modern industrial societies today. Other points on the continuum have been singled out for study by economists, particularly:

[1] This condition is sometimes stated in terms of the "concentration ratio," the ratio of the output of the largest firm (or acting group of firms) to total production of the industry. If one firm produces, say, 25 per cent of the total output, then this is plainly not a purely competitive industry. Nor would it be if four firms produced 40 per cent or 50 per cent of the total. There is, however, no universal agreement on just where to place the limit of the ratio and still have the industry classified as being purely competitive. Parallel statements are possible with respect to buyer concentration.

*Oligopoly:* a few firms, any one of which produces enough of the particular product to influence product price.

*Monopolistic competition:*[2] Chamberlin's term for an industrial structure wherein products are somewhat differentiated so that the producer has some, usually quite limited, influence on price (Chamberlin, 1933).

Oligopoly has been called the characteristic market structure of the United States, with the automobile industry providing an oft-cited example: each firm can influence the price of its own product, and in so doing will probably also influence the sales of the other firms. Because price competition is costly to oligopolistic producers, once a price pattern has become established in such industries the prices tend to be "sticky." Producers prefer to "let well enough alone"—competition tends to be expressed less by price and more by the services offered along with the product, and (perhaps primarily) by competitive advertising. During the late 1960's, for example, much of the competition between international airlines was in advertising that stressed food and beverages, or the friendly smiles of hostesses—neither unimportant, to be sure, but scarcely a sound basis for choosing an airline!

Tooth paste provides a good example of monopolistic competition: the many varieties are distinguished by flavor, color, size, and shape of tube, various additives, as well as by their name. The degree of price control that the producer can exert depends essentially upon the degree of differentiation, either real or imagined, that can be induced in the mind of the potential buyer. Such industries, therefore, tend toward highly competitive advertising.

Other market structures sometimes identified are duopoly (two producers) and what Bain once called "oligopoly with a competitive fringe," a few large firms that produce most of the output, and a substantial number of much smaller firms that produce the rest (Bain, 1948). Such industries are characterized, price-wise, by price leadership: one of the larger firms, normally the largest, sets its own prices and the others follow. There is no prima facie reason to assume collusion in such cases, for no collusion is necessary: it's just common sense to let the big fellow set the price—and most successful businessmen have a lot of common sense.

[2] This same market structure is sometimes called "imperfect competition," the term used by Joan Robinson in her book covering much of the same ground as Chamberlin's (Robinson, 1934).

Monopoly, duopoly, oligopoly, etc., refer to market structure on the *sales* side: monopoly is a market form characterized by a single *seller* of the product. Buyer's markets can be described in a parallel fashion. A market characterized by a single buyer, for example, is called a *monopsony*, and the single buyer is a *monopsonist*. The terms *monopsony, oligopsony*, and *monopsonistic competition* describe markets characterized by varying degrees of buyer concentration just as their counterpart terms describe seller concentrations. Because forest land is immobile the actual number of potential buyers is often relatively small and buyer concentration can be an important factor in the pricing of primary forest products.

### The Demand Concept

Few areas could be better chosen to illustrate the "tyranny of words" than that of demand. Most businessmen equate the term with "sales"; an increase in the demand for lumber means an increase in the quantity of lumber sold. To the economist, however, demand has a very specific connotation. Demand is the functional relationship between the price of a given commodity and the quantity of that commodity that will be sold in a market specified as to time and place. Demand conceptualizes the buyer's response to price changes; it is a "buyer's response curve."

Being a functional relationship, demand can be described by a schedule, a graph, or an equation. Suppose, for example, a door-to-door survey revealed that the households in Woodstock would, under the anticipated conditions, consume the amounts of fireplace wood shown in Table 2.1.

**TABLE 2.1 SCHEDULE OF QUANTITIES AND CORRESPONDING PRICES OF FIREPLACE WOOD THAT COULD BE SOLD IN WOODSTOCK.**

| Price per cord (dollars) | Quantity sold (cords) |
|---|---|
| 18 | 10,000 |
| 16 | 12,000 |
| 14 | 14,000 |
| 12 | 16,000 |
| 10 | 18,000 |
| 8 | 20,000 |
| 6 | 22,000 |
| 4 | 24,000 |

This relationship could then be described by the data in the table, by the curve of Fig. 2.1 or by the equation:

$$P = 28 - .001Q$$

One of the initial distinctions to be made is that between *shifts* of the demand curve and *movements along* the demand curve. A

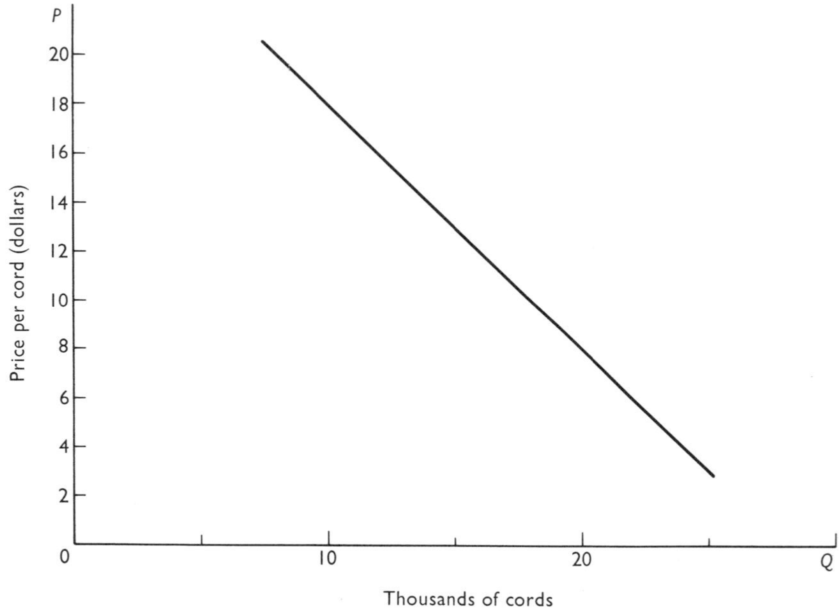


**Fig. 2.1** **Curve of the demand for fireplace wood in Woodstock.**

*shift* signifies a change in the position or shape of the curve itself, while *movement along* the demand curve describes the changing location of a point on the same demand curve. Shifts of a demand curve normally mean that, after the shift, a different quantity will be bought at every price. In contrast, a movement along the curve indicates the effect of a price change on product sales.

The demand concept is an instantaneous concept. Theoretically, every time a sale is made the curve should be reconstructed. Practically, however, time intervals are usually specified. Thus, one may meaningfully speak of the demand for (say) paper during 1950 or during 1980. When comparing the demand schedules for different time periods one usually speaks of demand "shifts"; demand curves shift from year to year or even from week to week.

Product specification causes real problems, particularly in quantitative research. We can be quite certain, for example, that the demand for black walnut is very different from the demand for southern pine or for redwood. And even within a product as carefully specified as "Southern pine construction lumber" one would find considerable variety, yet the phrase "demand for lumber" does make sense and is useful in analysis.

We also know that the price of most commodities is a function of many things in addition to the quantity of the commodity offered in the market. The price of butter would almost surely be influenced by the price of margarine, to a somewhat lesser extent by the price of bread, jams, and jellies, of rice and potatoes, and ultimately by the price of almost everything. Similarly, one would expect the price of hardwood flooring to be affected by the cost of installation, by the price of carpeting and asphalt tile, by the price of land upon which houses are built, etc. Generally these effects are held constant, in a conceptual sense, by the phrase *ceteris paribus* or by its approximate English equivalent "other things being equal."

The demand relationship as usually pictured graphically (Fig. 2.1) shows price as the dependent and quantity as the independent variables. Unlike many other economic expressions this relationship can be interpreted in two ways: (1) if the price is changed the quantity taken in the market will vary according to the curve shown; or (2) if the quantity offered in the market is changed, the price at which it will be taken will vary according to the schedule described. In other words, the demand relationship is "implicit."

The demand curve is also an average revenue curve for whatever part of the industry it represents. Given the market condition described by the particular curve, the unit price at any point $(p,q)$ will necessarily be the average revenue obtained for each unit sold as the number of units exchanged in the market is varied.[3]

## Elasticity

Given a demand curve such as that in Fig. 2.1, there will be some quantitative response in the quantity sold if price is changed. The relationship between the percentage changes in price and quantity is called "price elasticity" and may be precisely defined as:

[3] Total revenue, or $TR$, is clearly $P \cdot Q$. Average revenue must therefore be $\frac{TR}{Q}$ or $\frac{PQ}{Q} = P$ under the conditions given here.

$$E_p = \frac{\%\text{ change in quantity}}{\%\text{ change in price}}$$

Alternative formulations may be derived as:

$$\frac{\frac{\Delta Q}{Q}}{\frac{\Delta P}{P}} = \frac{\Delta Q}{Q} \cdot \frac{P}{\Delta P} = \frac{\Delta Q}{\Delta P} \cdot \frac{P}{Q}$$

where $\Delta Q$ and $\Delta P$ indicate, respectively, the change in quantity and the change in price.[4] If we now let $\Delta P$ and $\Delta Q$ tend to zero we can replace the ratio $\frac{\Delta Q}{\Delta P}$ with $\frac{dQ}{dP}$, and are left with the formula for point elasticity:

$$E_p = \frac{dQ}{dP} \cdot \frac{P}{Q}$$

Since demand curves are almost always negatively sloped it follows that elasticity will normally be negative. In practice, however, the sign is usually ignored and elasticity is said to vary from zero to infinity. If $E_p$ is less than one the curve is described as being *inelastic,* and if zero it is called *perfectly inelastic:* if elasticity is greater than one, then demand is termed *elastic,* and if $E_p$ tends toward an infinite value the curve is said to be *perfectly (or infinitely) elastic.* The special case of $E_p = -1$ is termed *unit elasticity.*

The elasticity concept has been particularly useful in empirical analyses of forest product demand and merits full understanding by foresters. The paragraphs which follow constitute considerably more than a simple review.

Elasticity is a pure number—a ratio of percentages—and completely free from any scale effect. Fig. 2.2 shows a hypothetical demand curve for Douglas fir lumber in Peru, with units converted to U. S. measurements.

As price falls from \$210 per MBM (thousand board feet) to \$200, the quantity sold in the market rises from 4,500 MBM to 5,000. The enlarged view permits a closer examination of the geometry involved. The change in quantity is $5{,}000 - 4{,}500 = 500$, so the per-

[4] Whenever the symbol $\Delta$ is employed it will signify a discrete change, while $\frac{dQ}{dP}$ for example will indicate the first derivative of $Q$ with respect to $P$, or the limit of the ratio $\frac{\Delta Q}{\Delta P}$ as $\Delta P$ approaches zero.

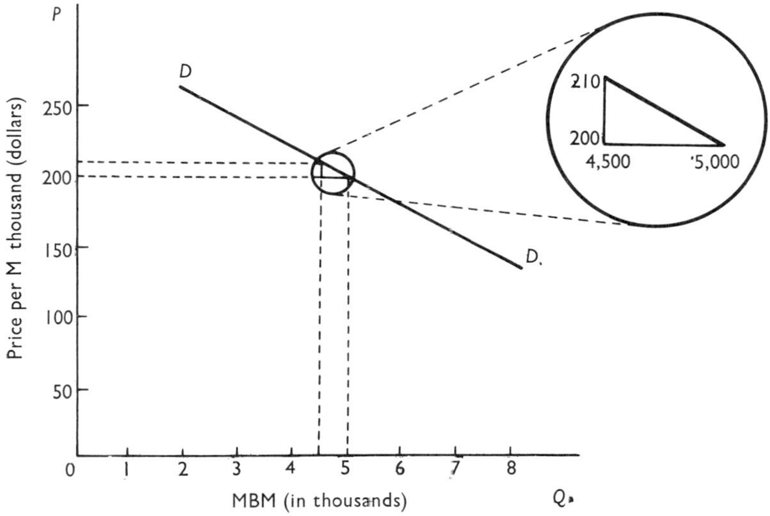


**Fig. 2.2** Arc elasticity.

centage change in quantity could be $\frac{500}{5,000} = 10\%$. Similarly, the corresponding percentage price change would be $\frac{-10}{200}$ or $-5\%$. The elasticity ($E_p$) might therefore be calculated as: $E_p = \frac{10}{-5} = -2.0$ (or 2.0).

Calculation of an "arc elasticity" such as this reveals a particular difficulty, for clearly, the percentage change might be based on either the beginning or ending values.[5] To eliminate this problem it is usual to make the averages of the two prices and quantities the bases for the percentage calculations. Consequently, arc elasticity becomes an "average" elasticity between the two selected points.[6]

[5] That is, the per cent change in quantity might be calculated as above, or as $500/4,500 = 11.11\%$. Similarly, the per cent change in price could be $-10/210 = -4.76\%$.

[6] In this example the calculations would be:

$$\frac{\frac{500}{4,750}}{\frac{-10}{205}} = \frac{10.53}{-4.88} = -2.16 = E_p$$

Boulding reduces this difficulty by defining elasticity as the percentage change in the quantity (demanded or supplied) which would result from a one per cent change in price (Boulding, 1948).

A more precise measure is obtained by permitting $p$ and $q$ to tend to zero, and treating elasticity as a point concept. $\frac{\Delta q}{\Delta p}$ then becomes $\frac{dq}{dp}$, or the reciprocal of the slope.[7] Hence, to calculate elasticity of demand with respect to price at any point on a demand curve one can simply multiply the reciprocal of the slope of the curve by the ratio $\frac{p}{q}$, using the values of $p$, $q$, and the slope at the desired point. Note the relationship between slope and elasticity: it is the *reciprocal* of the slope that must be multiplied by $p/q$. *Slope* and *elasticity* are not the same entity. In fact, for any straight line demand curve such as that of Fig. 2.2, *slope* is necessarily constant, but *elasticity* is different at every point on the demand curve. Since we have the equation for this particular demand relationship we can readily calculate the *point elasticity* at any point we wish. Demand being $P = 300 - .02Q$, the first *derivative* (or slope) is $-.02$. Point elasticity at a price of \$200 is $-\frac{1}{.02} \times \frac{210}{4,500} = -2.33$.

## Elasticity and Revenue

The relationship between average revenue, total revenue, and elasticity is of particular interest. It is illustrated in Fig. 2.3(a) and (b) with the Woodstock fuelwood demand curve (from page 24). This demand curve, $P = 28 - .001Q$, is also the average revenue curve, and is so labeled in Fig. 2.3(a). Total revenue ($TR$) we know is $P \cdot Q$, and therefore:

$$TR = P \cdot Q = Q(28 - .001Q) = 28Q - .001Q^2$$

This curve of total revenue has been drawn in Fig. 2.3(b), directly below the curve of average revenue. From the demand equation, elasticity can also be calculated directly and is given for three points on the demand curve.

The relationship between elasticity, average revenue, and total revenue is now apparent. The point $E_p = 1$ (unit elasticity) divides the demand curve into two significantly different portions. When

[7] This statement assumes the demand concept to be of the form $P = f(Q)$. For various reasons nearly all demand analyses are aimed at quantifying equations in which quantity is made the dependent variable, or equations of the form $Q = f(P)$. Since the demand relationship is implicit this does no violence to theory. In this latter form the derivative $dq/dp$ is obtained directly. The simple equation of the above example could be written $Q = 15,000 - 50P$, and $dq/dp = -50$, which is the same as the reciprocal of the slope of the equation $P = f(Q)$.

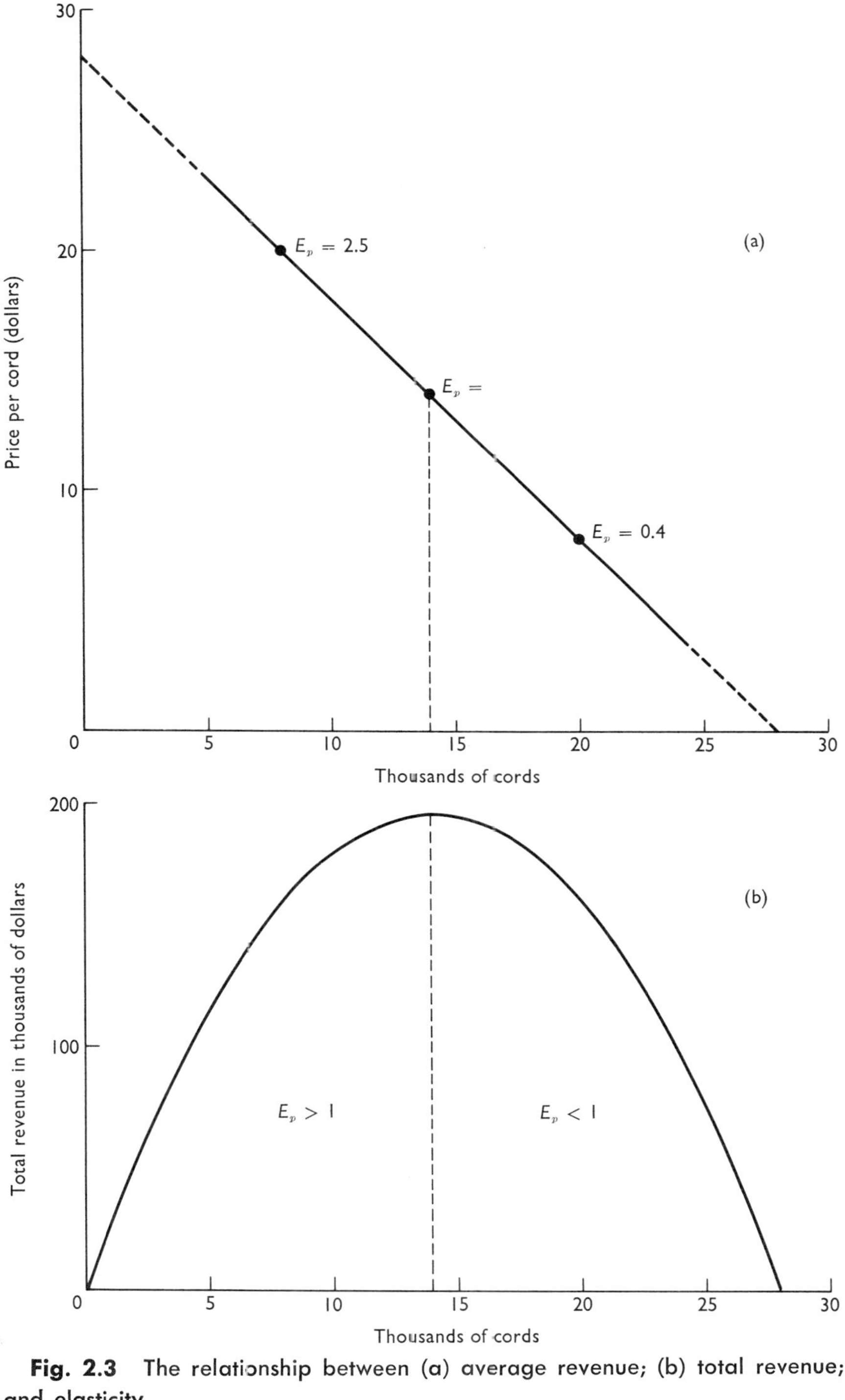


**Fig. 2.3** The relationship between (a) average revenue; (b) total revenue; and elasticity.

price is above this point any *lowering* of price brings increased sales and *increased total revenue*, as can be seen from the total revenue curve below. And below this point of unit elasticity, any further lowering of price brings about a *decrease* in total revenue. If with falling price, revenue increases up to this point and declines thereafter, then revenues must be at a maximum where $E_p = 1$.

This same relationship can be seen in a slightly different form from Fig. 2.4(a). Let price and quantity be, respectively, $p$ and $q$. Then $op \cdot oq$ (or $p \cdot q$) is the area of the rectangle $opaq$. But $p \cdot q$ also equals the total revenue. Hence the area of a rectangle such as

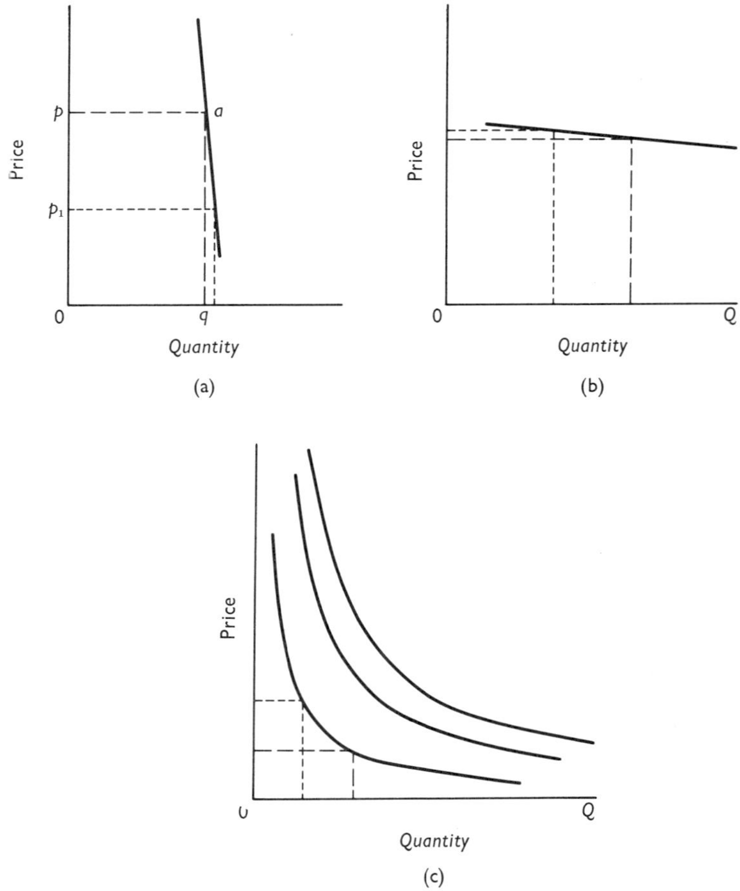


**Fig. 2.4** A curve (a) of relative inelasticity; (b) of relative elasticity; and (c) of unit elasticity.

*opaq*, inscribed under a demand curve (average revenue curve), is also the total revenue. Fig. 2.4(a) illustrates a demand curve that is relatively inelastic. Here a price rise *increases* total revenue—the area of the rectangle *opaq* is substantially larger than the corresponding one formed at the lower price $p_1$. When demand is inelastic, then, a price increase means an increase in total revenue to the industry.

Fig. 2.4(b) demonstrates the case of a relatively elastic demand. Quite the opposite of the preceding example, an increase in price will necessarily *decrease* the industry's total revenue.

Fig. 2.4(c) presents a series of demand curves, all of which have a constant unit elasticity. Such curves are called "rectangular hyperbolas"—all rectangles formed by connecting a point on the curve to the two reference points on the axes will have equal areas. With such demand curves the product $p \cdot q$ is a constant, and it follows that if $E_p = 1$, then total revenue cannot be affected by changing price.

The effect of changing price on total revenue under varying conditions of elasticity is summarized in Table 2.2.

**TABLE 2.2 EFFECT OF PRICE CHANGES ON TOTAL REVENUE UNDER VARYING CONDITIONS OF PRICE ELASTICITY.**

| | Effect on total revenue | |
|---|---|---|
| | **Price rises** | **Price falls** |
| Unit elasticity | No change | No change |
| Elastic | Revenue falls | Revenue rises |
| Inelastic | Revenue rises | Revenue falls |

## Income Elasticity

Price elasticity implies a *movement along* the demand curve, and is useful in describing the demand relationship and its implications with respect to price-quantity-revenue relationships. But we know that many things other than price play an important role in determining the amount of a commodity that consumers purchase. A change in any variable other than price implies a *shift* of the demand curve rather than a *movement along* the curve itself.

The effect of any third variable on the demand relationship can be shown graphically only if a third dimension is added to the demand picture. Fig. 2.5 illustrates one possible relationship between per capita income and the demand for industrial wood. As seen here, increased income results in an outward shift in the demand curve.

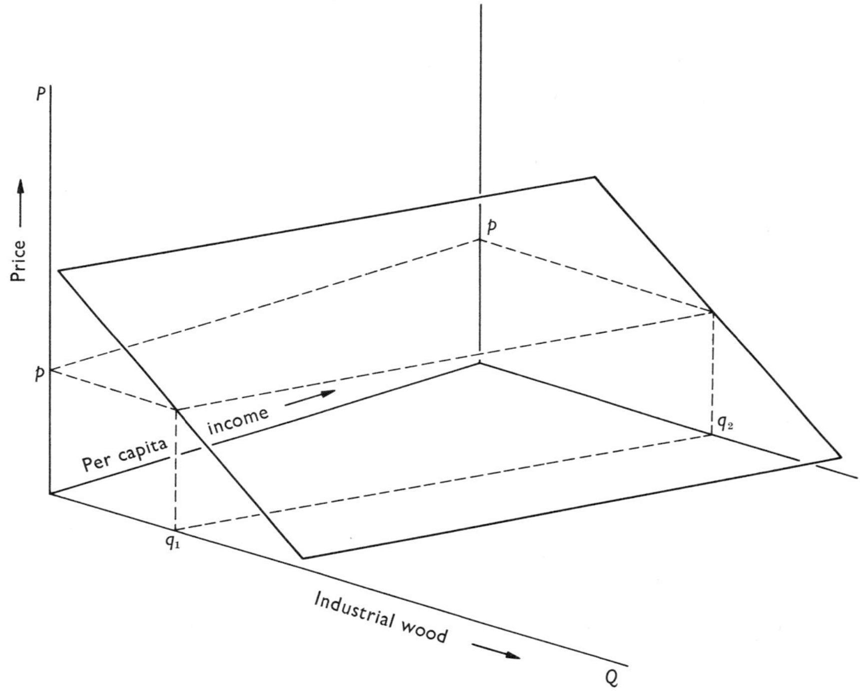


**Fig. 2.5** Per capita income may produce a shift in the demand for industrial wood.

Increasing incomes mean increased purchases of industrial wood at any constant price level.[8] Under these conditions (shifting demand, constant prices) the relationship of consumption to income is shown on the base of the diagram (Fig. 2.5) as $q_1 - q_2$. To describe this relationship between income and demand the term *income elasticity* is used. It is defined in a manner closely analagous to that of price elasticity:

[8] Observe that the price elasticity is also changed by the shift in the demand curve.

$$\text{Income elasticity} = \frac{\%\text{ change in quantity demanded}}{\%\text{ change in income}}$$

Historical price data usually reflect both *movements along* the demand curve and *shifts* in the curve itself, and may in addition involve substantial changes in the shape of the curve. For such reasons, and because the determination of the location and shape of demand curves is a difficult research task, studies of changing wood use often employ a somewhat different concept, perhaps best referred to as the *elasticity of income with respect to consumption.* Strictly speaking it should be defined as:

$$\frac{\%\text{ change in quantity consumed}}{\%\text{ change in income}}$$

and the *ceteris paribus* assumptions should be vigorously maintained. Because significant income changes usually occur only over substantial time periods the *ceteris paribus* assumption is most often ignored.

For clarity of illustrating, all the relationships of Fig. 2.5 were made linear. It is generally believed that the income-consumption relationship is probably curvilinear—perhaps more like that of Fig. 2.6. Note also that wood consumption is shown on the vertical axis and income on the horizontal. A declining rate of increase in wood consumption as incomes increase has important implications for forecasting future wood needs in any country with rising incomes.

In the literature of forest economics, graphs such as Fig. 2.6 are often used to present the results of investigations involving use of either historical or cross-sectional data. The elasticity coefficients, when presented, are usually calculated without holding other variables (such as price) constant. Such calculations are usually called simply "income elasticity," and occasionally are even more erroneously labeled "income elasticity of demand."

## Cross Elasticity

"Cross elasticity" is another term that may be encountered in forest economics readings. It may be defined as:

$$E_y^x = \frac{\%\text{ change in quantity of product } X}{\%\text{ change in price of product } Y}$$

and will be further mentioned in Chapter 8.

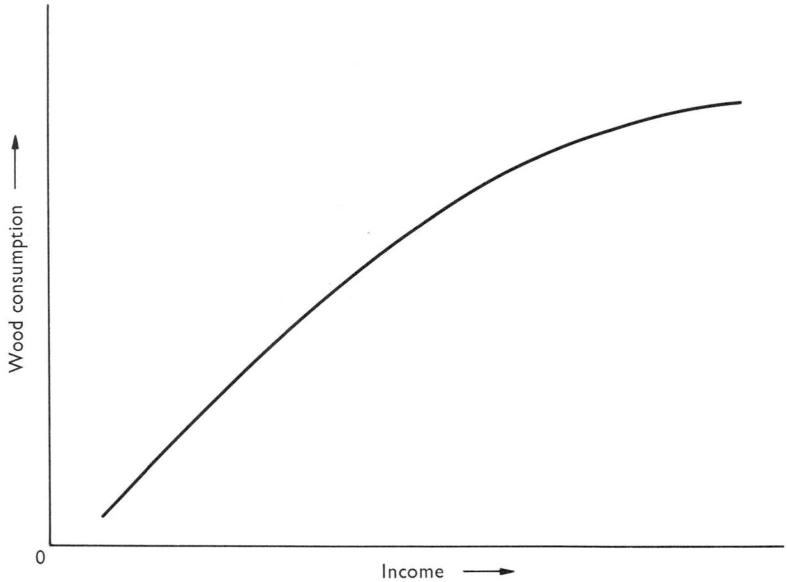


**Fig. 2.6** The generally accepted income–consumption relationship.

## THE RELATION OF INDUSTRY AND FIRM DEMAND CURVES

The market structure of an industry determines the relationship of the firm's demand curve to that of the industry. Suppose, for example, that the demand curve for lumber in the United States could be represented by *DD* in Fig 2.7. This relationship would indicate that at a price index of $300 per M, 15 billion board feet could be sold.

Now consider the extent of the price change that could be induced by a change in the production of a single one of the more than 30,000 sawmills of this country. Even a change in production of 500,000 board feet—half of the annual production of the average-sized mill of this country—could scarcely be identified on such an industry demand curve.

To the producer—the individual mill operator—the only relevant part of the industry demand curve is the part *he* affects. His demand curve, therefore, must show the range of his output and the relevant change in market price that will be induced by his production decisions. The demand curve for the individual producer, therefore,

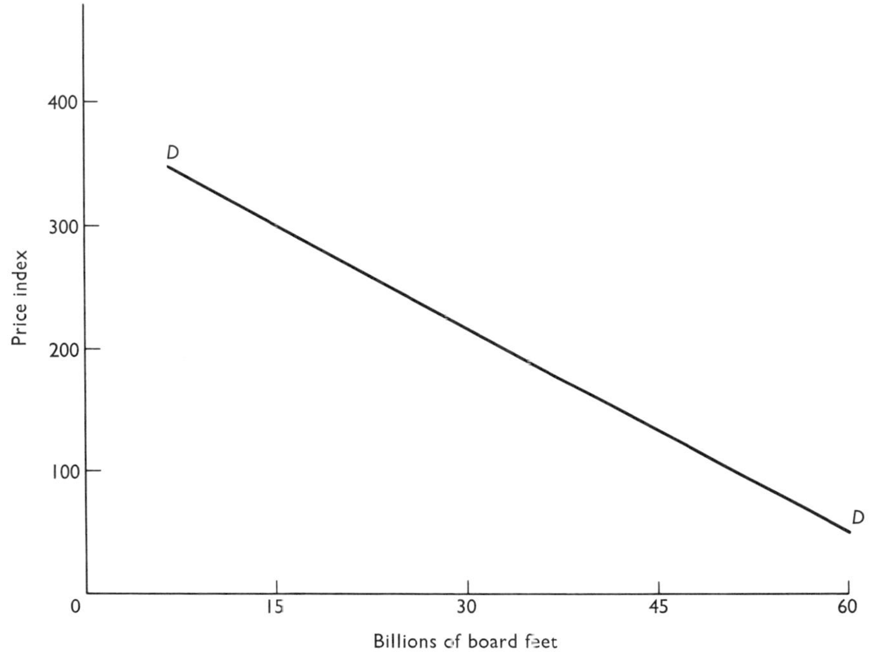


**Fig. 2.7** A hypothetical demand curve for lumber in the United States.

will appear almost perfectly elastic: it will be a horizontal line at the level of market price. The average sawmill operator, like the average farmer, must take the market price as something given to him by forces over which he has no control. He can produce or not produce—sell or not sell—but he will not be able to change the market price.[9]

Now suppose instead that the producer is the nation's sole manufacturer of some wood product: he is a perfect monopolist. In this case the demand curve of the firm is the demand curve for the industry, for the firm *is* the industry. Such a producer *can* influence price by changing his production pattern: his market demand curve is not perfectly elastic. And for competitive structures that lie be-

[9] Naturally, if he were to raise his price only slightly above the market average he would probably maintain some of his more loyal customers along with those of his clientele that were too slow-witted to observe the price differential, or too lazy to shift their buying patterns. But this condition would not persist long. And remember the *ceteris paribus* assumption, which rules out quality differences, service differences, etc.

tween monopoly and pure competition one would find varying degrees of market control over price.

One very significant difference brought about by the market structure is its effect on the kinds of decisions the producer must make. Under pure competition a producer must decide how much to produce at the given market price. But if some degree of market control exists the producer must make two decisions. He must decide upon his output pattern, and he must also consider the effect this will have upon the price he will receive for his output. And unless he is a monopolist—a rarity in this country—he must also consider the effect his production and pricing decisions will have upon his competitors, and his reaction to *their* decisions (and their's to his reaction, etc.).

To the public, of course, there are also very important differences in the market structure: differences in the possibility and necessity for market controls, differences in the resulting allocation of resources, and differences in the societal implications. These differences, however, are best brought out by specific example after the theory of production has been reviewed—a task to which we now turn.

# 3

# Production, Costs, and Supply

Economic demand is essentially an abstraction; no one has ever "seen" a demand curve taking shape in a market. Production, however, conveys a greater sense of reality. One can see timber growing, or lumber being pulled from a green chain, or a sheet of veneer being peeled from the bolt as it turns in the lathe. Hence, production concepts tend to be easier to describe and, on the whole, easier to understand, than are the concepts of demand. But in forestry one should not carry this argument very far, for forests also produce wildlife, water, and recreational opportunities. Some of these latter goods and services are as difficult to lay hold of as a demand function! For the moment, though, we can stay with the familiar and the tangible.

In the review which follows a few assumptions underlining production theory must be kept firmly in mind. Initially it will be assumed that the product is sold and all productive services are bought in a purely competitive market. Hence the price of the product and of all productive services are independent of the quantities sold or purchased. For supply curve formulation we will assume that all operators are independent in their production decisions, and that the goal of each operator is simply that of maximizing net revenue, where net revenue is defined as total revenue less total cost.

## THE INPUT–OUTPUT RELATIONSHIP

All production ties back to some set of physical relationships between inputs and outputs. To produce plantable seedlings, for example, one needs seeds, technical know-how, fertilizer, water, labor, land, and (at least when done on a commercial scale) capital in the form of tractors, sprayers, plows, and other equipment. These factors (inputs) are applied to a particular land area and the resulting product is what we plant in the fall or spring. Or, inputs of labor and capital may be applied to some particular forest tract and the product measured in board feet of logs, or cords of pulpwood, or numbers of campsites, or perhaps in the less well-defined units of increased game cover and wildlife, or millions of gallons of usable water.

The physical relationship can be described by a formula or equation—a recipe—for making the particular product and is called a *production function.* The basic principles of production, including the input–output relationships, are best seen if we initially confine our attention to the most simple types—those involving the production of a single product by combining a single *variable factor* with a set of *fixed factors.*

If (for example) one adds nitrogenous fertilizer (the variable factor) to a planted nursery bed already supplied with all other necessary nutrients (the set of fixed factors) the result might be as shown in Fig. 3.1. As successive increments of fertilizer are added to the nursery bed, the total physical product (*TPP*) rises, first at an increasing rate (part *a*), then at a decreasing rate (part *b*), and finally, as chemical burning occurs, yields begin to fall off (part *c*).

This is the type of production function usually encountered in textbooks, although many other kinds are found in real life.[1] It

[1] It is often helpful to express production functions as equations, and many curve forms have been used. The "general" equation for a production function is usually $Q = F(x_i)$, or $Q = F(x_1 \ldots x_n)$—that the quantity of product is a function of "*n*" inputs. Sometimes the equation form may be specified, as (for example)

$$Q = F(C + b_1X + b_2X^2 + b_3X^3)$$

a simple polynomial of the third degree that often takes a form similar to that of Fig. 3.1. Logarithmic types are much used in research—partially because they are

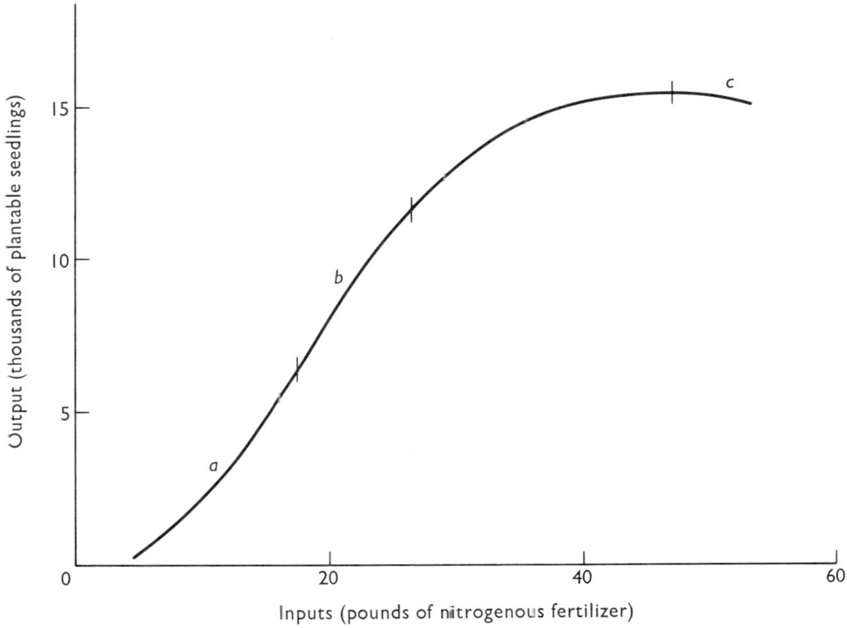


**Fig. 3.1** A production function; the relationship between inputs and output.

happens also to be the kind most usually encountered in forest production.

## Average and Marginal Physical Product

Once the basic output relationship has been specified, two additional functions can be derived, the *average physical product* curve ($APP$) and the *marginal physical product* curve ($MPP$). Both are shown in Fig. 3.2. The average product curve shows the average amount of product produced per unit of input for any particular input level. Arithmetically, it is found by dividing the output by the number of units of input required to produce that particular quantity, or

$$APP = \frac{\text{total physical product}}{\text{input required}}$$

easily fitted by ordinary regression techniques. Finally, a fully specified equation might be written—such as will be done with the example of total and average cost curves later in this text.

The maximum point on the average product curve (or the level of inputs required to produce the maximum average physical product) can be found graphically from the basic production function by drawing a line from the origin just tangent to the total physical

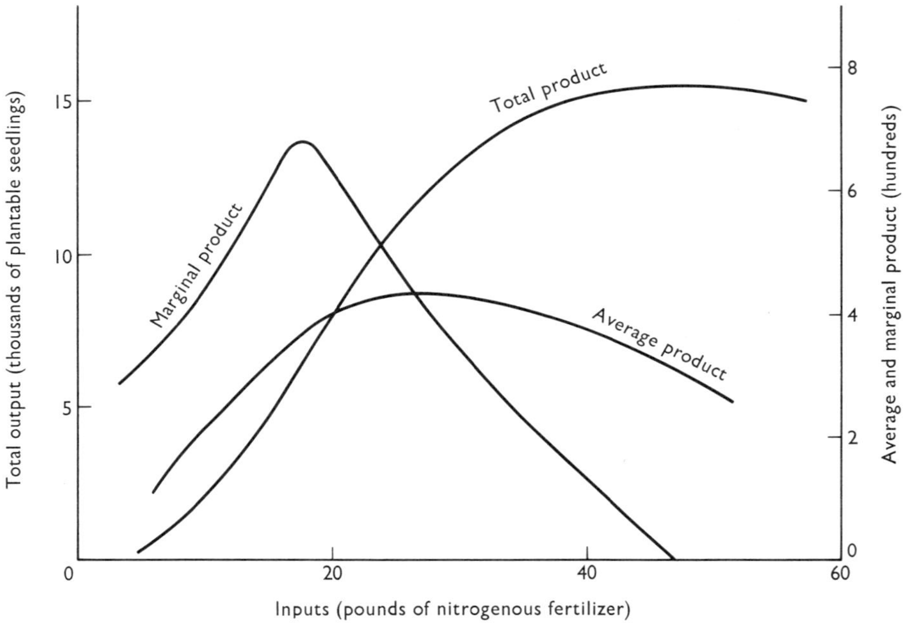


**Fig. 3.2** The three physical product curves: total, average, and marginal products.

product curve. Up to this level of input, every additional increment of the variable service increases the return to *every* input unit used. Once this maximum point is attained, successive increments of the variable factor *lower* the average return per unit of input. The average product curve will reach zero only if the total product curve is somehow also reduced to zero—something that certainly *can* happen, but would scarcely be desired.

The second curve, marginal physical product, involves a somewhat different concept. This curve shows how the quantity of product being produced *changes* as the amount of variable factor used is also *changed*. Mathematically, it is the ratio of the change in output to the change in input, or:

$$MPP = \frac{\text{change in output}}{\text{change in input}}$$

and as the change in input is allowed to approach zero, the ratio becomes the first derivative of the production function, or the slope of the total physical product curve. Accordingly, the maximum point on this marginal curve is the point of inflection of the physical product function. Dropping from this maximum, the marginal curve necessarily passes through the maximum point of the average product curve, and becomes zero at the input level that produces the absolute maximum on the total product curve (where the slope is zero). The relation between these three physical product curves is shown in Fig. 3.2. Note that the scale of the average and marginal curves is different from that for the total curve, and is shown on the right side of the diagram.

## THE SHORT RUN

The production function is peculiarly a "short run phenomenon." Frequently the "short run" is defined as a period of time in which no changes can be made in plant capacity. Strictly speaking, short run and long run are pure intellectual concepts, totally divorced from time in the temporal sense. A general definition, and one more appropriate to forestry problems, relates the terms to the character of the production factors; short run is always characterized by the presence of a fixed factor, while long run is devoid of fixed factors. Herein lies the first real peculiarity of forest economics, for in forestry the production period tends to be much longer than in most productive endeavors. One frequently finds short runs of fifty years or more—a time period that most analysts would scarcely wish to consider a short run. And of course, the *time* period is *not* short. Nonetheless, a fifty-year rotation for sawlog production from a forest area can only be called a short run if the area of land is fixed.

It should now be clear why the production function is essentially a short run phenomenon; this function describes the quantitative response (output) generated by varying the amount of certain inputs (the variable factors) that are applied to a fixed factor or to a group of fixed factors, and although the function may specify only the variable factors, a fixed factor is always implied.[2]

[2] Basically this is the reason for the usual assumption that production functions are "homogeneous and of the first order." A doubling of the right side of the equa-

## LAW OF VARIABLE PROPORTIONS

The graph of a production function is also a pictorial representation of the "law of diminishing returns," often called the "law of variable proportions." A complete statement of this "law" would be something similar to the following: if to a fixed factor one adds successive increments of a variable factor, output will first increase at an increasing rate, then increase at a decreasing rate, and finally may decrease absolutely. Clearly, this is simply a description of the curve of Fig. 3.1. The "law of variable proportions" is a restatement of this same law in terms of changing proportions of variable to fixed factors.

It may be worth observing that so long as the assumptions implicit in this law are fulfilled, the law has no exceptions in the physical world. As has been facetiously (but accurately) observed, the law of diminishing returns simply explains "why we cannot grow the world's food supply in a single flowerpot."

## COSTS AND REVENUES

Knowledge of the physical response of forests to man's managerial efforts (the inputs) is the technical basis for professional forestry, but this knowledge is only a starting point for *management* of a forest. Silvicultural training permits us to say, for example: "Loblolly pine will grow best on this site, and slash pine on that one," or "If a thirty-year rotation is adopted for this plantation, the yield will average around 100 cords per acre." Such knowledge is indeed important and valuable, but it is still necessary to answer the economic (managerial) questions of "*Should* I plant the loblolly pine (or slash pine) site?" and "*When should* I plan to cut this plantation?" A production function describes the input–output relationship but does not identify the best level of inputs to use. The best level necessarily depends on the objective, and the objective of management (at least as a first approximation) is that of maximizing net revenue. Because net revenue involves both costs and revenues, both must be brought into the picture.

---

tion (the inputs) results in a doubling of the left (output) side, which of course is only true if the fixed factor is also doubled. The same idea is sometimes expressed by assuming constant returns to scale. Thus a function such as that of Fig. 3.1 is applicable to a 1/10 acre plot, a fifty acre woodlot, or a 500,000 acre forest. If the inputs are scaled to the fixed factor, outputs will be proportionately scaled.

## Costs

Costs are easily introduced. In the example of Fig. 3.1 the input was measured in pounds of fertilizer. Now suppose that this fertilizer costs ten cents per pound. A new scale measuring inputs in dollars instead of pounds can be established by multiplying the original pound scale by .10. This converts the physical production function to an economic function—the *total variable cost* curve.

In economics it is customary to make cost a function of output, although cost obviously arises more directly from inputs. To so express the cost curve derived above requires only a slight rearrangement of the axes. As a first step the whole diagram can be rotated ninety degrees in a counterclockwise direction; that is, starting with (a) in Fig. 3.3, "spin" the diagram to the position shown

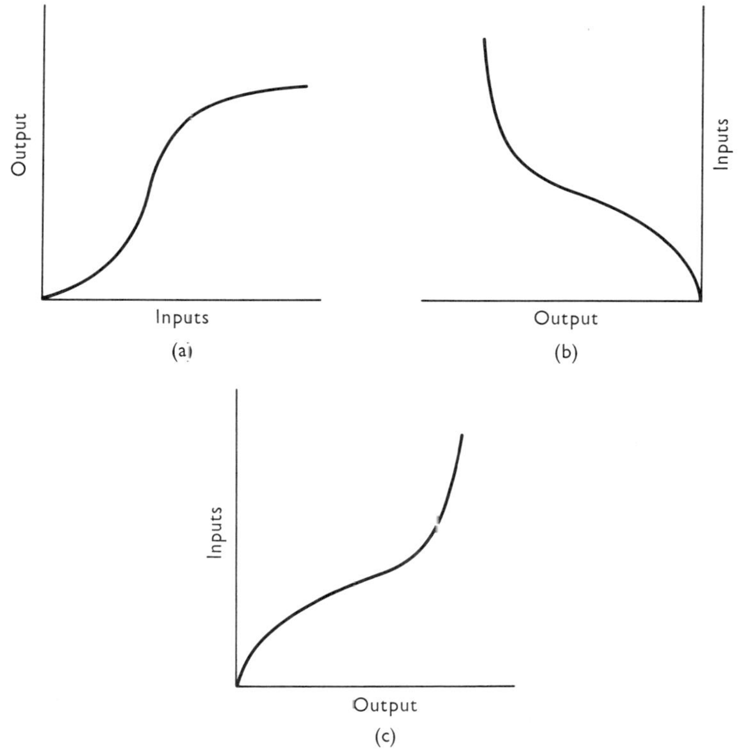


**Fig. 3.3** The relationship between the production function and the variable cost curve; (a) depicts a production function; (b) shows this rotated 90 degrees; and (c) is the mirror image of (b).

in (b). Now look at its mirror image, or look at the diagram from the back side of the page (through the paper). The result is that shown by (c), the familiar form of the total variable cost curve seen in nearly all elementary texts of price theory. Before proceeding further, however, be sure to note that the form of this variable cost curve stems directly from the physical relationships. Any change in the production function—a new technique, a new genetic strain, a new chemical—will change the cost curve.

Cost curves can also be expressed mathematically, the specific equation for the total variable cost curve of Fig. 3.4 and subsequent figures in this chapter being[3]

$$TVC = 9.75Q - .225Q^2 + .0025Q^3$$

In order to keep the example in general terms we will simply refer to the output as "product," measured in "units." The product could be almost anything—pallets, Christmas trees, boxes, or whiskey or boat trips.

Using an equation of this type, it is no longer necessary to think of a single input factor. Whatever factors have been used, and however many have been used, they have been converted to costs and these costs related to product output by the above equation.

To the total variable cost must now be added the cost of the fixed factor (or group of fixed factors). By definition these factors are not affected by output so they can be represented graphically by a horizontal line. In the example used here, $TFC = 100$. It is plotted, along with the $TVC$, in Fig. 3.4.

These two costs, $TFC$ and $TVC$, are now added together to form the total cost curve ($TC$) which is also shown in Fig. 3.4. Because $TC$ is derived by simply adding the constant to $TVC$, the whole $TVC$ curve is shifted up by the amount of fixed costs. Algebraically, then, the equation for total cost in the example being used here is

$$TC = 100 + 9.75Q - .225Q^2 + .0025Q^3$$

So far as the problem here is concerned, the cost picture is now complete. But we cannot yet determine the optimum level of output or of inputs, for we still lack the revenue side of the picture.

[3] In the specific equation used here, both quantity of product ($Q$) and total variable cost ($TVC$) are expressed in thousands. That is, an output of 40,000 units of product is entered in the equation as 40, etc.

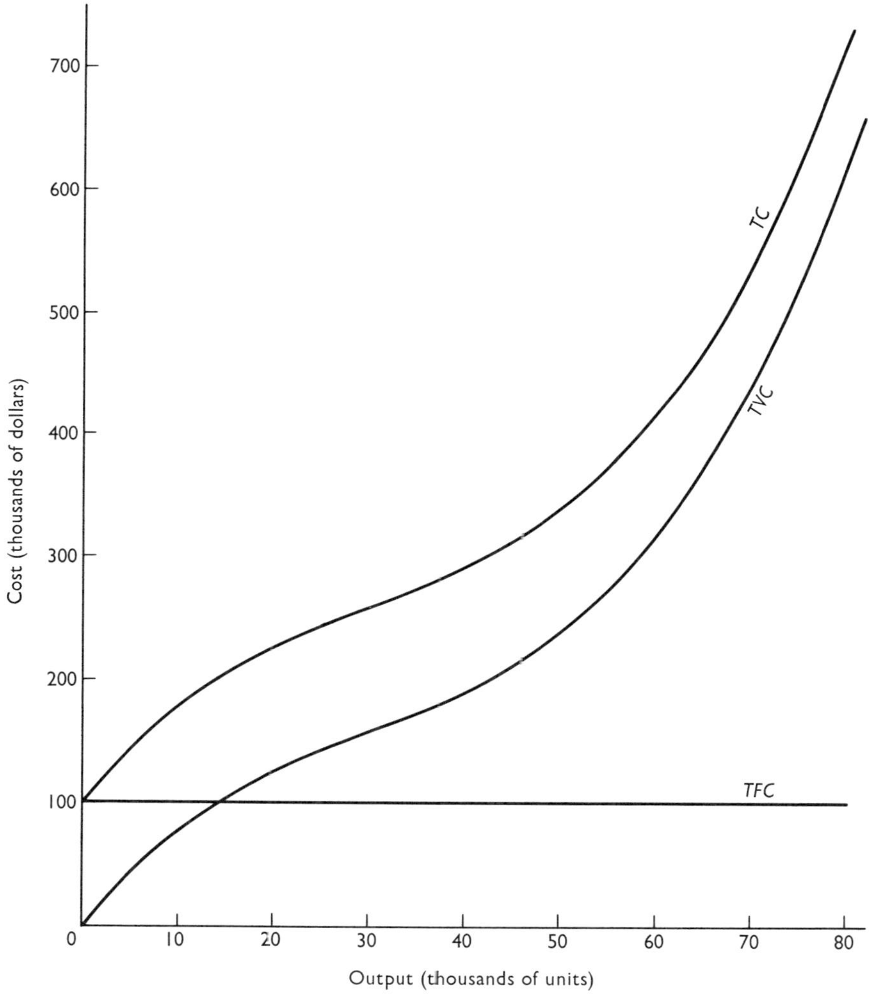


**Fig. 3.4** A set of total cost curves.

### Revenues

Total revenue can also be represented as a function of output, for it is by sale of products that revenue is obtained. Under the purely competitive conditions assumed here, total revenue will simply be a straight line passing through the origin. Algebraically, $TR = P \cdot Q$, where $P$ is the (constant) price and $Q$ is the quantity sold: in our example a product price of \$8 has been assumed and the equation could be written

$$TR = 8Q$$

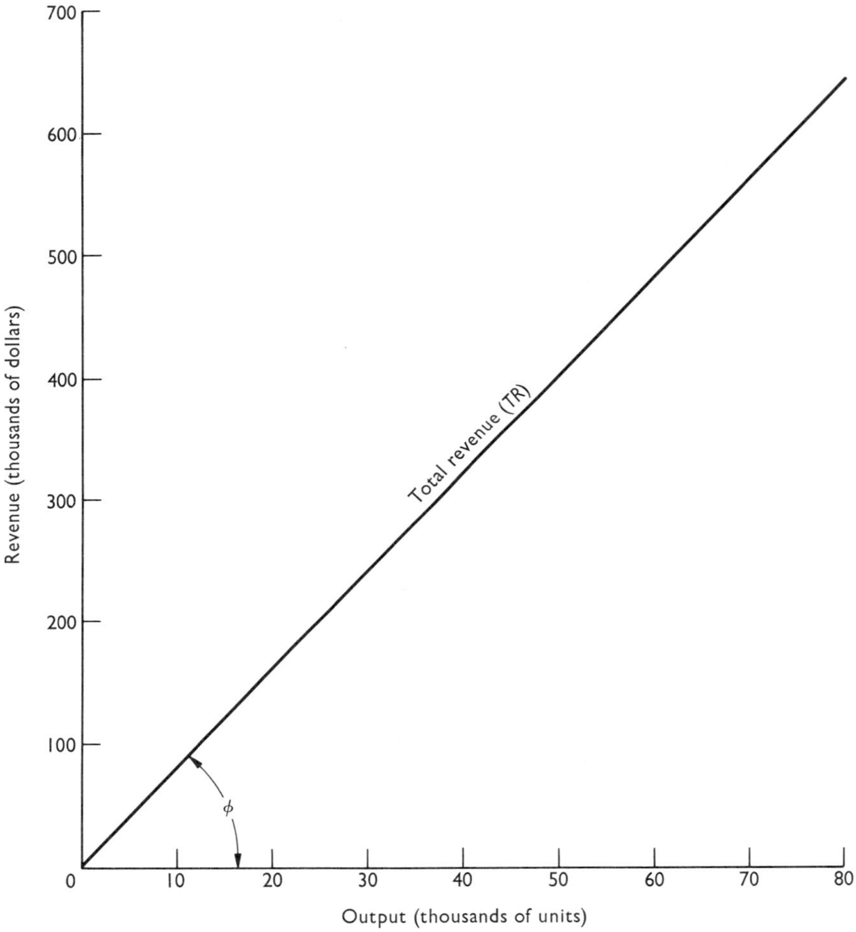


**Fig. 3.5** The total revenue curve of the firm under pure competition.

Under purely competitive conditions the slope of the total revenue curve is always the product price, and is the tangent of the angle $\phi$ in Fig. 3.5.[4]

[4] Since $TR = P \cdot Q$, the tangent will be $\frac{d(TR)}{d(Q)} = \frac{d(Q)}{d(Q)}$. $P = P$, wherever $P$ is a constant (as in pure competition).

Both the cost and revenue elements of the problem can now be combined into a single diagram by superimposing the revenue curve on the set of total cost curves, as in Fig. 3.6.

Our objective of maximizing total net revenue requires that we find the output with the greatest vertical distance between the total revenue and total cost curves in the area of positive net revenue, the shaded portion of the diagram. The point of maximum vertical

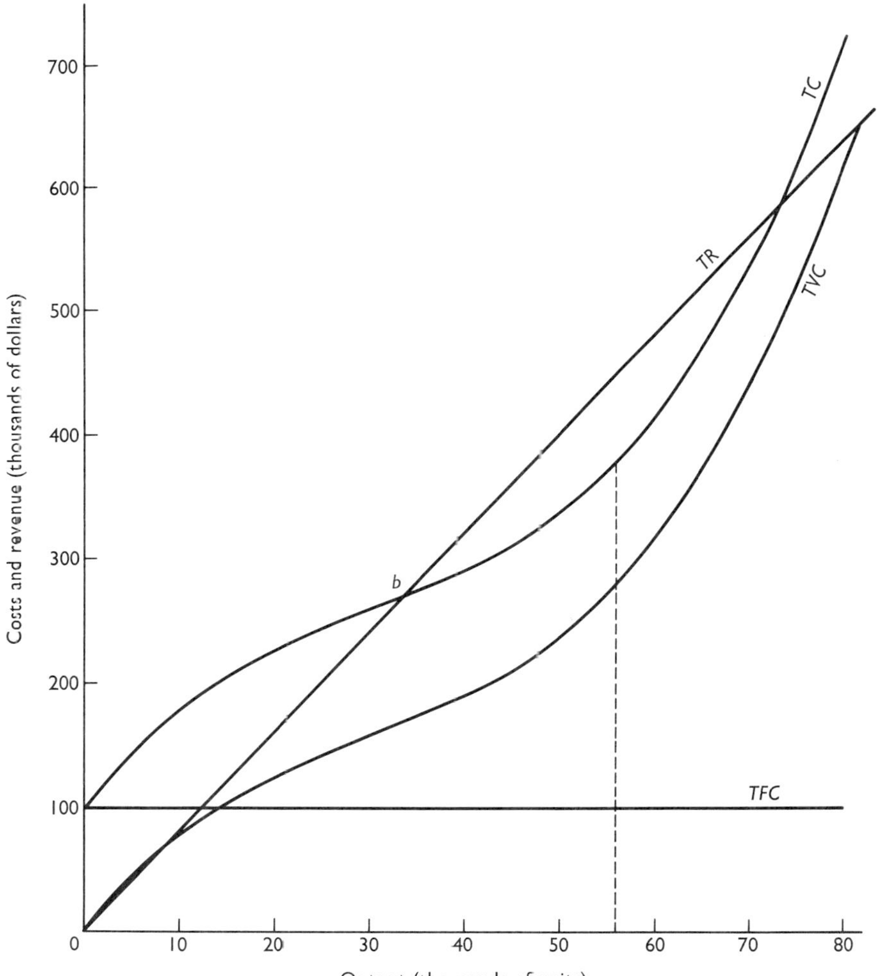


**Fig. 3.6** Total cost, total variable cost, total fixed cost, and total revenue. (For equations of curves see text.)

distance between any two curves is located where the curves are parallel at a given value of $Q$. Since $TR$ is a straight line, that point can be found by constructing a line parallel to $TR$, and tangent to $TC$. This point of tangency identifies the output that maximizes total net revenue—in the example, an output of approximately 56,000 units. The "break-even" point, often used by engineers, is at $b$ in this figure. If the cost function were linear, as is the case with constant cost machinery, the then straight line of total cost would intersect revenue at only one point.

## NET REVENUE AND PROFIT

The word "profit" is used for so many different quantities that the term "net revenue" will be used in this text wherever there is possibility of confusion. When studying a diagram such as that of Fig. 3.6 one should remember that fixed costs include whatever return on invested capital is deemed necessary to keep that capital so employed. These fixed costs also include whatever salary the entrepreneur decides he must earn if he is to continue his entrepreneurship. Hence the *net revenue* of Fig. 3.6 is a return over and above that necessary to pay the operator's wage or salary, to return the required rate on all invested capital, and to make the necessary payment for any resources owned by the entrepreneur. It is, in fact, true "economic profit." Profit will be reviewed in Chapter 4 as a topic of distribution theory.

## AVERAGE OR UNIT COST CURVES

Very frequently it is helpful to view costs and returns on a unit basis—per M board feet, per M square feet, or cubic feet, cubic meters, etc. The transition from total to average (or unit) costs can be made quite directly.

With three *total* cost curves we will be able to derive three average cost curves; one for fixed costs ($AFC$), one for variable costs ($AVC$), and one for total costs ($ATC$). The relationship between them, and their derivation from total costs, can be seen in Fig. 3.7.

Average fixed cost is determined by $AFC = \frac{TFC}{Q}\cdot$ The ratio, $\frac{TFC}{Q}$, viewed geometrically, is simply the tangent of the angle $\phi$ (Fig. 3.7) where $Q = q$. If we visualize the behavior of this tangent as $q$ is shifted from the origin outward, it becomes clear that this ratio

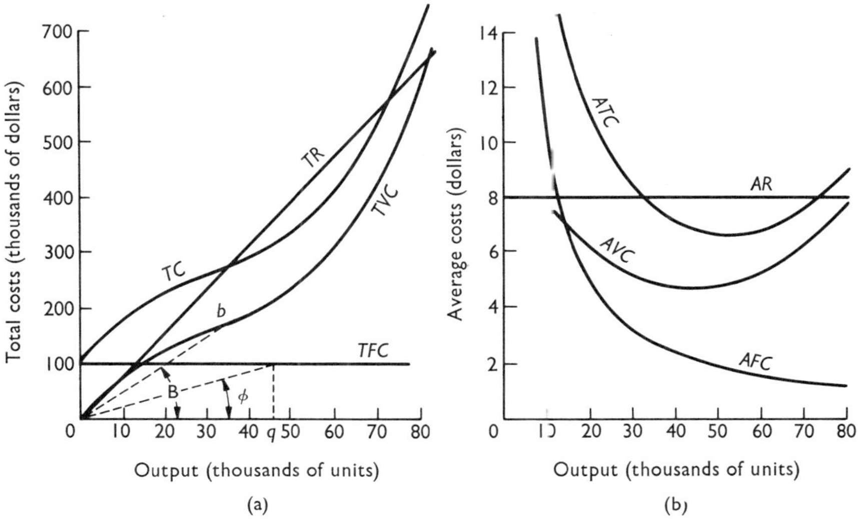


**Fig. 3.7** Derivation of average cost curves from total cost curves. (For equations of curves see text.)

(or tangent $\phi$) will drop from a high value to a low one, falling ever less rapidly. The average fixed costs, therefore, must be a constantly declining curve, as in Fig. 3.7(b).

In similar fashion the average variable costs and the average total cost curves can be defined as

$$AVC = \frac{TVC}{Q} \qquad \text{and} \qquad ATC = \frac{TC}{Q}$$

respectively, and these two ratios also represent tangents of angles. In these latter cases, however, the angle is formed by the horizontal axis and a line connecting the origin to a point on the *TC* or *TVC* curves—line $0b$ in Fig. 3.7(a). It is helpful to visualize the behavior of these tangents as the point moves along the particular curve: at first as the point moves out (as $Q$ increases) the angle (for example, $B$) and hence its tangent, declines. Eventually a point is reached on the *TVC* and *TC* curves where, as $Q$ increases, the angle and its tangent *increases*. Since the two ratios $\frac{TVC}{Q}$ and $\frac{TC}{Q}$ are at once the tangents and the value of *AVC* and *ATC* for the particular levels of $Q$, the two average costs curves also at first decline, then eventually increase, giving rise to the familiar U shaped average cost curves of Fig. 3.7(b).

Mathematically the derivation is equally simple. The equations of the three total cost curves of Fig. 3.7 can be readily transformed to the equations of the three average cost curves by simply dividing through by $Q$. Hence:

$$AFC = \frac{TFC}{Q} = \frac{100}{Q}$$

$$AVC = \frac{TVC}{Q} = \frac{9.75Q - .225Q^2 + .0025Q^3}{Q} = 9.75 - .225Q + .0025Q^2$$

$$ATC = \frac{TC}{Q} = \frac{100 + 9.75Q - .225Q^2 + .0025Q^3}{Q}$$

$$= \frac{100}{Q} + 9.75 - .225Q + .0025Q^2$$

Examination of these equations will show that these divisions accomplish precisely the steps completed in the graphic derivation of the average cost curves of Fig. 3.7(a) and (b).

Similarly, total revenue curves can be converted to average revenue curves and added to the unit cost and revenue picture. In this case, starting with the known relationship $TR = P \cdot Q$, we can derive average revenue by again dividing by $Q$:

$$ATR = \frac{TR}{Q} = \frac{P \cdot Q}{Q} = P = \$8 \text{ (in our example)}$$

## MARGINAL COST

Marginal cost is defined (at least for a start) as a ratio ofchanges:

$$MC = \frac{\Delta TC}{\Delta Q}$$

This ratio is seen in Fig. 3.8 to be the tangent of the angle $\theta$ in the lower left corner of the small triangle, and it is also the tangent of the angle $\phi$ (formed by extending the hypotenuse of the small triangle until it intersects a prolongation of the base of the graph). Note that this ratio is also the average slope of the total cost curve as costs rise from $c_1$ to $c_2$, and output increases from $q_1$ to $q_2$.

As $\Delta q$ is allowed to approach zero the hypotenuse of the triangle approaches the true slope of the total cost curve at that particular point. Hence, as we move from a discrete to a point concept we find that:

$$MC = \frac{\Delta TC}{\Delta Q} = \frac{d(TC)}{d(Q)}$$

Marginal cost is equal to the *change* in total cost that accompanies a *change* in output. It also equals the *slope* of the total cost curve at any particular point, and equals the first derivative of the total cost function. Looking at marginal cost in this light permits clearing up several points that often prove bothersome.

1. When discrete changes in cost are used to calculate marginal cost (as in most classroom problems and many real-life situations), it should be plotted at the midpoint of the two quantities involved.

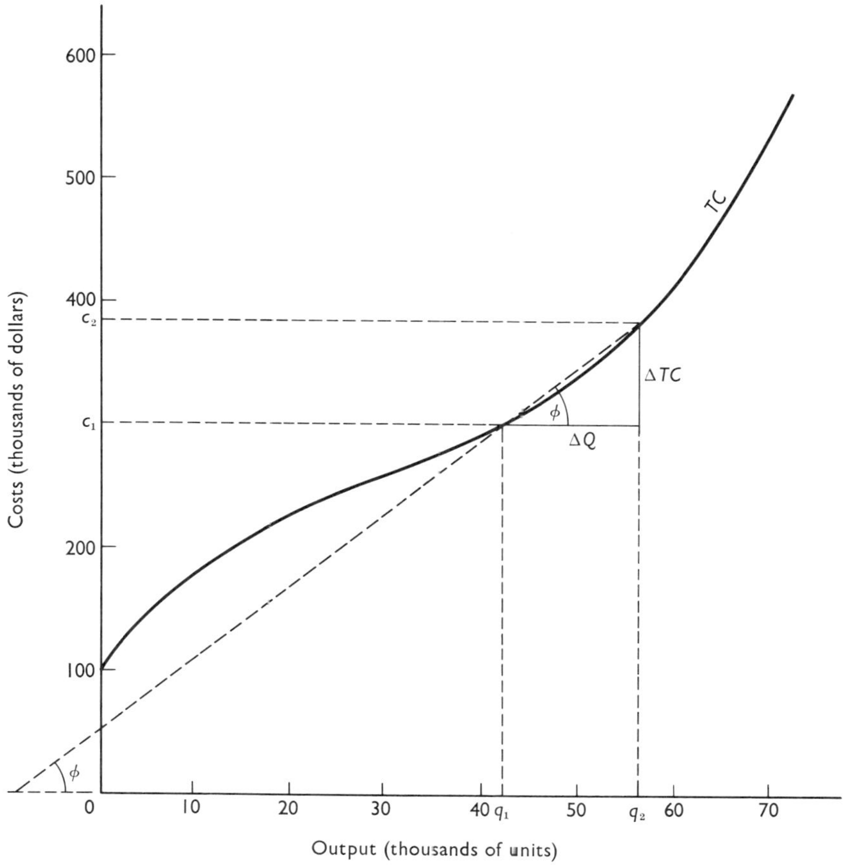


**Fig. 3.8** The derivation of the marginal cost curve from the total variable cost.

To take a simple numerical example, if costs were to change from (say) \$200 to \$210 as output changed from 45 to 50 units, marginal cost could be calculated as follows:

$$\frac{210 - 200}{50 - 45} = \frac{10}{5} = \$2$$

Since this would represent the average slope (or change) between the outputs of 45 and 50 units it would be plotted over 47½.[5]

2. The marginal cost curve can be derived directly from either the total cost or total variable cost functions. Clearly, since $TC = TVC + TFC$, and $TFC$ is a constant, both equations will have identical first derivatives. Returning to the example used earlier:

$$MC = \frac{d(TC)}{d(Q)} = \frac{d(100 + 9.75Q - .225Q^2 + .0025Q^3)}{d(Q)}$$
$$= 9.75 - .45Q + .0075Q^2$$

and to find the values for $MC$ one needs only to insert the desired values for $Q$ and solve. The resulting $MC$ curve, together with all the average cost curves, is shown in Fig. 3.9.

3. The $MC$ curve always cuts the two average cost curves exactly at their lowest points. A glance at the geometry already developed shows why; average total cost, for example, may be represented as the tangent of the angle formed by connecting the origin to some point on the total cost curve, and the average total cost curve was derived by permitting this point to move along the curve as $Q$ changed, then plotting this tangential relationship. Marginal cost, on the other hand, is the actual slope of the total cost curve—the tangent of the angle formed by the horizontal axis and a line just touching the total cost curve at the particular output. As the point connected to the origin begins to move along the total cost curve, the angle formed and hence its tangent (which is, numerically, $ATC$) is becoming smaller and smaller. Eventually, however, it starts to increase. This point of transition from a drop to an increase marks the low point of the average total cost curve, and here the line from the origin is *just tangent* to the total cost curve. Hence the slope of this line is also the slope of the total cost curve, and $ATC = MC$.

[5] Note that this is, at best, an approximation. Only in special cases would this average represent the actual slope of the total cost curve at its midpoint. It is, however, considerably more accurate than plotting over either the beginning or ending output level.

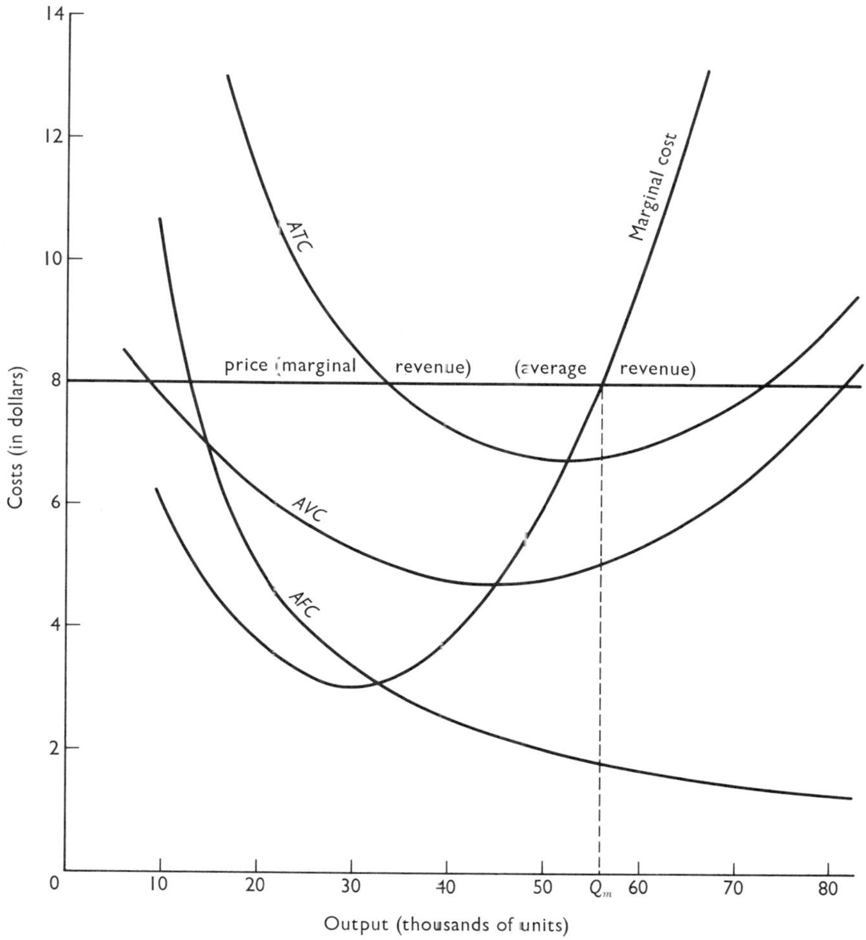


**Fig. 3.9** Marginal cost curve, together with the average cost curves and average revenue (or price).

## MARGINAL REVENUE

Marginal revenue, conceptually identical to marginal cost, may be defined as *the change in revenue that accompanies a change in output*. It is therefore a ratio of changes, and may be expressed mathematically as

$$MR = \frac{\Delta TR}{\Delta Q} = \frac{d(TR)}{d(Q)}$$

Hence, marginal revenue is the *slope* of the total revenue curve at any selected output level. Under assumptions of pure competition the market price of the product is a constant, unaffected by output decisions of the individual firm. And since $TR = P \cdot Q$, one can also see that

$$MR = \frac{d(TR)}{d(Q)} = \frac{d(P \cdot Q)}{d(Q)} = P \cdot \frac{d(Q)}{d(Q)} = P$$

Moreover, because price is also equal to the *average* revenue, a three-way equality always exists under pure competition for the firm in equilibrium; $AR = MR = P$. Under other market structures this equality does not necessarily hold. Marginal revenue under non-pure competition will be treated in the following chapter.

## MAXIMIZATION OF NET REVENUE

With both marginal revenue and marginal cost plotted on the set of average cost curves (Fig. 3.9) the point of net revenue maximization can be readily identified by the intersection of the rising marginal cost with price. The profit maximizing output of $Q_m$ is, of course, exactly the same as the output determined from the set of total cost curves, for in Fig. 3.9 the *slopes* of the total cost and revenue curves were plotted and the point of equality determined by their intersection, while in Fig. 3.6 the output corresponding to equal slopes of the total cost and revenue curves was determined by constructing a parallel tangent.

### Mathematical Maximization

It is often helpful (and at this point also serves as a quick review) to approach the problem of net revenue maximization in a strictly mathematical framework. Given the equations for total cost and total revenue as functions of output,

$$TC = f(Q) \qquad \text{and} \qquad TR = f(PQ)$$

we can certainly define total net revenue as a function of output also, since

$$TNR = TR - TC$$

Our objective, locating the output level (the value of $Q$) which will maximize $TNR$, can be attained by taking the first derivative of the equation with respect to $Q$, setting it equal to zero, and solving for the desired $Q$ value (remembering that this procedure establishes *either* a maximum or a minimum). Hence,

$$\frac{d(TNR)}{d(Q)} = \frac{d(TR)}{d(Q)} - \frac{d(TC)}{d(Q)} = 0$$

and

$$\frac{d(TR)}{d(Q)} = \frac{d(TC)}{d(Q)}$$

These derivatives, $\frac{d(TR)}{d(Q)}$ and $\frac{d(TC)}{d(Q)}$, are respectively the slopes of the total revenue and total cost curves and are, mathematically, the marginal cost and marginal revenue functions. Hence, to find the point of either maximum (or minimum) net revenue one must locate the point at which the slope of the total cost curve equals the slope of the total revenue curve. This was done geometrically in Fig. 3.6 and the equation system was solved graphically in Fig. 3.9, both solutions establishing the maximizing output at approximately 56,000 units.

The corresponding mathematical operation consists of setting the marginal cost equation equal to that of marginal revenue. In the example used here, since marginal revenue is equal to price (pure competition has been assumed) the equality becomes:

$$8.00 = 9.75 - .45Q + .0075Q^2$$

Since this is a quadratic equation there are two roots, in this case both positive, and the two corresponding solutions are

$$Q = 55{,}820 \qquad \text{and} \qquad Q = 4{,}780$$

The first solution, $Q = 55{,}820$, corresponds to that found graphically in Figs. 3.6 and 3.9. The second solution, $Q = 4{,}780$, represents the output that would *minimize* net revenue. That such a point exists is also apparent from the graphical presentation; in Fig. 3.9 this becomes apparent if one extends the left arm of the (parabolic) marginal cost curve until it intersects price; in Fig. 3.6 the equality of the two slopes at the lower output is readily seen. The existence of these two solutions requires that an additional condition be attached to the statement regarding maximization of net revenue. Net revenue is maximized when marginal cost equals marginal revenue *provided the marginal cost curve is rising* (when the *second* derivative of total cost is positive).

## MARGINAL COST AND REVENUE IN INPUT TERMS

The preceding review of marginal cost and revenue has been in terms of *output* units, and is the conventional presentation. Some problems, however, are more easily approached if *inputs* are made the primary analytical unit. The starting point, as before, is the production function, but instead of using this relationship as a basis for making cost a function of output (as previously) the total physical product curve is converted directly to revenue. Under assumptions of pure competition this conversion can be accomplished by simply rescaling the vertical (output) axis in dollar terms—a process that requires nothing more than multiplying output by the price per unit. The curve itself remains unchanged, and total revenue thus becomes a function of inputs, as in Fig. 3.10. In this example the production function appropriate to Fig. 3.4 has been utilized, and to aid comparison with the preceding development, a product price of eight dollars per unit has again been assumed.

Similarly, the relationship between the physical quantity of inputs and their cost can be shown in dollar terms, using the same scale as that depicting total revenue (Fig. 3.10). Costs and revenues are thus both made a function of input levels, and the problem of determining the optimum level of inputs (or the optimum expenditure level) can be directly tackled. Several major differences between this approach and the more conventional one of making costs and revenues a function of *output* should be emphasized.

1. Total fixed cost (*TFC*), since it is not related either to the level of input or output, remains a constant. The cost of the fixed factor is, by definition, unchanged as either input or output changes. In a graphic representation such as Fig. 3.10, therefore, fixed cost will show as a horizontal line, its level being fixed by the total amount of the fixed costs (here at $100,000).

2. Variable cost (*TVC*) becomes a straight line extending from the origin, with a positive slope equal to the cost per unit of input. In Fig. 3.10 the total variable cost was obtained by assuming a cost of one dollar per unit of input ($1,000 per 1,000 units).

3. Total cost (*TC*) will necessarily also be a straight line, parallel-

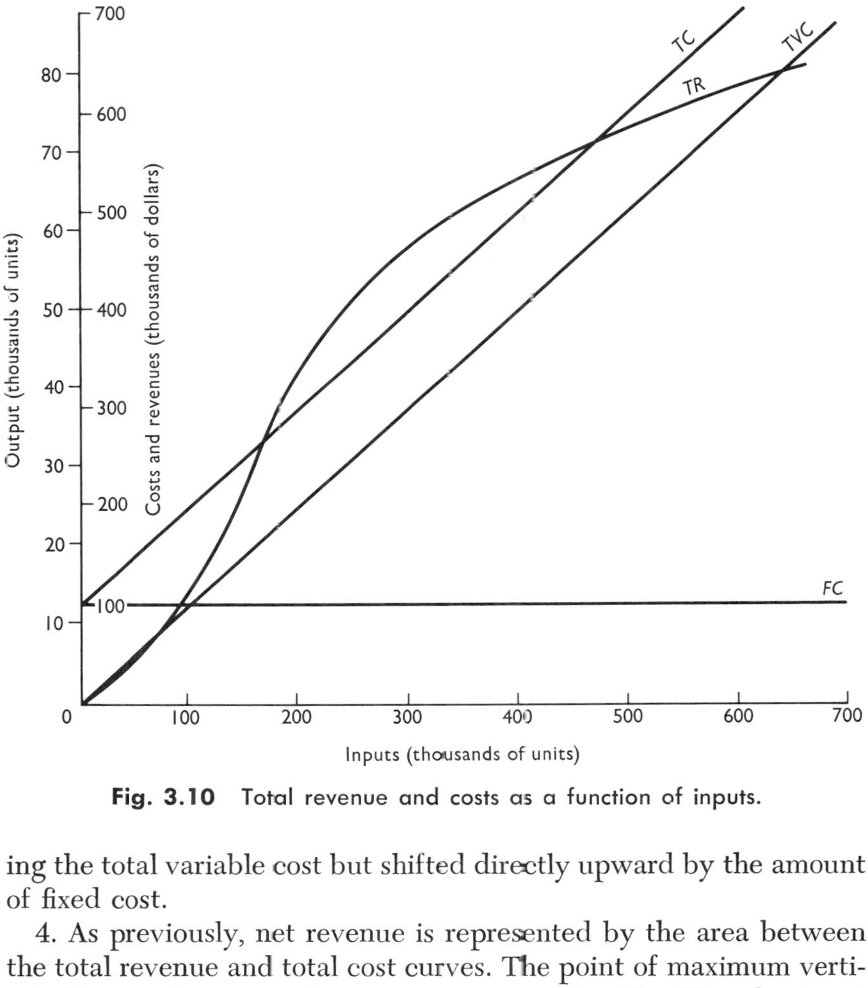


**Fig. 3.10** Total revenue and costs as a function of inputs.

ing the total variable cost but shifted directly upward by the amount of fixed cost.

4. As previously, net revenue is represented by the area between the total revenue and total cost curves. The point of maximum vertical distance between these two curves again identifies the input level maximizing net revenue, and can be located by finding the input level at which the slopes of the two curves (*TR* and *TC*) are identical. A line parallel to the total cost curve, and just tangent to the total revenue curve (Fig. 3.10) will identify this point.

5. A complete set of average cost and average revenue curves can also be constructed in terms of input units, as can corresponding marginal cost and marginal revenue curves (Fig. 3.11). Average variable cost per unit of input (*AVC*) and the marginal cost (*MC*)

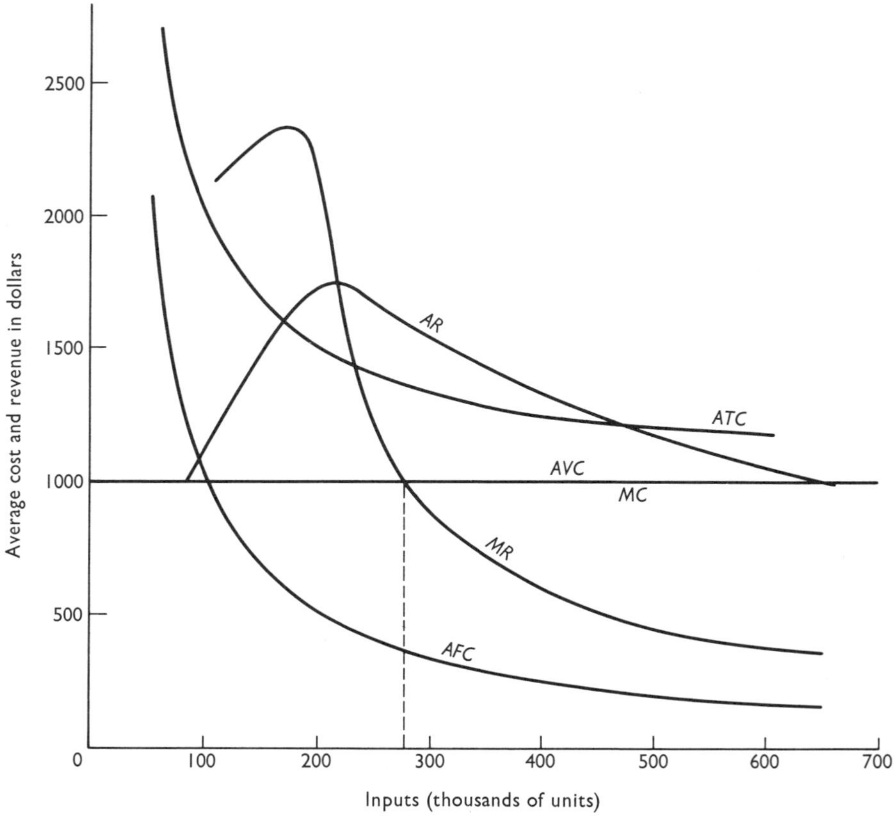


**Fig. 3.11** **Average costs and revenues, and marginal costs and revenues, per thousand units of *input*.**

will, in this formulation, be identical horizontal lines, their level determined by the cost per unit of input. Average fixed cost will be a constantly declining curve, as will average total cost, since it is the sum of the average variable and average fixed costs. Note that the average total cost will always lie *above* the marginal cost.

6. In terms of input units, average revenue $\left(\frac{\text{Total revenue}}{\text{input units}}\right)$ and marginal revenue (the change in total revenue accompanying a change in inputs) now become curvilinear, their shape depending directly upon that of the original production function. As long as the product price is assumed to remain constant, these curves (providing the scale is maintained) will be identical to the curves of average and marginal physical product, and are sometimes referred

to as the curves of "average value product" and "marginal value product."

7. The input level maximizing net revenue is again indicated by the intersection of the marginal cost and marginal revenue curves (Fig. 3.11). The indicated solution (approximately 275,100 input units) is, of course, identical to that obtained earlier and can be confirmed by checking with the total revenue.

8. In using this approach the production function is sometimes assumed to be parabolic in shape. With such a total revenue curve, and with constant product prices, average revenue and marginal revenue curves are both straight lines, with negative slopes, and with the marginal revenue curve always having exactly twice the slope of the average revenue curve.

Most forestry problems are more easily fitted into the output framework. An exception, and an important one, is that of rotation determination, which will be treated in Chapter 15.

## SUPPLY CURVES

A supply curve, like a demand curve, is a functional relationship between price and output. Whereas the demand relationship represents a "buyer's response curve," the supply concept represents a "producer's response curve," showing the output schedule for a firm under a variety of possible prices. And again paralleling the demand concept, this functional relationship of supply can be represented by a schedule, by a graph, or by an equation. Table 3.1 is the supply

**TABLE 3.1 SUPPLY SCHEDULE OF THE FIRM (ROUNDED TO NEAREST 100 UNITS OF OUTPUT).**

| Price (dollars) | Quantity (units) |
|---|---|
| 4.70 | 45,000 |
| 5.00 | 46,300 |
| 5.50 | 48,300 |
| 6.00 | 50,000 |
| 6.50 | 51,600 |
| 7.00 | 53,100 |
| 7.50 | 54,500 |
| 8.00 | 55,800 |
| 8.50 | 57,000 |
| 9.00 | 58,300 |
| 9.50 | 59,400 |

schedule for the firm we have been following during the production review. The same data can be presented in graphic form (Fig. 3.12) or by an equation, in this case $P = 9.75 - .45Q + .0075Q^2$, where $Q \geqq 45$ (outputs in 1,000's, and output of minimum $AVC = 45{,}000$ units).

### The Firm's Short Run Supply Curve

In the short run, purely competitive case, the supply curve of the firm is that part of the firm's rising marginal cost curve that lies above the minimum average variable cost, identified in Fig. 3.12 by the heavier portion of the *MC* curve. Fig. 3.12 illustrates the

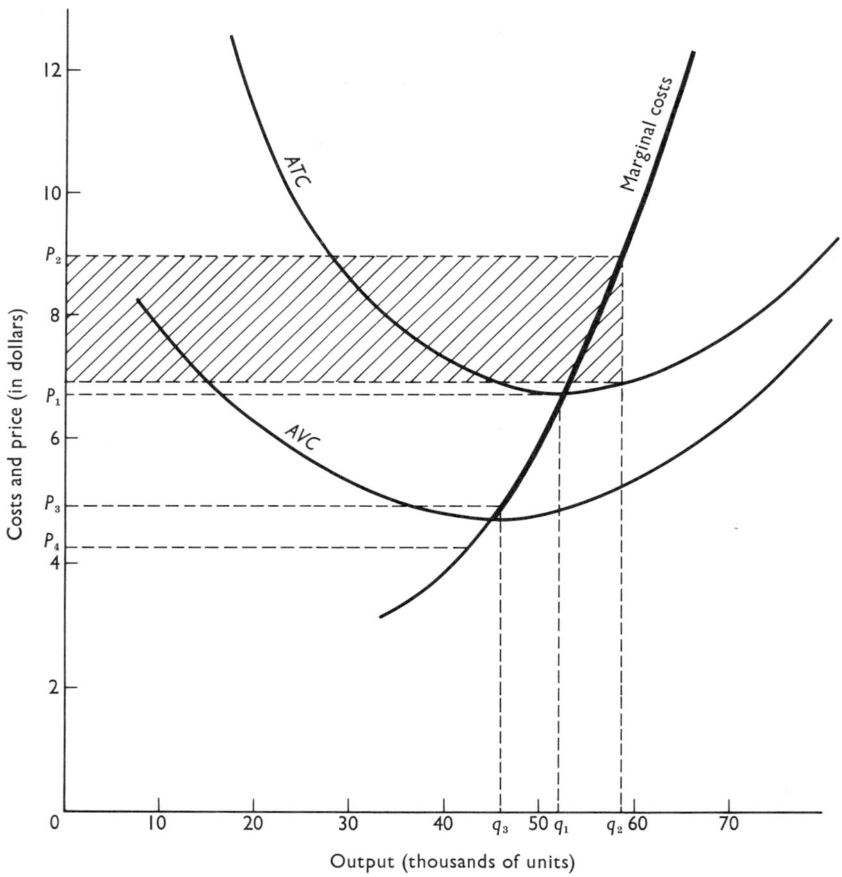


**Fig. 3.12** The short run supply curve of the firm.

situation with the cost curves of Fig. 3.9. If price is at $P_1$ then all costs are being covered. This is the "equilibrium output" of the short run: the producer is making an adequate return on his investment, is covering his own salary, all fixed costs, and is also meeting all other expenses. He is making no *profit,* but will be content to remain in business. He *is* maximizing his net revenue; it just happens to be zero at this particular output level, but at any other output his net revenue would be negative. If price rises to $P_2$ he will increase his output to $q_2$, and here he will not only cover all expenses but will also make a profit, the amount of which is indicated by the shaded rectangle in the figure. Note that his average total costs have risen from that of the equilibrium output; his production is now less *efficient,* but because of the increased price he can more than cover the increased costs.

If price should drop to a level below $P_1$ the producer would be far less satisfied. At a price of $P_3$, for example, he would continue to produce, but at the reduced output of $q_3$. By operating at this production level he will still cover all his variable costs and will be making some return on his fixed costs—not as much as he would like, but *some* return is better than none at all!

If price should drop to a level below his average variable cost curve—say to a level of $P_4$—he would stop production entirely, for he thereby eleminates the necessity of meeting his variable costs. True, he makes nothing on his investment: in fact, he earns no production revenue whatever. But now he has no production *expenses* that must be met by cash outlay. His fixed costs continue, but he loses less by stopping production than he would if he were to continue operating.

In summary, then, when operating to maximize net revenue in the short run, that part of the marginal cost curve above the average variable cost curve is the producer's response curve, and is his short run supply curve.

### Industry Short Run Supply Curves

The industry short run supply curve is derived directly from the marginal cost curves of the firms making up the industry. To illustrate this derivation: suppose the entire industry were composed of the four firms, *A*, *B*, *C*, and *D*, whose average total and marginal cost curves are given in Fig. 3.13. (Four firms would hardly meet the specifications for a purely competitive industry, but we will

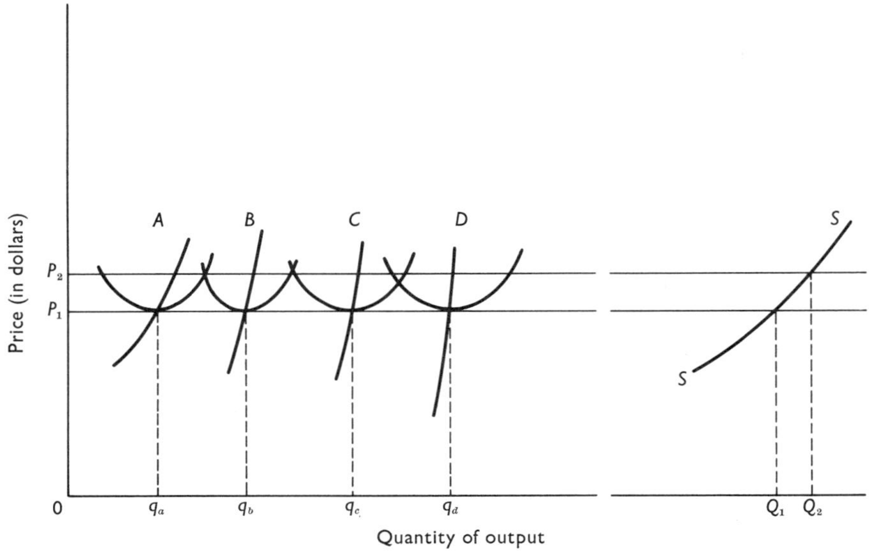


**Fig. 3.13** The industry supply curve (*SS*) is formed by aggregating the marginal cost curves of all the firms in the industry.

assume that each of these four producers *acts as if* he were in a purely competitive industry.) Now suppose that market price is at the level $P_1$. To maximize net revenue, firm $A$ will produce the quantity $q_a$, and firms $B$, $C$, and $D$ will produce $q_b$, $q_c$, and $q_d$ respectively. Since the output of all four firms will comprise the total market offering, the supply on the market will be $q_a + q_b + q_c + q_d = Q_1$.

If price should increase to $P_2$ each of the four firms would expand their output, moving up on their own marginal cost curve until $MC = P$. The new quantity offered on the market by the industry would be $Q_2$, again the sum of the individual firm's outputs. In the same fashion the response of producers to any price can be determined, so the short run market supply curve represents the horizontal aggregation—the "summing across"—of the short run marginal cost curves of all the firms in the industry.

## Long Run Supply Curves

Once restrictions with respect to fixed factors are removed the firm is able to adjust its *scale* of operation to whatever seems most desirable (within limits of capital, labor, abilities, etc.). Reductions in unit costs brought about by a larger scale of operation are called "economies of scale" and are usually divided into two types: *technological* (bigger machinery, faster equipment, more efficient management, better organization, etc.), and *pecuniary* (reductions in cost due to bulk purchasing, etc.). Such economies of scale give rise to the so-called "envelope" curve, or the *long run average cost curve* (*LRAC*).

This envelope curve is strictly a planning concept, not a chronological description of the development of any particular firm. For example, suppose one were planning a new firm and that plans were drawn for plants of five different scales of production. *If* these plants were built they would have operating characteristics that would result in the five short run average cost curves of Fig. 3.14. The marginal cost curve (short run supply curve) of each proposed plant is also shown, but remember we are still in the process of deciding *which* plant to build.

Suppose plant No. 1 were selected. As long as price remained at or above the level $p_1$ profitable operation could continue satisfactorily. But this is a competitive industry, and clearly the owner of

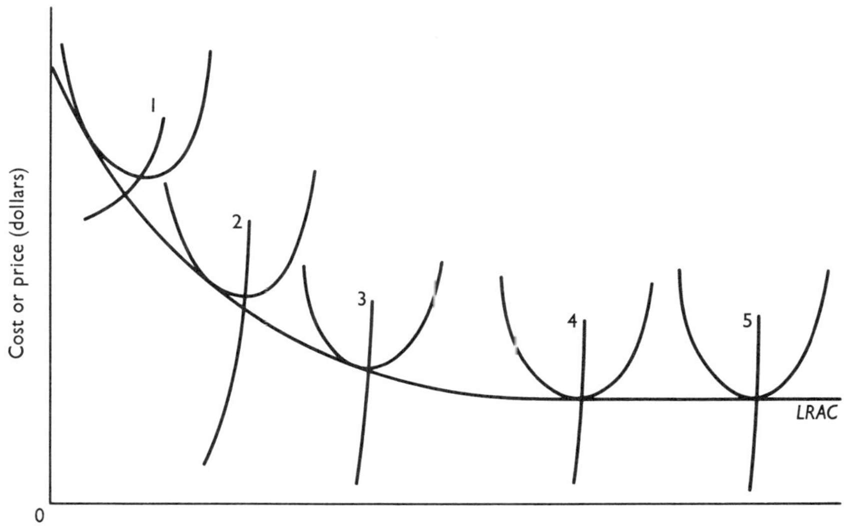


**Fig. 3.14** When economics of scale exist, the long run average cost curve (LRAC) is an "envelope" curve.

this plant would be hard pressed to compete with the owner of plant No. 2, for plant No. 2 can operate profitably as long as price remains at or above $p_2$. Similar reasoning leads one not only to reject plant No. 1, but also No. 2 and No. 3. Only when the scale represented by plant No. 4 is reached can there be any real hope of success, for in a purely competitive industry it is certain that plants at least as large as No. 4 will be built. A plant as large, or even larger, than No. 5 might be selected, but there would be no cost advantage to be gained by such a decision, and because each prospective producer would go through a similar analysis, we can be sure that all firms in the industry will be at least as large as plant No. 4, where all the advantages of scale are present.

If a curve is drawn so that it is tangent to all the possible short run average cost curves, this "envelope" (*LRAC* in Fig. 3.14) will indicate the trend of *long run average costs* as scale is increased. If a long run equilibrium is to be reached it will necessarily come only when all firms in the industry are large enough to have achieved all the economies of scale—when all firms are at the "bottom" of the long run average cost curve. In this happy state of affairs all firms in the industry will have identical long run average costs, short run average costs, marginal costs, and product price. What is more, all four of these will be equal. This is, of course, the familiar condition of *long run equilibrium.*

But what happens if, before all the economies of scale can be attained, the firm becomes so large that it can influence product price? In this case it simply means that the technological facts of production rule out the purely competitive model. It also illustrates well the idea that competition tends to create the very forces which eventually result in its own demise: the drive toward lower costs imposed by a purely competitive industrial structure may literally force the firms to get so large that they can influence price. Once this latter position has been reached the purely competitive model is no longer applicable.

### Constant Costs and Increasing Costs

If the *LRAC* curve flattens out at a scale such that the industry remains purely competitive, the industry long run supply curve will be perfectly elastic—a horizontal line. In such an industry output is expanded or contracted by firms entering or leaving the industry, as has occurred in sawmilling in the United States.

If diseconomies of scale exist, and if they outweigh the possible economies, then costs could increase with increasing scale. Increasing costs, however, can usually be avoided by duplicating the plant of most efficient size. For example, one doesn't *have* to build a bigger sawmill to double output; he might instead build *two* sawmills. Foresters should hold the possibilities of these diseconomies in mind, however, for forest land cannot be moved about and it is often difficult to assemble exactly the pattern of ownership desired.

### External Economies and Diseconomies

In contrast to the *internal* economies discussed so far—the economies arising from the action of the firm itself and which to a large extent give rise to the declining average costs observed as firms grow larger—are the so-called *external* economies and diseconomies. These latter economies are those dependent on the general development of the industry. Such economies are numerous in most industries and the forest-based industries are no exception. In the Pacific Northwest, the South, or the Northern Lake states, for example, the supply of woods labor is large and the tradition of woods work strong. Hence, if one crewman leaves his job another can be picked up with relative ease—an economy certainly not attributable to any single firm but rather to the size of the industry itself, and hence an "external" economy. The network of woods roads, many of which are used in common, provides another example. Perhaps most important is the fact that any industry supporting a substantial number of employees can attract the attention and support of political leaders. Through the efforts of such men, the forest industries have been given valuable aids such as federal and state forest experiment stations, forest product laboratories, forest fire protection systems, etc. These are outstanding examples of *external* economies of scale.

The opposite side of this coin must also be examined. The general public is often aware of the *external diseconomies* which plague forest industries. Such diseconomies are often divorced from the question either of firm or industry scale of operation: some writers refer to these as *social costs* (Kapp, 1950) while the more modern term is *externalities.* As used today, *externalities* are either costs or benefits that do not enter into the financial accounting system of the firm or the industry. Logging roads and landings (for example) have been located along stream bottoms in northern California mountains because they are easier to build here and considerably less expensive to the logging company. Because of this practice the state is losing high-quality forest sites, is silting up its reservoirs at a more rapid rate than can be warranted, etc. (Wallis, 1965). This latter is a large negative externality of the logging industry. Water pollution by pulp mills, damage to roads by logging trucks, the destruction of waterfowl nesting sites, the silting over of salmon spawning beds, the destruction of scenic values—all are examples of negative externalities associated with timber production.

Other forest industries are equally guilty, but the debit is not all on one side of the ledger. The timber industry, for example, presently shoulders substantial costs of outdoor recreation—increased fire prevention and protection charges, highway maintenance, hunter patrols, etc. It also bears the increased costs of plantation due to damage by deer and bear and other costs introduced into forest management by silvicultural changes that wildlife maintenance might necessitate.

Public resource policy is often directed toward either preventing externalities harmful to the public, or providing some means of recovering the costs from those who create the problem. Such actions are referred to as "internalizing the externalities." Taxes may be levied according to the amount of effluent discharged into the river, for example, or subsidies paid for "proper" forest management, plantations exempted from property taxation, or various other means of "internalizing" the costs or benefits may be employed.

Externalities such as these are obviously far too important to be passed over in a paragraph or two. They are introduced at this point only to show how and where they fit into the general scheme of economic theory, and will be treated more fully in future chapters.

# 4

# Prices in a "Free Enterprise" Economy

It is difficult to overestimate the role of price in a market-oriented economy such as that of the United States today. On the one hand the price system permits each consumer to view the many thousands of items offered for sale, compare them on a basis of their relative cost *to him* in terms of his own purchasing power, and make his own decision as to what and how much he should purchase. On the other hand, these same prices reflect to a considerable degree the relative cost to society of making the items available. It is this balancing of societal costs and consumer wants that makes a free market price system such an efficient means of allocating resources.

But this is not the entire picture. Imperfections in knowledge, in market organization, or in production know-how, may make some price signals less accurate than others. Furthermore, some firms may be able to control or influence price to a degree that makes price comparisons a distorted reflection of comparative production costs. Even more significant to our study and to society is the existence of some very real and important externalities—benefits and costs that do not enter into the accounting system of the firm at all. Quite obviously an allocation system based on prices established under any of these latter conditions might be very inefficient.

These reasons, together with the fact that price formation necessitates an integration of supply and demand theory, makes it appropriate to end this brief review with a look at some of the more simple pricing models of economic theory.

## PRICES UNDER PURE COMPETITION

### Short Run

In the short run, and under purely competitive conditions, the individual producer will maximize his net revenue by adjusting production until the cost of producing one more unit just equals the price received for that unit: until marginal cost equals price. Since every producer acts in this fashion, the industry supply curve can be visualized as the horizontal aggregation of all the individual firm's marginal cost curves. On the demand side, consumers adjust their purchases until the satisfaction gained from purchasing one more unit is just equaled by the price they must pay for that unit. Market price will be determined by the intersection of the supply curve with the market demand curve.

### Long Run

In the long run, market price is always moving toward the elusive, constantly shifting equilibrium of supply and demand. Fig. 4.1 depicts this situation under conditions of constant costs. Here the short run supply curve ($mc$) of one of the firms in the industry

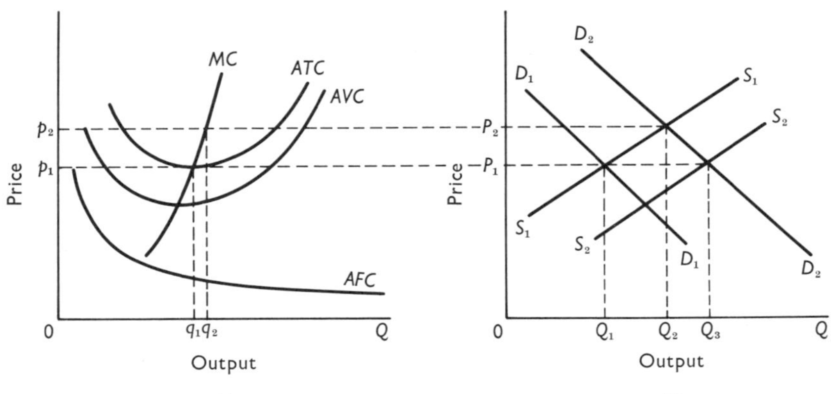


(a) (b)

**Fig. 4.1 Short run adjustment of the firm to demand change in an industry characterized by constant costs.**

is shown on the left, while on the right are the industry demand and supply curves ($D_1D_1$ and $S_1S_1$). Under the equilibrium conditions shown, the industry is producing the output $Q_1$, price is at $P_1$, and the particular firm shown is producing $q_1$.

Now suppose the demand curve shifts outward to $D_2D_2$. In the *short run* this will result in a rise in price, causing the firm to expand its output to $q_2$. The similarly increased output of all firms will raise the industry's output to the new level of $Q_2$, where price will stabilize at $P_2$.

At this price the firm will obviously be making a substantial profit (in the economic sense; see Chapter 3, page 48). Hence new firms will be attracted to the industry and old firms will expand plant capacity. Providing there are no limitations on the supply of production factors, this increased capacity will eventually shift the supply curve outward to $S_2S_2$, where a new equilibrium price will be established at an output of $Q_3$, with price restored to exactly its former level of $P_1$. Once again we will have the four-way equality of $MC = P =$ short run average cost $=$ long run average cost.

## PRICES UNDER NON-PURE COMPETITION

Now how about *non*-pure competition? (Bear in mind that a monopolist is not necessarily a big, fat cigar-chewing individual who delights in foreclosing mortgages on helpless widows in the dead of winter, but is just someone sufficiently important to affect the price of things he buys or sells.) We can be very glib and simply say that producers will still maximize net revenue by equating marginal cost to marginal revenue. But this is not very informative; the *shape* of the marginal cost curve becomes important here, and in the case of a *buyer* monopolist (a *monopsonist*) the shape of *his* marginal cost (or marginal outlay) curve.[1]

### Monopoly Pricing

As an example we can take the Woodstock fuelwood market data as given previously (Fig. 2.1 and Table 2.1, on pages 23 and 24), but now we must assume that a single producer has established a monopoly. Since an industry demand curve is also its average revenue curve, total revenue can now be derived directly by multiplying the aver-

[1] Monopsony will be treated in Chapter 17.

age revenue (or price) by the quantity marketed. Hence, since average revenue is given by the equation $P = 28 - .001Q$, and total revenue equals $P(Q)$, the equation for total revenue will be:

$$TR = P(Q) = 28Q - .001Q^2$$

and marginal revenue, the first derivative of total revenue, will be

$$MR = \frac{d(TR)}{d(Q)} = 28 - .002Q$$

Note that the intercept (28) remains unchanged, while the slope coefficient is exactly doubled. Hence, if the demand curve is represented by a straight line as in the example used here, the marginal revenue curve will also be a straight line, will intersect the price axis at the same point as the demand curve, and will have exactly twice the slope of the demand curve. Under these particular (and peculiar) conditions, therefore, the marginal revenue curve will always bisect a horizontal line drawn from the price axis to the demand curve (Fig. 4.2).

The supply curve of our Woodstock monopolist will still be the

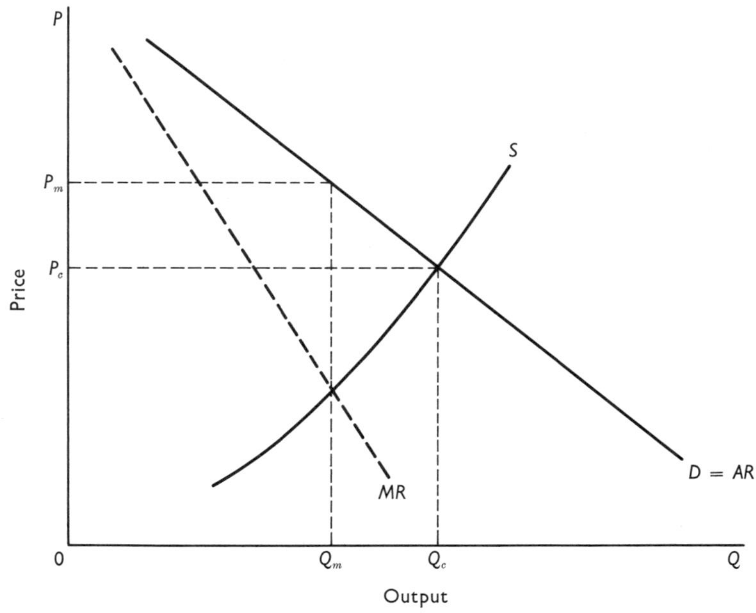


**Fig. 4.2** Price and output decisions under monopoly.

same as previously calculated: the *production* aspects are unchanged. But now the supply curve will intersect the *marginal revenue* curve at an output level of $Q_m$, a decidedly reduced output! And this output can be sold at the price $P_m$. These price and output levels can be contrasted in Fig. 4.2 with the competitive price and output levels of $Q_c$ and $P_c$. Clearly, the tendency of monopoly will always be to restrict production and to increase prices. How much production will be restricted, and how much prices will be increased, will depend upon the shape and the slope of both the demand and supply curves involved, as a little experimentation on scrap paper will readily confirm.

## Oligopoly Pricing

Oligopoly poses a more difficult problem. Suppose there are four firms, all identical in every way, producing some product for which demand is as shown on the right side of Fig. 4.3. The average and

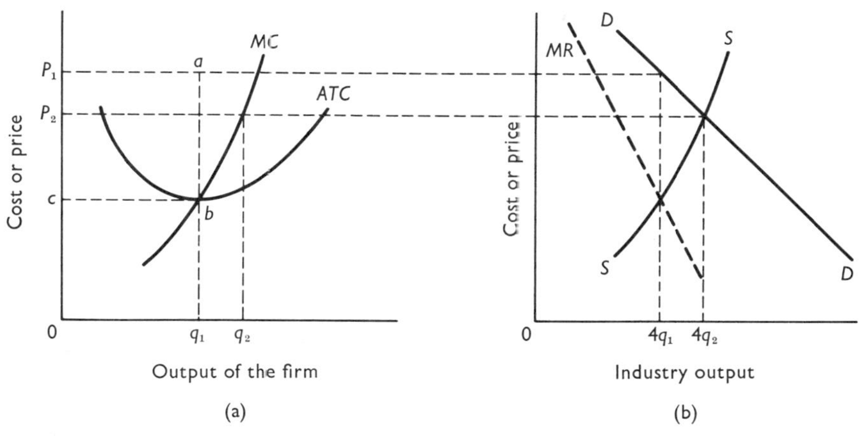


**Fig. 4.3** Price and output decisions under oligopoly.

marginal cost curves of one of the firms is shown on the left side to facilitate comparison.[2]

As shown in the figure, if each firm produces at minimum average cost it would produce an output of $q_1$ units, and the combined out-

[2] Since the four firms are identical, the supply curve (the horizontal aggregation of the four *MC* curves) will have exactly the same slope as the *MC* curves if the output scale of the industry is exactly four times that of the firm. These conditions are maintained in the above figure.

put of the four firms would be $4q_1$ units. If full "cooperation" of the four producers could be assured, this output could be sold at the price $P_1$, and each firm would have a net revenue of $P_1abc$.[3]

But why would any firm want to sell at this particular price? Clearly, by lowering its own price very slightly below $P_1$ any one firm would be able to expand its sales very substantially: it would, in fact, capture some of the market of the other producers, reducing *their* profit, but substantially increasing its own. But if each one of the producers acts in this same fashion, price would drop to $P_2$, with the new industry output of $4q_2$ and each firm realizing a smaller net revenue than previously.

In making production decisions the oligopolist must consider not only the effect of his own output on product price but also the effect of his decision on the production policy of his competitors. If he increases his own output he may increase his net revenue—providing his competitors make no change in their output patterns. But if they follow his lead he will lower his profit—unless he were to lower his price still further. Since he cannot be certain of his competitors' reaction the oligopolist tends to take one of three positions:

1. He does not vary his price. He (and all the others) go along with the existing price pattern since any attempt to change it may worsen his position.
2. He tries to bring about a merger of the competing companies. It is clearly to the benefit of all four to cooperate since then they can maximize their net revenues by operating all four plants at the optimum level.
3. The four producers can make "subrosa" arrangements.

But (2.) is often prevented by anti-trust legislation, and (3.) is punishable as a criminal offense. Hence if (3.) is accomplished, it will be at best temporary, for the arrangements cannot be enforced. Sooner or later one of the producers will "cheat" on the agreement by shading a price to a particularly large buyer. Then a price war will break out (although a price war should not be taken as evidence of prior collusion) and all four producers lose.

In oligopoly, therefore, the usual course is that of (1.). Prices tend to be "sticky," and competition takes the form of offering

[3] In legal parlance such cooperation is spelled "collusion."

better credit terms, better delivery service, higher quality, and often a high degree of competitive advertising. Unlike either the purely competitive or the monopoly cases, there is no precise theoretical solution.

### Monopolistic Competition Pricing

In a very real sense, pricing under monopolistic competition is simply an extension of the oligopoly case. With many producers, each with some small measure of control over the price because of some real or fancied difference in their particular product, there is far less drive toward mergers simply because there are too many firms. And for similar reasons, collusive attempts are patently useless. Hence, each producer strives to intensify the degree of product differentiation in the buyer's eyes. Brand names and trademarks are a common means of product differentiation. The Weyerhaeuser and Georgia Pacific brands of Douglas-fir lumber provide an easily observed example from the timber industry, while writing-paper watermarks and tent trademarks illustrate a similar approach by other industries.

One way of expressing this diagrammatically is to say that each producer has a demand curve that is *slightly* sloped; it is not perfectly elastic, as it would be under pure competition. The greater the degree of product differentiation the more inelastic to price changes the demand curve for his product becomes.

### Need for Market Control

This last permits bringing up a final review point concerning monopoly pricing under all degrees of market imperfection: monopoly, in the sense of controlling production, is not particularly significant unless it is accompanied by the power to control price. This latter power is dependent upon the *demand* situation as well as that of *supply*.

Suppose, for example, that by some stroke of fate all southern pine lumber production were to come under the control of a single producer. To be sure, price would be affected, but increases would be limited substantially by the fact that Douglas fir, ponderosa pine, white pine, and other species could be—and *would* be—substituted for southern pine if prices were raised very far out of line. Monopoly *position* and monopoly *power* are not the same thing!

## DISTRIBUTION THEORY

Distribution theory is concerned with who gets how much of whatever our economic system produces. Ricardo, Marx, Pigou, Pareto, and other famous early economists were primarily concerned with these distribution issues. Small wonder that in earlier days the term "political economy" was used to describe the field now usually referred to as "economics." Clearly, no matter how vital or interesting distribution theory might be, a text in forest economics is no place for a detailed treatment of distribution theory, but three of its topics are of great importance to forest economics: profit, interest, and rent.

### Categories of Productive Factors

When speaking in general terms economists usually divide productive factors into at least three, and often four, broad categories: land, labor, and capital are almost universally recognized, and management is frequently cited as the fourth. The return to land (or "natural resources") has traditionally been called *rent;* the return to labor is called *wages*; and the return to capital is usually termed *interest* (to be covered in Chapter 10). Management is the most recent addition to this classic trio of land, labor, and capital, for without management, land, labor, and capital are often combined very ineffectively. Few would agree, however, that *profit*—the fourth type of return—is properly defined as the return to management. This reluctance stems in part from the knowledge that management, ownership, entrepreneurship, and risk bearing are all closely interrelated but quite distinct functions. Perhaps even more important is the growing awareness that management in the modern corporate sense of the term is better regarded as a highly specialized form of labor.

**Profit.** Profit has already been quantitatively defined as the difference between total cost and total revenue and is synonymous with "net revenue" (see Chapter 3, page 48). The latter term will generally be used in this text to avoid confusion with "accounting profit" which is defined as sales receipts less the sum of payments for hired factors plus depreciation. This residual, or accounting profit, generally goes to owners of the enterprise and may include interest, wages, and rent, for owners often supply much of the capital, furnish at least part of the management, and may also own resources used by the firm.

As long as this residual consists entirely of wages, rent, and interest (whether they are identified or only imputed to the owners) there is no particular problem of distribution. But after the owners have received payment for services supplied (even though not contracted) there may still be something left over. *This* remainder is true *profit* in the economic sense.

In Chapter 3 it was seen that the purely competitive equilibrium position of the firm is a "no profit" position. There will be an accounting profit, since all costs are covered and total costs include payment to *all* productive factors including those to owners for wages, interest, and rent. But there is no profit in an economic sense.

Equilibrium conditions, however, are seldom achieved, and if achieved are not long lived: demands shift, factor costs change, technologies alter, new products appear, competitive products are invented. In this dynamic situation profits can arise whenever market price exceeds the firm's average total cost. Profit can also arise from market imperfections, particularly from monopoly, oligopoly, and similar non-perfect market structures. There is a distribution problem with respect to such profit: who should get it, and why?

Profits arising from dynamic disequilibrium in a purely competitive industry are usually regarded as a reward for innovation, or as a payment to enterprise for particular astuteness in assessing the true risk of some specific venture. The opportunity to earn such rewards has been regarded as the mainspring of our capitalistic economy.

There is much more ethical concern over profits arising from monopolistic positions. A great deal of effort is expended by industries to secure and hold some market advantage, even though its benefit may be temporary. Yet monopoly profits arising from discovery of new products or new processes are also a reward for innovation, and the possibility of establishing such a position is surely a major incentive for technological progress in any capitalistic system.

The source and magnitude of rent, interest, profit, and wages is the proper study of distribution theory, but the ethical problems of distribution are not to be settled by economics. Essentially, a capitalistic system distributes rents, interest, and profits on the basis of private ownership of resources and capital, and includes these payments as part of personal income along with any wage payments that might be due the individual. A socialistic system makes personal income primarily dependent on wages. The choice between the two systems must be based on a very complex system of value judgments: there is no *a priori* "justice" that can be accorded either.

**Wages.** This term is defined as any payment made for services rendered by humans. Wages thus include many kinds of payments—to singers and other artists, to doctors and school teachers and bank presidents and ditch diggers and garbage collectors—and even those made to professional foresters. It is thus somewhat more broadly defined here than in some other uses.

**Rent.** Rent plays a peculiarly significant role in forestry. Fig. 4.4(a) gives the basic short run average total cost curves for a firm

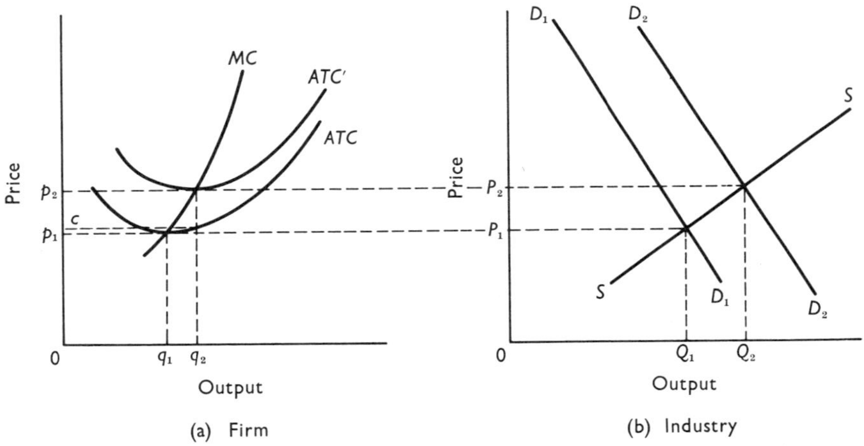


**Fig. 4.4** "Ricardian Increasing Rent," rent arising from an increase for product demand when one productive factor is absolutely fixed.

operating in a purely competitive industry. Price is at the equilibrium level $P_1$, established by the intersection of the industry demand and supply curves, shown in Fig. 4.4(b).

Now introduce a couple of basic changes: (1) suppose the product involved requires a very special type of land and that all this land is already in use by the industry to produce $Q_1$; (2) suppose the land in use by the particular firm illustrated in Fig. 4.2(a) is being *rented* from a landlord.

If for any reason whatever the demand curve shifts out to $D_2D_2$, price will rise; our firm will increase its output (as will all other firms in the industry) and eventually industry price and output will reach $P_2$ and $Q_2$. At this price the firm will be producing $q_2$ and will not only be covering all fixed and variable costs (including the rent) but will also be making a very handsome profit of $q_2 (p_2 - c)$.

Under normal conditions one would expect new firms to enter the industry, or old firms to enlarge their scale of production (or both) and the former equilibrium to be restored. But this normal reaction to shifting demand cannot take place, for there is no more land of the required type. Now, observing the large profit being enjoyed by the firm, the landlord may well decide to increase the rental charge. The firm will have to meet the increase or go out of business, for there is no more of the needed type of land. The only remaining question concerns the amount of the rent increase, and this can be settled quickly; the landlord can take the entire profit of the firm as rent for the scarce factor of production he owns. After including this new, increased rent in his production costs the average cost curve of the producer will be moved up along the marginal cost curve to the new position indicated by $ATC'$—differing from the former $ATC$ by precisely the increased rental payment.

This is an example of what is often called "Ricardian Increasing Rent" (Viner, 1932). Ricardo defined rent as the payment "to the indestructible powers of the soil." It has since become obvious that these powers are not entirely indestructible, and the concept of rent has been generalized to include all factors of production rather than confining it to land. In modern terminology economic rent is defined as "any payment to a unit of a factor of production, in an industry in equilibrium, which is in excess of the minimum amount necessary to keep that factor in its present occupation" (Boulding, 1948).

## NORMATIVE ASPECTS OF PURE COMPETITION

The following question is sometimes raised. "Since a purely competitive price will always gravitate to a no-profit level, and since profits are the life-blood of any enterprise economy, why is so much attention paid to the competitive price? Is it any better, socially, than a monopolistic price?"

The answer is not a simple one, but the purely competitive pricing mechanism of $MC = P$ does have certain normative qualities. To see, if not to understand, these qualities is in itself a capsule review of production theory.

Demand theory relies upon the assumption that each consumer, when free to choose as he wishes, will maximize his own personal well-being. The last unit of any good or service purchased by a consumer yields a "marginal utility" just equal to its purchase price. Overlooking the fact that many things must be purchased in sizable "lumps," it can be shown that the consumer will adjust his purchases so they will all yield the same *marginal* utility—that there is no shift in his purchasing schedule, given the amount he has to spend, that would increase his well-being.

On the production side we have another sort of balance. By producing that quantity which just makes his marginal cost equal to market price each producer will maximize the net returns to his firm.

The interplay of these two forces of consumption and production establish a market price at the intersection of the demand and supply curves. When prices are so established we can say:

1. Each consumer can purchase as much or as little as he wants of any good or service offered *as long as he is willing to pay its cost of production.*
2. Each producer will be happy to produce enough of his product to make the total quantity (the industry supply) just sufficient to assure consumers of the above.
3. *Given the existing distribution of income,* there is no shift in the allocation of resources (or in the relative amounts of goods being produced) that will permit any consumer or producer to be made "better off" without lowering the welfare of some other person.

But note carefully the assumptions which must be met; pure competition in all production, complete knowledge on the part of consumers and producers, perfect mobility of all production factors, and acceptance of the given income distribution.

In an enterprise economy price is at once the "carrot" dangled in front of the producer which encourages him to produce just the right amount, and on the other hand serves as a brake on consumption, requiring consumers to choose between alternative quantities as well as between alternative goods. Much of the strength of a competitive economy stems from the fact that price acts as an auto-

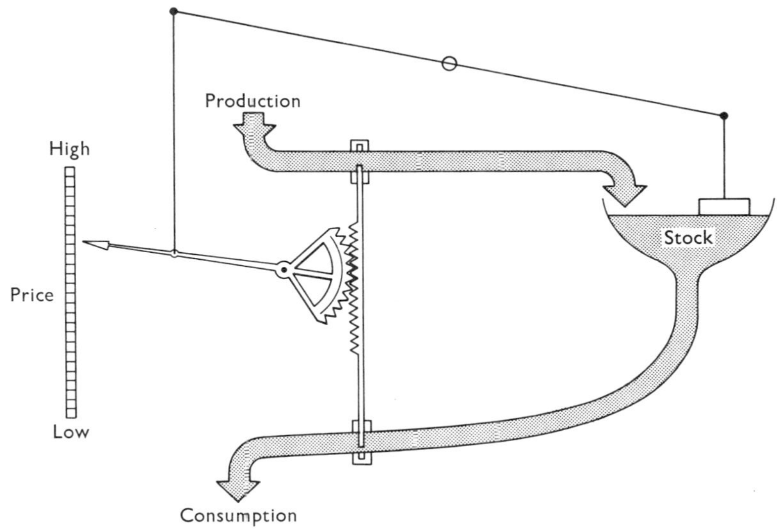


**Fig. 4.5** The regulatory nature and function of price in an enterprise economy.

matic regulator of production and consumption. Fig. 4.5 represents an attempt to depict this regulatory feature in physical terms.

When the stock of some good begins to get low, price rises. The rise in price is accompanied by an "opening" of the valve controlling production, and an increased stream of products flows into the consumer's pool. Simultaneously the volume of consumer buying is checked and the outflow from the pool correspondingly reduced. On the other hand, if stocks become high, prices fall, production is checked, and consumption encouraged.

While illustrating the role of price in a competitive economy, one must also recognize that such a diagram is a very naive view of our complex economy. Many goods and services are not subject to the automatic control of the market, and among these are important forest products. Recreation, water, wildlife, sports fishing, aesthetics, air purification—none of these is subject to market decisions in the usual sense. The problems created by such non-market items will be treated more fully in Chapter 19.

Even if attention is confined to products subject to the market system, one must agree that many things could happen to throw off the regulator. The float might stick, or fail to respond to changes in the level of consumer wants, or producers might conceivably combine so as to restrict the flow of goods into the pool, thus keeping prices high and consumers relatively unsatisfied. Or the gears on the control box could get rusty or dry, thus preventing the signals from the pool from reaching the producers. And one must point out that there is also the possibility that some plumber might come along and try to "adjust" the mechanism so that a particular group would have an advantage. Governments, for example, do establish price controls for various purposes—occasionally even for very *good* purposes.

Perhaps the biggest danger in presenting such a diagram lies in the possibility that it might lead some students to believe "if we just let the price system alone everything will work out satisfactorily." *If* the conditions stated earlier describe reality this might conceivably be the case. But industries are *not* always purely competitive; in actuality very few could be so described. Consumers do *not* have perfect knowledge (any college sophomore who still thinks the advertising business exists solely to educate consumers has failed to take advantage of his opportunity to learn). The factors of production are *not* perfectly mobile: capital is *not* equally available throughout the economy, labor is *not* completely transferable, and land (natural resources) is *not* equally or evenly distributed. Very few people are willing to accept the distribution of income as given: income taxes, taxes on luxury goods, and a host of other devices have been designed to promote a "better," more equitable distribution—and even those devices sometimes work in opposite directions. And finally, nearly everyone recognizes that in our society the private sector tends to emphasize the rewards going to the individual, which automatically means a lesser emphasis on things that benefit the group as a whole. Hence education, highways, recreational systems, postal departments, police departments, etc., all tend to be slighted by the profit-motivated economy. Unless compensated by society itself through government this could lead to disaster for all.

At this point in the study of forest economics a real dilemma arises. Forest economics deals with the economic problems that arise in managing *forests,* yet the teacher who confines himself to forestry may fail to train men competent to deal with the economic problems of tomorrow's forestry. Professor Fusfeld has observed that the teacher of economics "faces a situation in which error must be unlearned before advances can be made"—that very often students entering college "have been indoctrinated with the folklore of economics, most of which is either unrealistic, or just plain wrong." High on the folklore list is the idea of "free enterprise," which term disguises the fact that "large segments of the United States economy have strong monopolistic elements and that much of the economy is strongly influenced by public policy" (Fusfeld, 1963). One can scarcely expect such "folklore" to be corrected by any single course, but the warning should be held in mind.

# PART II

# TIMBER PRODUCTS ECONOMICS

# 5

# Timber in the United States Economy

The fact that timber production constitutes only a part of forestry and that it may well decline during the coming decades was emphasized in Chapter 1. For simplicity, however, the review drew only on timber products for illustrative material. It is well, therefore, to remind the reader again that *forest* production and *timber* production are not synonymous. The importance of *forestry* to the economy of the United States cannot be measured by the dollar value of the timber produced, or even of the lumber, plywood, paper, piling, poles, fuelwood, and other wood products used by the nation's consumers. At the same time one should recognize that timber is by far the major source of dollar revenue from forests, and for a forester or any other responsible citizen to overlook this fact would be little short of absurd.

Few would deny that we, or any country, could far more easily get along without timber than we could without water. Yet water is a product of the forest. Virtually every river in the United States has its origin in forested areas, and without the protection that forests afford these watersheds there would be little hope for maintaining either the quantity or quality of water needed by our industries, our cities, or our farms. Unlike lumber and plywood values, however, the value of water is not determined in a national market.

The contribution of water to our economy has not yet been quantified in a way that permits either comparison or inclusion with timber values.

Much the same things could be said about the two closely related products, forest-based recreation and forest wildlife. Every year increasing numbers of people visit forests for recreational purposes of one kind or another, leading one to accept the simple statement that the opportunity to participate in forest-based recreational activities is highly valued by the average citizen. Yet wildlife and recreation are not measured by the same value scale as timber, and perhaps not by the same value scale as water, for that matter. Aesthetic and many other values of forests are only now being identified; the task of measurement has not yet begun. At the present time there is simply no way to measure the total contribution of forests to the economy.

Still, the fact that many forest values cannot presently be measured should not prevent discussion of those values that can be identified and measured. Moreover, if the difficulties of estimating the contribution of timber to the economic well-being of the nation can be understood, perhaps a better appreciation of the problems involved in including non-market values can be gained. But this treatment requires an almost constant awareness that we are examining only part of the values produced by forests. With this caveat firmly in mind a quick look at *timber* values in our economy may be ventured.

## ECONOMIC INDICATORS

Anyone who reads *Time, Newsweek,* or the financial page of any newspaper knows that Gross National Product, the GNP, represents the market value of goods and services produced by our economy.[1] As an important indicator of how the economy is functioning, estimates of GNP formation are studied by businessmen almost as avidly as by economists. The contribution of an industry to the Gross National Product is one of the measures that can be used to indicate that industry's importance to the general economy.

The *Census of Manufacturers,* published every five years by the Department of Commerce (Bureau of the Census), reports various

[1] A good many items are excluded. One important example of such excluded items is the value of services contributed by housewives—certainly a significant item.

statistics collected from all manufacturing units. Of these data, three items of key significance are reported for each industry class: "employment," "payrolls," and "value added by manufacturing." Employment is reported monthly, and all full time or part time employees on the payrolls of operating manufacturing establishments who "worked or received pay for any part of the pay period which included the 12th and ended nearest the 15th" were included in the 1963 census. "Payrolls" include gross earnings paid in the reporting year to "all employees on the payroll of operating manufacturing establishments." The third item, "value added by manufacture," is obtained by subtracting all costs of materials, supplies, fuel, electric energy, sales, and miscellaneous receipts, from the value of shipments (including resales) and other receipts. The result is adjusted for changes in inventory and work in process. "Value added by manufacture" represents the amount available for wages and salaries, interest payments, profits (whether distributed or not), taxes, and depreciation and depletion charges. If these amounts are totaled for all manufacturing units of the nation, the result will be the contribution of manufacturing to Gross National Product; totaled for any particular part of manufacturing the result will be the contribution of that segment to GNP.

Unfortunately, the *Census of Manufacturers* does not distinguish between timber products industries and non-timber products industries, so we must face the problem of how to separate timber products industries from the others. Some cases are so clear as to create no difficulty; timber management, along with sawlog and pulpwood production, would probably be called timber production industries by nearly everyone, although one might well hesitate to include them in a "timber products" group. Similarly, few would argue about lumber, veneer and plywood, or poles and piling, or perhaps even wood pulp. But what of the furniture industry? Very possibly its wooden furniture segment would meet with general acceptance, but unfortunately (for present purposes) many furniture companies manufacture not only wood furniture but also metal, plastic, and upholstered products as well. And what of paper, or the construction industry, or paper products, or textiles, or transportation? In each of these timber provides an important input, yet timber is clearly only one of several important raw materials.

The situation is far more complex with the non-timber products of the forest. There is no industrial classification or grouping within the *Census of Manufacturers* that can be identified as "water industries," or as "forest recreation industries," or "wildlife industries," a fact explaining in part why these important forest products cannot be included in this chapter.

Actually, there is no neat way of determining just where to draw the line between "timber products industries" and "non-timber products industries." Almost all industries use some wood, just as all timber products industries use other raw materials; it is primarily a matter of degree. Consequently, there is no way of determining precisely the contribution of timber or of timber products to GNP, to employment, or to payrolls—just as there is no way of determining precisely the contribution of oil, agriculture, or steel, or that of any other raw-material-centered industry. If one is to attempt extension of the analysis beyond the value of the raw material itself, some arbitrary decision must be made as to how the particular measure being considered (employment, value added, etc.) is divided or allocated among the various materials that go into the manufacturing process. This allocation might conceivably be based on proportionate cost of raw materials, on raw material volumes, on sales receipts, on revenues, or on almost any other basis. And of course, the same problem of allocation arises with the so-called primary industries as well as with the others, for even lumber manufacture uses other raw materials besides wood.

## ECONOMIC IMPORTANCE OF TIMBER

In 1963 the United States Forest Service, using data from the 1958 *Census of Manufacturers,* attempted an assessment of the economic importance of timber in the United States, in a publication bearing that title (Hair, 1963). The study began with an estimate of the values produced and the employment furnished by *forest management,* which was defined to include all activities involved in protecting and managing forest lands for the production of timber and related crops. Under this definition, forest management was estimated to "produce" nearly one billion dollars worth of stumpage —some 9,980 million cubic feet of timber—in 1958. To produce this quantity required the equivalent of some 82,000 people working full time, including the efforts of slightly under 11,000 professional foresters.

No comparable estimate is available for later dates, but production has increased considerably since that time. Table 5.1 shows production in several product catagories for each year beginning with 1950. Note particularly the changing proportion of the total contributed by each major use. Sawlog production, the basis for lumber manufacture, has fluctuated considerably and there is some evidence of an overall decline. In the last five reported years, however, production was only slightly under that of the early 1950's. Veneer logs and pulpwood have both increased dramatically; by 1967 production of veneer logs was fully three times that of 1950, while production of pulpwood had more than doubled. Fuelwood continued the decline that has characterized this product since the turn of the century, and was less than ⅓ of its 1950 production in the last reported years. Timber used for other products declined to less than five per cent of the total, even though this category includes such important items as mine timbers, poles, piling, cooperage, fence posts, shingles, excelsior, and a host of other products in common use.

Some question should be raised about using the term "timber production" to describe volumes such as those of Table 5.1. The figures of the table indicate, as correctly as possible, the total harvest of the products itemized, but "total domestic production" is a somewhat misleading terminology. Strictly speaking, *timber produced* should be equivalent to *timber growth*. In 1962 (the last year for which growth estimates are available) growth was estimated to be 16,260 million cubic feet, as compared to the total domestic production in that same year of only 10,180 million cubic feet (Forest Service, Timber Trends, 1965). How much of this total growth should be attributed to efforts of those who protect and manage forests in this country is another unanswered question; presumably some growth would occur even if no conscious effort were devoted to forest management.

## Primary Manufacturing

*Primary manufacturing* is concerned with the initial treatment of the harvested raw material. As applied to timber products it includes three major industrial groupings of the *Census of Manufacturers*: lumber and wood products; pulp, paper, and board; and veneer and plywood. Before these primary manufacturing plants can begin to produce their products, however, the timber must be harvested. "Harvesting" is, therefore, included in Table 5.2, which presents the measures of economic activity described earlier (employment, payrolls, and value added) for each of the three primary manufac-

**TABLE 5.1 PRODUCTION OF TIMBER PRODUCTS, BY MAJOR PRODUCT GROUP, SINCE 1950. (IN MILLIONS OF CUBIC FEET ROUNDWOOD EQUIVALENT). FROM *THE DEMAND AND PRICE SITUATION FOR FOREST PRODUCTS* (U.S.D.A., 1971).**

| Year | Total domestic production (all products) | Total industrial roundwood | Sawlogs | Veneer logs | Pulpwood | Other products (apparent consumption) | Log exports | Fuelwood (apparent consumption) |
|---|---|---|---|---|---|---|---|---|
| 1950 | 10,795 | 8,525 | 5,905 | 345 | 1,500 | 770 | 10 | 2,270 |
| 1951 | 10,970 | 8,740 | 5,780 | 300 | 1,825 | 730 | 15 | 2,230 |
| 1952 | 10,780 | 8,775 | 5,820 | 435 | 1,810 | 700 | 10 | 2,010 |
| 1953 | 10,710 | 8,790 | 5,710 | 475 | 1,910 | 675 | 20 | 1,920 |
| 1954 | 10,585 | 8,750 | 5,635 | 480 | 1,960 | 655 | 25 | 1,835 |
| 1955 | 10,950 | 9,205 | 5,785 | 575 | 2,185 | 630 | 25 | 1,745 |
| 1956 | 11,280 | 9,625 | 5,920 | 590 | 2,480 | 605 | 30 | 1,655 |
| 1957 | 10,190 | 8,625 | 5,100 | 560 | 2,360 | 580 | 25 | 1,565 |
| 1958 | 9,985 | 8,505 | 5,160 | 615 | 2,135 | 560 | 30 | 1,480 |
| 1959 | 10,750 | 9,360 | 5,745 | 720 | 2,330 | 535 | 35 | 1,390 |
| 1960 | 10,210 | 8,910 | 5,080 | 705 | 2,565 | 510 | 45 | 1,300 |
| 1961 | 9,940 | 8,725 | 4,945 | 765 | 2,450 | 490 | 75 | 1,215 |
| 1962 | 10,155 | 9,035 | 5,120 | 800 | 2,565 | 465 | 85 | 1,125 |
| 1963 | 10,615 | 9,560 | 5,355 | 870 | 2,670 | 515 | 150 | 1,055 |
| 1964 | 11,175 | 10,190 | 5,635 | 960 | 2,885 | 540 | 170 | 985 |
| 1965 | 11,465 | 10,550 | 5,670 | 1,030 | 3,100 | 500 | 190 | 915 |
| 1966 | 11,525 | 10,680 | 5,645 | 1,030 | 3,220 | 565 | 220 | 845 |
| 1967 | 11,185 | 10,405 | 5,360 | 995 | 3,220 | 515 | 310 | 775 |
| 1968* | 11,785 | 11,035 | 5,630 | 1,025 | 3,405 | 570 | 405 | 750 |
| 1969* | 11,725 | 11,000 | 5,500 | 940 | 3,630 | 550 | 375 | 725 |
| 1970* | 11,825 | 11,125 | 5,285 | 960 | 3,925 | 525 | 430 | 700 |

* Preliminary estimates

**TABLE 5.2 MEASURES OF ECONOMIC ACTIVITY IN THE UNITED STATES PRIMARY FOREST PRODUCTS INDUSTRY, 1958, 1963, AND 1965.**[1]

| Sector | All employees (thousands) | Payroll ($ millions) | Value added by manufacture ($ millions) |
|---|---|---|---|
| *Pulp, paper, and board* | | | |
| 1958 | 223.3 | 1,239.4 | 2,801.6 |
| 1963 | 220.2 | 1,484.0 | 3,492.4 |
| 1965 | n.a. | n.a. | n.a. |
| *Veneer and plywood* | | | |
| 1958 | 59.3 | 240.8 | 390.1 |
| 1963 | 66.2 | 324.7 | 559.2 |
| 1965 | 71.7 | 385.7 | 638.1 |
| *Lumber and wood products* | | | |
| 1958 | 288.9 | 904.2 | 1,413.6 |
| 1963 | 253.8 | 981.0 | 1,665.8 |
| 1965 | 250.1 | 1,061.1 | 1,738.3 |
| *Harvesting* | | | |
| 1958 | 71.7 | 226.7 | 390.4 |
| 1963 | 73.1 | 281.8 | 520.5 |
| 1965 | 73.5 | 292.1 | 585.4 |
| *All industry total* | | | |
| 1958 | 643.2 | 2,611.1 | 4,995.7 |
| 1963 | 613.3 | 3,071.5 | 6,237.9 |
| 1965 | n.a. | n.a. | n.a. |

Source: Bureau of the Census. United States Department of Commerce, 1961. Census of Manufacturers, 1958, 1963, 1966, 1968. Annual Survey of Manufacturers, 1964, 1965.

[1] Public Land Law Review Commission, 1969.

turers of timber products—and for harvesting. Bear in mind that these data apply to the industries described and should not be taken as representing timber's contribution to the economy. Raw materials other than wood are used in all three of these industry groups, especially in pulp, paper, and board.

Altogether, these industries employed over 600,000 people in the United States during the years for which data are available. Employment in the lumber and wood products industries dropped by fully 35,000 during the 1958–63 period (by almost 39,000 by 1965) but only slightly in pulp, paper, and board. During this same five years, employment in veneer and plywood increased nearly 12 per cent, while the number employed in harvesting registered a much smaller rise. This increase continued through 1965 in the latter two industry groups, but no data are available for pulp, paper, and board. In total, employment in United States primary forest products industries, including harvesting, dropped by about 30,000, or slightly under five per cent, between 1958 and 1963, but probably remained fairly constant over the next two years.

In spite of this decline the total payroll increased very substantially—to over three billion dollars in 1963. In real dollar terms this represented an increase of 11 per cent over the five-year period. Value added by manufacture showed an even larger increase, rising to more than 6.2 billion dollars. Using the wholesale price index as a deflator this means a real dollar increase of some 25 per cent. A substantial further rise probably occurred during the next two years, but without data from the important pulp and paper sector no comparable figures can be given.

Some idea of the varying regional importance of timber and the timber based industries may be gained from Table 5.3. Here, the regional employment, payroll, and value added are shown for the four census regions into which the United States is usually divided. Note the heavy concentration of these industries in the West and South, and the differences between these two regions. The South, for example, has by far the largest employment in these primary manufacturing industries, but its payroll is less than that of the West, thus reflecting the decidedly higher wages of the latter region. On the other hand, the value added by manufacturing is highest in the South, which leads in the production of pulp, paper, and board.

**TABLE 5.3 REGIONAL EMPLOYMENT, PAYROLL, AND VALUE ADDED FOR THE PRIMARY FOREST PRODUCTS INDUSTRY, 1963.**[1]

| Region | All employees (thousands) | Payroll ($ millions) | Value added by manufacture ($ millions) |
|---|---|---|---|
| Northeast | 92.1 | 535.6 | 990.8 |
| North Central | 84.9 | 476.5 | 958.8 |
| South | 249.7 | 998.0 | 2,251.9 |
| West | 186.8 | 1,061.5 | 2,036.3 |
| U. S. | 613.5 | 3,071.6 | 6,237.8 |

Source: Bureau of the Census, 1966. Census of Manufacturers, 1963.
[1] Public Land Law Review Commission, 1969.

## Other Manufacturing

Once primary manufacturing has been covered, the employment, payrolls, and value added that can be attributed to timber become more and more subject to question. *Secondary manufacturing* of timber products is usually said to include the manufacture of wooden containers, furniture, paper, and paperboard products, and similar industries. That these industries use wood as an important component of their raw material mix is an undeniable fact; none of these industries could produce their products if for any reason the forests stopped producing timber for their use. Yet clearly, timber is only one of the inputs to their production, and any means of apportioning employment or value added, or any other indicator of economic importance, may be challenged. To go beyond secondary manufacturing to industry groups such as textiles, transportation, marketing, or even construction makes the allocation problem even more difficult.

The Forest Service study mentioned earlier attempted to estimate the contribution of timber to all industry groups, using as the primary basis of allocation the proportionate cost of raw materials consumed by each industry class (proportionate freight revenues were used to allocate transportation totals, and the ratio of total sales of timber products to all sales was used for marketing). When the amount of the value added by manufacturing and allocated to timber was totaled over all industry groups it was estimated that nearly 25 billion dollars could be attributed to timber in 1958, or about 5.6 per cent of the total GNP. This is an impressive figure—decidedly more so than that of the value of stumpage, or the value added by primary manufactures—but it must be remembered that it follows from using a very arbitrary method of estimating timber's contribution to our economic life.

Even after recognizing these many difficulties, however, it should be clear that timber production and the primary timber manufacturing industries of this country are significant to our economy. Stumpage alone was worth something approaching two billion dollars annually by 1969, and together with the primary manufacturing industries dependent on timber, furnished employment to well over a half million people. The addition to GNP in 1963 by primary forest product industries was over six and one quarter billion dollars, and has been rising. Against this general background we can proceed to the major primary users of wood and inquire into the economic aspects of lumber, plywood, and pulp and paper.

# 6

# Lumber

If the nation's forest lands are to produce the products needed in future years then we must have some idea of both the relative and absolute quantities of those products the nation will demand. While today's uses may not be good indications of future demands, the changes that have occurred in these uses may help us understand some of today's difficulties and could help avoid problems of the future.

Ideally, we should examine what has happened in each timber industry, then construct an analytic model that would show how the important economic, social, and political factors have interacted. These models could then be used to forecast the quantities of each timber product consumers might demand in the coming decades. This objective, however, is both overly ambitious and exceedingly narrow. Furthermore, the pattern of national wood use has been presented elsewhere, with the series of national timber trend studies providing the major example of such work (U.S.D.A., 1946, 1958, 1965). Most such studies also project past consumption trends into the future. In this section we will examine the consumption of four major wood products in an economic framework of demand, supply, and price formation and thereby provide a different, perhaps a better, insight into some of the problems that forest managers should be anticipating.

It would undoubtedly be helpful to examine the declining use of fuelwood (still the major use of wood if viewed on a global basis), of cooperage, cross ties, fence posts, shingles, and other timber products now considered minor in the United States. Only the major timber products, however, will be covered, for while the data pertaining to minor uses are available, changes in these uses will not significantly affect forest management decisions over the foreseeable future. In contrast, changes in use of lumber, pulp and paper, plywood, and particle board *could* be significant to forest management in the United States and throughout the world. We begin with the most traditional of all wood products, *lumber*.

## HISTORICAL DEVELOPMENT

Since the first settlement on the Virginia coast, lumber has played an important and often dramatic role in the development of this nation. According to shipping records, a cargo of clapboards was carried from Plymouth to England in 1623, but the first well authenticated, commercial sawmill was established at Berwick, Maine in 1631 (Dana, 1956). Since then the lumber industry has gone through all the stages that characterize the industrial development of this country; periods of pioneer daring and ruggedness, of dramatically rapid growth, of sustained progress, of precipitous decline and frequent failure, of comeback and reestablishment, and even, perhaps, of reasonable maturity—all are present in lumber's history. A vast quantity of raw material has been processed; estimates place the total volume of lumber produced since the first mill down through the present at over three trillion board feet, a quantity of lumber that would be sufficient to build a boardwalk 1,200 feet wide and two inches thick from the earth to the moon!

In 1830 Maine led all other states in lumber production. As settlement moved westward, the raucous, rambunctious, hard-drinking loggers from Bangor found equally rugged companions among the men of Saginaw, AuSable, and Menominee in the Michigan of the 1880's, a time when the Lake states assumed production leadership. Michigan hit its production peak in the decade 1880–90. A twenty-foot high pile of white pine lumber stretched for over 20 miles along the Saginaw River when the shipping season opened. After "daylight had been let into the swamps" of Michigan, Wisconsin and then Minnesota became the largest producer. The lead passed to the Southern states during World War I, then went West to the Pacific coast, where the production leadership is still maintained.

Today it is customary to decry the "rape of the Lake states forests," overlooking entirely the economic role thereby fulfilled as well as the economic climate in which the alleged crime occurred. Immediately following the Civil War there was a great surge of western migration into the very fertile but virtually treeless Great Plains. At this same time the Lake states pine stands were opened and connected by barge and rail to the transcontinental railroads just being completed. This conjuncture in time of a tremendous market for home-building, a new and large supply of excellent construction material, and a transportation network able to bring the two together—all in an economic climate favoring a rapid expansion of big business (in fact, placing a blessing on those strong enough and ruthless enough to seize economic opportunity)—encouraged the rapid exploitation of what is usually recognized as the finest stand of softwood timber this nation possessed. The fact that most people believed the cut-over lands would pass into agricultural use after logging provided still further encouragement.

To a considerable degree it was this very "stripping" that enabled the rapid settlement of the Great Plains and also provided capital upon which much of the industrial might of Michigan and Wisconsin was eventually built. It also created the Paul Bunyan legends and enough ballads of frozen loggers and drunken river drivers to stock the folk singers of a later age. While in no way condoning the wasteful cutting and other destruction of that period, it is worth observing that, in developing economies, nature's stored wealth may be converted into capital to build roads, school houses, homes, factories, and churches, and that these may represent as much or even more of a contribution to the welfare of a nation's people than would an undisturbed and beautiful forest.[1]

During this century lumber has been by far the largest consumer of timber in this country, and until recently, lumber companies owned more forest land than all other industries combined. In 1970 approximately 48 per cent of all industrial roundwood production in the United States was destined for sawmills. In this same year the West produced nearly 57 per cent of the nation's lumber, the South

[1] It was not the *cutting* of these forests that did the damage. Properly protected, most of the area would have quickly regenerated. The major destruction came from the repeated and uncontrolled burning that followed the logging. Unfortunately the same sequence of events is being repeated today in many of the developing nations. Brazil and Chile provide outstanding examples of such destruction.

some 31 per cent, and the North the remaining 12 per cent (U.S.D.A., 1971).

As defined by the Forest Service, lumber producers now own only five per cent of the nation's commercial forest land, though consuming nearly half of all United States timber harvested. Sawmill owners, therefore, must obtain much of their stumpage from other sources, making stumpage prices and supply of major significance to these owners. In this connection it is important to recognize that most major lumber producers also manufacture pulp, paper, plywood, and are often heavily engaged in recreational home site development and in other kinds of operations. The term "lumber corporation" is probably as outmoded as big wheel logging, but will probably be with us for many more years.

Lumber's position as the most important use of timber is being seriously challenged today by pulpwood, but there seems little doubt that for at least the coming decade sawlogs will continue to be the most significant single timber product, volume-wise, from our forests.

## Production

An important part of the lumber industry's recent past and quite possibly an indication of its future is revealed by the production record since 1900 (Fig. 6.1, page 99). Note the difference between the peaks and valleys of hardwood and softwood lumber, the long decline in production from 1906 to 1932, and the upward rise from 1932 through 1956. It is this latter change that has given rise to much difference of opinion with respect to future lumber markets. As can readily be seen, by exercising a little care in the choice of the time span considered, one can use these statistics to substantiate a forecast of either rising, declining, or stable lumber production.

The lumber import–export balance has shifted considerably over the recent past (Fig. 6.2, page 100) with most of the change on the import side. Up to World War II the United States was a lumber exporting nation. Since 1941 imports have exceeded exports. Imports have risen almost constantly since the end of World War II, and since 1960 have been averaging about one-sixth of the United States production.

The 1930's and late 1940's saw a very large increase in the number of mills in this country, with almost all the expansion being in the so-called "portable" mill category. The 1950's witnessed a dramatic (and, for many owners, traumatic) reversal of this trend—a fact well documented by the data on number of active mills by size classes in Table 6.1. There is little to suggest that lumber production has important scale economics beyond very modest mill sizes. The unusually wide size variation of existing operations goes far toward

**TABLE 6.1 UNITED STATES SAWMILLS BY PRODUCTION SIZE CLASS[1]**

| | 1947 | | 1958 | | 1963 | |
|---|---|---|---|---|---|---|
| Size class | No. mills | Production (per cent) | No. mills | Production (per cent) | No. mills | Production (per cent) |
| 50,000 M ft. + | 43 | 11.0 | 67 | 18.1 | 80 | 19.3 |
| 25,000 M to 50,000 M ft. | 122 | 12.0 | 169 | 18.3 | 200 | 20.5 |
| 15,000 M to 25,000 M ft. | 150 | 8.0 | 219 | 12.9 | 436 | 20.0 |
| 10,000 M to 15,000 M ft. | 181 | 6.1 | 326 | 8.7 | | |
| 5,000 M to 10,000 M ft. | 608 | 11.8 | 564 | 12.1 | 571 | 11.0 |
| 1,000 M to 5,000 M ft. | 4,467 | 25.4 | 3,235 | 20.1 | 2,879 | 18.3 |
| 500 M to 1,000 M ft. | 6,429 | 13.0 | | | | |
| 50 M to 500 M ft. | 25,340 | 11.8 | 27,065 | 9.8 | 28,031 | 10.9 |
| 1 M to 50 M ft. | 15,769 | 0.9 | | | | |
| Total: | 53,109 | | 31,645 | | 32,197 | |

Source: United States Department of Commerce, Bureau of Census 1949, 1960, and 1965.

[1] 1947 data for number of mills; 1958 and 1963 data for number of establishments.

proving the point, and even in the Douglas fir region, where large mills are most numerous, the optimum output range probably lies between 60,000 and 140,000 board feet per 8 hour shift (Mead, 1966).

## Employment

For several reasons, employment data for the lumber industry is difficult to interpret. In modern integrated operations, pulp and paper production is frequently combined with lumber production and plywood manufacture under the same corporate structure, making it difficult at best to allocate woods workers between the two industries. Moreover, woods work tends to be quite seasonal in many areas and labor to be somewhat migratory. If one looks at the whole industry, however, it is the retailing phase that raises the most problems in determining employment levels, for lumber is often

**Fig. 6.1** United States lumber production since 1900. Sources: *Historical Forestry Statistics of the United States* (U.S.D.A., Statistical Bull. No. 228, 1958). *The Demand and Price Situation for Forest Products—1964* (U.S.D.A., Misc. Pub. No. 983, 1964). *The Demand and Price Situation for Forest Products.—1970–71* (U.S.D.A., Misc. Pub. No. 1195, 1971).

sold along with a great variety of construction materials, including many that compete directly with lumber.

According to the United States Department of Commerce (1969) the lumber industry in 1950 was the fifth largest employer of the nation, furnishing employment to some 808 thousand workers.[2] By 1960 this number had been reduced to 627 thousand, and in 1970 only 500 thousand were so employed. In 1954 it was calculated that it took, on the average, about 28 man hours to log, manufacture, and distribute each thousand board feet of lumber—a figure often quoted by analysts as indicative of a labor intensive industry. These figures include not only mill workers, but also woods workers, truckers, and retailers as well.

[2] These figures are adjusted for part time workers by placing on a calculated "full time equivalent." If two workers, for example, each worked a half year they would be counted as one full time equivalent.

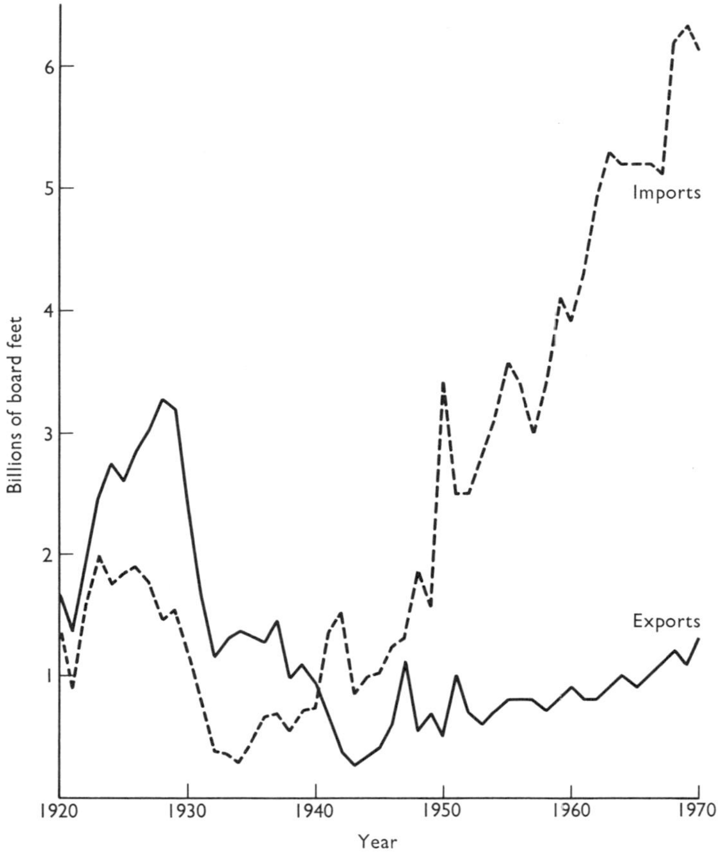


**Fig. 6.2** United States lumber trade since 1920. *Sources: Historical Forestry Statistics of the United States.* (U.S.D.A. Statistical Bull. No. 228, 1958). *The Demand and Price Situation for Forest Products—1970–71* (U.S.D.A., Misc. Pub. No. 1195, 1971).

If one measures progress by increased productivity (and there is good reason for so measuring it) then it appears that the lumber industry is in fact less progressive than most United States industries. Furthermore, research and development expenditures offer little promise for improving this picture. Of 15 manufacturing industries studied, only the textile industry ranked below the wood product group in terms of company financed research and development costs expressed as a percentage of net sales (Mead, 1967). Zivnuska pointed out that the Standard Oil Company of California alone had a research budget three and one half times that of the entire lumber industry (Zivnuska, 1963).

In summary: lumber takes nearly half of the total cut from United States forests, is produced by thousands of independent, highly individualistic operators in mills ranging from units producing a hundred thousand board feet of badly made, undried, ungraded lumber to those manufacturing annually a hundred million board feet of carefully graded, sized, kiln dried, and packaged construction material.[3] The industry grew steadily with the nation until 1906, producing at that time a record 46 billion feet. It then went into a long decline with production dropping to 13 billion board feet in 1932. Since then it has made a considerable comeback and recently has averaged about 35 billion board feet per year. Imports have made substantial inroads in domestic markets and now about one-sixth of our lumber is from foreign sources. Lumber has not fully shared in the increased productivity so characteristic of United States industry. It is a significant employer of labor, owns little of the forest land upon which it depends for raw materials but purchases its raw material from millions of forest land owners, the largest of whom is the United States Forest Service.

## DEMAND

Two approaches could be taken to a broad study of United States lumber demand. As with any product such as bread or automobiles one might approach it directly; alternatively, one might recognize that lumber is a "producer's good" and reason that its demand should be treated as "derived" from the demand for the final products into which lumber is manufactured. In this chapter, because a broad quantitative and historical viewpoint is believed the first requisite, the initial possibility will be explored. In Chapter 8 the concept of derived demand will be reviewed.

Annual lumber production, imports, and exports were given in Figs. 6.1 and 6.2. The corresponding trend of lumber prices (in constant dollars) is shown in Fig. 6.3. It takes only a glance to see that lumber prices rose substantially from 1920 through 1950, but then remained relatively constant until 1968.

[3] This does not mean that all small mills produce poor lumber, or that all large mill lumber is correctly manufactured.

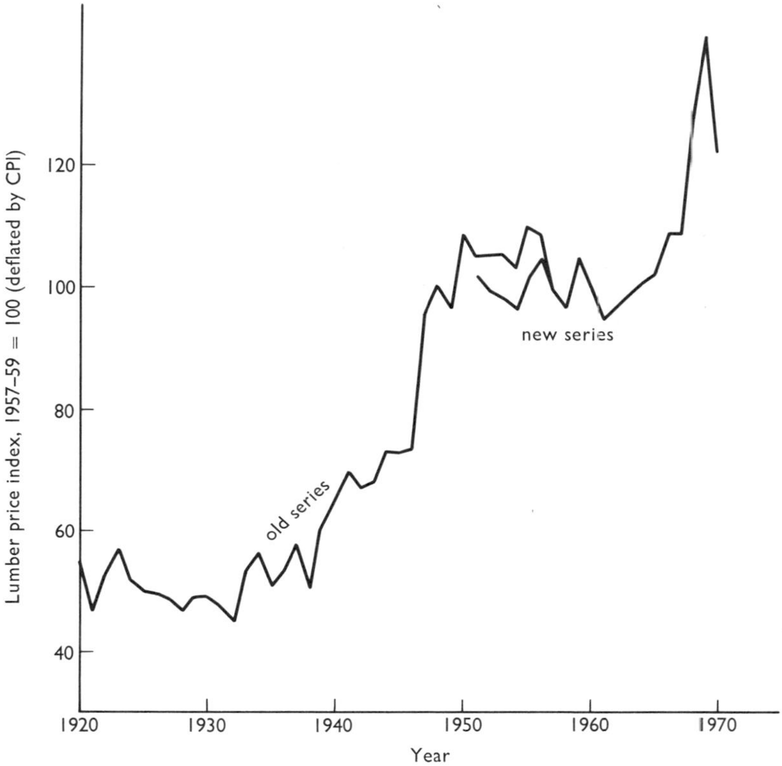


**Fig. 6.3** Trend of United States lumber prices. Sources: *The Demand and Price Situation for Forest Products 1970–71* (U.S.D.A., 1971). *Historical Forestry Statistics* (U.S.D.A., 1958).

For demand analysis it would be preferable to have market sales data rather than production figures, but unfortunately no such information can be obtained directly. As an approximation to sales we can add imports to production, subtract exports, and then add or subtract the changes in stocks of lumber at mills and in retail and wholesale yards. The result is called "apparent consumption." A comparison of the figures in Table 6.2, compiled from statistics assembled by the Forest Service, will help establish the order of difference between apparent consumption and production.

The adjusted price is plotted over apparent consumption in Fig. 6.4. Obviously the results bear little resemblance to the neat demand relationship that theory might lead us to anticipate. Nor should

**TABLE 6.2 PRODUCTION AND APPARENT CONSUMPTION OF LUMBER IN THE UNITED STATES, 1920 TO 1970.***

| Year | Lumber production | Apparent consumption |
|---|---|---|
| 1920 | 35.0 | 34.6 |
| 1922 | 35.3 | 34.9 |
| 1924 | 39.5 | 38.5 |
| 1926 | 39.8 | 38.8 |
| 1928 | 36.8 | 34.9 |
| 1930 | 29.4 | 28.2 |
| 1932 | 13.5 | 12.7 |
| 1934 | 18.8 | 17.8 |
| 1936 | 27.6 | 25.7 |
| 1938 | 24.8 | 23.6 |
| 1940 | 31.2 | 34.3 |
| 1942 | 36.3 | 43.8 |
| 1944 | 32.9 | 34.6 |
| 1946 | 34.1 | 33.5 |
| 1948 | 37.0 | 36.4 |
| 1950 | 38.0 | 40.9 |
| 1952 | 37.5 | 39.2 |
| 1954 | 36.4 | 38.7 |
| 1956 | 38.2 | 40.9 |
| 1958 | 33.4 | 36.1 |
| 1960 | 32.9 | 36.0 |
| 1962 | 33.2 | 37.3 |
| 1964 | 36.6 | 40.8 |
| 1966 | 36.6 | 40.8 |
| 1968 | 36.5 | 41.5 |
| 1970 | 34.0 | 38.8 |

* Source: For data prior to 1950: Historical Forestry Statistics of the United States (U.S.D.A., 1958).
For data 1950 to 1970: *The Demand and Price Situation for Forest Products 1970–71* (U.S.D.A., 1971).

such a relationship be expected, for the specifications of the demand model have not been met. The market area has been reasonably constant over the time period being considered, but in 1920 the United States contained about 106 million consumers, while in 1967 the population was over 200 million. A virtual doubling of the population should bring about substantial changes in the market!

Per capita lumber consumption is plotted over time in Fig. 6.5. The all-too-apparent trend of declining per capita lumber consumption has been a matter of deep concern to everyone interested in lumber markets. While far from uniform there can be little doubt that a continuing reduction in lumber use by the average United

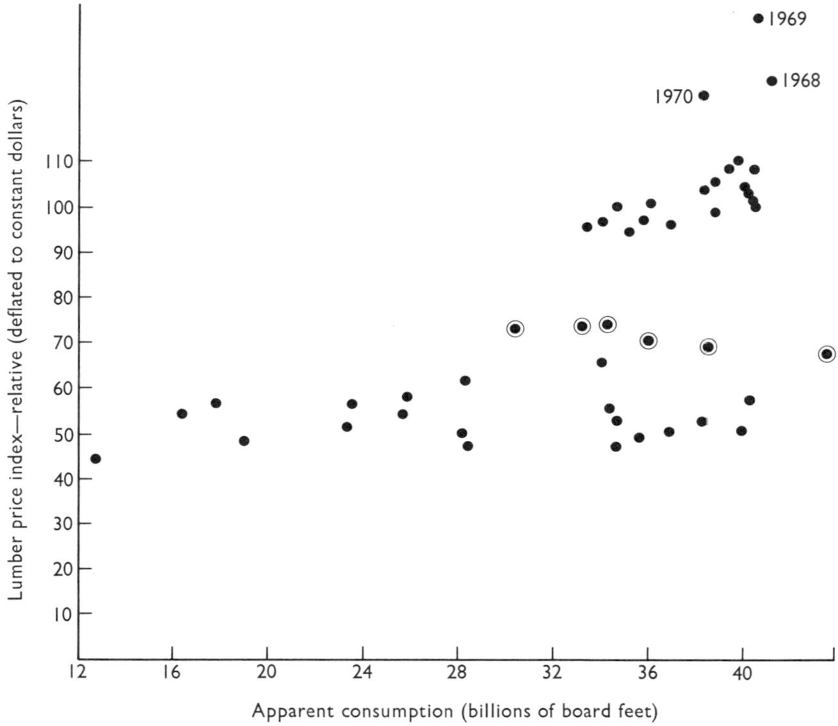


**Fig. 6.4 Lumber price index plotted over apparent consumption. Source:** ***The Demand and Price Situation for Forest Products 1970–71*** **(U.S.D.A., 1971).**

States consumer has been underway since 1920, with the all time low of 189 board feet being recorded in 1970.

The changing population, however, is not the only deficiency in our model. The time span used includes a period during World War II, when lumber prices were controlled. These years, circled in Fig. 6.4, form a population quite distinct from other years. In fact, these war years seem to divide the data into two distinct populations, pre- and post-World War II. Perhaps our model should be reexamined, with an eye to determining the effect of departures from the *ceteris paribus* assumption.

## Shift Variables

The model we have been attempting to fit began with the simple two-dimensional price/quantity relationship of (a) in Fig. 6.6. To

**Fig. 6.5** Trends in per capita consumption of lumber in the United States. Source: *The Demand and Price Situation for Forest Products 1970–71* (U.S.D.A., 1971).

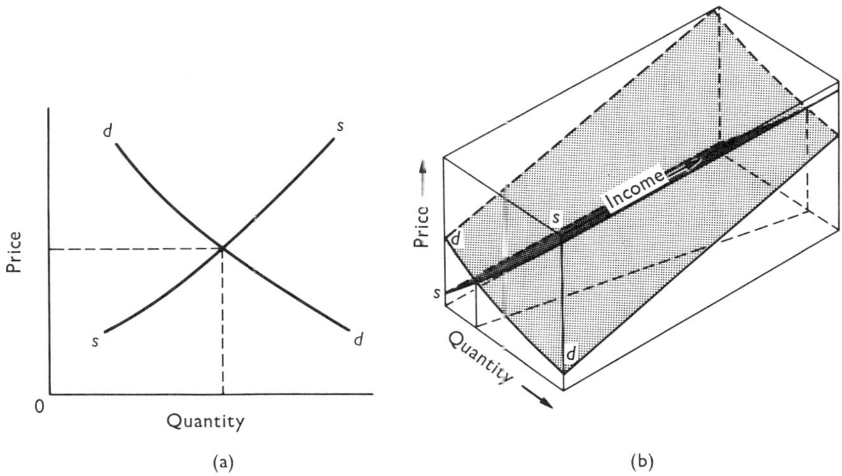


**Fig. 6.6** An increase in income tends to shift the demand curve upward.

counter the effect of population increases we might use *per capita* consumption data, but not all factors can be handled in so simple a fashion. Most of the factors producing a "shift" in the demand curve (and therefore called *shift variables*) require that a new dimension be added to the model. Income, for example, might be included as shown in Fig. 6.6(b); as incomes increase the per capita demand curve shifts upward. Other dimensions are needed for factors such as mortgage rates, per cent of population living in urban areas, etc., but a graphic presentation is necessarily limited to three variables.

Actually, our data consists of a series of points, each one representing the (average) intersection of two curves—demand and supply—that have been subjected to constantly changing forces. So far we have been looking at factors that might produce shifts in the demand curve, yet clearly a shift in supply could produce an equally pronounced effect on the observed price/quantity intersection. The situation is illustrated in Fig. 6.7. Each observation of price/quan-

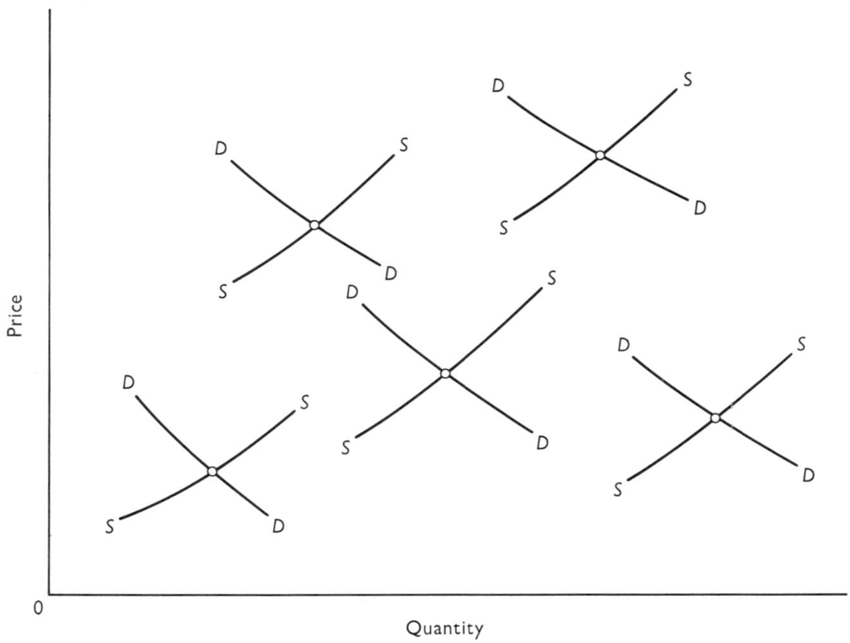


**Fig. 6.7** The identification problem—each observed point represents the intersection of a demand and a supply curve, both of which may be shifting.

tity is a point of intersection of demand and supply functions. The analyst of such an array of points is faced with the task of *identifying* whether demand or supply factors, or both, are responsible for the shifting point location.

### The Identification Problem

This is the essence of the "problem of identification" that has bothered econometricians since research in quantitative economics

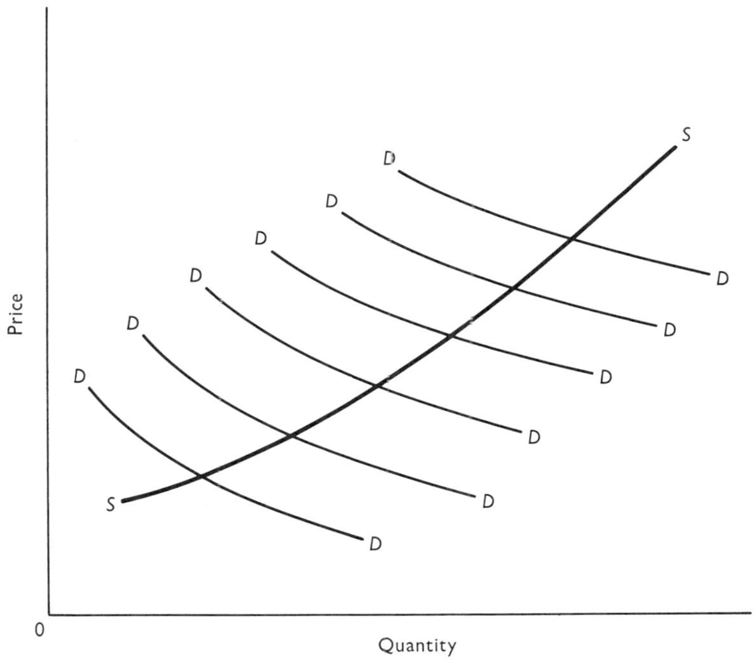


**Fig. 6.8** If supply is stable while demand shifts, the observed points indicate a supply curve.

began. With some products the problem is not as severe as with others. Suppose, for example, one were reasonably certain that supply conditions had been unchanged over (say) a five-year time span. If this were the case, then the changing points of intersection caused by shifts in the *demand* curve (provided, of course that it did shift) would trace out the location of the *supply* curve, as in Fig. 6.8. And, of course, the reverse would also be true; during periods of constant demand conditions a shifting *supply* curve, by its intersections with the demand curve, would trace out the *stationary* demand curve.

What if both shift? Then things get difficult! To solve this problem requires more sophisticated analytical techniques and additional information; from these data alone there is no means of "identifying" which curve shifted how much to produce the observed effect. Such methods do exist, but they lie considerably beyond the scope of this text (see, for example, McKillop, 1967).

### A Chronological Approach

But we are able to learn something about shifts of the demand (and supply) curves with a reasonably simple model. In Fig. 6.9, per capita consumption data are plotted against the wholesale lumber price index in current dollars (1957–59 base) and the resulting points connected serially—from 1920 to 1921 and 1922 to 1923, etc. The pattern revealed by this figure would seem to indicate that during 1921, 1922, and 1923 demand was increasing (the curve was shifting outward), that it decreased from 1923 until 1932 (a really tremendous decrease on a per capita basis), and from 1932 through 1940 (with the exception of the 1938 recession) demand was again increasing. Only this sequence of demand shifts and relatively stable supply schedules could possibly produce the quantity/price pattern shown during this period.

What happened between the two periods (prewar and postwar) is less clear. A large upward shift in supply coupled with a very inelastic demand curve would conceivably produce the big price rise indicated by the postwar grouping of the figure, and especially the 1965 to 1970 configuration. But with a supply curve shifting upward in this fashion one would expect to find many more substitutes being used, and high substitutability is characteristic of highly *elastic*, not *inelastic*, demand. If the demand for lumber is elastic with respect to price, the postwar group could only be produced by a distinct upward shift or increase in demand over the past two decades, coupled to be sure, to an upward shift in supply. The data are not sufficiently precise to establish shifts in demand at all definitely, but certainly the indications are strong. The supply aspects revealed by these analyses will be discussed in a following section.

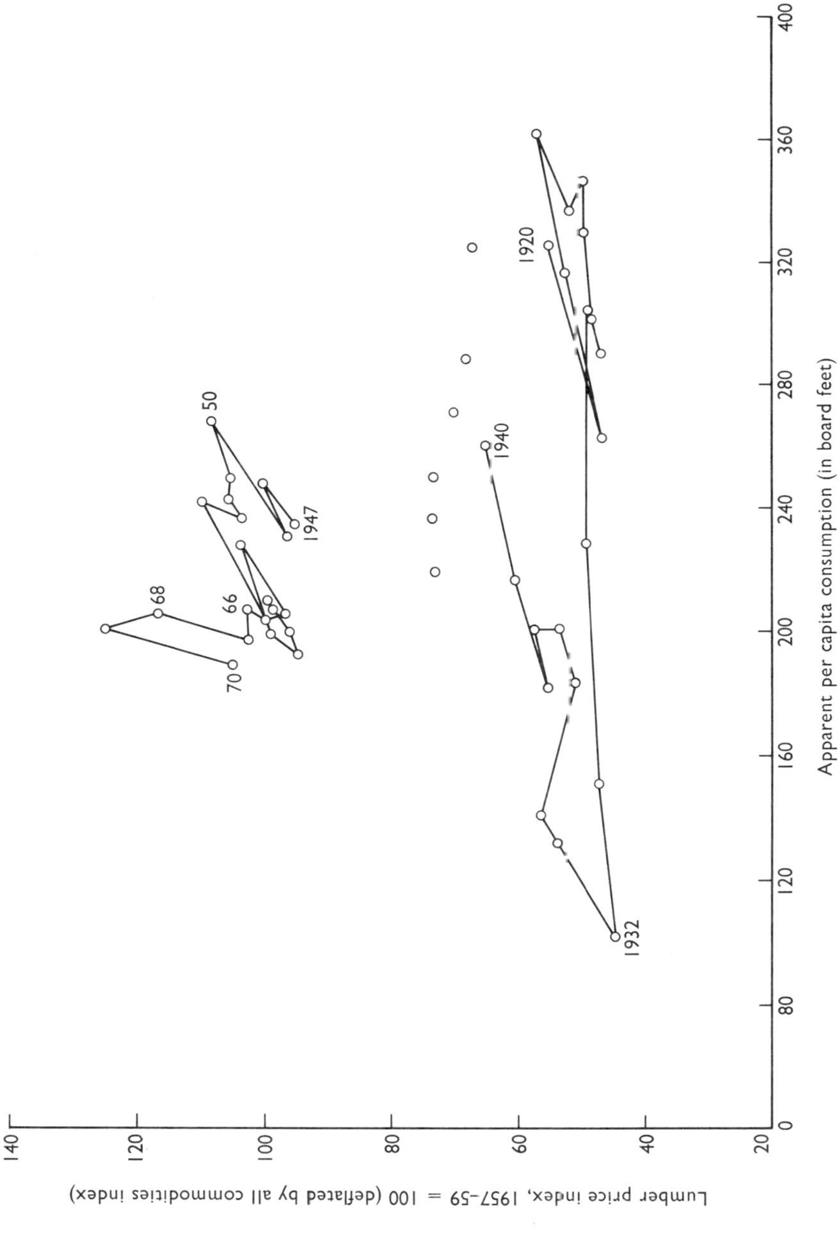


**Fig. 6.9** **Lumber price index plotted over apparent per capita consumption. Source: *The Demand and Price Situation for Forest Products 1970–71* (U.S.D.A., 1971).**

## Cross-Section Analysis

There are other types of analyses that throw light on demand behavior. Zaremba, for example, interviewed 63 New Haven families, determining their income and wood consumption, and related these graphically. The resulting curve indicated that as incomes rise, use of wood in house construction also rises, but at a declining rate. The small number of families sampled as well as the lack of usual statistical tests throw the specific results under doubt, but the method itself seems promising (Zaremba, 1963).

A similar cross-section approach was used in an attempt to identify factors affecting industrial wood consumption on a world wide basis (Gregory, 1965). According to this analysis, income and wood availability accounted for over 90 per cent of the observed variation in saw wood consumption. The two variables were almost equally important at average levels—again indicating the importance of considering both the supply and the demand aspects.

## Other Shift Variables

Before turning to supply it is well to think for a moment of the factors that could shift the demand schedule, even though one can only hypothesize the level of their effect. *Income* almost certainly has an influence—probably strong at low-income levels, but decreasing (as indicated by the two cross-section studies cited) with increasing income. At high-income levels people might prefer houses made of steel, glass, aluminum, ceramics, brick, or stone to those made of wood. On the other hand, since income and the degree of urbanization are closely associated in most nations, it is possible that income has relatively little effect on lumber consumption while urbanization is very important.

The extent to which *consumer taste* can be separated from income is a moot point; in any event, taste is extremely difficult to quantify objectively. Most analysts tend to believe that income is a reasonably good "proxy" variable for taste.

*Home financing* is undoubtedly an important factor influencing new residential construction and is not a simple variable. Interest rates (both short and long term), the period of time over which the loan is to be repaid, and the size of the required down payment—all have effects on home purchase, and through this on the demand for lumber.

*Construction methods* have had a very significant effect over the past. The change from two-story to ranch style houses, and the increasing numbers of houses built on concrete slabs (hence requiring no floor joists, and influencing the type of flooring to be used) have been particularly important. In a somewhat similar fashion, *building codes* that restrict the use of lumber—regardless of the reason—have significantly reduced the use of lumber in the past. In the long run the ongoing change from rural to urban living is undoubtedly an important demand shifter. As our cities grow ever larger, our way of living inevitably changes. The percentage of the population living in multifamily dwelling units has increased substantially since World War II, as has the percentage of new multifamily residences constructed. Because high-rise apartments use little lumber, this change would have a very deleterious effect on lumber markets. Any forester who is given the opportunity to travel in Europe should observe the adaptations to urban living that have been made there, and ponder their significance.

A discussion of lumber demand would be incomplete without reference to the United States Forest Service and F.A.O. studies of timber requirements (U.S.D.A., 1946, 1958, 1965) (F.A.O., 1963, 1964, 1966, 1967). Although using the term "demand," these are in reality analyses of consumption, and have been designed primarily to forecast future wood consumption under quite rigid supply conditions. By examining trends in, for example, wood used in framing, floors, and siding for residential construction, and in shipbuilding, mining, railroads, etc., a better understanding of factors influencing the changing consumption patterns has been achieved. Almost all the factual information presently available for demand analysis has been produced for these requirement studies. It is unfortunate that neither agency has seen fit to sponsor rigorous work in demand analysis.

## SUPPLY

All supply curves represent some kind of cost curve, hence any economic analysis of supply is necessarily an implicit or explicit analysis of cost. The specific kinds of costs to be investigated will depend on several factors—on the competitive structure of the industry, upon whether one is speaking of an industry, a region, or of some particular firm, and upon the time period under consideration.

There seems little question that the purely competitive model is directly applicable to lumber markets as long as the view is national. The number of operating firms is large (over 30 thousand in 1968) and the number of buyers is very large; the degree of product differentiation is small, and there are few barriers to exit or entry.[4] In 1958 the 4 largest producers of "sawmill and planing mill" products accounted for only 8 per cent of the total value of shipments, while the 8 largest produced 12 per cent. The corresponding figures for gypsum building materials and for primary aluminum (to give two examples of concentrated industries) were, for the 4 largest companies 87 per cent (gypsum) and 82 per cent (aluminum), and for the 8 largest companies 97 per cent and 90 per cent respectively (Mead, 1966). These low concentration ratios indicate the lumber industry is still one of the most competitive in the entire United States economy, although the degree of concentration has been rising significantly.

On a regional basis the purely competitive model may not be applicable. In the Douglas fir industry, for example, the "Big Four" of Douglas fir lumber production accounted for 23.9 per cent of the value of shipments in 1958. By 1963 this percentage was only 19.7 per cent.[5]

The discussion here will concern the overall lumber supply situation, since the review of lumber demand indicated that supply changes had been a decided factor in declining markets. In this case the purely competitive model is appropriate, and the industry short run supply curve will represent the horizontal summation of the marginal cost curves of the various firms composing the industry. Long run supply curves would vary in shape, depending upon

[4] Species differ greatly, to be sure, and this does offer opportunity for non-competition in some instances. Substitution possibilities are so great, however, that in almost all instances the purely competitive model seems appropriate for lumber markets. This does not, of course, carry over to purchase of stumpage, where monopsony may occur. And in some regions the lack of available forest land may limit entry.

[5] The four largest producers of Douglas fir lumber and their shares of regional production in 1963, were according to Mead:

| | |
|---|---|
| Weyerhaeuser Timber Co. | 14.2% |
| Pope and Talbot, Inc. | 2.0% |
| Simson Timber Co. | 1.8% |
| Georgia Pacific Corp. | 1.7% |

Between 1960 and 1963 Georgia Pacific dropped from second to fourth place, and U. S. Plywood Corporation from third to ninth place, in regional production.

whether costs to the industry were increasing, decreasing, or constant.

## A Chronological Approach

The approach will parallel that used with demand, although production rather than consumption data provide the starting point. The result of plotting lumber price (in constant dollars) over production for the 1920–70 period is shown in Fig. 6.10. By again connecting the points serially, six distinct periods became obvious: 1920–32; 1932–36; 1937–40; the war years; 1947–67; and 1967–70. If we suppose the straight lines show approximate industry supply curves for the first four identified periods (war years omitted), it would appear that from 1920 to 1932 the supply curve for lumber was essentially horizontal, that it remained horizontal but slightly higher from 1932 through 1936, and that it was sloping upward from 1937 until the war. Following World War II the industry curve shifted upward, and either became virtually inelastic from 1967 through 1970 or (more probably) shifted rapidly.

A competitive industry characterized by constant costs would vary its output as demand shifted with very little change in price *even though output was altered by large amounts.* Short term production increases would, probably, be accompanied by a price rise but the price rise would not be long maintained. Demand could shift from $D_1D_1$ to $D_2D_2$ to $D_3D_3$ (Fig. 6.11) with little long term price reaction. Change in the industry output would come primarily from changes in the number of active firms.

What actually happened during this period can be traced out in Fig. 6.10. Immediately following World War I there was a substantial boom in construction, followed by a very sharp but short recession in 1921. Lumber price dropped and production fell off by some 6 billion board feet. Recovery was rapid, however, and by 1922 production had fully recovered while price was rising: both price and output continued to rise through 1923. Then, as the industry adjusted to relatively stable output levels through 1924, 1925, 1926, and 1927 price settled back to almost exactly its former level. During this time the industry was gradually shifting from the south to the west. Although the western mills were larger there was no surge of technological innovation and mills of very different size competed for the market. There was a decided over-capacity in the

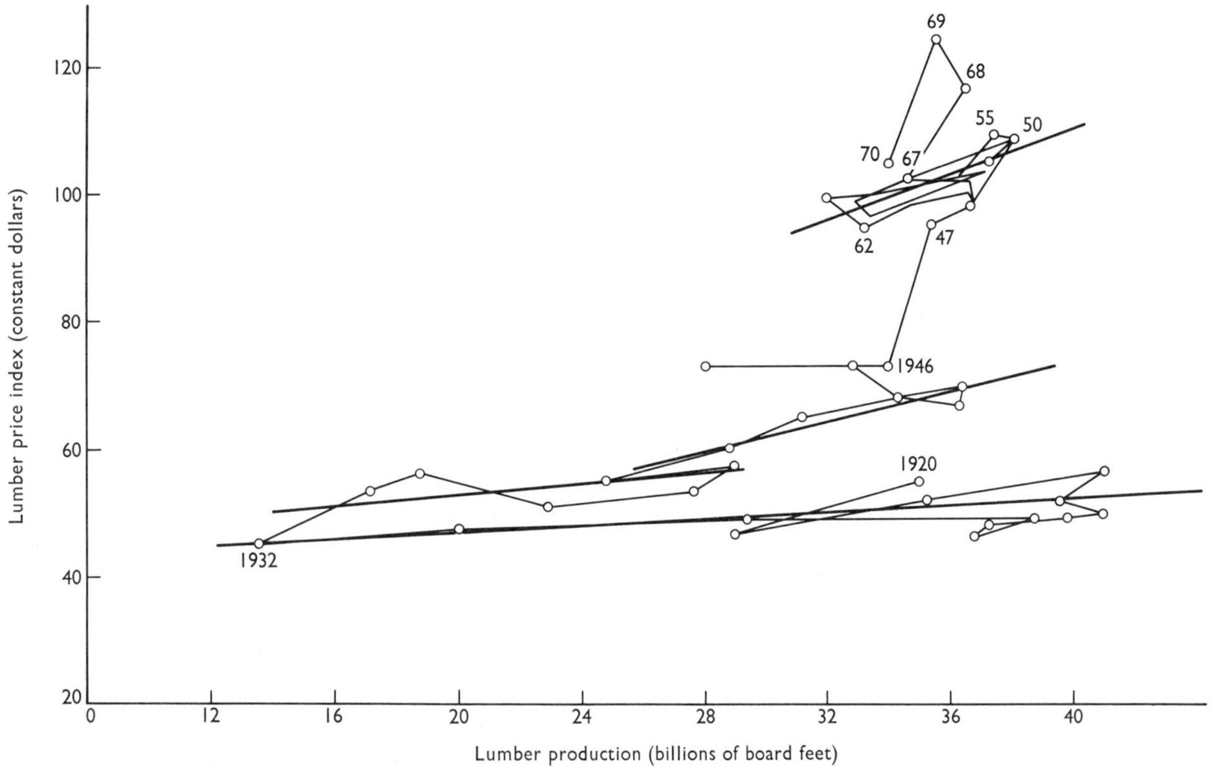


**Fig. 6.10** Lumber price index plotted over production. Source: *The Demand and Price Situation for Forest Products 1970–71* (U.S.D.A., 1971).

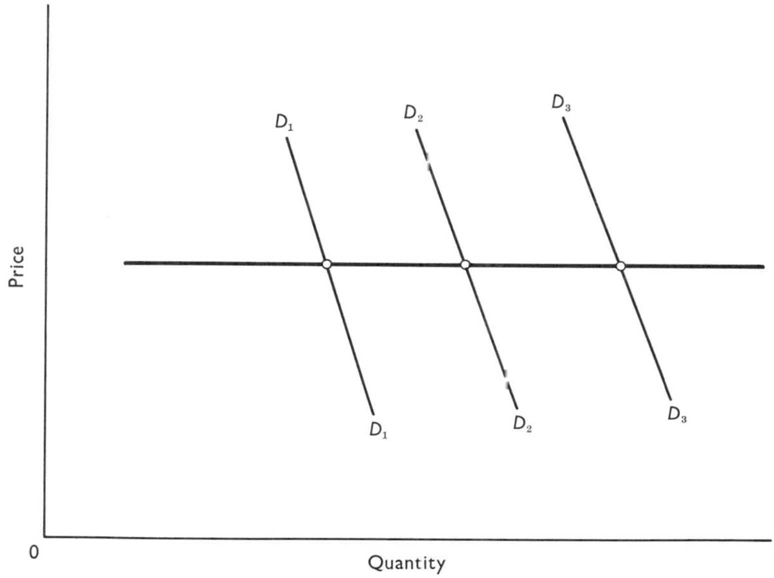


**Fig. 6.11** A "model" of a competitive industry under conditions of constant cost. Shifts in demand trace out by price/quantity indications, a horizontal supply curve.

industry and competition was keen. There was no effective shortage of any productive factors.

Then came the great depression. Production dropped to the lowest point in the century. Mill after mill went bankrupt as production slowed to a trickle—*but price fell very little,* as is seen in Fig. 6.10.

If during this period the lumber industry was in fact one of constant costs, these large shifts in demand *should* trace out an industry supply curve that is virtually a horizontal line. *And it does!* Taking the 1920 and 1932 period as a whole, the apparent industry supply curve is virtually a straight line (within the range of production observed over the period) with a very slight upward inclination. It is almost a classic example of an industry operating under conditions of constant cost, with changes in output brought about by changes in the number of firms.

After 1932, prices and production staged a comeback, but production did *not* decline when the NRA's production code was abandoned by the industry. Instead, production increased with almost no change in price until 1936, exactly as one would anticipate under virtually constant costs.

### Shifts in Supply

The pattern from 1936 through 1940 indicates that something was changing. Demand declined in the 1938 recession, but prices did not drop back to former levels. More significant, the increased output of the next two years was accompanied by substantially rising lumber prices. After centuries of being able to go "over the next ridge" to find more timber, the industry had reached the end of the virgin stands. Logging new stands meant going much further up the mountains or valleys than previously, and both of these cost money. Forestry suddenly became possible—*and profitable.* The supply curve for lumber took on a decidedly sharper slope. One might almost claim to see evidence of the upward shifting supply curve as the change in timber supplies became apparent: two quite distinct curves are traced by the 1935–37 and the 1938–40 periods.

The same pattern is evident in the post-World War II data. The industry supply curve apparently increased its slope still further and shifted upward very decidedly. Demand shifted outward, but this increase was more than offset quantitatively by the sharp upward shift in supply. And since 1967 the industry has apparently been unable to increase supplies to any appreciable extent.

In a competitive industry an upward shifting supply curve means an increase in production costs. Since the data used here are in constant dollars, they indicate increases in real cost of very close to 100 per cent—the difference in level of the indicated supply curves at an output of around 35 billion board feet. Two factors would seem to be of overwhelming significance here: labor productivity and stumpage costs.[6]

**Labor productivity.** Productivity in the lumber industry simply has not kept pace with other United States industry during this century. Zaremba by an ingenious method [7] has shown that produc-

[6] Stumpage price determination is the subject matter of Chapter 17.

[7] Reasoning that the ratio of annual changes in price of lumber to price of "all commodities" should equal the ratio of annual changes in general economic productivity to productivity in lumber manufacture, he substituted available estimates for the first three variables and solved for the fourth—productivity in lumber manufacture.

tivity in the lumber industry has *declined* relative to the rest of the economy since 1900 (Zaremba, 1963). Other efforts at measuring productivity in the lumber industry have, in general, produced similar results. Actual man hour requirements in lumbering have undoubtedly declined substantially since 1900, but because the rest of American industry has increased productivity so very rapidly, relative lumber manufacturing costs have risen sharply. It has been shown that while output per man hour in the lumber industry increased an average of 1.1 per cent annually during the 1899 to 1954 period (Kendrick, 1961) the average for all 80 industries studied was 2.2 per cent—just twice as fast.

There have been real gains in the technology of sawmilling, and in felling and hauling. Many of the equipment changes that contributed so importantly to other industries have been adopted by sawmills as well—particularly the mechanized handling of lumber and the hundreds of applications of electric, gasoline, and diesel power. But unlike most other American industries the lumber industry has suffered offsetting decreases in productivity—distances that logs must be hauled have increased, log quality has undoubtedly declined, logs are now much smaller—and freight rates for finished products have increased as the industry has moved further from population centers. The latter would decrease consumption by raising delivered price, even though the price indices used in the diagrams are based on f.o.b. mill prices. The end result is, as we have seen, a decline in productivity *relative* to the rest of American industry, and a marked increase in lumber price when price is measured in constant dollars.[8]

**Stumpage costs.** The cost of stumpage could also be a major element in the lumber price increase. Fig. 6.12 shows average stumpage prices, in constant dollars, of Douglas fir and of southern pine since 1920. That the big jump in stumpage prices came immediately after World War II and coincides almost precisely with the big increase in lumber prices (in time, not necessarily in amount) is not mere happenstance!

[8] As the reader will undoubtedly observe, the increase in lumber price is even more marked if price is measured in *current* dollars. This, however, is due to the severe inflation that has marked the United States—and world—currency over the past 50 years. Had *deflation* been the pattern, the increase would be *less* obvious in current dollars than in constant dollars.

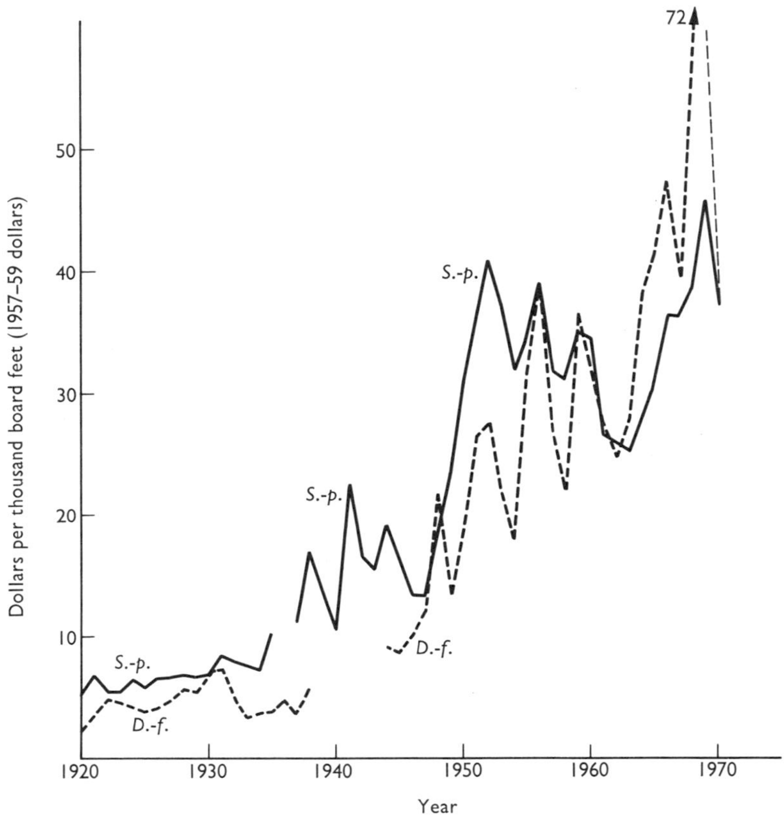


**Fig. 6.12** Average stumpage prices in constant dollars. Source: *The Demand and Price Situation for Forest Products 1970–71* (U.S.D.A., 1971).

How much of the supply shift should be attributed to declines in relative productivity, how much to rising stumpage prices, and how much to other items cannot be determined precisely, but there can be little doubt that both labor productivity and stumpage price have contributed substantially to lumber price rises.

## LUMBER PRICE

It would be redundant to treat price at length after having traced shifts in both demand and supply through the changing price/quantity relationships. Two points, however, are worth making. First, of all the natural resource price series studied by Resources for the Future, only timber displays an almost constant increase over time. Fig. 6.13 shows wholesale lumber prices in constant dollars since

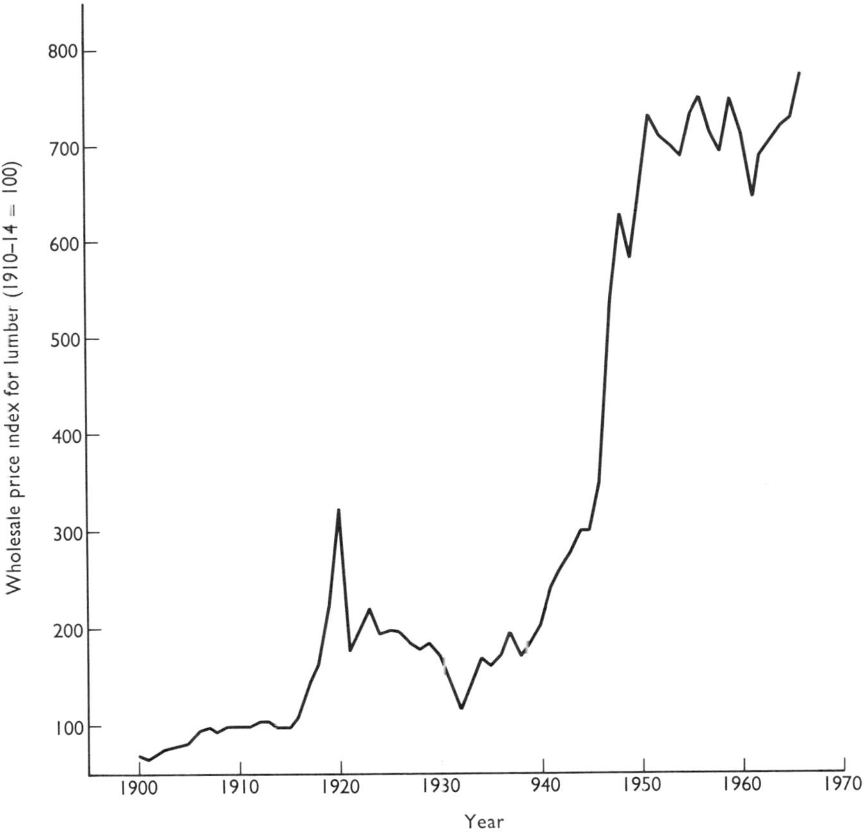


**Fig. 6.13** Lumber price trends since 1900. Sources: N. Potter and F. T. Christy, *Trends in Natural Resource Commodities* (Baltimore: Johns Hopkins Press, Resources for the Future, 1962). *The Demand and Price Situation for Forest Products 1970–71* (U.S.D.A., 1971).

1900. Second, lumber prices have also increased very substantially relative to products competing with it directly, and relative to the general price level. This comparison, made in Fig. 6.14, emphasizes

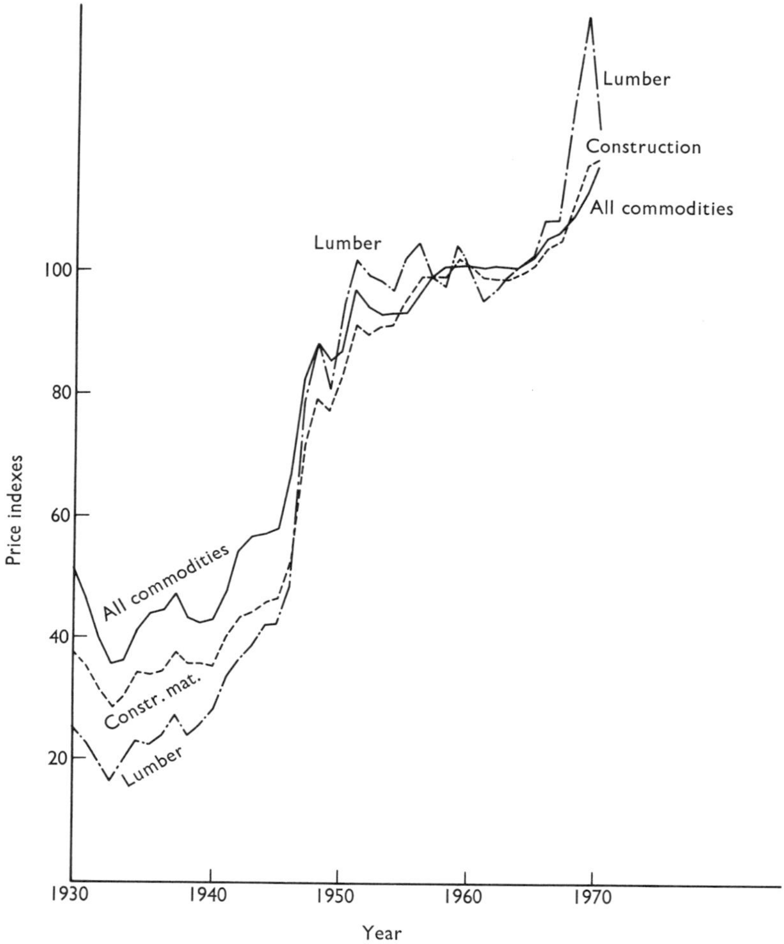


**Fig. 6.14** Lumber price index compared with price index of various other building materials. Source: *The Demand and Price Situation for Forest Products* —1970–71 (U.S.D.A., 1971).

the fact that lumber prices have risen far more rapidly than have prices of almost any other major product in our economy. Foresters should remember this relative price rise whenever they hear someone bemoaning the "declining lumber demand."

# 7

# Other Wood Product Industries

The three industries to be discussed in this chapter present quite different problems from both technological and economic viewpoints. They follow lumber in quantitative significance to forestry, and are generally called the *pulp and paper industry*, the *plywood industry*, and the *particle board industry*. Their technical aspects will be treated only to the extent necessary to permit taking an analytical approach to their timber requirements.

## PULP AND PAPER

Usually the Chinese are given credit by historians for discovering the art of paper making, probably about 2,000 years ago. The industry was introduced to the American colonies about 1690 when the first papermill began operating near Germantown, Pennsylvania (Stevenson, 1940; Guthrie, 1950). By 1776 the industry was firmly established in Connecticut and Maine. Cylinder and fourdrinier machines were introduced in the early half of the next century, making it possible to produce a continuous sheet of paper rather than having to produce it sheet by sheet.

In very early days papermills were located on streams in or near cities, thus assuring access to fresh water, labor, and to markets for both their end product and their primary raw material, rags. Once continuous production of paper became possible, power also became important, again favoring location along streams.

The use of straw for making wrapping paper shifted production towards the Midwest. Wood was not used on a commercial scale for paper until after the soda process was introduced in Europe around 1855. The sulfite process was discovered in 1886, but nearly 25 years passed before it was adopted by United States manufacturers. Once wood came into common use the industry began to follow in the wake of lumber mills, a tendency which persisted as long as non-resinous woods were available.

Today's paper industry provides an outstanding example of how technological advances and mass production techniques can change the pattern of a nation's living. Contrast the status of packaged goods in a modern supermarket with those of the old-time grocery store when the cracker barrel was a major attraction. As you thumb through the *New York Times*, think of the handbill-sized, four-sheet newspaper of 100 years ago. Pause before a display of "paperbacks." Low-cost paper may well be the most significant single development in news dissemination and education that the world has yet seen.

### Economic Importance

How important to the United States is the paper industry? Again, there is no simple answer, but there are some significant indicators. Net sales of the industry amounted to 11.8 billion dollars in 1960 and to 14.8 billion in 1964. In all but three of the years between 1950 and 1969 the industry has paid over a half billion dollars in Federal taxes alone. The Department of Commerce reported that in 1950 some 485 thousand workers were employed in manufacturing "paper and allied products" at an average wage of $60.53. Similar data for 1970 showed 710 thousand workers employed at an average weekly wage of $144.14 (U. S. Department of Commerce, 1971). The industry's contribution to national income amounted to over 5 billion dollars in 1964 and in the calculations made by the Forest Service (Hair, 1963) 60 per cent of timber's total contribution to the United States economy was attributed to the pulp and paper industry.

From forestry's standpoint the most significant aspect of the paper industry is the manufacture of pulp. Many materials have been and are being used to produce paper pulp; the residue from sugar cane refining (bagasse) is the primary raw material in some countries, and other materials include esparto grass, bamboo, cotton waste, and rags. Wood, however, generally proves to be the most economical fiber source, if land is available for its production. In the United States wood has steadily strengthened its hold over the last 25 years as the major raw material for paper making, and there seems little likelihood of any sudden change from this preference. Fig. 7.1 shows how wood has gradually increased its dominance in the paper making processes.

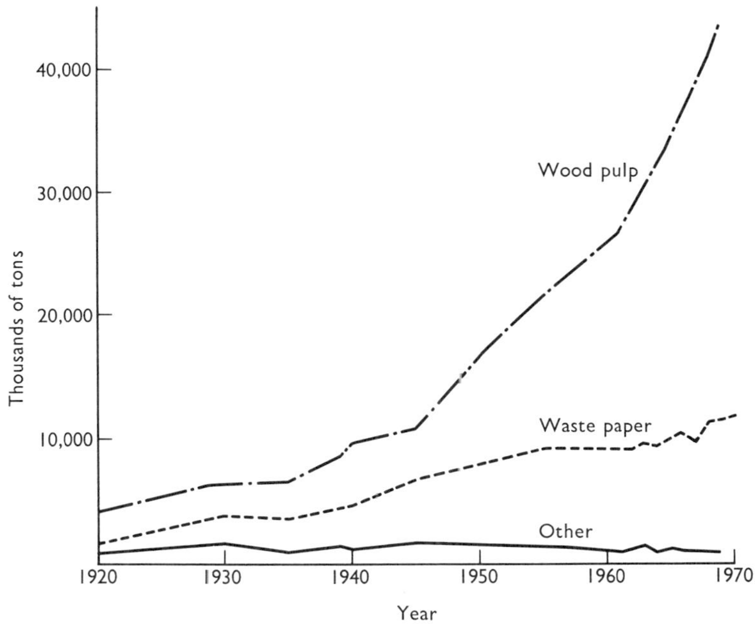


**Fig. 7.1** Raw materials used for making paper in the United States. Note increasing importance of wood pulp.

Because of this dependence on wood supplies the pulp mills of the country are well scattered throughout the major forested regions of the country (map, Fig. 7.2).

Two technological developments triggered the development of large southern and western pulp mills. The move southward was preceded by discovery of methods for pulping the highly resinous hard pines. Utilization of sawmill waste through the chipping of slabs and edgings further aided the southern expansion and made

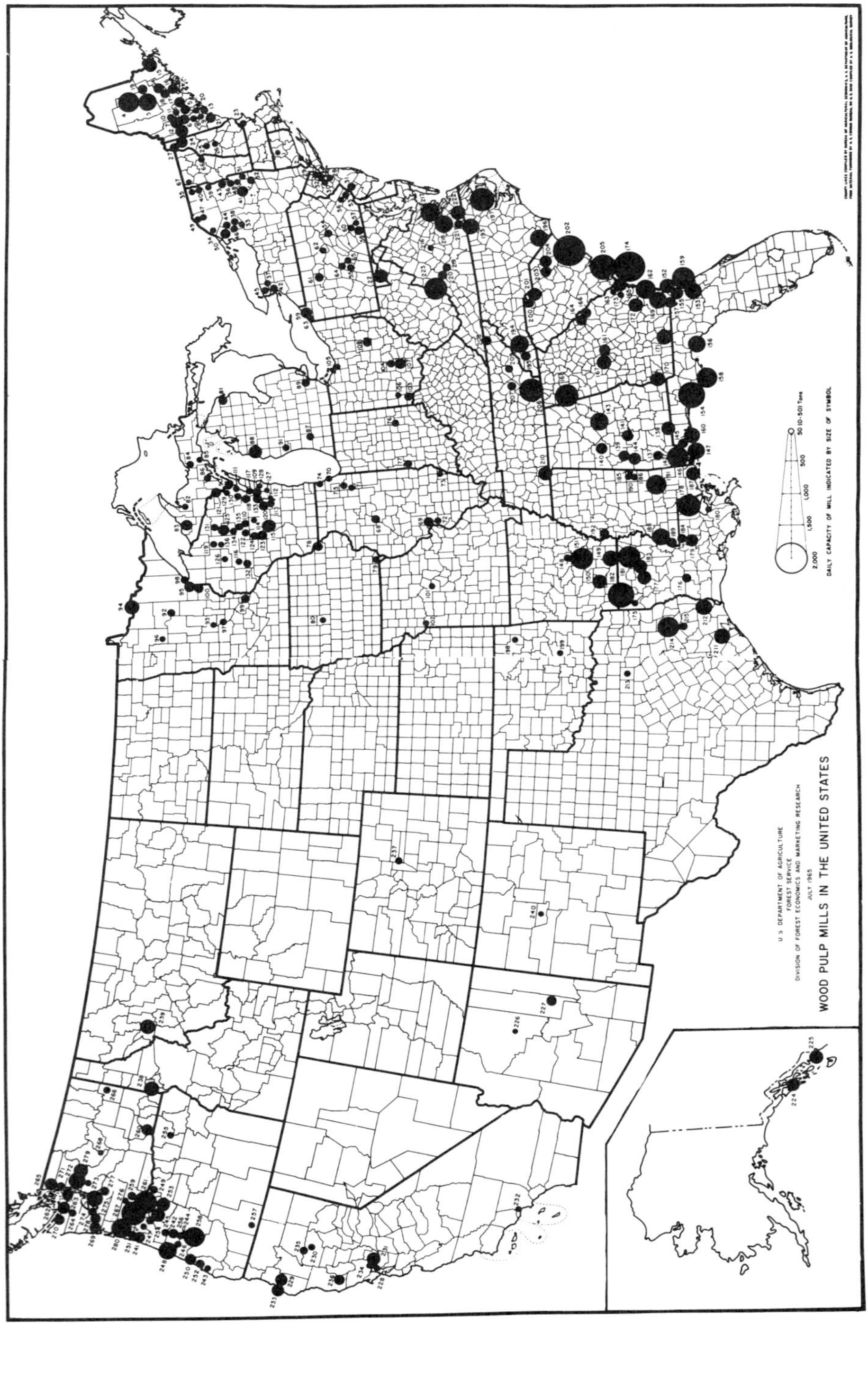


**Fig. 7.2** Location of wood pulp mills in the United States in 1965 (U.S.D.A., 1965). Courtesy of United States Forest Service.

possible the use of Douglas fir mill residues by western mills after World War II. Prior to that time the western industry used primarily spruce and hemlock. The increased utilization of dense hardwoods has also been an important factor in expanding the total wood fiber supply.

Table 7.1 presents United States pulpwood production data,

**TABLE 7.1 SOURCE OF DOMESTIC PRODUCTION OF PULPWOOD. TAKEN FROM *THE DEMAND AND PRICE SITUATION FOR FOREST PRODUCTS, 1970–71* (U.S.D.A., 1971).**

| Year | Total | Roundwood | | | Chips |
|---|---|---|---|---|---|
| | | Total | Softwoods | Hardwoods | |
| | | (thousands of cords) | | | |
| 1950 | 20,715 | 19,465 | 16,680 | 2,785 | 1,250 |
| 1951 | 25,130 | 23,720 | 20,070 | 3,650 | 1,510 |
| 1952 | 25,045 | 23,475 | 20,000 | 3,475 | 1,570 |
| 1953 | 26,320 | 24,785 | 20,705 | 4,080 | 1,535 |
| 1954 | 26,970 | 25,470 | 20,945 | 4,525 | 1,500 |
| 1955 | 30,950 | 28,600 | 23,365 | 5,235 | 2,350 |
| 1956 | 35,195 | 32,145 | 26,210 | 5,935 | 3,050 |
| 1957 | 34,420 | 30,535 | 24,525 | 6,010 | 3,885 |
| 1958 | 33,240 | 28,090 | 22,445 | 5,645 | 5,150 |
| 1959 | 36,715 | 30,580 | 23,380 | 7,200 | 6,135 |
| 1960 | 40,010 | 33,465 | 25,450 | 8,015 | 6,545 |
| 1961 | 40,270 | 32,115 | 23,995 | 8,120 | 8,155 |
| 1962 | 42,770 | 33,330 | 24,315 | 9,015 | 9,440 |
| 1963 | 44,710 | 34,665 | 25,120 | 9,545 | 10,045 |
| 1964 | 48,600 | 37,450 | 26,920 | 10,530 | 11,150 |
| 1965 | 52,320 | 40,290 | 29,250 | 11,040 | 12,030 |
| 1966 | 56,070 | 41,840 | 29,590 | 12,250 | 14,230 |
| 1967 | 57,470 | 41,810 | 30,100 | 11,710 | 15,660 |
| 1968 | 61,670 | 44,250 | 32,130 | 12,120 | 17,420 |
| 1969 | 66,970 | 47,130 | 33,600 | 13,530 | 19,840 |
| 1970 | 70,300 | 51,000 | 37,000 | 14,000 | 19,300 |

showing quite clearly the changing sources of wood fiber since 1950. Continued expansion of chip production is a matter of some doubt, for chip output is presently associated with lumber and plywood production. An increased use of dense hardwoods, however, seems almost certain.

One would be hard pressed to find a better example of contrasting trends than that shown by the two major timber-using industries —lumber and paper pulp. Production of paper and paper board has increased almost constantly since 1920, on both a total (Fig. 7.3) and per capita basis (Fig. 7.4). Here, indeed, would seem to be an industry with a brilliant past and an even more promising future. For there is much to indicate that changing rates of paper consumption in this country may well be indicative of changes that

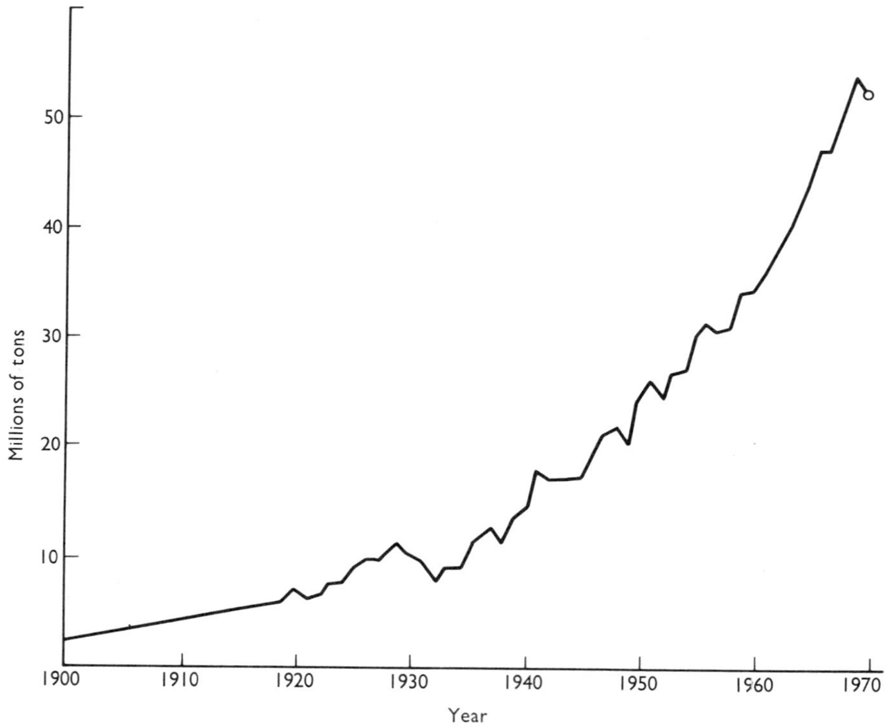


**Fig. 7.3** Production of paper and board (total) in the United States. Source: "Statistics of Pulp and Paper," American Pulp and Paper Association, 1965 supplement. Subsequent data from Survey of Current Business, Department of Commerce (April, 1971).

should be anticipated in other areas. In the lexicon of the space age, all signals are go and one of the brightest indicators is found in the relatively high expenditure for research and development.

## Competitive Structure of the Paper Industry

Since 1920 the number of paper mills and of pulp mills in the United States has remained relatively constant, with over 800 paper

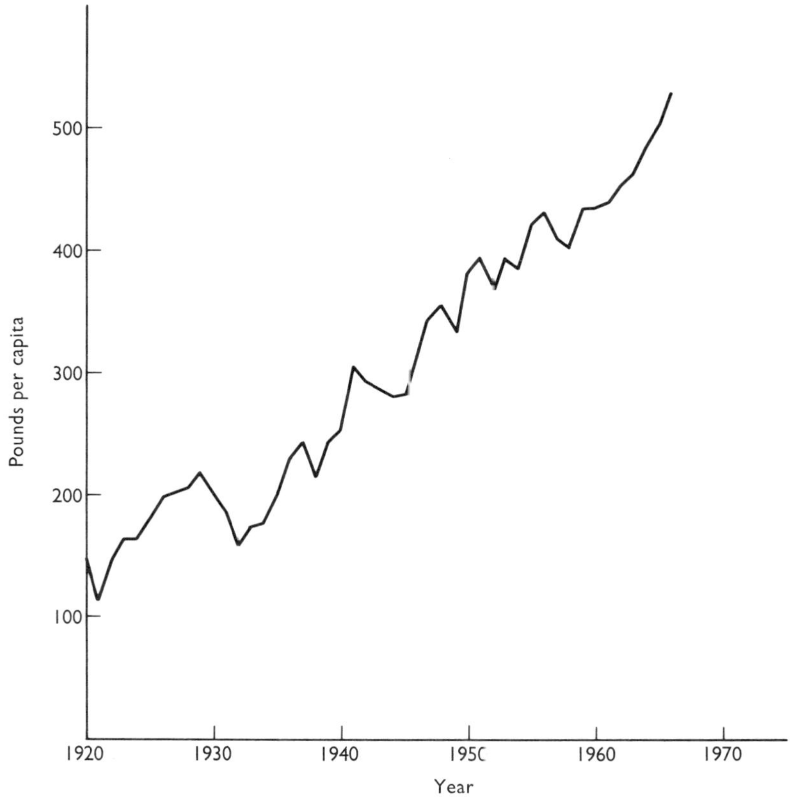


**Fig. 7.4** Per capita production of paper and board in the United States. Sources: *Business Statistics,* U. S. Department of Commerce, 1969. *Survey of Current Business,* U. S. Department of Commerce, 1971.

mills and around 350 pulp mills in active production. During the same period production has increased about 700 per cent, an expansion made possible largely by the construction of new, very much larger sulfate mills in the South and West as mills in the Northeast and the Lake states became overaged.

The industry is probably as good an example of oligopoly with a competitive fringe as one can find. Several corporations produce enough to influence product price through their production decisions. The industry is intensely competitive, though not in a price-wise fashion. Product differentiation by trade name, by watermark, and by finish, is a prime objective of advertising and especially so among the producers of fine paper. Service is probably the major variable factor among producers of Kraft.

Concentration within the pulp-producing sector is quite heavy. The five leading pulp producers in the United States have for several years produced around 30 per cent of the total United States production (U.S.D.A., Woodpulp Mills, 1965), while the largest of the corporations produced nearly 17 per cent of all the sulfate pulp in 1965. Eight paper making corporations were listed in the 1970 "Fortune Directory" among the 100 largest firms in the United States.

## Demand

The same problems encountered in discussing lumber demand are encountered in the paper industry; product identification, changing price series, and changing product groups, all complicate the statistical information. As with all aggregates, data describing the general United States production pattern (Figs. 7.3 and 7.4), tend to mask as well as to reveal, and the problem of disaggregation posed by the paper industry is not easily solved. The tremendous diversity of paper products is a point of pride to paper producers!

> Special papers are being fabricated for the manufacture of bags and beach slippers, cigarettes and cartridges, diapers and Dixie cups, filters and facial tissues, polishing cloths and patterns, roofing and roof trusses, soda straws and seat covers, towels and twine; wallpaper, sandpaper, building paper; meat wrappers, butter wrappers, candy wrappers; cartons, coat hangers, and condensers . . . all the multiplicity of products which aid transportation, sanitation, refrigeration, and . . . "You name it, we got it!" (Ainsworth, 1958)

To the manufacturer this diversity represents tremendous opportunity for expanding the markets for his products. To the economic analyst, however, it represents trouble. In an analysis of paper consumption (Hair, 1963), for example, separate analyses of 14 different types of paper and board were made. The necessity for treating each type individually is illustrated by the fact that although over the 1950 to 1960 decade less than one per cent of the tissue paper consumed in the United States was imported, nearly 80 per cent of our newsprint came from abroad over the same period (76 per cent in 1966). Newsprint is, in fact, one of our major import items, and the annual purchase of this product plays an important role in the Canadian economy.

This product diversity alone would pose severe problems for

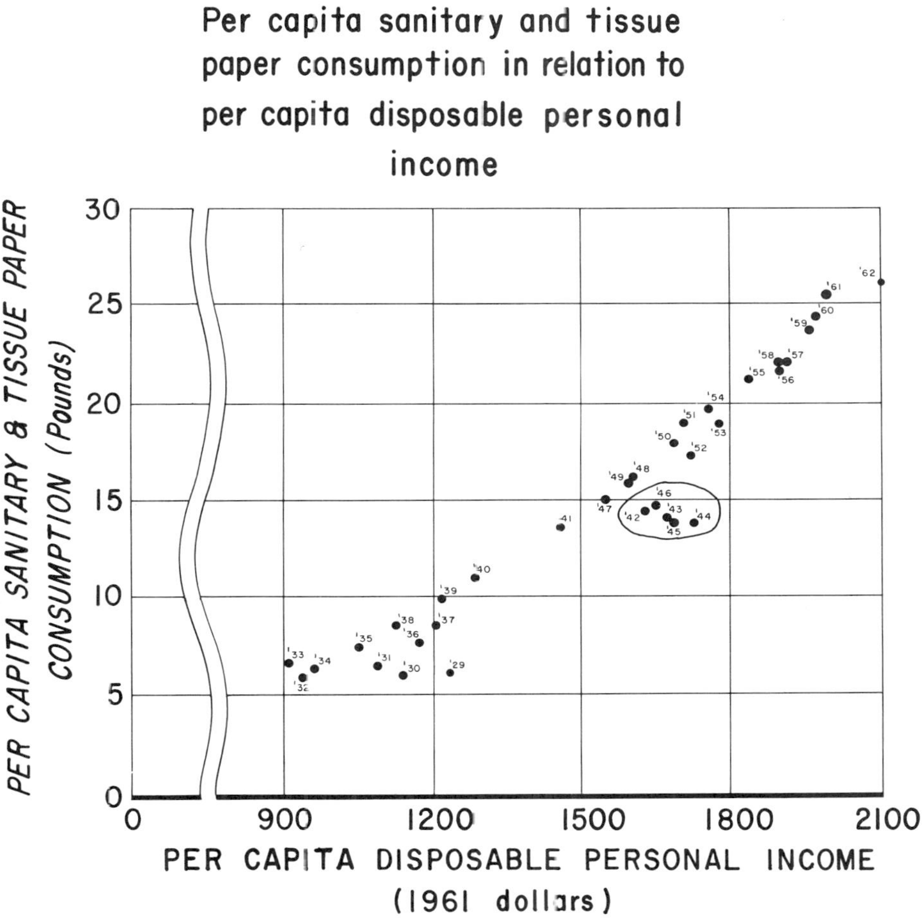


Source: U. S. Department of Commerce

**Fig. 7.5(a)** Per capita consumption of sanitary and tissue paper in relation to per capita disposable personal income. Forest Service Chart, Forest Resource Report No. 18 (Hair, 1963).

anyone analyzing product demand, but the lack of even reasonably good price information has been a further deterrent to demand analysis. The authors of the *Timber Resource Review* (U.S.D.A., 1958) after commenting on the inadequacy of price information, offered an index of "real price of wood pulp products." The index, however, "showed no discernible upward or downward movement of any significance."

There are, of course, many factors besides price that influence

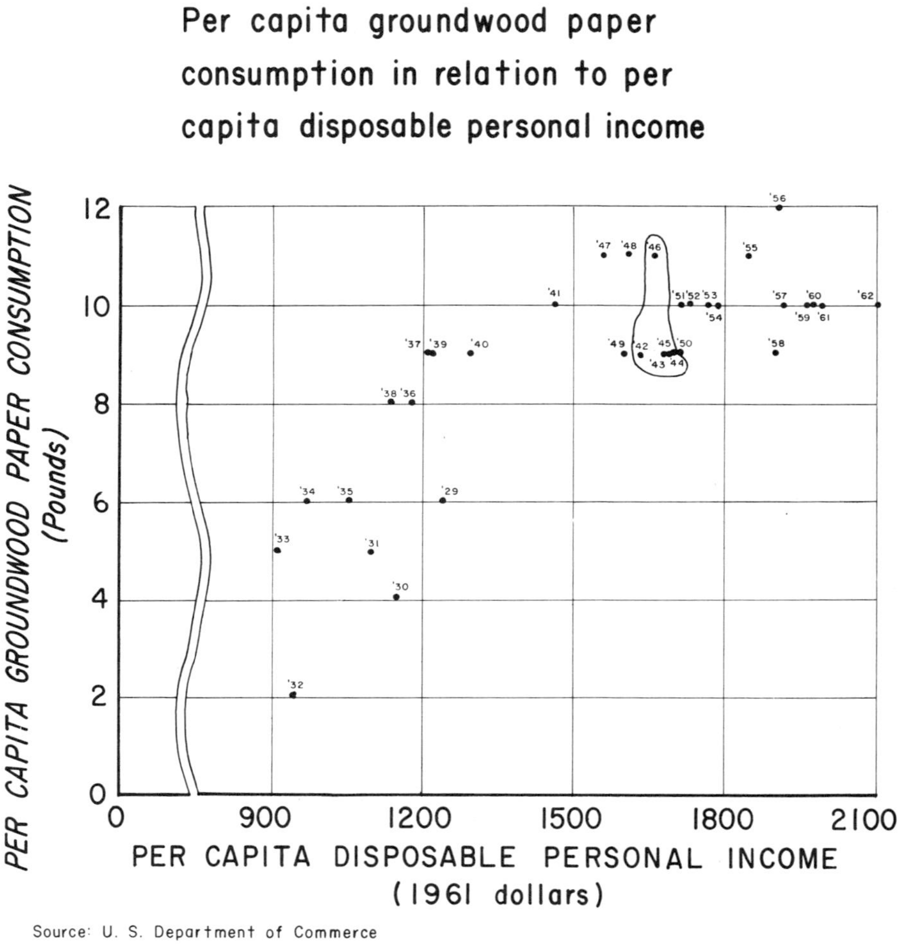


**Fig. 7.5(b)** Per capita consumption of groundwood paper in relation to per capita disposable personal income Forest Service Chart. Forest Resource Report No. 18 (Hair, 1963).

paper consumption. One would expect, for example, to find per capita paper use of many paper products increasing with income, and this is indeed the case. Fig. 7.5(a) and (b), taken from Hair's study, shows that not all paper types have been affected equally as United States incomes have changed over time. The correlation between per capita consumption of sanitary and tissue papers, Fig. 7.5(a) is extremely close, far closer than that for ground wood paper consumption, Fig. 7.5(b) although the relationship is positive in all cases. When the period analyzed was confined to post-World War II, however, per capita consumption of sanitary and tissue papers showed a high correlation with per capita income, while ground wood paper per capita consumption was virtually constant for the entire postwar period (1946–66).

As would be expected, factors other than income are also important in per capita paper consumption. For projecting consumption Hair (1963) found that population, number of households, Gross National Product, Disposable Personal Income, and the Index of Industrial Production, would all yield high correlations with past per capita consumption, but would result in substantially different projected values. If analyses of international consumption were made, the urban/rural ratio, literacy per cent, and similar variables might be found significant.

Various equation forms were fitted to cross-sectional data of per capita consumption by FAO analysts (FAO, 1960). High correlations between income and consumption were obtained and the relationships thus established were used for forecasting probable consumption of paper by nations anticipating improvement in national income.

These forecasts point to a worldwide expansion in paper consumption if world economic conditions continue to improve. This expansion is particularly meaningful for developing economies in tropical and semitropical areas—in Asia, Africa, and Latin America. If this increased consumption is to be realized it will necessarily be preceded by a tremendous expansion of paper production facilities in the tropical countries or in the presently established centers of paper production such as the United States, Canada, and the Scandinavian nations. Almost all the demand shift variables seem to be pushing the demand curve outward rapidly, whatever its shape. Levels of consumption and of price will apparently both be under decided upward pressure; their behavior will depend significantly upon supply conditions.

## Supply and Price

It would be inappropriate in a text such as this to develop at any length the technical side of the paper industry. Three aspects of supply, however, should be noted here—productivity, economies of scale, and change in raw material needs. Productivity in the paper industry has apparently kept pace with that of United States industry generally. The index of paper price relative to prices of all commodities has fluctuated around 100 since 1920, and although a similar index of board prices shows a significant rise between 1920 and 1940, it has remained virtually constant since that time. Hair (1963) found no functional relationship between price and consumption. Since production has increased some sevenfold during this same time span, it appears that technological developments have occurred at a rate at least sufficient to keep the supply curve moving outward as fast as demand has shifted.

In spite of existing variation in plant size, there are real economies of scale in pulp and paper manufactures, and the minimum size of efficient plant appears to be increasing as larger and faster paper machines come into use. The increase in plant size has been most pronounced with sulfate mills, and least pronounced with those using the soda process. The latter process has, in fact, almost been abandoned by United States producers.

Economies of scale are particularly important to developing nations. If the total market to be served by a proposed mill is small then it often proves inadvisable to build a mill. Figs. 7.6 and 7.7

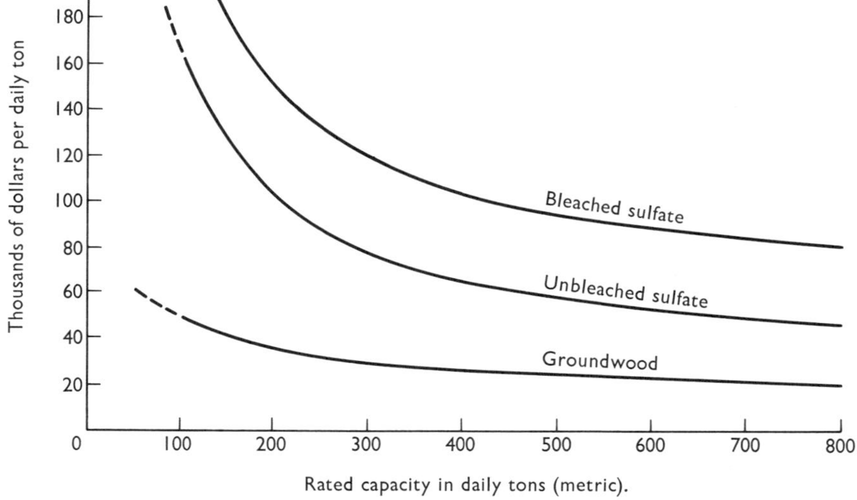


**Fig. 7.6** Pulp mill investment requirements as a function of mill size (after Eklund).

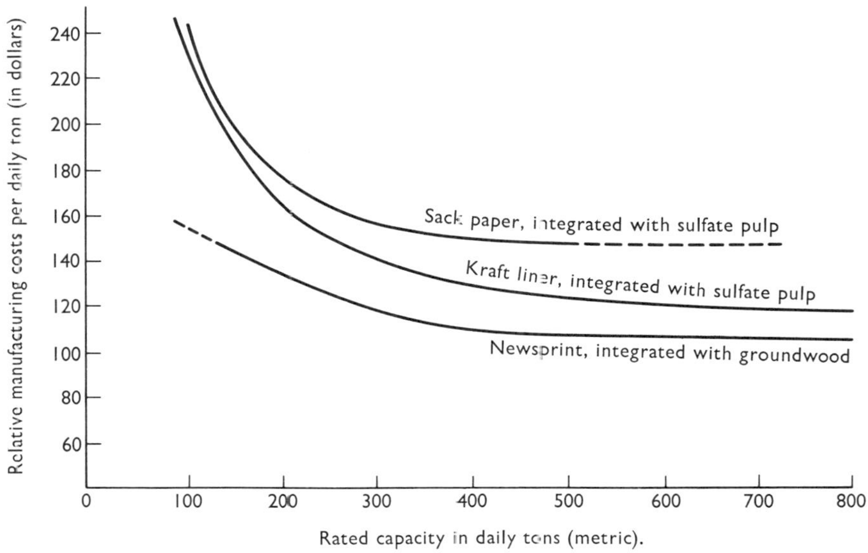


**Fig. 7.7** Costs of paper and paperboard manufacture as a function of mill size (after Eklund).

show two aspects of the scale question (Eckland, 1967). The three curves of Fig. 7.6 show how estimated investment requirements for pulp mills vary as mill size is increased, while Fig. 7.7 gives manufacturing costs of paper and paperboard as a function of mill size for these different kinds of operations. Suppose, for example, an integrated newsprint mill with a rated daily capacity of around 125 daily tons can produce at about $150 per ton. If the market can support an output of 400 tons per day, costs could be reduced to $110 per ton, certainly a sizable reduction. Yet if transportation costs to a developing country are as much as $40 per ton, then even the small mill might be highly economic.

Kraft mills present quite a different picture. Costs run around $230 per ton of output at a scale of 100 tons per day, are down to $140 per ton at outputs of 300 tons per day, and continue to drop (although much more slowly) as output is further increased, reaching $120 per ton at 600 tons of daily output. It is highly unlikely that cost savings of this magnitude can be achieved on transport.

There have also been important changes in the raw material side and some of the more recent developments may prove particularly significant over the coming decade. Until recently, conifers have been heavily favored for pulping because of the strength afforded by their long fiber length, and for their ease of management under pure, evenaged silvicultural systems. This has been a major factor in determining the location of the paper industry, both nationally and internationally. Hardwood forests are usually characterized by species variation, and this has made for substantially higher pulp production costs (and hence paper costs) because different species require different cooking procedures, and log sorting is expensive.

The semi-chemical process, newest of the major paper making processes, is particularly well suited to hardwoods. Because of their higher specific gravity, hardwoods yield more paper per cord and the increased costs of wood procurement are offset to a considerable degree. A change in the kinds of trees most favored for pulping may be just over the horizon. When one couples this possibility with a recognition of the vast quantities of tropical hardwoods that could be available, and additionally considers the likelihood that plantations of fast growing species are being established in many of the developing countries, there seems room for very substantial shifts in paper making centers.

## PLYWOOD

The individual who first decided to use a thin sheet of figured wood as a decorative cover for other material had a brilliant idea. His name is unknown, but he made his discovery a long time ago for beautiful examples of the art of veneering were found in the tombs of the pharaohs. A mural now credited to the reign of Thothmes III, who lived around 1500 B.C., shows the steps in the veneering process used at that time, the tools required, and even a pot of glue heating on a fire (Perry, 1948). And in those very ancient times successive sheets of veneer were combined to construct built-up units of high strength: plywood, very much as we know it today, was in use 3,500 years ago.

In this country some of our most famous relics were constructed of veneers and plywood. George Washington, William Penn, and Daniel Webster all used plywood-built desks, while the desk at which Thomas Jefferson wrote the Declaration of Independence is veneered.

Such older uses of veneer (and plywood) were attributable primarily to the beauty, strength, or resistance to effects of changing moisture content such materials possess. Produced by hand, veneers were extremely costly, and it was not until the late 1800's that mechanical means of cutting veneer were introduced. The manufacture of "veneered stock" remained a minor industry until after World War I. Then in 1919 the Plywood Manufacturer's Association was organized in Chicago and the name "plywood" replaced the older term "veneered stock." There are many people who would claim that this seemingly simple step was the key to the industry's later success, for it eliminated the unfavorable connotation the public had associated with the term "veneer." At any rate, from that time on the industry prospered, and plywood was increasingly accepted.

Two segments of the industry have long been recognized, one producing hardwood plywood, and the other softwood plywood. The two products are usually manufactured in different plants and because of the substantial differences in their physical properties there is little substitutability between them. There is, however, considerable substitutability within these two major subdivisions. Hardwood plywoods and veneers have a much longer history, but in recent years softwood plywood has become by far the largest segment of the industry.

## Softwood Plywood

Since 1900 (and probably earlier), softwood plywood has been manufactured in the United States. In 1906 about 15 per cent of all plywood was made from coniferous species. The present day industry can trace its origin directly to the introduction of mechanized veneer cutting techniques to the Douglas fir of the Pacific Northwest, and to the discovery of moisture resistant and waterproof glues—the latter a product of the 1920's and 1930's. With these ingredients at hand, it required only the spur of the market to create an almost wholly new industry. A glance at Fig. 7.8 demonstrates the effectiveness of this spur. Almost completely lacking during the early 1930's, markets began to develop just before World War II. Production nearly tripled from 1938 to 1941, then dropped from the severe impact of wartime restrictions and did not regain the level then established until 1948—well after the conclusion of the war. From that point on, though, the industry has had a growth that is deservedly called "phenomenal." Beginning in 1945, virtually every year (possibly excepting 1967) has seen a new production record established.

**Fig. 7.8** Production of softwood plywood in United States (millions of square feet) and number of mills.

The importance of the development of moisture resistant adhesives cannot be overstressed in this growth. The expanding construction industry of the 1950's was quick to recognize the advantages of exterior grade plywood for sheathing, sub-flooring, cabinet work, repair, maintenance, etc., and plywood makers responded by expanding existing facilities, building new plants, and developing means for better utilizing the materials at hand. The end result is documented in Fig. 7.8; in the 17 years from 1947 to 1963 the number of plants increased from 43 to 161—about 275 per cent. During this same period softwood plywood production increased almost sevenfold, truly a remarkable record.

**Competitive structure.** Prior to World War II the industry consisted of less than 35 plants, nearly all in Washington and Oregon.[1] Without far better information than is currently available one can only observe that 35 mills would scarcely meet the requirements of a purely competitive industry, especially since some firms operated more than one plant. Capital requirements probably acted as a moderate barrier to entry, but certainly not an insuperable one. The product was, essentially, undifferentiated, and there were no monopsonistic concentrations of buyers. From the little evidence at hand, one can observe that this was apparently an industry in which economies of scale were such that the existing market could be supplied by a small number of mills and that very probably the structure was oligopolistic. Just how important trade association activity was at the time is not now known. With the producers so geographically concentrated the associations could have played a significant role.

The rapid expansion of production facilities following World War II would indicate little real restriction on entry; in four years (from 1946 to 1950) the number of mills doubled and in less than another decade it had quadrupled. The flurry of new mill construction slowed after 1959, but it will take a few more years to see whether or not this slower expansion is to be maintained. With over 160 mills in operation a more competitive situation probably exists. The number of companies involved, however, is considerably less than the number of mills, and some of the so-called independents are essentially captive mills since they market their output through the larger companies. One would hesitate to cite this industry as an example of "pure competition."

**Demand for softwood plywood.** The usual data deficiencies are encountered in analyzing softwood plywood demand as in other forest based industries. The relationship between per capita consumption and the softwood plywood price index in constant dollars (1957–59 = 100) is seen in Fig. 7.9 with the individual price quantity points connected chronologically. The graph differs markedly from that shown earlier for lumber (Fig. 6.10). With softwood plywood the supply curve apparently was shifting outward far more rapidly than demand until 1968, although demand was surely shift-

[1] The *Facts for Industry Release* (Series M13B, supplement 1), of August 22, 1946, states "In Sept., 1941, there were thirty-four active mills." Only two mill locations outside of Washington and Oregon were cited, both in California.

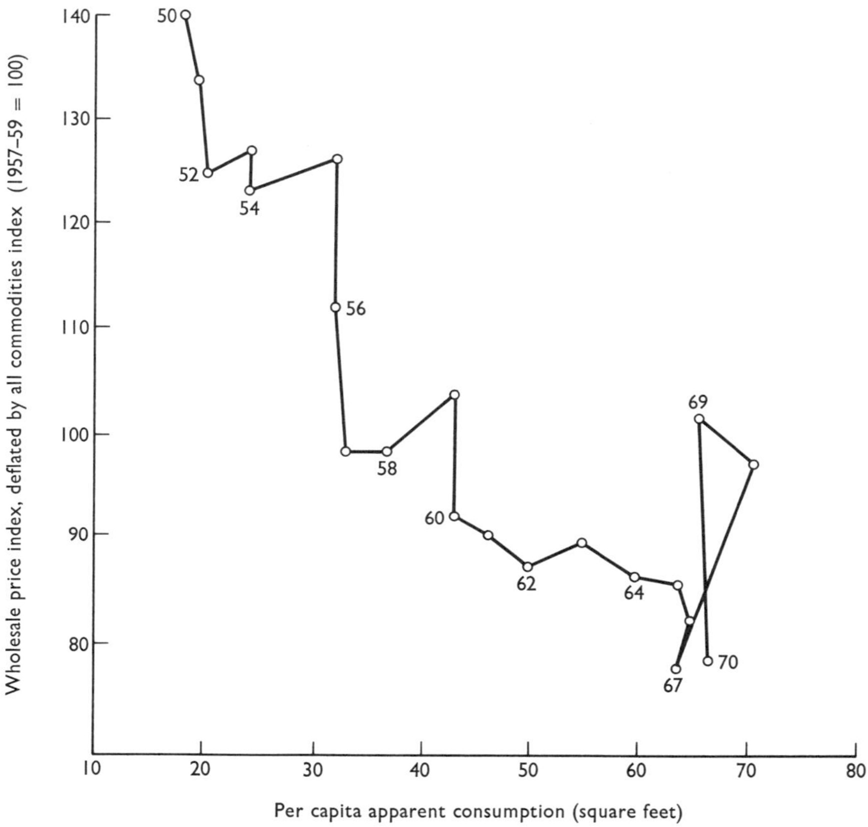


**Fig. 7.9** Per capita apparent consumption of softwood plywood and wholesale price (index, 1957–59 = 100). Points connected chronologically. Source: *The Demand and Price Situation for Forest Products 1970–71* (U.S.D.A., 1971).

ing at this same time. At least two factors have been operating to shift the demand curve outward, and both are important.

First, labor is one of the most important cost elements in construction. Labor costs have been rising steadily since 1950 and any material that promises to reduce labor cost is given a chance to prove itself by the construction industry. Second, plywood can be substituted for lumber in a good many construction applications: for sub-flooring, for sheathing, for roofing, and for concrete forms. A considerably higher material unit cost is offset by the reduced labor required for installation. It is the cost *in place* that is the real test.

**Supply and price.** Figure 7.10, in which price is plotted over total United States production, shows a pattern very similar to that of

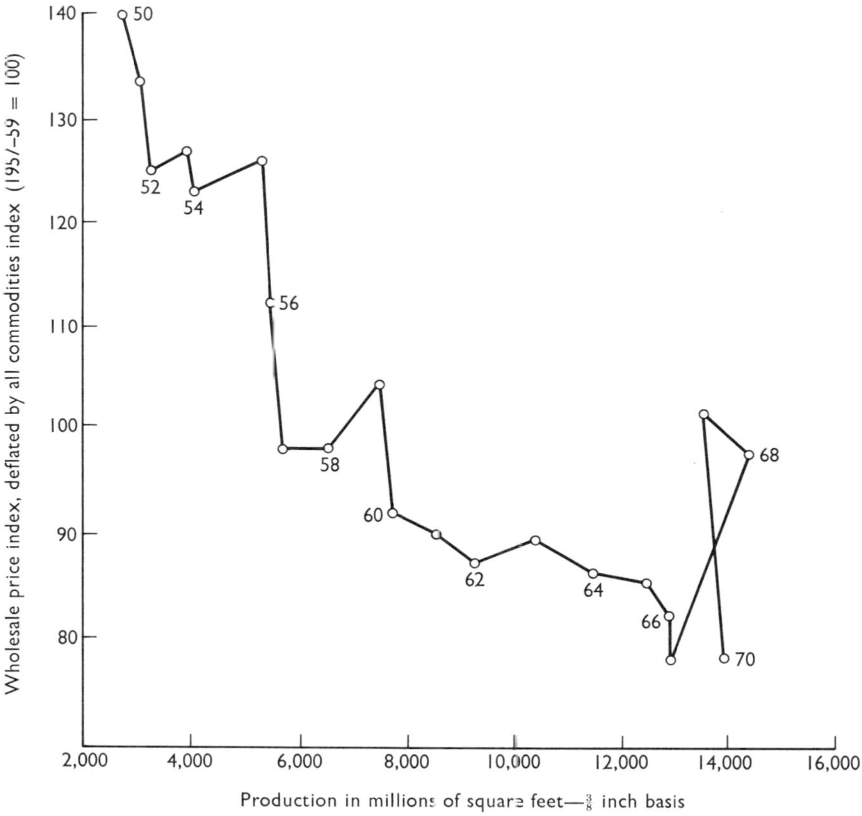


**Fig. 7.10** **Production of softwood plywood and wholesale price index (1957–59 = 100). Points connected chronologically. Source: *The Demand and Price Situation for Forest Products 1970–71* (U.S.D.A., 1971).**

per capita consumption. The short run supply curve apparently shifted outward at a rapid rate until 1968. Part of the price decrease undoubtedly represents lower production and marketing costs. One could easily argue that, up to 1968, Fig. 7.10 is an illustration of a long run average cost curve in an industry having substantial economies of scale, for average plant size has increased from about 40 million to over 70 million square feet annually. But more than scale is involved; there have been real gains in productivity as well. Figure 7.10 was constructed with deflated price data, and from these it seems quite certain that productivity increases in this wood-using industry have far exceeded those of United States industry generally.

Finally, during the decade following World War II softwood plywood manufacturing was a highly profitable business—sufficiently profitable to attract large numbers of new plants. As the new plants added their production to the industry total, prices dropped. Competition acts as a sharp spur to increased efficiency and with this one often sees price reductions. Moreover, there have been changes in the "mix" of plywood grades; sheathing grades have expanded relative to the higher priced sanded grades, and sheathing types involve lower labor requirements.

### Hardwood Plywood

As organized in the United States the hardwood plywood industry is usually divided into two branches, one producing container and packaging plywood, the other making "ordinary" plywood. Attention here will be focused entirely on the latter.[2] Plywood of this type is identified according to the species used on the panel face, although other species—even softwoods—may be used as an interior core.

The beauty and unique character of hardwood plywood varies greatly with the species but it depends almost as much upon the manner in which the face veneer is cut: rotary cutting, for example, yields one type of figure while sliced, sawn, and half-round veneers show quite different patterns. Most plywood veneer is rotary cut, for this yields by far the lowest cost material. Once cut the successive sheets of veneer can be arranged (matched) in various ways, yielding an almost infinite variety of designs.

The hardwood plywood industry was well established in this country prior to World War I, but it was World War II that really pushed this material into prominence. New adhesives permitted far stronger, more durable plywoods to be made and techniques were

[2] As late as 1950 container plywood production amounted to over one-fourth of the total United States hardwood plywood output. By 1964, however, this percentage had dropped to around five per cent.

developed for "molding" plywood into various curved forms. Under pressure of wartime shortages, hardwood plywood was quickly put to a variety of new uses. The dramatic success of the PT boats and of the Mosquito bomber provided superb publicity and, as would be anticipated, new applications came rapidly in the postwar period. Today hardwood plywood is used in widely differing ways. In construction it is used primarily for interior paneling, cabinet work, and doors.[3] In manufacturing, hardwood plywood finds its major use in furniture construction, store fixtures, radio and television cabinets, boats, etc.

By 1951 there were some 262 plants engaged in the production of this material, nearly all of them east of the Mississippi River. Most of the firms were small, but as is usual in such cases, production was heavily concentrated in the mills of larger size.[4] Apparently there are substantial economies of scale in this industry; man hour requirements per 1,000 square feet of production almost halved as size increased from a mill employing fewer than 100 workers to one employing over 200 men.

**Demand, Supply, and Price.** The relationship between consumption and the price index of hardwood plywood is shown in Fig. 7.11 over the period for which reasonably reliable data exist (1951 to the present). As in previous analyses, the points are connected chronologically. Despite the 300 per cent increase in consumption there has been no dramatic change in price, only 13 per cent between 1952 and 1969. The use of per capita consumption data (not shown) yields almost identical results—per capita consumption increased from 8 square feet in 1951 to 19 square feet in 1969.

The evidence provided by these data throws little light on the shape of the demand curve. From 1951 through 1958 demand was apparently fluctuating, but there is little evidence of any real shift, and prior to 1951 there is simply no price/quantity information

[3] "Door skins," a natural for hardwood plywood, became a major item with many plywood manufactures immediately following World War II when flush doors became standard for interior construction.

[4] According to a Tariff Commission report (United States Tariff Commission, no date) the seven largest mills produced 16 per cent of all United States output in 1951, while four large firms operating 11 large mills produced fully 27 per cent of the total. Unfortunately, no later data are available for comparison or for examining trends toward concentration. The information on economies of scale is contained in a report by the Bureau of Labor Statistics (U. S. Bureau of Labor Statistics, 1953).

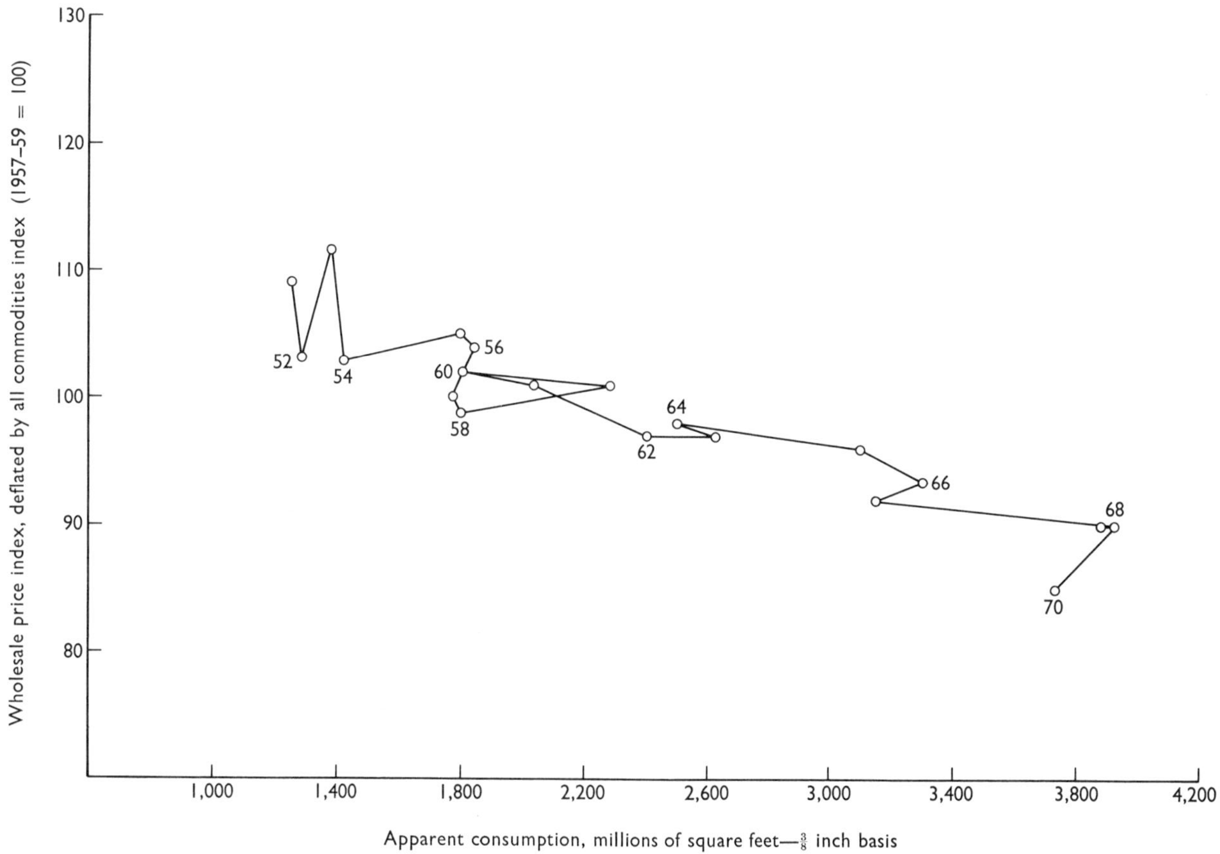


**Fig. 7.11** Apparent consumption of hardwood plywood in United States and wholesale price index (1957–59 = 100). Source: *The Demand and Price Situation for Forest Products 1970–71* (U.S.D.A., 1971).

available. From 1958 on, however, it seems clear that the demand curve has been shifting outward, at least until 1969.[5]

The relation between United States *production* and the deflated wholesale price index is shown in Fig. 7.12—again with the separate

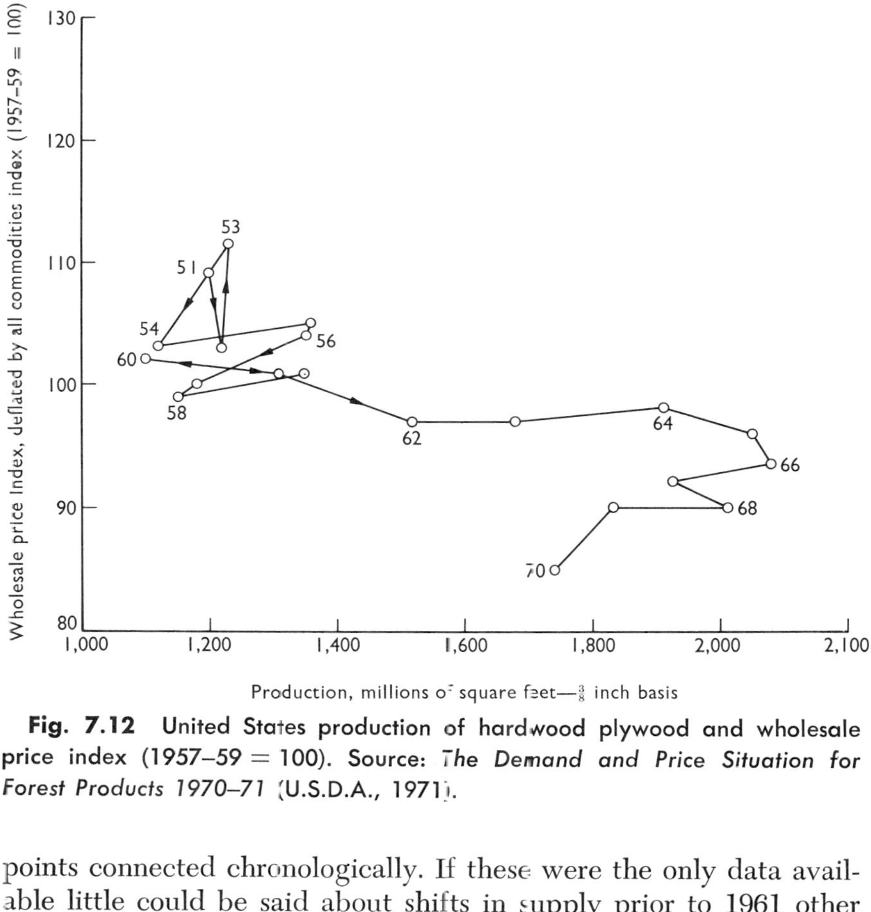


**Fig. 7.12** **United States production of hardwood plywood and wholesale price index (1957–59 = 100). Source: *The Demand and Price Situation for Forest Products 1970–71* (U.S.D.A., 1971).**

points connected chronologically. If these were the only data available little could be said about shifts in supply prior to 1961 other than that the supply curve during this period was certainly not steeply sloped. If, however, we look at the entire supply offered to consumers in the United States by including imports with domestic production, then we can use the consumption series of Fig. 7.11 as evidencing a "total available" supply curve. At the same time, by

[5] The drop in consumption between 1959 and 1960 is probably accounted for by the very large drop in housing starts, from 1,554,000 in 1959 to 1,296,000 in 1960.

looking at domestic production and imports individually (Fig. 7.13) we can see whether the increased quantities came from home or abroad, since from 1952 through 1960, supplies were increasing while prices held up reasonably well (Fig. 7.11). *Domestic* supplies

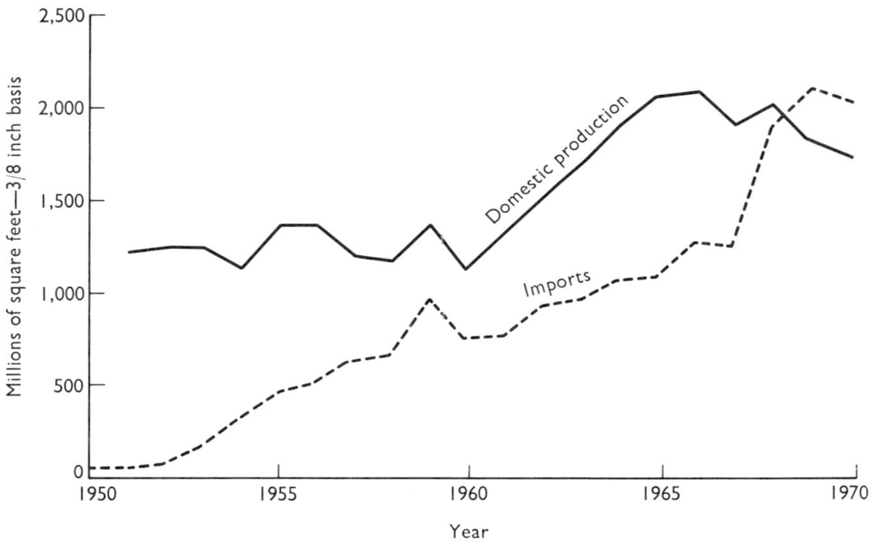


**Fig. 7.13** **Hardwood plywood imports and domestic production. Source:** ***The Demand and Price Situation for Forest Products 1970–71*** **(U.S.D.A., 1971).**

were almost constant, with the increased quantities coming from imports. Apparently the price was high enough to encourage imports very substantially, but not high enough to encourage domestic producers to expand. And from 1960 through 1966 both imported and domestic supplies increased as the demand curve shifted outward with little change in price—an increase in consumption of 118 per cent while prices dropped only 12 per cent. During this period domestic production increased at a faster rate than imports; domestic producers nearly doubled their output while imports continued to increase substantially. One could scarcely use these data as proof of a declining supply curve for domestic producers, since a very substantial part of the total supply was being imported. It is plain, however, that by 1968 supplies available to consumers in the United States had increased over 300 per cent (since 1951) while real price had actually declined. The consistency of the price decline seen in the domestic production (Fig. 7.12) does lead one to suspect that a real cost saving had been accomplished.

Imports of such a substantial amount deserve a closer look. From 1940 through 1950 imports of hardwood plywood consisted primarily of Canadian birch. Other plywood, mostly from Japan and the Philippines, had been imported at around one-third the level of that from Canada, and total imports had held well below 10 per cent of total domestic production. In 1953 these "other" imports increased almost 600 per cent and have increased nearly every year since that time (Fig. 7.13). Japan was the early leader, and before long other nations had entered the market. At the present time over 90 per cent of our hardwood plywood imports are from Asian countries—primarily Japan, Korea, Taiwan, and the Philippine Republic (in order of quantitative importance in 1968)—yet the total number of suppliers is much larger—34 different nations in 1964. Total imports of hardwood plywood finally passed domestic production in 1969.

How much of this expanding market could have been supplied by domestic producers alone is a moot point. From 1930 on, the tariff on most hardwood plywood was set at 40 per cent ad valorem. It was reduced in 1950 to 20 per cent, where it has remained until the present.[6] During the middle and late 1950's, and after the reduction in the tariff, domestic producers were vociferous in their demands for higher protection. The statistics, however, would seem to indicate that imports have been necessary to supply the market and simultaneously hold price levels reasonably steady. United States production has expanded; in fact, it nearly doubled during the period here considered, but consumption increased much faster.

The gap between United States consumption and United States production has been filled by imports, and undoubtedly at great advantage to the American consumer, since the imports have apparently been in sufficient quantity to keep real prices in line with increasing productivity. Producers, quite naturally, take a somewhat different view toward trade policies which permit importation in quantities sufficient to hold prices down.

[6] Birch plywood was accorded different treatment. Its tariff rate was 50 per cent in 1930, 25 per cent in 1936, and 15 per cent in 1950. Strangely, rates for Spanish cedar have remained unchanged at 40 per cent since 1930.

## PARTICLE BOARD

The newest of wood-based sheet materials is an outstanding illustration of how technical developments might change the wood use picture of the future, and how economic considerations influence the introduction of new technologies. The techniques of particle board manufacture, including chipboard, flakeboard, and other types, had been developed in Europe before World War II. It took some time to establish production after the war, but by 1950 particle board—defined by the Department of Commerce as "an engineered product, mass formed, consisting of machined fiber particles such as granules, chips, slivers, flakes, shavings, etc., of a controlled moisture content and size, bonded together with a synthetic resin or other binder into panel form under controlled heat and pressure" —was readily available. West Germany, the leader in particle board technology, produced about 4.5 billion square feet in 1960, and by 1969 had increased this to 27.3 billion square feet annually.

Particle board was relatively slow in becoming established in this country, but once started it grew rapidly. In 1957, the first year a census of particle board plants was undertaken by the Census Bureau, the 46 known plants produced a total of 183 million square feet (Wright & Phelps, 1967), and in 1969 FAO estimated United States production to be 26.2 billion square feet.

Particle board can utilize a wide variety of species and lends itself to relatively small plant operation. Currently, most production comes from plant waste and is primarily used for core stock in furniture manufacture. New uses, however, are continually being promoted.

A major deterrent to particle board manufacture in this country has been the ready availability of Douglas fir plywood. Particle board is substitutable for plywood in many uses and when waterproof particle board is eventually developed, this substitutability, already large, will be greatly enhanced. Moreover, as with any new process, unit production costs can be expected to drop rapidly with improved techniques, and as costs drop further competition with plywood will develop. New market possibilities are of course present, but most observers agree that continued expansion of the particle board industry will inevitably result in a further loss of lumber and plywood markets.

The eventual impact of such a development will depend upon

the extent to which such substitution occurs, a matter more fully developed in Chapter 8. For now it suffices to point out that the possible impact on forestry is large. Almost any species of nearly any size can be utilized for particle board construction. For many years foresters have joked about "mowing machine forestry," and particle board technology makes this a real possibility. Obviously, particle board would not substitute for all plywood and lumber uses, but it could have an impact strong enough to push forestry into the mowing machine era before many foresters are ready for the change.

# 8

# Derived Demand and Demand for Aggregates

One speaks of the demand for lumber or the demand for paper, assured that most people will understand what is intended. Yet we know that the demand for most lumber, paper, and plywood is not direct but is derived from the sale or the expected sale of finished houses, furniture, boxes or toys, or of newspapers, magazines, books and grocery bags, or of the hundreds of other things into which these products are fabricated. One subject to be explored in this chapter, therefore, is that of "derived demand," a topic seldom given more than passing mention in introductory economics courses, but of singular significance to anyone who would understand timber products markets.

We will also take a closer look at the demand for aggregates of products. Lumber, pulp and paper, plywood, and particle board are well identified commodity classes, yet within each class there is a wide range of specific products. An examination of the demand for such aggregates is also warranted.

## DERIVED DEMAND

The classic approach to derived demand may be illustrated as follows: In Fig. 8.1(a) let $D_h$ be the demand curve, and $S_h$ the supply curve, for new residential housing in the United States, measured in terms of square feet of floor space. $Q_h$ and $P_h$ are, respectively, the equilibrium quantity and price for residential housing floor space in the nation. $S_o$ is the aggregate supply curve for all factors of housing production other than lumber—factors such as paint, glass, bricks, labor, concrete, capital, nails, etc.

The demand curve for lumber used in residential construction may now be "derived" by subtracting this supply curve for all other productive factors ($S_o$) from the housing demand curve ($D_h$). The result of this subtraction is shown in Fig. 8.1(b), and is the demand for housing lumber ($D_L$); the shaded portions of the two diagrams are equal.[1]

$S_L$, in Fig. 8.1(b), is the supply curve for construction lumber. Like all supply curves this represents the horizontal summation of the producer's marginal cost curves.[2] The equilibrium price of lumber ($P_L$) is determined in the usual fashion, by the intersection of the lumber demand and supply curves.

Note that any movement along the derived demand curve for lumber necessarily implies something about the prices for all other productive factors used in house construction. Because of the assumptions required for its derivation the derived demand curve for lumber implies that the prices of all other productive factors will change as indicated by the supply curve ($S_o$) if there is any movement *along* $D_L$. A derived demand analysis has a built-in assumption of fixed proportions of all input factors. In the present illustration this means that house construction is assumed to require certain fixed proportions of lumber, paint, nails, cement, glass, steel, etc. In the short term this may be a reasonable assumption, but it would scarcely be realistic over longer time periods.

Four propositions are usually advanced respecting the elasticity of derived demand curves (Marshall, 1947). The derived demand for a product will be *more inelastic:*

[1] The scale used for the lumber demand curve must also be derived from that for the final product if a direct relationship is to be maintained.

[2] In Fig. 8.1 the supply curve for lumber is also necessarily the difference between $S_o$ and $S_h$, for the supply curves of lumber and of all other productive factors must add to the supply curve for housing; there are no other factors.

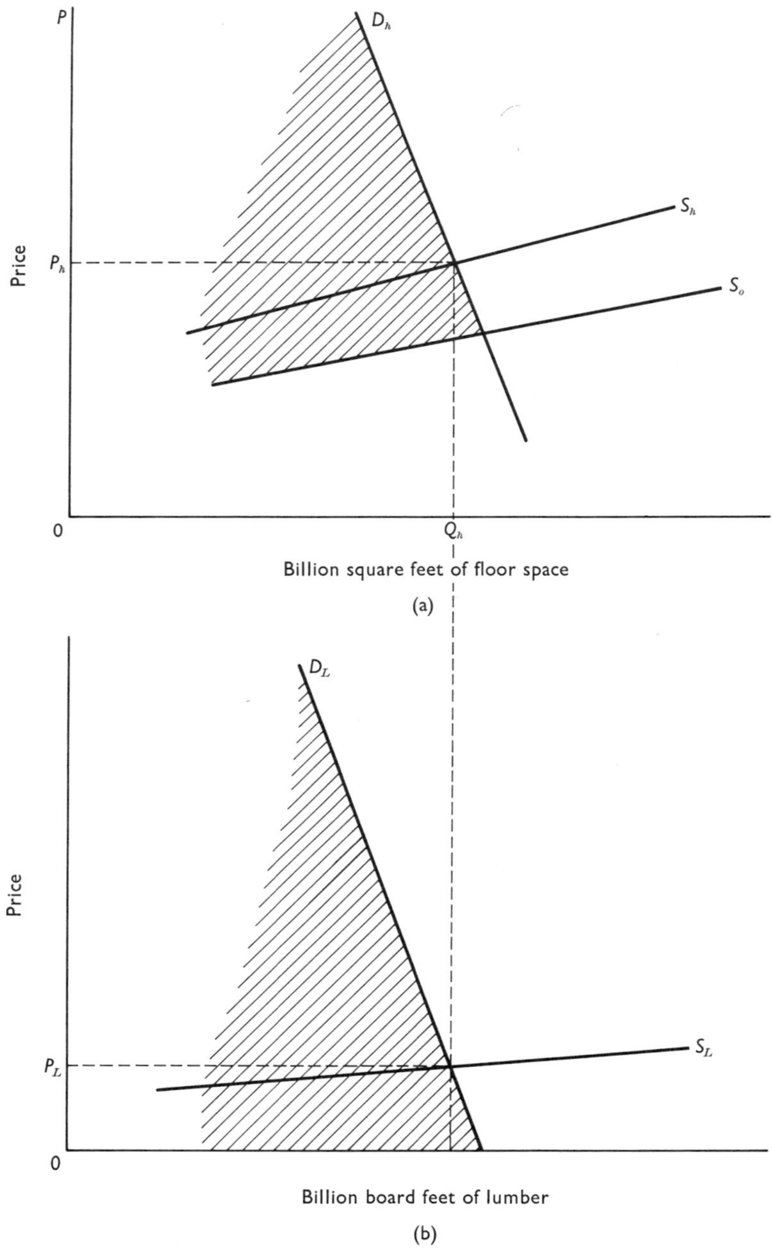


**Fig. 8.1** Derived demand.

1. the more essential the factor being examined;
2. the more inelastic the demand for the final product;
3. the smaller the fraction of total cost that goes to the factor;
4. the more inelastic the supply curve of the other factors.

Citing these four propositions, some authors have concluded that the demand for lumber must indeed be quite inelastic (Mead, 1966). An examination of the four points as they apply to lumber would certainly lead to agreement; with current techniques of construction, lumber is quite essential to houses; the demand for the final product (housing) is probably inelastic at given income and population levels; lumber makes up no more than 15 to 20 per cent of the total cost of a house; and the supply curve for other productive factors (labor, capital, paint, glass, etc.) could be no more than moderately elastic. Both Holland and McKillop concluded as a result of their econometric studies that indeed, lumber demand was inelastic, even though the analyses *per se* were inconclusive (Holland, 1955; McKillop, 1967).

What is frequently overlooked is the fact that most of the conclusions regarding the inelasticity of lumber demand stem from analyses treating lumber as a single commodity. The findings are analagous to those one would expect from an analysis of the demand for food. Certainly in the United States the demand for "food" would be highly inelastic; after all, the human stomach is of limited capacity. Yet a different conclusion about demand elasticity should be anticipated if one were to analyze the demand for frying chickens instead of the demand for "food." Other meats are readily substituted for chicken in our diet, and a relatively small change in fryer prices can and does have a decided effect on sales. Similarly with the demand for a specific wood product produced in a given region, for example, redwood siding, or southern pine box lumber, or black cherry paneling, or most of the 2,000-odd items that together are called "lumber."

The demand for siding, for example, would be derived from the demand for housing and would probably be quite inelastic; it is difficult to build a house without some sort of siding. But there are many siding materials—plywood of several types, shingles of cedar or asbestos, brick, aluminum, pine and fir materials, red cedar, redwood, etc. At any given point in time each of these materials will have a specific price. Given the choice of siding materials that exists, a relatively small change in the price of, say, redwood siding lumber *could* have a very substantial effect on the amount of redwood siding sold. The demand for many, perhaps most, lumber products may well be quite *elastic* when viewed from a specific product standpoint. An over-reliance on an inelastic derived demand for "lumber" could have very serious effects on lumber markets.

## AGGREGATE DEMAND

One aspect of the diversity in the demand for many timber products is illustrated in Fig. 8.2. Here the total demand curve for lum-

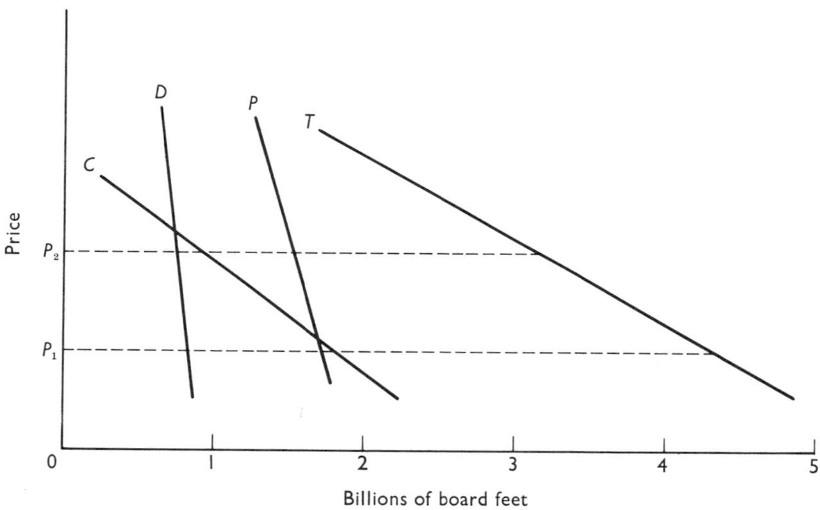


**Fig. 8.2** Total demand for shipping lumber (*T*) is an aggregate of the demands for containers (C), pallets (*P*), and dunnage (*D*).

ber used in shipping ($T$) is represented as the aggregate of three major components—containers ($C$), pallets ($P$), and dunnage ($D$).[3] The demand curves of the figure are hypothetical, though the

[3] Dunnage is the term applied to wooden bracing used when shipping by rail or truck.

quantities used at price $P_1$ are those reported for 1962.[4] With a demand schedule as shown, a price rise to $P_2$ would result in a substantial change in the proportion of lumber taken by each product. To understand thoroughly just what was affecting the change in total demand it would be necessary to investigate each of the components, for it is the elasticity of demand with respect to price that is important, and each component has a different elasticity.

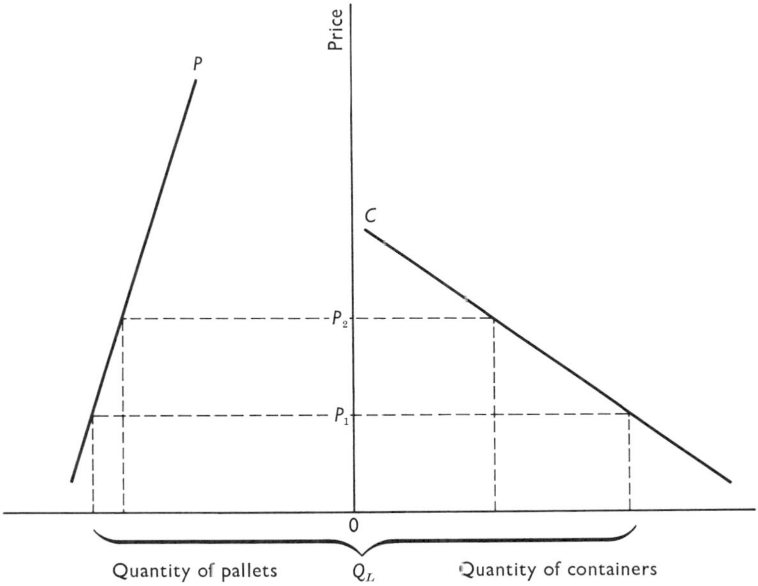


**Fig. 8.3** If two demand curves are plotted "back to back" the distance between them at any given price represents the sum of the quantities demanded at that price.

A better understanding of the response of any two markets to such changes may be facilitated by plotting the two curves "back to back." As an illustration, the demand curves "Containers" ($C$) and "Pallets" ($P$) are shown in Fig. 8.3.

[4] In 1962 boxes and crates (called containers in the illustration) used 1,790 million board feet, pallets used 1,710 million board feet, and dunnage used 840 million board feet (U.S.D.A., 1958; Table 29, page 39).

One could argue about the elasticities of the demands for the three products shown, in either absolute or relative terms. No claim for accuracy on this point is made. The curves are for illustrating the *effect* of elasticity when deriving aggregate (or composite) demand curves.

Plotted in this fashion the distance between the two curves ($Q_L$) represents the total amount of lumber used for the two products. A price rise from $P_1$ to $P_2$ reduces total lumber consumption, but by a far greater amount in the more easily substituted (more elastic) use for containers. This "allocation" of available resources between competing uses is one of the most important functions of our price system.

Shifts in the demand for one product can also have a substantial effect on the availability of a limited resource for use in a second product. Suppose, for example, that the demand for pallets increased very substantially over time—as has been the case—and that the demand for containers diminished somewhat—again in accord with reality. If the shifts were as indicated in Fig. 8.4, and if

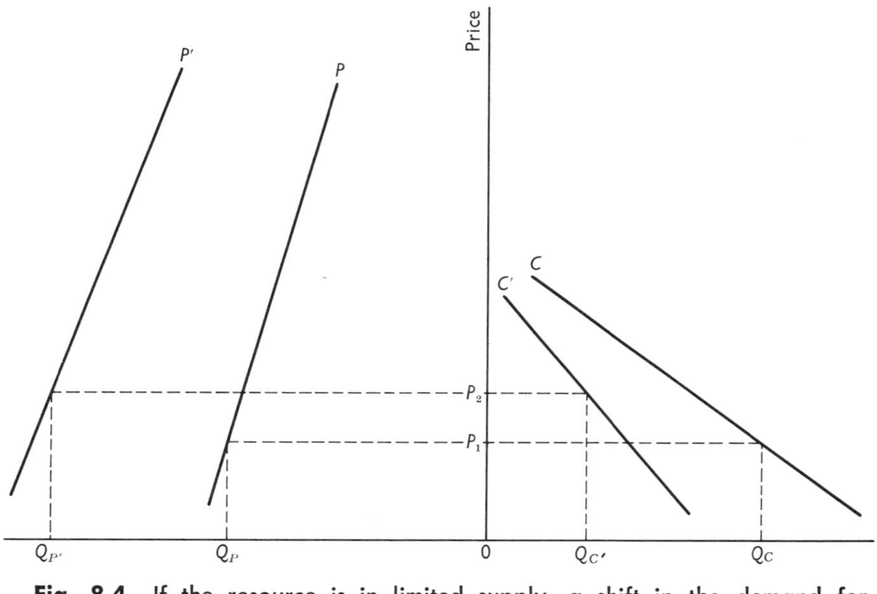


**Fig. 8.4** If the resource is in limited supply, a shift in the demand for one product can adversely affect the supply available for the second.

the amount available for the two uses remained about the same, price would rise slightly, from $P_1$ to $P_2$, the amount entering the more "essential" use of pallets would increase very substantially (from $Q_P$ to $Q_{P'}$), while the quantity consumed by the container industry would slump drastically (from $Q_C$ to $Q_C'$).[5] Similar illustrations could be evolved for many wood products.

Graphic representations such as these are not intended as a means for quantitatively analyzing problems of the real world where many different uses may compete for the same resource. Complex problems are usually better handled in terms of elasticities of the various demands, as indicated in earlier paragraphs. Unfortunately, few quantitative studies for determining demand elasticities for timber products have been undertaken.

## Cross-Elasticity

In discussions of markets where two or more products are in direct competition—either wood or non-wood—one may encounter the term *cross-elasticity of demand.* This refers to the ratio between the relative change in the price of one product and the relative change in quantity of the second product sold. One might have been interested a few years ago, for example, in determining the relationship between a change in sale of wooden storm sash and the price of aluminum storm sash. The cross-elasticity between the two products would be (using $a$ and $w$ to indicate aluminum and wood):

$$E_a^w = \frac{\%\text{ change in quantity of wooden storm sash}}{\%\text{ change in price of aluminum storm sash}}$$

Various formulas for calculating cross-elasticities—point or arc—may be devised by referring to the earlier treatment of elasticity in Chapter 2. The concept itself should pose no problems. There is, however, one significant difference between the usual demand elasticity and cross-elasticity: cross-elasticity can be positive, negative, or zero, whereas elasticity of demand with respect to price is almost always negative. Positive cross-elasticities characterize substitutes: if the price of cellophane bags *decreases,* sales of paper bags *decrease.* A negative cross-elasticity indicates the goods are complementary: if the price of glass goes *up,* sales of wooden sash go *down.* A zero cross-elasticity indicates the two goods are independent:

[5] It appears that the demand curve for pallets actually did shift outward rapidly, as indicated in Fig. 8.4. However, the demand curve for containers apparently shifted also, adversely affected by the substitution of paperboard and containerboard, so the price actually declined somewhat and total consumption remained very close to the original amount.

prices of airplanes may rise or fall with little or no effect on sales of red oak flooring.

## AN INTEGRATED EXAMPLE

For a final look at the complex of demand, supply, and price of timber products we will return to the demand for lumber. Derived demand theory provides a good takeoff point for attempting to bring together timber product markets. It makes markets for housing, shipping containers, newspapers, and packaging the starting point of a long reaction chain that stretches all the way from the grocer placing his order for 20-pound bags to the pine tree growing in a southern forest plantation, or from a building contractor in Ohio to the Douglas fir standing in Oregon.

The construction industry provides the most important single market for lumber. Construction uses wood in many different ways, often substituting one timber product for another. Here we will focus on just one of these many uses—sheathing. Nearly all residential construction requires sheathing material of some kind; in former days lumber was used almost exclusively, but now sheet products such as plywood and particle board are finding more favor.

### Housing

The starting point for demand analysis is the demand for new housing, shown in Fig. 8.5. Several factors could be identified as influencing the position and shape of the demand curve for new housing: population, age class distribution, the marriage rate (or rate of new family formation), per capita income, number of existing houses, degree of urban concentration, number of rental units available, and so on. The supply curve would be influenced by a quite different set of factors: cost of raw materials (including wood products), wage rates for construction workers, mortgage rates, cost of building sites, etc. In passing it should be noted that some of these factors are quite stable over time periods likely to be chosen for an analysis, while others might fluctuate rapidly. Most factors would probably vary more as the time span lengthened.

Figure 8.6 is a set of related diagrams, beginning in the lower right hand corner, labeled (a), with a reproduction of Fig. 8.5.[6]

[6] Note that the scale has been changed to square feet of floor space and cost per square foot.

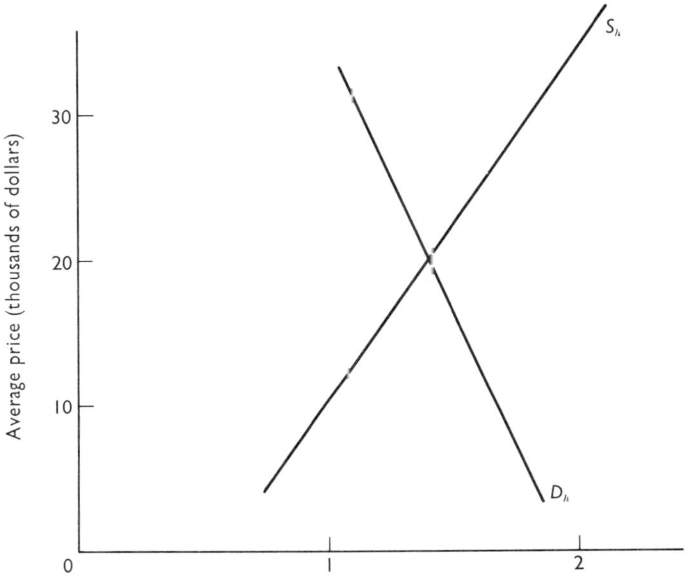


**Fig. 8.5** A hypothetical demand and supply relationship for new housing.

The demand for many products is derived from this demand for new housing, and among these is the one in which we are especially interested, that for sheathing material.

### Sheathing Material

The demand and supply curves for sheathing material are shown in part (b), directly above those for housing. The amount of sheathing required will vary with the total square feet of new floor space to be built; larger houses, or more houses, require more sheathing material. There is, therefore, a direct tie-in between the two diagrams, (a) and (b). If either the demand or the supply curve for housing shifts there will be a corresponding shift in the demand for sheathing material.

Sheathing material accounts for only a small part of the total cost of a new house—perhaps around one and one-half per cent of the total. In line with what we know of derived demand, therefore, we would scarcely expect sheathing material to be very sensitive to price changes. A 20 per cent increase in the price of sheathing, for

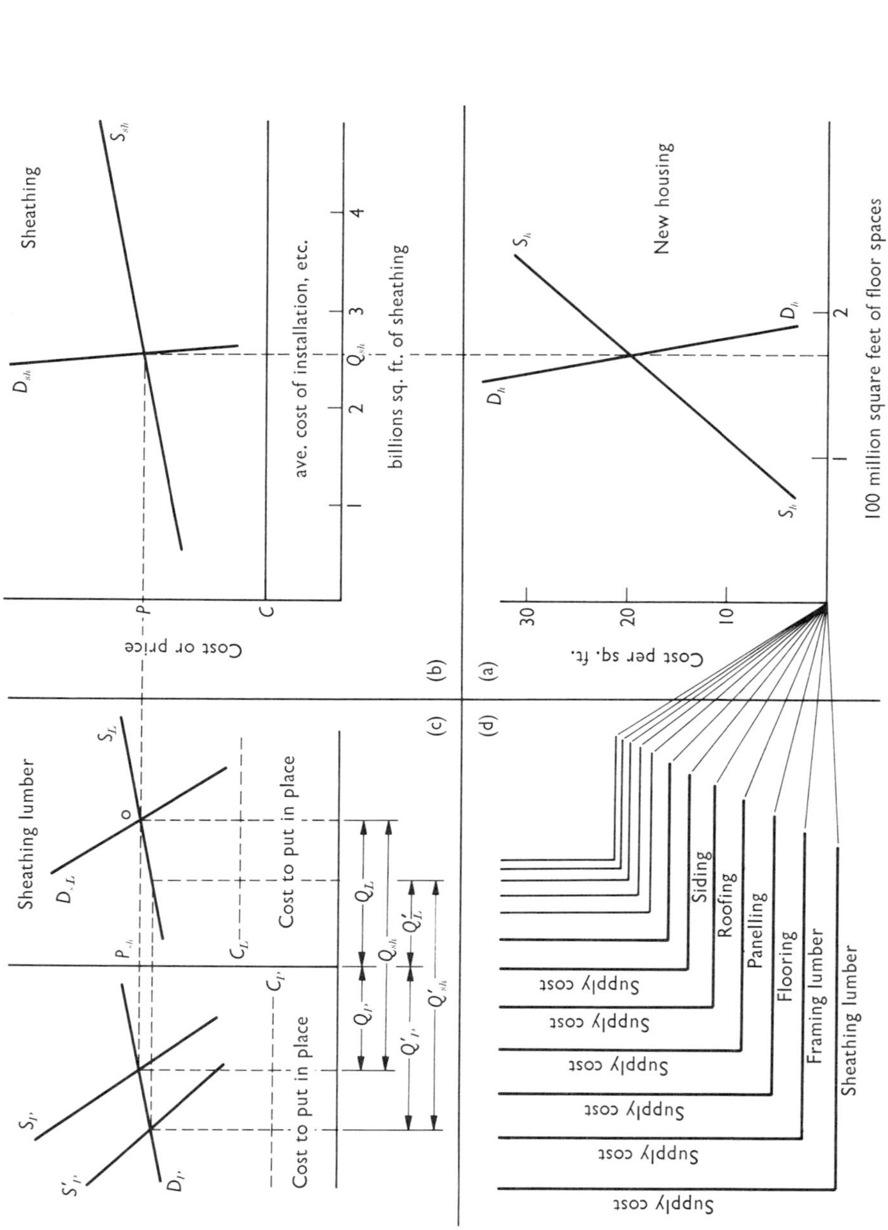


**Fig. 8.6 The relationships between demand for new housing, sheathing lumber, lumber and plywood, and other factors of production—and their relation to housing prices.**

example, would (*ceteris paribus*) cause a $20,000 house to cost perhaps $20,060, hardly enough to cause a builder to change his mind about building the house! Over the short term, then, the demand for sheathing ($D_{sh}$) would probably be very inelastic, just as it is for most of life's necessities.[7]

Conversely, the supply curve ($S_{sh}$) for sheathing material would probably be very elastic. Why? Because sheathing material is not very exacting in its technical requirements and, therefore, a good many species and grades of lumber could be used in addition to a variety of plywood, particle board, etc. Moreover, sheathing makes up only a part of the total production of any of the products so used; the demand for *lumber of sheathing quality*, as an example, is a composite of several demand curves, *one* of which is the demand for sheathing lumber.

Both the demand ($D_{sh}$) and supply ($S_{sh}$) curves are shown in Fig. 8.6(b) so they will show the cost (or price) *in place*. This means that the 1,700 million square feet of new housing floor space built in the selected year required $Q_{sh}$ square feet of sheathing material at an average cost in place of $P$, of which $P$ $C$ was the "price" (or cost) of the sheathing material, and $C$ was the cost of transportation to the site, installation, wastage, etc.

## Substitution Effects

Now focus more sharply on two particular wood products, lumber and plywood, both of which can be, and are, used for sheathing. And to simplify our example a little we will assume that these are the only two materials that are important sheathing materials. The supply curve for sheathing lumber ($S_L$) and for sheathing plywood ($S_p$) are drawn "back to back" to facilitate comparison, part (c), upper left. The quantity of the two materials supplied (under the assumptions) must be $Q_{sh}$, as given in part (b). As long as these two materials are technically equivalent one would suppose builders would be indifferent as to the raw material used; if it "does the job" the material with the lowest cost in place would be selected by the rational builder. We know, therefore, that $Q_P$ of plywood, and $Q_L$

[7] The usual method of deriving such a demand curve—by subtracting from the demand curve for housing the supply curve for all productive factors other than sheathing—is of course appropriate here, but probably inaccurate, since the "remainder" is only one and one-half per cent of the total. Hence a small error in establishing the supply curve of other factors would make a very large error in the derived demand curve.

of lumber would be used for sheathing material; the distance $Q_{sh}$ must be fitted into the space between the supply curves.

Note, however, that the installation cost for the two products is not the same. Lumber must be put in place piece by piece, and each piece must be sawed to length. Plywood, on the other hand, goes up in 4 × 8 or 4 × 10 panels. A reduction in labor cost can easily offset a higher material cost. Consequently, at the installed price of $P_{sh}$ the price of lumber ($P\ C_L$) must be less per square foot than the price of plywood ($P\ C_p$). The two supply curves, $S_p$ and $S_L$, must of course "add up" to $S_{sh}$, the supply curve for sheathing material shown in the upper right-hand diagram.

What about the *demand* for sheathing *lumber*? It is, we know, derived directly from the demand for sheathing materials. A brief digression seems in order.

## A "Kinked" Demand

Under usual *ceteris paribus* assumptions both the price and the quantity of plywood must be held constant. If this assumption is maintained then the demand for sheathing lumber must resemble the curve ($D_L$) shown in Fig. 8.7. An increase in the price of sheath-

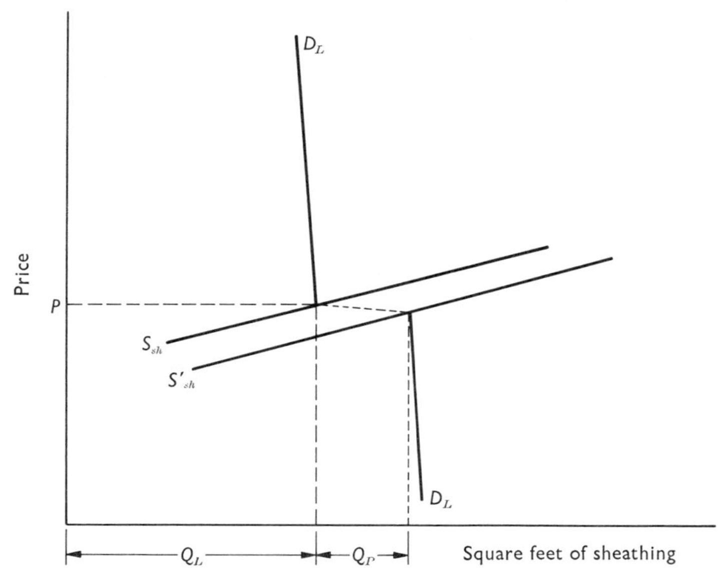


**Fig. 8.7** Under certain conditions the effective demand curve for sheathing lumber could be "kinked."

ing lumber would have little effect on the quantity sold, for the demand curve for sheathing lumber above price *P* is the demand curve for sheathing material—shifted to the left by the (fixed) quantity of plywood consumed ($Q_L$).

But suppose the supply curve for sheathing *lumber* were to shift outward as indicated in the figure ($S'_{sh}$). Now, providing builders really had no preference between the two materials at price *p*, demand for sheathing lumber would be very elastic. The price differential between plywood and lumber would have to be large enough to compensate builders for the trouble of changing orders and techniques, but this would not amount to any great sum. Once the plywood market had been taken over completely the demand curve would again become inelastic; it would now be *identical* to the demand curve for sheathing material. This "kinked" demand curve[8] is not particularly rare among goods having close substitutes, but it should be remembered that such a curve arises as much from the *ceteris paribus* assumptions as from any other factor.

## Interdependence in Demand

A more rational (or at least more useful) approach to this problem would seem to be to drop the restrictive assumption that price and quantity of plywood would remain constant and instead try to recognize the interdependence of the two commodities. We would still regard the demand for lumber as derived from the demand for sheathing, but it should also reflect the *supply* situation for plywood. In fact, *the demand curve for sheathing lumber will be exactly the same as the supply curve for plywood*, transferred so as to pass through the point *o*. It is shown in Fig. 8.6 as $D_{sL}$. This comes about because the total quantity of sheathing material must come from these two materials. Given $S_p$ as the supply curve for sheathing plywood, any increase in lumber price would result in a loss of lumber markets to plywood; the "mix" of plywood and lumber would change while the total quantity of sheathing material going into housing would remain virtually constant. As a consequence, the demand for sheathing lumber could be very *elastic*, reflecting the supply elasticity of plywood. In similar fashion a demand curve for

[8] The "kinked" demand curve is generally discussed as a reflection of oligopolistic conditions, but seems at least equally appropriate here. Kaldor suggested such curves should be called "imagined demand curves" (see Sweezy, 1939).

plywood could be constructed from the *supply* curve for sheathing lumber, and would be at least equally elastic.[9]

The moral to this part of our exercise seems fairly clear. Price elasticities for wood products with acceptable substitutes tend to be high so long as supplies of the substitute materials are readily available.

## Effect of Shifting Supply

Returning again to Fig. 8.6, suppose that for some reason the supply curve for plywood were to shift outward to the new position $S'_P$. As long as the demand for sheathing material remains unchanged the total quantity of sheathing will remain at least close to $Q_{sh}$. But again the composition of the "mix" would alter, this time plywood substituting for lumber. The new equilibrium would be reached with $Q'_P$ of plywood and $Q'_L$ of lumber continuing to make up $Q'_{sh}$ of sheathing material. The installed price would drop slightly to $P'$, followed by a *very* small increase in the total amount of sheathing material used. $Q'_{sh}$ would consequently be slightly larger than $Q_{sh}$, but almost negligibly so.

Clearly, any outward shift in the supply schedule for one product spells trouble for its competitor. If the plywood industry is subject to economies of scale, or if technological advances bring about real gains in productivity which permit lowering of plywood prices, this would be expected to have almost immediate repercussions in the lumber industry. With very little qualification we can picture the supply schedule for plywood and the demand schedule for lumber as being "tied" together, but with a rope that pulls in only one direction (an outward shift in demand for lumber will not necessarily pull the supply schedule for plywood along with it). And one can generalize this finding still further. An outward shift in the supply schedule of aluminum, or steel, or glass, or plastics, will bring decided changes in the markets for wood products *even though this is not accompanied by any drastic price change.* As a matter of fact, the previous chapter presented enough quantitative information to make one strongly suspect that plywood is a case in point when coupled with lumber; the supply schedule for plywood *has* been shifting outward. The following specific facts should come as no

[9] This assumes that the two supply curves (for lumber and plywood) are independent. Strictly speaking, of course, they are not independent, but this lack would not produce major practical changes in the conclusions reached here.

surprise. In 1959 the average new FHA-inspected single family dwelling used 985 board feet of lumber and 1,010 square feet of plywood for sheathing purposes. Three years later (in 1962) the corresponding figures were 617 board feet of lumber and 1,342 square feet of plywood, and by 1968 use of sheathing lumber had declined to 465 board feet while 1,707 square feet of plywood were used (U.S.D.A., 1971).

Before leaving this particular section of Fig. 8.6 it may be observed that an increase in the cost of installation would also produce a substitution of plywood for lumber. An increase in carpenters' wage rates, for example, would increase the installed price of sheathing. For reasons already mentioned this increase would probably not produce any substantial change in the total quantity of sheathing material used, but $S_L$ would rise more than $S_p$ (since installation costs are higher for lumber than for plywood). The product mix will again be altered, with plywood substituting for lumber. And again it may be pointed out that carpenters' wage rates did rise during the 1955–65 period, very probably accounting for some of the change referred to in the preceding paragraph.

## Aggregating Supply

Lumber used for sheathing purposes accounts for only part of the total lumber use when building a modern house. The supply schedule for sheathing lumber, therefore, is only one of a large array of such schedules, as shown in the lower left (d) of Fig. 8.6. Conceptually, one must aggregate these supply curves (accounting for lack of independence wherever necessary) in order to arrive at the supply curve of "lumber used in residential construction." Along with this set of lumber supply curves one can envisage a whole series of similar arrays, each representing a particular industry group. And finally, these all come together in the supply curve for new housing, the diagram in the lower right-hand corner, and the starting point for this discussion.

Since the supply of sheathing material is an element in the supply schedule of new housing it is clear that any shift in sheathing material supply—plywood, lumber, or whatever—will indeed cause *some* response in the housing market. If the supply curve for plywood moves outward, as suggested here and in the previous chapter, the market for sheathing lumber would drop. But the induced outward shift of the housing supply curve would increase the number of houses built (*ceteris paribus*) and the total market for sheathing materials would expand, though ever so slightly. This expansion would automatically be transmitted around the diagram to all the industries affected.

One final point should be made. A diagram such as Fig. 8.6 serves, hopefully, to illustrate some of the complexities and interrelations that are involved in the economics of wood product markets, but is nevertheless a drastic simplification of reality. Of particular importance is the fact that market responses are far from instantaneous. The time lags between price and production changes can be quite substantial. The connections between the various curves of Fig. 8.6 should be imagined as ties possessing varying degrees of "stretch," while the movements of the curves are subject to all sorts of delay switches, triggers, and rachets.

In spite of these time lags and other friction, however, the curves do shift and responses to market changes do occur—often with dismaying rapidity. In the minds of many it is this very "sensitivity and complexity" which makes the market such an impressively efficient mechanism. Without relatively free markets man must somehow develop other means for identifying the need and transmitting the signals to increase or diminish production. The "market" is a staggeringly complex mechanism, but it works with amazing efficiency. One can toy with the idea of establishing vast systems of complex equations to be fed into huge computer systems which would then grind out simultaneous solutions, thus generating a set of production guides. But this is exactly what the price system already does for us!

# PART III

# TIMBER PRODUCTION ECONOMICS

# 9

# Factors Influencing Timber Production

One of the genuine attractions of our profession is the range of skills and knowledge that a successful forester must command. Though only a part of forestry, timber production provides an excellent example of this variety. The peculiarities of stumpage production and the biological limits imposed by species, soils, patterns of rainfall, and temperature variations, as well as the likelihood of fire, insect attack, or storm damage, must first be understood. These must be worked into a framework of specific but everchanging economic limitations: land ownership, interest rates, credit availability, taxes, logging, and, ultimately, manufacturing. Some of these factors, particularly interest, have often been regarded as essentially mechanical in nature. Yet important theoretical considerations also exist, and in order to assure a full understanding we will necessarily examine interest theory. Others, like ownership, credit, and taxes, are highly institutionalized. Finally, an analytical framework for fitting together the direct and indirect costs of timber production must be established so that stumpage supply, demand, and price formulation can be discussed as an integrated whole. Before considering these economic aspects, however, we should look briefly at certain peculiarities of the timber producing process.

## PECULIARITIES OF TIMBER PRODUCTION

If timber production were identical in every way to other businesses there would be little reason for special courses in forest management or forest economics. Perhaps the differences that do exist tend to be exaggerated, but there are some which make the process of timber production quite distinct in an economic sense from that of other production endeavors. Seven "peculiarities" are worth singling out for attention.

### Immobility of Standing Timber

Once a forest is established it is fixed in space; it becomes essentially immobile. Purely from a production standpoint of combining inputs this fixity of location differs little from that of most complex commodities—automobile factories and forests are both difficult to move! But in timber production the product itself is immobile. Stumpage must be sold where it stands, a fact of great significance in pricing the final product of the forester's efforts. True, the owner may enter the logging business, cut and haul the tree to a distant market for sale. But the return he receives is no longer attributable to his timber production effort; part of this latter return must be credited to his logging activity.

### Time Period Involved

Stumpage production requires far longer time periods than are usually encountered in productive endeavors. Under the very best conditions pulpwood plantations may be ready for cutting in ten years or less, and fuelwood plantations in some parts of the world may be producing in even shorter periods. In this country a 25-year time span between plantation and harvesting is more usual, and for sawlog or veneer log production the time periods are normally much longer. One can admittedly point to fruit crops which require time periods approaching those of pulpwood production, and it is probably correct to say that the time period involved in timber production does not make forestry unique—that *all* production requires time and that forestry differs only in degree. The point is that the degree is so large the analytical techniques and decision making processes are changed. Hence, the long time period can be called a "peculiarity" of production.

The extent to which this peculiarity influences forest decisions is well demonstrated by the fact that it is the long production period that makes interest charges so important to forest management decisions.

### The Dual Nature of Standing Timber

When a forester goes into a stand to mark timber for harvesting he makes a decision which is almost unique in productive enterprises; he must decide not only if that particular product is ready for harvesting but also if he should destroy that portion of his timber producing factory. The standing tree is at once the final product and the factory which produces that product.

### High Ratio of Inventory to Annual Production

Few if any productive enterprises have the high ratio of inventory (growing stock) to annual production (growth) encountered in stumpage production. Measured in volume terms the ratios may vary from as little as 17 to 1 to as much as 50 to 1 or even more. This high ratio influences many expenditures (protection for example) since not only is the factory involved but also the entire stock (inventory) of saleable product existing in a productive forest.

### One-Way Flexibility of Production and Marketing

The time period involved coupled with the dual nature of standing timber produce a peculiarity of stumpage production that can be described by the phrase "one-way flexibility of production." Because of the time factor it is difficult to change output (growth) significantly without investing long time periods. But production can be reduced—in fact, eliminated entirely—in no more time than it takes to clear-cut the stand. Growth rates can be raised only over fairly long time spans, but they can be reduced to zero very quickly.

Fortunately, there are offsetting flexibilities in product and in time. Standing timber, once it has achieved a certain size, can be sold for a variety of products: pulpwood, veneer logs, sawlogs, Christmas trees, etc. This interjects a marketing flexibility that is quite rare. But even more important is time flexibility. Suppose, for example, one has grown sawtimber with the expectation of harvesting the crop at 50 years. If for any reason the harvesting is delayed there is usually little financial penalty involved. Because of its dual nature the stand will go on producing sawtimber at nearly the same rate as before. From a marketing standpoint, timber possesses great time flexibility, even though a few years delay in harvesting might have a major effect on the management organization of the whole property.

### Aggregative Nature of Forests

Anyone who has been involved in timber cruising will readily appreciate this peculiarity of the stumpage production process; forest trees grow in a community and cannot be treated entirely as individual entities. Each tree is related to those immediately surrounding it and, in fact, forms a part of a total forest environment which influences, and in turn is influenced by, a host of conditions. The fact that forests are aggregative in nature means that management must recognize the ecological forces that are involved, but it means far more. This peculiarity has resulted in the creation and adaptation of an entirely new (and expensive) set of management tools. The very act of determining the extent, volume, and quality of timber on an extensive tract is an expensive undertaking and one which requires quite distinct skills and methods. In almost all cases sampling techniques are involved; one simply cannot "inventory" an extensive forest in the usual sense of "going out and counting the trees at the end of each year."

Because of its aggregative nature modern computing techniques may well offer dramatic opportunities for advancement in forest management. Biolley (the French forester who instituted the "method du control" almost a hundred years ago) could not foresee that modern computers would make his intensive methods applicable over large areas. One wonders, however, if Biolley himself might not be quite disturbed were he to return and see some of the problems created by what has sometimes been called "IBM Forestry." Modern computers do indeed permit the rapid accumulation and sorting of literally fantastic quantities of inventory information, but computers do not satisfactorily frame questions nor do they sift out the information management needs unless compelled to do so by the admittedly slower but infinitely more capable human mind. Just where computer technology will take forest management is indeed a moot question at this time. It has already made possible more complete information than ever before on extensive forest properties. It has also resulted in the collection of large quantities of information we have not yet learned to use.

### Externalities

Simply because it occupies so much space, timber production necessarily involves many externalities. Forests produce products and services other than wood (recreation, scenery, etc.) which private owners cannot always convert to revenues, while timber harvesting often involves costs which are not borne by the operator (silting of streams, water pollution, etc.). This places forest management in a somewhat different category than management of many other enterprises. In some ways forests are a public utility, and in fact there have been many suggestions that, like public utilities, timber lands should be subject to public regulation as well as benefitting from public investment. The identification and measurement of some of these externalities will be treated at considerable length later on; for now we should be aware of their existence and make sure we leave room for their consideration in the forest management decision process.

## OWNERSHIP OF COMMERCIAL FOREST LAND

Perhaps the most important institutional reality of commercial forestry in the United States is the present pattern of forest land ownership. So far as timber production and sale are concerned, this nation is a "mixed economy." Part of our timber is produced and owned by private industry operating under competitive market forces, and part is publicly owned and managed. Of the 510 million acres of commercial forest land in this country, some 113 million acres are federally owned, with 97 million acres (86 per cent) administered as National Forests. More detailed information is given in Tables 9.1 and 9.2. Note the very heavy concentration of sawtimber ownership in the western region, and the overwhelming predominance of the western National Forests in sawtimber ownership. The areas of public and private ownership will change from year to year in both absolute and relative terms, but if past trends continue, Federal lands will dominate the supply of sawtimber in the coming decades even more than today.

## INTEREST

Interest charges are particularly important to forestry because of the time periods involved. The very fact that production periods are

**TABLE 9.1 OWNERSHIP OF COMMERCIAL FOREST LAND IN THE UNITED STATES, AS OF JANUARY 1, 1968. (IN THOUSANDS OF ACRES)***

| Ownership class | Total | North | South | West |
|---|---|---|---|---|
| Public | | | | |
| Federal | | | | |
| National Forests | 96,870 | 10,331 | 10,589 | 75,949 |
| BLM | 4,913 | 79 | 15 | 4,820 |
| Indian | 6,222 | 1,040 | 230 | 4,953 |
| Miscellaneous | 4,741 | 930 | 3,529 | 281 |
| Total Federal | 112,746 | 12,380 | 14,363 | 86,003 |
| State | 21,321 | 12,820 | 2,427 | 6,074 |
| County and Municipal | 7,900 | 6,775 | 677 | 447 |
| Total Public | 141,967 | 21,975 | 17,467 | 92,524 |
| Private | | | | |
| Forest Industry | 65,554 | 15,512 | 35,532 | 14,510 |
| Farmer | 140,913 | 57,014 | 68,627 | 15,272 |
| Miscellaneous Private | 161,790 | 71,492 | 77,173 | 13,126 |
| Total Private | 368,257 | 144,018 | 181,332 | 42,908 |
| All ownerships | 510,224 | 175,993 | 198,798 | 135,432 |

* From information furnished the Public Land Law Review Commission Timber Study Team by the United States Forest Service.

long has made some foresters in the not too distant past argue that "forestry should not be sacrified to the laws of compound interest." In our economic world, however, interest is a fact of business life, and forestry can no more escape the necessity of meeting interest payments than can the house buyer or the student purchasing his car "on time."

Interest is sometimes defined as the price of money; one who in-

**TABLE 9.2 VOLUME OF SAWTIMBER ON PRIVATE AND PUBLIC COMMERCIAL FOREST LAND, 1968. (BILLIONS OF BOARD FEET, INTERNATIONAL ¼ IN. RULE)***

| Region | Total | Public | Private |
|---|---|---|---|
| North | 239.5 | 38.5 | 201.1 |
| South | 474.0 | 57.6 | 416.4 |
| West | 1,811.5 | 1,326.0 | 485.5 |
| Total U.S. | 2,525.0 | 1,422.1 | 1,103.0 |

* From Timber Study, Public Land Law Review Commission, Table 4.8a, Appendix G. Study by G. Banzhaf & Co. (Public Land Law Review Commission, 1969)

vests "gives up" whatever the invested sum might purchase as long as the investment is maintained. The compensation for this abstinence is "interest" (although to regard interest as only a payment for abstinence would be far too simple). Because the investment period is a major factor in the price of the money, interest necessarily becomes a rate concept, a payment of so much per unit of time. The conventional time unit is a year, and when one quotes an interest rate of, say, six per cent it is normally understood that the "price" will be 6/100 of the capital amount for each year the loan is outstanding.[1]

Interest may also be defined as the rate of growth of capital. If I obtain needed funds by selling a share in my corporation for $1,000 and in exchange promise to pay the holder of this security $60 annually, the nominal rate of interest is 60/1000 or six per cent. But clearly, if you purchased this same security for $1,000, received the $60 dividend twelve months later, and then immediately sold the share for $1,040, you would have earned $60 + $40 or $100 on your investment—a yield of 10 per cent. This is the real rate of growth of your capital, for you would now have $1,100. (Tax-wise, these two sources of revenue should be separated. The dividend is "ordinary income," the other "capital gain," but both represent income. They are as available for spending as wages received, profits, or any other revenue.)

### Compounding and Discounting

The somewhat confusing arithmetic of interest will not be developed here since the subject is well treated in several texts (Davis, 1966; Forbes, 1955). Before looking at some of the factors influencing interest rates, however, it is necessary to make sure that the mechanics of compounding and discounting will not hinder understanding.

To students of biology, and especially to students of forestry, the basic principles underlying compounding and its counterpart, discounting, should be almost self-evident. In discussing the volume growth of a tree, or the growth of a deer herd, or the population of a fruit fly colony, percentages are commonly used and create no particular problems. Asked to predict the deer population, for ex-

[1] The price is not necessarily paid each year since the interest charge can, and often is, allowed to accumulate. When this is done the charges also accumulate on the unpaid amounts, giving rise to "compound interest."

ample, five years hence for an area having a present herd of one thousand deer, and told that the herd is increasing 10 per cent annually with no harvest, most students would probably set up a table much like that in Table 9.3. If a tree with a volume of 1,000

**TABLE 9.3 GROWTH OF 1,000 HEAD DEER HERD OVER FIVE-YEAR PERIOD AT NET INCREASE OF 10 PER CENT PER YEAR.**

| Year | Population | Calculation |
|---|---|---|
| present | 1,000 | |
| after one year | 1,100 | 1,000 + (1,000 × .10) |
| after two years | 1,210 | 1,100 + (1,100 × .10) |
| after three years | 1,331 | 1,210 + (1,210 × .10) |
| after four years | 1,464 | 1,331 + (1,331 × .10) |
| after five years | 1,610 | 1,464 + (1,464 × .10) |

board feet were assumed to grow 10 per cent annually this table would be equally valid for predicting its volume over a five-year time period. When applying the growth rate to a tree no forester would worry about removing the 10 per cent increase *each year:* the increase is simply added to the previously existing volume and provides an increased base for growth over the following growing period.

Exactly the same reasoning applies to the growth of any capital fund. If $1,000 were invested at 10 per cent and the proceeds allowed to accumulate as part of the investment, then the preceding table would serve to "predict" perfectly the capital value of the investment as it grew each year. This is really about all there is to compounding. Naturally, if the growth of the capital were paid out each year (simple interest) it could not be "compounded," nor would the deer herd increase as predicted if 100 deer were harvested each year. But as long as the increase is not removed it serves to increase the base for growth over the following period.

The results can be graphed over time (Fig. 9.1) to yield a quite typical form of growth curve. Two interest rates (10 per cent and 6 per cent) are shown in the figure, and two initial starting values ($1,000 and $500). Both of these factors influence the position and the shape of the resulting curve—the rate, and the amount at the beginning of the investment. (Incidentally, there is a simple and convenient rule-of-thumb for estimating growth of any sum under com-

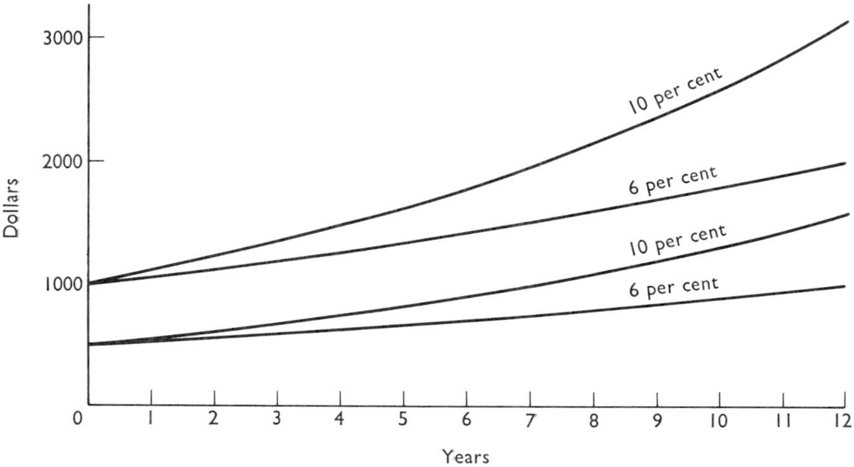


**Fig. 9.1** Growth of two amounts ($500 and $1,000) at two rates (6 per cent and 10 per cent) of compound interest.

pound interest. Divide the number 72 by the selected rate, and the result is the number of periods required for any sum to double at that particular rate. For example: at 6 per cent it takes 72/6, or twelve years, for an invested sum to double.)

*Discounting* is simply the reverse of compounding. As an example, a rate of 10 per cent must be earned by $1,000 invested today if this amount is to grow to $1,610 at the end of a five-year period. At 10 per cent, therefore, the present value of $1,610 to be received five years in the future is $1,000. Why? Because if $1,000 were to be invested today at 10 per cent it would "grow" to $1,610 at the end of five years. Because discounting is nothing more than "compounding in reverse," curves appropriate to one can be used equally well for the other. In Fig. 9.1, for example, the present value of a future sum is found by moving "backward" along the curve—downward to the left.[2]

[2] An ingenious device for illustrating the relationship between compounding and discounting was developed for a graduate seminar by Dr. Frank Shirley while studying at the University of Michigan. Take any transparent material (acetate drafting material works well) and construct a set of curves resembling those of Fig. 9.1, using the desired interest rate or group of rates and a beginning sum of 10 dollars. Make the horizontal axis to cover a period of at least 50 years, scaled conveniently, but unnumbered.

Now construct a graph (a set of axes) on regular (opaque) paper, having the

### "Pure" Rate of Interest

The "pure" rate of interest is, essentially, a risk-free rate. In the early, post-World War II literature, this was often taken as the rate on United States Government long term bonds and was usually thought to be around 2¼ or 2½ per cent. Since that time the long term rate has had its ups and downs—but mostly ups—and the recurring inflationary pressure on the economy has made the whole idea of a "pure" rate much less acceptable. Some economists have pointed out that if one reduces to an annual rate the inflation that has more or less characterized the United States economy since 1950, and adds this to the 2¼ per cent that was formerly taken as the "pure" rate, the result is not far from the average rate on long term government bonds. This implies that the general public simply adds to the "pure" rate a factor to allow for the risk of inflation, and that the factor has been derived from past experience. If this is really so—if a constant inflationary trend is to be anticipated—the idea of a "pure" rate becomes little more than a convenient synonym for "risk free," and it will be used in this sense here.

### Time Preference

*Individual* time preference is the rate at which an individual discounts the value of future income to make it comparable to present income.[3] Individual time preference will therefore be directly affected by an individual's present and expected income, his present and expected needs, his attitude toward saving versus consumption, his age and general outlook on life, and so on. The proverbial story

---

same vertical (dollar) and horizontal (years) scale as that used on the transparent material. The *future value* of any amount on the vertical axis can be found by placing the transparency over the graph and shifting the transparency horizontally (keeping the horizontal scale lines superimposed) until the starting value (in terms of 10's) is under the appropriate curve. This curve will now describe the growth of the amount over time, with the amount of time being indicated on the graph, and read through the transparency.

The same set of curves can also be used to determine the *present value* of any sum. Again, slide the transparency horizontally as above, this time until the desired point (the future value, plotted over the proper year) lies under the desired curve. Present value of this sum is now indicated by the intersection of this curve with the vertical axis of the graph.

[3] The rate of individual time preference may more precisely be defined as "a ratio between the marginal utility of . . . money in more distant future intervals and the present marginal utility of the same amount of money in intervals nearer the present and reduced by unity" (Ciriacy-Wantrup, 1952).

of the ant and the grasshopper may be explained in terms of the individual time preferences held by these two creatures of the fable. The grasshopper had a very high time preference rate, heavily discounting all future income in favor of present high living. On the other hand, the ant displayed all the virtue mankind expects of intelligent creatures, working all summer long, slaving and saving for the hard winter he could foresee in his future. The ant clearly had a relatively low time preference.[4]

But far more than personal attitude affects individual time preference. An individual's income is especially important; people with high incomes tend to discount future values at a low rate because their need for additional income today is slight (the marginal utility of additional income today is small). Future values tend to be more nearly equivalent to present values, hence time preference rates are low. Those with low incomes often require their income today. They discount future values at high rates, and have a high time preference rate.

*Social* time preference requires little additional explanation. It is the rate at which society (rather than the individual) discounts future values as compared to present values. Society (at least before the bomb and environmental crises) has a much longer life expectancy, and tends to take the "long view" toward utilization of resources. As a consequence, social time preference rates tend to be low relative to those of most individuals.

A high time preference generally rules out undertakings that are long in maturing, while low time preferences favor such investments. One should anticipate, therefore, that in developing regions of the world where present incomes are low, time preference rates are high and traditional forestry finds little acceptance. For similar reasons almost identical conclusions might be reached by low income farmers or small sawmill operators in our own nation. At the same time public agencies, representing society more as a whole, could find these same forest investments quite attractive.

[4] It may be worth observing here that traditional moralistic interpretations of this fable may well have to undergo some changes in our increasingly affluent society. The fable offers a real challenge to the modern economist. For the moment, consider that: (1) it is quite possible that the grasshopper had enough fun during the summer to more than compensate for his having to beg from the ant; and (2) the economics of private thrift is being increasingly questioned in times when social security and medicare are established facts and a guaranteed annual income is being seriously considered.

### Risk, Uncertainty, and Expectations

A distinction is sometimes drawn between *risk* and *uncertainty*. In such cases the term *risk* is reserved for events of a nature such that the probability of their occurrence can be calculated, while *uncertainty* refers to events whose likelihood of occurrence cannot be quantified in probabilistic (stochastic) terms. In this chapter no such distinction will be made; *risk* will be used to cover both kinds of events.

A businessman holds an *expectation* with respect to some future event. His expectation may, for example, be that prices will fall over the next five-year period, then rise substantially. An expectation is "how one feels" about a future event, and depends on a large complex of factors—the present situation, past and present trends, the individual's interpretation of their significance, and even how he happens to view life on a particular day. Expectations can and do change very rapidly.

### Interest Theory

Since World War II, a major concern of national economic policy has been the maintenance of full employment. There has been, and remains, much debate as to the relative effectiveness of monetary and fiscal remedies. To a very large extent these questions rest upon the nature of the *demand for money*.[5] The demand and supply of money determine its price, and the effective price of money is the interest rate.

To present day economists, interest theory is much more than an intellectually stimulating exercise in logic, but it can scarcely be regarded as a direct topic of *forest* economics. Still, because interest is of such prime significance to the long term investment decisions that characterize forestry, foresters need a general understanding of how various factors affect interest rates in order to appreciate the role interest plays in allocating private and public investment funds, as well as to see why interest rates change with time.

**Factors in money demand.** The demand for money is usually attributed to three major needs: *transactions, precautionary purchases,*

[5] Money is defined here in the usual manner—the sum of all currency and bank deposits *other than savings deposits* held by the non-bank public. It thus corresponds to the "stock of non-interest bearing liquid assets held by the non-bank public" (Dernberg, 1968; Kreps and Pugh, 1967).

and *speculation.* Transaction demand arises because people need money to make day-to-day purchases. Precautionary demand stems from the need to maintain some balance for unforeseen needs. Speculative demand includes the entire range of other needs, most of which are tied to investment of some kind or other.

It is unnecessary here to go into how the present *supply* of money came into being. Control over the supply of money is exercised by Federal authorities through (1) changes in the re-discount rate (the rate at which the Federal Reserve System lends to member banks); (2) changes in the legal minimum reserves required (the percentage of their deposits a bank must keep on hand at all times); and (3) the purchase and sale of government securities in the open market. These will be taken as given for present discussion purposes.

Both transaction demand and precautionary demand were for some time believed to be primarily a function of income level and were regarded as inelastic with respect to interest rates (price). Recent developments in monetary theory, however, suggest that both of these demands might be quite elastic with respect to interest. As the price of money rises (as interest rates rise), the cost of holding money for transactions or for precautionary purchases also rises.

To avoid the connotation many people attribute to "speculation" the term "investment demand" will be used here, but the reader is cautioned to bear in mind that investment can be in uranium stocks or oil wells (which most would call "speculative"), in so-called "blue chips" (bank stocks, AT&T), or in cash itself (the most easily convertible, or "liquid," of all commodities).

**Securities and interest rates.** In any economy, industrialized or developing, one can visualize a demand and a supply of money for investment purposes, as in Fig. 9.2(a). The demand and supply curves, $D_c$ and $S_c$, intersect at some point to yield a specific rate of interest $r$ (the price of investment funds) and a specific amount of investment capital, $q$. In a simple economy, this relationship alone may provide an explanation of interest rate behavior if both transaction and precautionary demands are inelastic and if money supply is held constant.

However, in the advanced economies which characterize most Western nations, the demand for investment money has led to the development of markets organized specifically for selling securities. In such markets, the demand for investment money, $D_c$ in Fig.

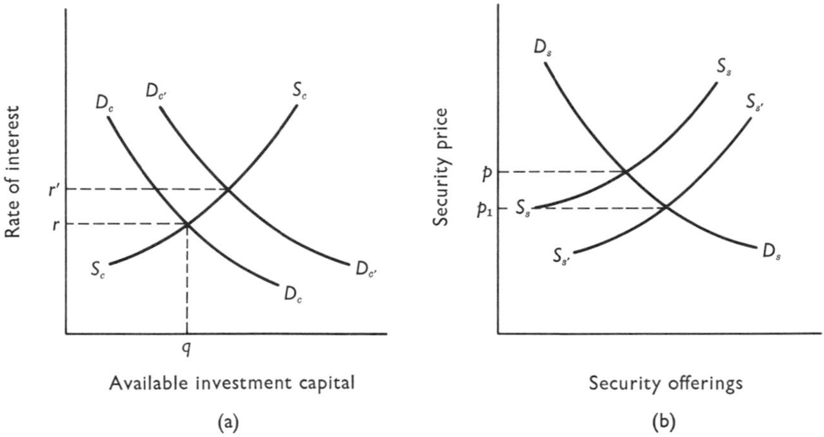


**Fig. 9.2** The relationship between interest rates and security prices.

9.2(a), is directly related to the supply of securities, $S_s$ in Fig. 9.2(b). As the demand for investment money increases to $D'_cD'_c$, the interest rate rises from $r$ to $r'$ while the supply of securities (which are offered to raise the desired investment money) also shifts out—to $S'_sS'_s$. If the supply of investment money remains constant at $S_cS_c$, then the demand for securities will also remain constant at $D_sD_s$ and the price of securities will fall from $p$ to $p_1$. The relationship between security prices and interest rates should now be reasonably clear. If security prices are high relative to their expected return, the interest rate will be low, and vice versa. But keep in mind that the expected rate of return will include not only the "dividend" but also any anticipated change in market price.[6]

**Aggregating demands for money.** Transaction demand ($D_T$), precautionary demand ($D_P$), and speculative demand ($D_S$) can now be added together (Fig. 9.3) to form a single demand curve for money ($D_M$) which will intersect with the supply curve for money ($S_M$) at some level (where $r$ = the price of money, or the interest rate). A shift in any one of the three demand components will clearly affect interest rates and any factor that tends to shift one of these demand curves tends also to influence interest rate levels.

It should also be observed that by taking appropriate action Fed-

[6] These two considerations, dividends and change in market price, are key items in classifying securities into two groups—the "high yield" and the "growth" types.

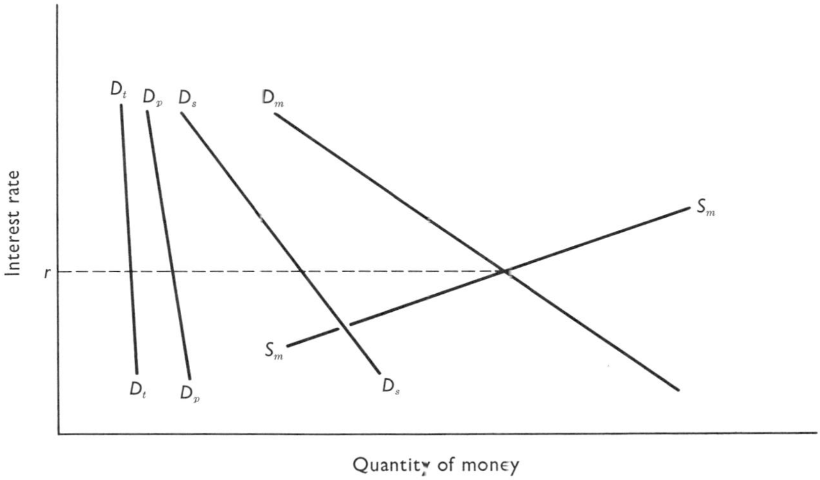


**Fig. 9.3** Aggregation of transactions demand, precautionary demand, and speculative demand to form the demand for money, together with the formation of the market rate of interest.

eral authorities may also influence interest rates through their effect upon the money supply. An increase in the rediscount rate, for example, will usually shift the money supply curve upward while raising interest rates, thereby reducing the amount of money available for investment purposes. This can have an almost immediate effect on forest management. New construction is a major form of investment and is also a primary user of lumber. An increase in the rediscount rate can consequently have an important effect on the sale of lumber through its effect upon the availability of mortgage money for new houses. The decision to attempt to control inflation through monetary measures in 1969 provided an excellent example of the possible Federal monetary policy on lumber prices and, as will be seen in Chapter 16, on stumpage prices as well.

**Effect of risk, expectations, and time preference.** How do risk, expectations, and individual or social time preference get into the interest picture? The pure rate of interest provides a starting point, representing a minimum price of money below which people will be as willing to hold cash as they are to invest. To this we must add the allowance for *risk*. A high risk venture must pay a higher rate of return if it is to attract investment capital. Relative to their possible rates of return, uranium securities tend to be lower priced than those of the blue chip corporations. Timber production (and forestry itself) is also subject to risks. It is an unfortunate fact that forests can be damaged and even destroyed by such things as windstorms, fire, ice, and insects. Investors view these as increasing the risk of a forestry investment, and the interest charged for forest investment capital (or the return required from forestry) is increased correspondingly. An increase in risk means an increase in effective interest rates, in forestry as well as in other businesses.

Risk is also a major reason for discarding the myth of "the market rate of interest." As almost everyone knows there are many market rates, and the actual rate applied in any particular situation depends partly on the risk involved. "The market rate" is, however, a convenient term, and as long as it is understood to mean "the market rate for an investment of this particular risk, duration, expected investment cost, etc.," no great harm is done.

Expectations act somewhat similarly to risk, but often in a more volatile fashion. If producers expect business to be good, prices to rise, profit margins to increase, etc., then the demand for investment funds (and offerings of securities) moves outward. But the people who loan money (who buy securities) are not the same ones who are producing (who borrow money or sell securities) and may hold different expectations. If the supply of investment funds moves outward less rapidly than the demand, interest rates will rise. Expectations are important to both the investor and the borrower, and play a particularly critical role at the onset of troubled times. A drop in investor's expectations may suddenly "dry up" sources of investment funds, while a decline in producer's expectations may cause them to cut back plans for new production, or even slow down existing production rates. Neither of these effects is necessarily serious, but such actions tend to be cumulative. Once shaken, confidence in business may take weeks, months, or even years to become reestablished. The "elasticity of expectations" (Hicks, 1946) changes quickly and can be high; a drop in prices may make producers expect a further drop, which leads to a still further drop, and so on. Expectations are believed to be an important element in the psychology of business booms and recessions, and the concept is receiving considerable attention in studies of dynamic price models.

Individual time preference, as an attitude of the individual, influences the interest rate an operator will assume when making calculations upon which management decisions are based. Table 9.4 illustrates why both the rate and the time period involved are important to forestry decisions. Clearly, high time preference rates are not conducive to long term investments of any kind. One would also

**TABLE 9.4 PRESENT VALUE OF AN INCOME OF $1,000 WHEN DISCOUNTED AT VARIOUS INTEREST RATES FOR SPECIFIED TIME PERIODS.**

| Rate | Period | | | |
|---|---|---|---|---|
| (in per cents) | 10 years | 25 years | 50 years | 100 years |
| 2 | 820.35 | 609.35 | 371.53 | 138.03 |
| 4 | 675.56 | 375.12 | 140.71 | 19.80 |
| 6 | 558.39 | 233.00 | 54.29 | 2.95 |
| 8 | 463.19 | 146.02 | 21.32 | 0.45 |
| 10 | 385.54 | 92.30 | 8.52 | 0.07 |

anticipate that individuals with high time preference would be reluctant to invest in timber production activities such as plantation establishment, pruning, or many types of thinning, while those with a low time preference might find the same practices quite attractive. In developing regions of the world where present incomes are low, time preference rates high, and interest rates very high, traditional forestry finds little acceptance among the small landholders. In contrast, owners of large estates may be enthusiastic forestry supporters. For similar reasons, almost identical conclusions might be reached by low-income farmers or small sawmill operators in our own nation, while public agencies, representing society as a whole, could find these same forest investments quite attractive.

The other side of the coin is the effect of interest on the costs of forest management activities. To illustrate, suppose you have inherited a 25-acre plot and decide to make this your first investment in forestry. You calculate it will cost 40 dollars an acre to prepare the area for planting and to purchase and plant the seedlings. This amounts to a $1,000 investment in your forestry enterprise. The growth of this initial cost at various levels of compound interest and for four different time periods is shown in Table 9.5.

Few investments can be expected to pay high rates of return for 50 or 75 years. If one is going to make investments in forestry that will be long in maturing, then funds must be available at low interest rates. Fortunately, interest rates do tend to decrease as the length

**TABLE 9.5 GROWTH OF $1,000 OVER FOUR TIME PERIODS AT FIVE INTEREST RATES.**

| Rate (in per cents) | Time period<br>10 years | 25 years | 50 years | 100 years |
|---|---|---|---|---|
| 2 | 1,218.99 | 1,640.61 | 2,691.58 | 7,244.65 |
| 4 | 1,480.24 | 2,665.84 | 7,106.68 | 50,504.95 |
| 6 | 1,790.85 | 4,291.87 | 18,420.15 | 339,302.08 |
| 8 | 2,158.92 | 6,848.48 | 46,901.61 | 2,199,761.26 |
| 10 | 2,593.74 | 10,834.71 | 117,390.85 | 13,780,612.34 |

of the loan period increases, since costs of loan management are decreased and reinvestment costs avoided.

### Interest and Resource Allocation

The most important role of interest in a market economy is that of allocating investment capital. In industrialized countries markets for securities are highly organized and closely integrated. Any change that might affect the supply of money, the return from an investment, risk, expectations, or time preference, is rapidly reflected in security prices and in interest rates. People having funds to invest balance the opportunities they see in forestry and in timber production against all other investment opportunities available to them; they weigh the rate of return from timber, recreation, and the other forestry production opportunities against their "alternative rate of return" (the opportunity cost of capital). Conceptually, each investor establishes his own schedule of investment alternatives, using the rates of return from each alternative to rank them in descending order. The result can be generalized as a curve formed by plotting the rate of return over the accumulating capital required for investing in successive opportunities (Fig. 9.4). If the individual has investment capital amounting to $q_1$ he will invest in those opportunities promising a return of $r_1$ or higher.

Applied to a corporation this scheduling of investments is called "capital budgeting." Some corporations restrict the investment possibilities to those within traditional corporate boundaries; a paper corporation, for example, might consider pulpwood production from several alternative areas, along with investment in a new plant, a new paper machine, etc. More and more, however, the traditional corporate lines are disappearing. Corporations holding extensive

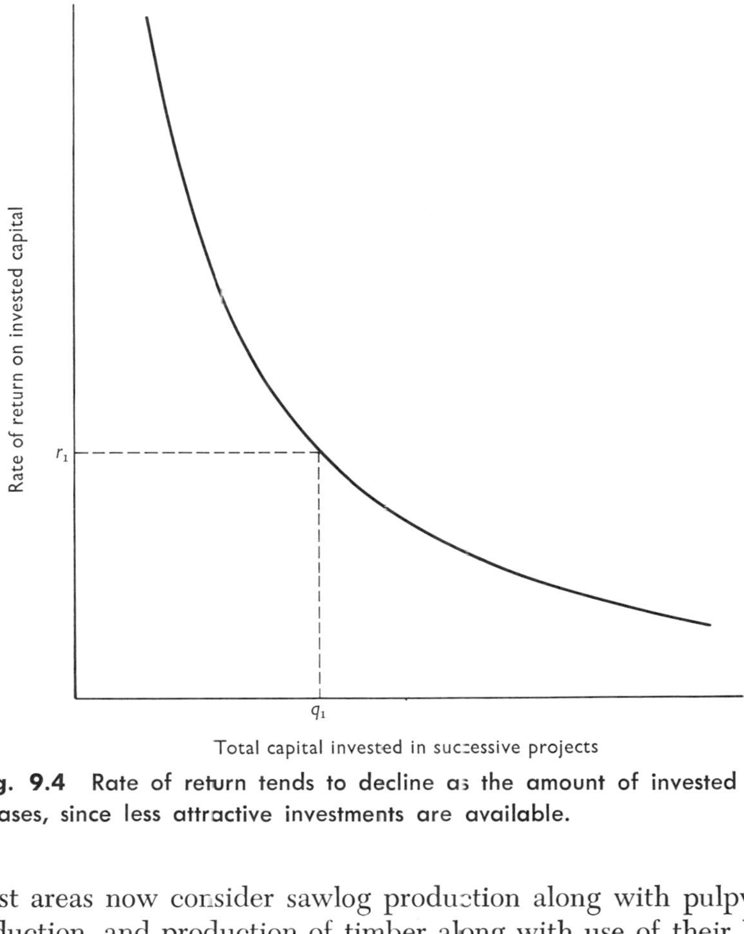


**Fig. 9.4** Rate of return tends to decline as the amount of invested funds increases, since less attractive investments are available.

forest areas now consider sawlog production along with pulpwood production, and production of timber along with use of their lands for residential purposes, vacation homes, campgrounds, or for hunting preserves. Even more broadly oriented are the so-called conglomerates, whose management may view chemicals, oil, transportation, or fertilizer along with recreation, timber production, and cattle when making up their list of alternative investments.

Public investments must also be made according to some schedule of priorities, for public funds are not limitless. Benefit/cost analysis is increasingly being used to determine rates of return from alternative public investments, and these rates provide one basis for allocating public funds. Benefit/cost analysis is a topic of Chapter 20; at present it is necessary to observe only that the interest rate appropriate to public investments has been, and still is, a major topic of debate.

## FOREST CREDIT

Prior to World War II the lack of forest credit availability was frequently cited as a major deterrent to forest management. Interest rates were high, loans were made for short periods only, and usually standing timber was not acceptable as collateral. Following World War II the situation changed. Stumpage prices rose dramatically, while the two major reports on the timber situation issued by the Department of Agriculture drew attention to the strong possibility of timber shortages (U.S.D.A., 1946, 1958). Progress in fire protection lessened the risk of loss, at least in the eyes of the general public, and loan agencies began to take a serious look at forests as a basis for long term loans. The result has been a very real change in the forest credit picture, and today few people regard credit as presenting any genuine problems except for the small owner.

The larger timber holding corporations, and particularly those integrated with processing plants, usually obtain funds through sale of stock or through retained earnings in much the same manner as do most large corporations in this country. Insurance corporations have recently become important lending agencies, and some of the more aggressive have actively sought to expand loans to forest-based corporations. Usually insurance company loans are restricted to those seeking sizable amounts.

National banks have only recently been able to loan on forest land as collateral. Until 1953 forest land was considered unimproved property, and as such was not acceptable collateral to national banks. In 1953 an amendment to Section 24 of the Federal Reserve Act was passed by the Eighty-third Congress (Public Law 285) under the sponsorship of the Pacific Northwest banking and timber industry groups. As a result of this ruling, national banks can now make mortgage loans secured by first liens[7] on forest tracts which are "properly managed in all respects." A national bank can loan up to 40 per cent of the appraised value of the merchantable timber offered as security. Ordinarily the loan is limited to two years, but

[7] If the borrower cannot meet the repayment schedule, such a *first lien* gives the lender the right to sell (or sometimes to take ownership of) the forest property and recover the amount of the loan.

loans can be made for periods up to ten years if secured by an amortized mortgage, scheduled for annual payments on the principle of at least 10 per cent. There are other restrictions which a bank must observe but these seldom prove significant to a borrower. In the years since passage of Public Law 285 the national banks have moved slowly but steadily into the previously untried and untested area of extensive forest credit. Most loans have been made in the southern states, with a few in the Pacific Northwest. There has been little activity in the Lake states and the Northeast, attributed by some to the relatively low growth rates which make sustained forest management less attractive financially (Resources for the Future, 1958).

Other than national banks and insurance companies the principal sources of credit are the Federal Land Banks and the Farmers Home Administration, the latter only extending credit for forest practices under the Soil and Water Conservation activities authorized by Public Law 597. As with the other institutions most of the activity has been in the southern states, with the Federal Land Banks much the more active of the two agencies.

Investigations made prior to World War II (U.S.D.A., 1940) suggested that forest credit banks, similar in nature to the land banks, might be established to solve the "forest credit problem." With the general easing of the picture since 1950 there has been little pressure for such a move.

# 10

# Taxes

It is doubtful if any single institutional factor is so all-pervading in its effect on modern economic society as our tax system. City, county, state, and Federal governmental units exist to provide services, and all must be financed. The impact of taxes on economic life is partially indicated by the fact that during the last few years between one-third and one-fourth of the gross national product has been taken by taxes of various kinds. Relative roles of Federal, state, and local governments have changed markedly along with the needs of the nation's people. These shifting patterns, the evident trend to greater reliance on Federal power, and both the absolute and relative level of the "tax burden" provide endless material for discussion by economists, political scientists, politicians, and almost every other group or discipline one can name.

## PRINCIPLES OF TAXATION

The problems to be considered here are less dramatic than those concerned with Federal–state–local issues, yet they are part and parcel of the same picture. Attention will be confined to four major tax types: the general property tax; the yield tax; the income tax, both corporate and personal; and estate and inheritance taxes. After defining and describing each tax, their influence on forestry, and especially on timber production, will be assessed. In keeping with one of the expressed purposes of this text, however, it is essential that consideration first be given to the *principles* of taxation, for only then can particular taxes be viewed within the context of the economy as a whole. Clearly, a tax system can be appraised only within the set of goals or objectives society has established for its government and for its economic system (Due, 1963; Musgrave, 1959). Societal goals are not derived from economic principles but are expressions of the value judgments of society at the period during which the goals are relevant. When and if these value judgments change significantly, then we can expect corresponding changes in the tax structure.

So far as the present United States economic system is concerned at least five goals may be identified:

1. Freedom of individual choice.
2. Attainment of satisfactorily high living standards.
3. Maintenance of a satisfactorily high rate of economic growth.
4. Economic stability.
5. A distribution of income which conforms to society's currently held standards of equity.

These goals may not always be compatible with one another, yet all must be considered when attempting to derive or develop principles appropriate to our tax structure.

The generally accepted *principles of taxation* which stem from these goals are three in number (Due, 1963):

1. A tax should interfere as little as possible with attainment of an optimum allocation and use of resources. Whenever possible it should assist in attaining this optimum allocation.
2. A tax should be *equitable;* the distribution of the tax load should help attain the desired pattern of income distribution.
3. The real costs of collecting the tax should be a minimum—and real costs include inconvenience to the taxpayer.

The relation of these principles to the goals of society will be illustrated when each of the specific tax types is considered. The general relationship is not difficult to see. A tax levied on a particular product, for example, may well alter freedom of individual choice, or interfere with the attainment of optimum living standards or rates of growth.

## Equity

Equity problems call for special comment since they may take on several dimensions. From our general commitment to, and belief in, the market system comes one quite generally accepted approach to equity in taxation: a tax system should be devised so that the burden is proportional to the benefits received. The second approach stems from the belief that on the one hand the tax system should provide that equal treatment be given to all individuals receiving equal incomes,[1] and on the other that persons of different circumstances should be given the "desired" *relative* treatment. Both beliefs are largely satisfied if the tax burden is made proportional to the ability to pay, which in practice is interpreted as being (roughly) proportional to income.

Neither the ability to pay nor the benefits received criteria can be applied at all widely, and they are frequently in direct conflict. Most governmental functions are in the public sector simply because the benefits accrue to society as a whole rather than to individuals. Hence the benefit criterion, the basis for commercial activities in the private sector, cannot be applied. Furthermore, strict application of the benefit approach would often result in those with low incomes bearing the same per capita burden as the rich—a violation of the "ability to pay" criterion. Yet some public activities can be financed on a benefit basis without adverse results. Provision and maintenance of public roads through taxes on motor fuel and by license fees is a classic example, and provision of forest recreation facilities by user fees may provide another.

The "ability to pay" criterion has no real economic basis. It stems directly from a value judgment of our current society and is in accord with the general ideas of desirable income distribution. Since ability to pay is interpreted in terms of measurable economic well-being, this approach to equity has the very real advantage of being "workable." It should be observed that taxes meeting this criterion (progressive individual income taxes, for example) may work against attaining the desired economic growth through their effect on incentive. At the same time such a tax may play an important role in attaining stability. The "trade-off" between economic stability and

[1] There exists the feeling that the "equal treatment for equal incomes" equity guide requires qualification by differentiating between incomes subject to different degrees of risk. Those with incomes subject to high risk should have a relatively lighter tax load.

economic growth creates a major policy issue whenever attempts are made to raise or lower the income tax rates.

Finally, the principle that costs of tax collection should be minimized points to the fact that resources used for collecting a tax add nothing useful to the national product of goods or services. Tax legislation should be clearly understandable to both taxpayer and collecting agency, and should be as simple as possible. Generally speaking, the marginal approach is accepted (in theory, though not in practice) as the proper guide for determining the level of expenditure for collection purposes. Expenditures should be increased until the additional cost is just matched by the added tax revenue thereby attained. Equity considerations, however, impose a real constraint on minimizing collection costs. Obviously, if individuals can avoid payment an inequity has been introduced.

## PROPERTY TAXATION

During the early part of the 1800's taxes on land, houses, stores, factories, carriages, inventories of goods, and similar items were grouped together and called the "general property tax." Today this tax supplies between 85 and 90 per cent of the tax revenue for local governments—those of the counties, cities, and villages of the nation. Broadly speaking the tax applies to two classes of property:

1. *Real* property: land, buildings, and permanent improvements. Usually real property has a fixed location.
2. *Personal* property: usually movable, personal property is itself of two kinds:
   a. *Tangible:* autos, farm machinery, personal effects such as clothing, jewelry, etc.
   b. *Intangible:* claims against tangible wealth, such as stocks, bonds, bank deposits, etc.

Real property is by far the more important of the two, accounting (in 1967) for about 85 per cent of the total assessed valuation of all taxable property in the United States. Almost all the rest is tangible property. Forest land is, quite obviously, classified as real property. Since forested land subject to the property tax makes up not less than one-fifth of the total U. S. land area and is valued (for tax purposes) at around ten billion dollars, the problems arising from forest land taxes are widespread, although only accounting for around two per cent of the total assessed property value in the United States.

### Administration of Property Tax

The operation of the property tax is best explained by a description of the five steps through which it is administered: assessment, review or appeal, determination of the tax rate, collection and distribution, and provision for delinquency.

**Assessment.** Assessment is the process of placing a dollar value on the real property concerned, or "assessing the property." This is normally the task of the local "tax assessor" who is often an elected official. Title records are usually used as a guide to ownership. Annual assessment is frequently specified by law but rarely occurs in practice. Instead, the figure of the previous year is simply carried forward to the next. Assessed value is usually related to market value, but the relationship is often distant.

**Review or appeal.** The taxpayer has the right to request a review of his assessment by a local board of review, often the county commissioners. If unsatisfied with these results he may contest the assessment in court—a step rarely taken, and when taken, rarely successful. Even the review by the local board is infrequently requested since knowledge of assessments on other properties is seldom at hand, assessments are typically at less than market value, and even if the individual sincerely feels his assessment is too high he may fear that his appeal might result in the raising of his neighbor's assessment rather than in a lowering of his own.

**Determination of tax rate.** In theory this is a simple operation. Each of the separate tax jurisdictions (county, township or town, school district, etc.) first totals the individual assessments in the jurisdiction. Then, after taking into account all other revenue sources, it determines the amount of revenue needed to perform the services, and divides this amount by the assessment total. If, for example, the school board of some school district whose boundaries embrace properties with an assessed valuation of 500 million dollars decides it will need 4 million dollars during the coming tax year, it will calculate the required tax rate as $4/500 = .008$, or 8 mills. This rate, applied to the assessed value of the property, will yield the needed money. In practice the process tends to be less simple. There may be legal limits on the rate, or more frequently political questions are raised by changes in the tax rate. But, if the planned expenditures exceed the legal or politically feasible limit, either the level of services must be reduced or new sources of revenue must be found. Neither of these problems has an easy solution.

**Collection and distribution.** The individual tax bill is calculated by summing the rates that apply to each taxable unit and multiplying the result by the assessed value of the property. With a school tax of 8 mills, a county tax of 4 mills, and a city tax of 12 mills, for example, a residential property assessed at $10,000 will pay $10,000 \times .024 = \$240$. Taxes are paid once or twice each year to a designated central office, usually the county or city treasurer, who distributes the collected funds to the proper taxing jurisdiction.

**Provision for delinquency.** Following the due date of the tax a period of grace is usually allowed during which the taxes, together with accumulated interest (and often a penalty) may be paid. After this period has passed the fact that taxes are levied *on the property* and not against an individual becomes important, for the property may be sold at auction and the realized funds used to pay the tax. Usually the buyer receives a temporary deed which may be redeemed by the delinquent owner at any time within some period of years by paying the overdue taxes and the penalties. Forest land in some states simply reverts to state ownership rather than being subject to sale.

## Deficiencies of the Property Tax

In general terms there are two reasons for criticizing the property tax as a means of raising public funds: (1) its administration is often poor, and (2) it does not meet the usual criteria for distributing the tax load.

The major administrative deficiency arises from the assessment step. In the past, and even today in some parts of the country, the assessor is an elected official, often without adequate training for what is inherently a difficult task. Poor assessment is most frequent in those places where the assessor is elected on a township basis, a unit scarcely large enough to permit hiring full time personnel.

There are more specific grounds for criticism. (1) The general property tax can and often does interfere with an optimum allocation and use of resources. Assessed unit values often vary by size of property and by use, are not necessarily related to market value, and the tax rates vary from county to county. Such variation influences the use of resources in ways not conducive to economic efficiency. (2) Property ownership is not necessarily related to an ability to pay nor is it an accurate indication of benefits received; by either approach to equity problems the property tax is found wanting. (3) Costs of collection would be large if the tax were to be administered at all equitably. All property would have to be identified and assessed each tax period, and at present the means of assessment, of rate determination, and even of disbursement are seldom well known to the average taxpayer.

Forest properties do not escape these drawbacks of property taxation. Indeed, many of the common faults are magnified in the case of forest land. Assessment of timber land is difficult at best. Very few assessors are able to appraise forest land values either accurately or consistently, and the frequent reappraisal that would be needed to achieve something approaching an accurate tax base could scarcely be expected.[2] Additionally, lands held by absentee owners or large corporations tend to be assessed at rates higher than similar properties that are locally owned.

**The time bias.** These factors all tend to make the property tax as usually administered somewhat inimical to forest land management. But the largest drawback of property taxation when applied to forest land stems from a quite different cause—the long time period involved in stumpage production. The property tax must be paid annually (in some cases, semi-annually). A forested acre, however, only rarely yields an income every year. Hence the property tax represents an annual cost that accumulates until the harvest cut, and for financial success the eventual harvest cut must yield at least an amount sufficiently large to meet these accumulated costs together with the interest charges thereon. An example, purely fictitious, will illustrate the effect of this "time bias."

Suppose you, as manager of a hardwood forest, had decided that unevenaged management with a ten-year cutting cycle would be the most appropriate silvical system to use. The site is reasonably good, and growth is expected to average about 300 board feet annually. Growing stock immediately following the cut will be 7,000

[2] This would require initial assessment on the basis of an accurate inventory, then annual adjustment (or reinventory) for all growth, all harvest, and all natural mortality.

board feet per acre. Average value of the growing stock is $30 per thousand board feet, while the harvested timber is sold for $50 per thousand. The forest is subject to a property tax of 40 mills. For illustrative purposes we will assume that the tax is levied accurately and on the full market value of the residual stand ($210), but is not reassessed annually. (In many states the assessor is legally directed to assess properties each year, but this is rarely done with forest properties). It is also assumed that growth is linear over the ten-year cutting cycle, and that all other costs amount to $1.60 annually.

The value of the forest growth, the accumulating values of both annual costs and tax payments (totalling $10 per year), and the accumulating carrying charges at a 6 per cent rate, are shown on a per acre basis in Figure 10.1. In the illustration the first payment

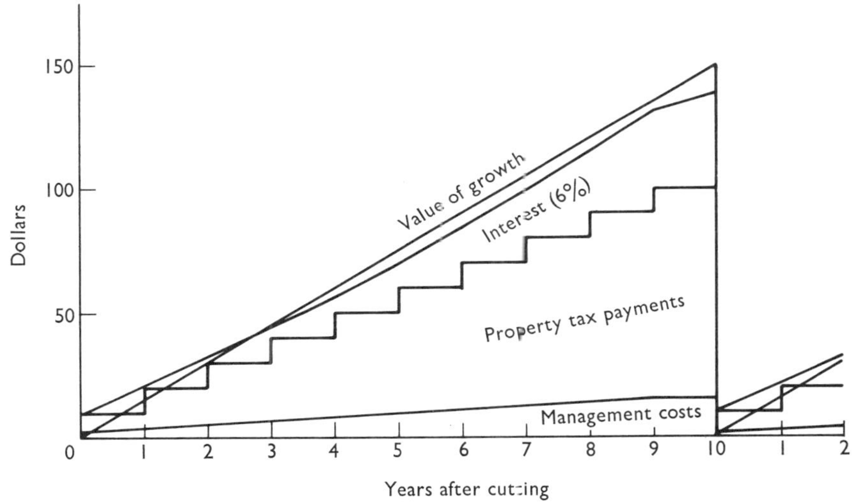


**Fig. 10.1** Per acre values of forest growing stock, management expenses, tax payments, and interest charges (at 6 per cent) as they accumulate over a ten-year cutting cycle. Stand under uneven-aged management.

is made immediately after the harvest cut, and the final payment charged against the cut of the ninth year, with interest accumulating to the end of the period. The harvest cut is made just prior to the tax payment in the tenth year.

Quite clearly, the tax alone would make timber growing a dubious financial venture. Even if there were no other expenses whatever, each acre, with an average investment of $250.50 in growing stock, yields only $32.65 after a ten-year wait. This means an average return on the investment of only slightly under one and one-quarter per cent, scarcely enough to encourage investment. Any managerial expenses cut this even further.

Fortunately for forestry the illustration is exaggerated. In most states today the property tax is made bearable by maintaining low assessments on forest properties, assessments far below the actual market value of the timber. But assessments do change, and in some areas reassessment problems are becoming acute. Normally, land is assessed at the value resulting from "its highest and best use," that use bringing the greatest value. For example, prime recreational land, or land suited to summer home construction usually has a high per acre value. Practically speaking, the example just given (Fig. 10.1) is well suited to show the effect of imposing high assessed values on forest land; $210 per acre is low for second home use in, say, the Lake states, but it would effectively rule out forestry. Zoning laws can help prevent fragmentation of management units, but over time those areas yielding higher returns from non-timber uses will drop out of timber production.

There is also another way out for the timber owner. Faced with a high tax situation, and if only timber values were involved, the owner could avoid the tax completely by clear-cutting everything of value—by liquidating the forest—then letting the land revert to the state for non-payment of taxes. And he would probably do just that under the circumstances illustrated here. This is what happened to much of the northern hardwood area during the 1910 to 1940 period. In the Lake states, forest land reverted to state ownership in such large amounts that it was called "the new public domain."

Basically, the problem exists because the land and timber are both real property. Land has value because of its income producing potential. Agricultural land rises in value as its productivity increases, and there is a close relationship between the annual income produced and the market value of agricultural land. Put directly, the value of land is the capitalized value of the net income stream it produces. Were forest land values calculated similarly[3] they

[3] Of course, they can readily be calculated in this same fashion. The calculation requires, however, that future yields and prices can be estimated—and future yields are directly dependent on levels of management and levels of growing stock.

would offer no particular problem, but because of the peculiarities of stumpage production they are not so calculated by assessors, and they *do* offer particular problems. The time period required for production is only the most obvious difficulty in applying the general property tax to forested areas. It is the dual nature of standing timber and its one-way flexibility of production and marketing that paves the way to liquidation, the high ratio of inventory to annual production that makes a property tax so burdensome, and it is the aggregative nature of forests that makes accurate assessment so difficult.

The problem created by tax delinquent forest lands became so acute in the 1930's that the Congress appointed a study commission to investigate "The Forest Tax Problem." The report of this group (Fairchild Commission, 1935), has become the classic study of its kind. The report proposed:

> That the property tax be modified in one of three ways in order to remove the excess burden inherent in that form of tax. The first is to reduce the current property tax on the deferred-yield forest property by an amount proportioned to the extent of income deferment in each individual case. The second is to defer the payment of property taxes until income is received. The third is to apply a flat-rate reduction to the property tax based on the degree of income deferment which may be considered typical or average for the State! (Hall, 1935).

Each of the three alternatives involves a conceptual separation of the land and the timber, and suggests modifying the property tax so that it would "impose a burden about equal to that of a tax on income or net yield at a rate equivalent to the property tax rate" (Fairchild, 1935). None of the alternatives has ever been tried in this country, most probably because the suggested methods are believed too complex to replace the relatively simple general property tax.

The many drawbacks associated with the property tax—to say nothing of the thorough report and the logical, though complex, alternatives proposed by the Fairchild Commission—might well lead one to ask why this tax has been permitted to survive so long. Why not do away with it entirely? The answer is purely pragmatic. The property tax has survived in the United States because it yields a large, stable, and predictable income to the taxing units; it is easily, although often badly, administered; and long use has made it familiar to the American public. Moreover, any drastic revision would result in large windfall gains to property owners, for to a very considerable extent this tax has been capitalized into property values.[4] The first factor (its large and predictable yield) is of great importance to local governmental units which must somehow be able to plan on future revenues. School houses, city halls, sewer and water systems, and fire stations must be built, and payment for them has to be scheduled over future years. The predictability and dependability of the property tax is a very real asset in planning, even though this very same characteristic may be a serious liability when it comes to promoting economic stability or aiding full employment.

## THE YIELD TAX

The Fairchild Report did not recommend the yield tax as a substitute for the general property tax, yet this tax type is the one most widely accepted as overcoming the deficiencies of the property tax. A yield tax is exactly what its name implies, a tax imposed on the *yield* of some particular operation. In forestry it is normally administered as an ad valorem tax, a tax according to value; hence it is a tax on the value of the forest yield. Because payment is usually made at the time of harvest, yield taxes effectively eliminate the "time bias" of the property tax when the latter is applied to timber-bearing properties.

Typically, yield taxes in the United States have required a conceptual separation of land and timber, even though land takes on a value only because it has a productive capability. The land is then subjected to the prevailing property tax while the timber values are, essentially, ignored until harvest. In a very real sense this represents a compromise between the general property tax and an income tax. As Marquis (1952) observes, this is a compromise "between two contradictory desires: the postponement of tax payments by the owner and the need for current revenue by government units." In addition, the yield tax has frequently been made a vehicle for regulating cutting practices. Clauses specifying some minimum

[4] That is, the market value of the land has been reduced by an amount which would yield, at market interest rates, an annual income just equal to the amount of the tax. (A tax of $4 per acre, at eight per cent, reduces the property value by $50).

level of forest practice before the owner can qualify his lands for classification are often made part of the law.

*Yield* taxes are sometimes distinguished from *severance* taxes on the basis of their intent. A yield tax is one intended to replace the general property tax, while a severance tax is a revenue measure levied in addition to the property tax. The difference between severance and yield taxes is really much deeper. Severance taxes are most often applied to mineral or other non-renewable resources that, once removed, are gone forever. They reflect the quite general notion that such resources are part of society's heritage. A severance tax is a kind of payment to society for the privilege of depleting the resource. Such a tax is almost always fixed on a unit basis—so much per ton of ore, barrel of oil, etc. A yield tax, however, is usually calculated as some percentage of the product value of a *renewable* resource. Moreover, timber yield taxes are often optional. If the land owner so elects he may file his property under the yield tax option and thus remove it from the property tax rolls, but he is under no particular compulsion to do so. In contrast, severance taxes are always compulsory.

At first glance the yield tax would seem to offer a logical means for eliminating some of the difficulties imposed on timber production by the property tax. And in fact it does, but new problems are created. Most important is the obvious difficulty imposed on the tax receiving units, for a yield tax by its nature simply cannot provide a predictable, dependable source of income. Even if the tax *rate* were completely fixed the income would vary by the amount of timber cut during the year and by its price. Conceivably even under a "sustained yield" program, the yield tax income to a local community might be very high during a year when the lands in the county or township were scheduled for harvest, but almost nonexistent during the following years when cuts were scheduled in forest areas outside the taxing unit. To counter this difficulty some legislation specifies that the yield tax be paid directly to the state, while the state makes an annual payment "in lieu of taxes" (usually fixed on a per acre basis) directly to the local taxing unit. The yield tax thus becomes a payment made at infrequent and irregular intervals into a revolving fund, out of which fund *annual* payments are made to local taxing units. Clearly, to inaugurate such a plan almost always requires that a substantial initial grant be made to establish the fund in the earlier years of the program.

Yield tax proponents often express surprise at the comparatively small percentage of eligible land owners who have taken advantage of available yield tax options. In its study, the Fairchild Commission found that of the total area of forest land eligible for classification under such options only 2.6 percent had been so classified (Fairchild, 1935). By 1967 this figure had increased to only 3.7 per cent, and in no state did the area reach 10 per cent. In a very real sense, however, the small number of acres grossly understates the effectiveness of these laws. Whenever such options exist the land owner can, if necessary, escape overly heavy property tax assessments, a possibility fully recognized by local assessors. Many owners who have not directly used the yield tax option still regard the legislation most favorably. It is a potent "club" to have in hand in case some overly ambitious assessor comes into office.

In the early 1950's a Forest Service survey (Marquis, 1952) disclosed several reasons for the apparent lack of interest on the part of forest owners. Among the more important reasons were: lack of knowledge on the part of owners; cumbersome methods of inspection, application, and approval (too much red tape); restrictions on forest practices, and the threat of future restrictions; lack of sympathy or downright opposition on the part of public officers; excessively high eligibility requirements; and an absence of any clear tax advantage.

## EFFECT OF PROPERTY AND YIELD TAXES

The very discouraging effect the property tax can have on an anticipated investment in forest land should be quite clear from the example already given. The discouragement will increase with either an increase in interest rate, in tax rate, or an increase in the interval between investment and harvest. It will become less as these same items are reduced. Some of the other effects, particularly the effect on management practices, are less easily seen.

To understand the differing effects of property taxes and yield taxes the analytic framework of production theory already developed may be called into play. Actually, this application of production theory to real forestry situations is also a good illustration of the absolute necessity for specifying the conditions if valid results are to be achieved, a point that will become quite apparent within the next few paragraphs.

## A Comparison

We can begin by taking the "ideal" forest, a fully established, operating, going forest concern. What effect would a property tax or a yield tax have on the *production* decisions of the forest manager? Fig. 10.2(a) gives a generalized total cost curve for such an

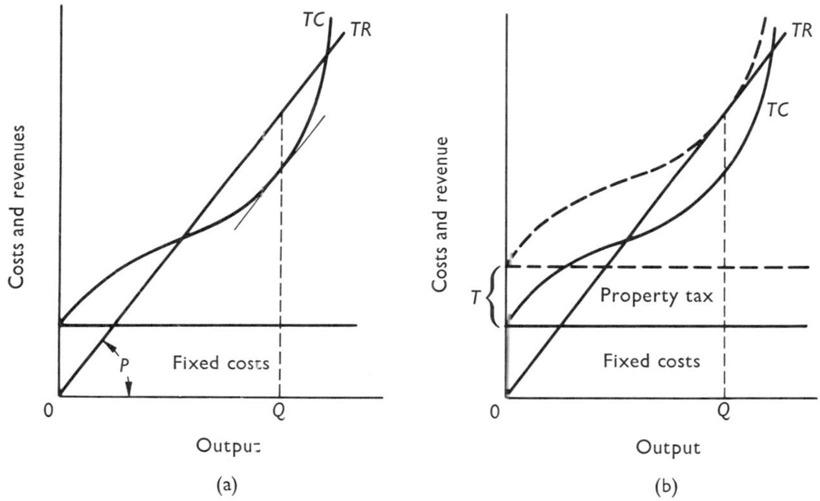


**Fig. 10.2** Effect of property tax on optimum output levels for a fully established "ideal" forest property.

operating forest, with both cost and output increasing as the intensity of management (and output) is increased. Note that the figure represents the entire operating forest area involved; it is *not* a "per acre" figure. At the given unit price, $P$, and under the cost/output relationships shown, the optimum annual output would be $Q$.

Now introduce a tax of $T$ dollars that is levied on the forest *property*, Fig. 10.2(b). Since the tax is not directly dependent on the productivity of the property or on its present or anticipated output it will have no effect on the production decisions of the forest manager.[5] Net revenue is decreased—in fact, it disappears—but all costs

[5] If the property were reassessed every year and the property tax levied on this changed assessed value reflected any change in level of growing stock, then the property tax would be dependent on the productivity of the property and would necessarily affect the production decisions of the manager.

are still covered and the optimum output level remains unchanged by the tax. Were the tax to be increased further it is clear that the long run outlook for the operation would not be very promising, but from the standpoint of normal production decisions regarding output levels, a property tax of *any* level on this kind of forest property would have no effect. The property tax, in this respect, is a fixed cost, unvarying with output. If the tax were to be increased *above* the level indicated ($T$), the long term continuation of the firm would be jeopardized.

Compare the situation of Fig. 10.2 with that of Fig. 10.3(a) and (b). The yield tax does have an impact on the production decisions,

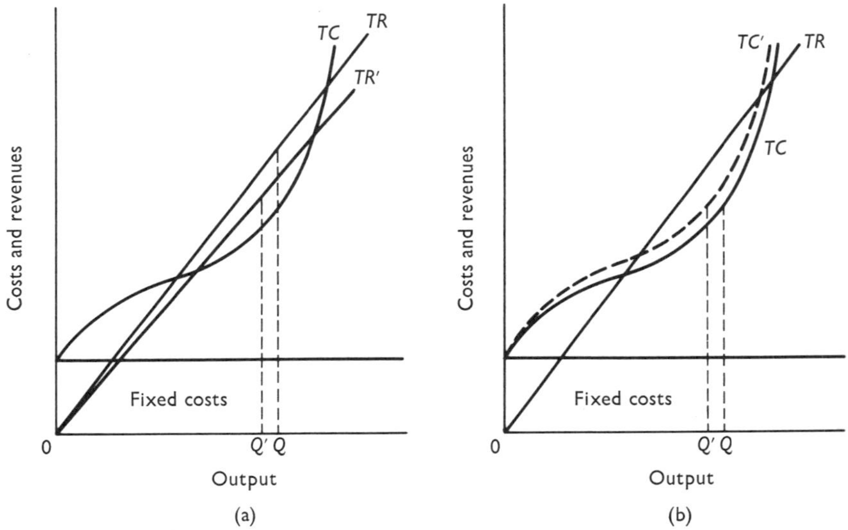


**Fig. 10.3** The effect of a yield tax can be treated (a) as a reduction in revenue, or (b) as a variable cost.

for a yield tax is a tax on value produced. One can view this tax as reducing the selling price of the produce, as in (a), or as a variable cost (b). If, for example, stumpage price were $10 per cord and the yield tax had been fixed at 10 per cent of the stumpage value, the effect of the tax would be that of reducing the price to $9 per cord. As a consequence the total revenue curve swings down (from $TR$ to $TR'$), its slope decreased (by 10 per cent), and the optimum level of output thereby reduced from $Q$ to $Q'$.

Exactly the same result is obtained if one views this tax as a cost.[6] Under the stated conditions the tax amounts to one dollar per cord. Hence the total amount to be paid varies directly with output and the tax is a *variable* cost. In (b) this is reflected by an "upward attenuation" of the total cost curve; the total cost curve is stretched upward and its slope quite obviously increased at every point. The net result is the same as that shown in (a); optimum annual output is reduced, along with, of course, total net revenue.

A going forest concern with a steady annual income is an unusual forest in this or in any country. Would the conclusion reached with respect to the property tax from Fig. 10.2 apply to a farm woodlot, or a 500-acre plot of timber owned by a retired postmistress (and managed by a consultant)? It would if the property yielded an annual income to pay the property tax, but this is not the usual situation. The tax must be paid every year, as in the example given for the northern hardwood stand, and interest charges accumulate constantly. The longer the period between harvest cuts the more burdensome the property tax becomes, and the more attractive the yield tax appears. The effect of either of these taxes is dependent not only on the rate but on the cutting cycle, on the rotation, and even on the kind of silvicultural system employed. One can even argue that the property tax favors all-aged management as compared with evenaged management and, since most of our commercial coniferous species are probably more effectively handled on an evenaged system the property tax, through this mechanism, has an unfavorable impact on timber supplies (Yoho, 1965). By far the more important influence on supply, however, is the direct effect of the "time bias."

**Conclusions.** From these diagrams several conclusions can be drawn.

1. The effect of either a property or a yield tax will vary with the tax *rate*. When property taxes, for example, are low almost no one bothers about their effect on timber production. When rates rise forest owners begin to think anew about "the forest tax problem."

2. The burden of a property tax increases as the income from the property is moved into the future, or as the periodic incomes from

[6] It is probably best to look on a yield tax as a reduction in revenue even though the same *result* follows from viewing it as a cost. Yield taxes are usually applied as a rate charge on the sales price, which is more in line with the revenue approach above.

the property become less frequent. With a fully regulated forest having a steady annual income a property tax will affect income levels but *not* input or output decisions. Under such conditions a yield tax offers no particular benefit (assuming that effective rates are comparable) and from a societal standpoint may in fact lead to a lower intensity of management.

3. With virgin forests, as presently exist in many of the developing countries, a property tax may be a very useful device to force utilization of presently standing timber, for a property tax cannot be avoided by non-utilization.[7] Under such situations (in developing countries where large areas are held under private ownership) the property tax can aid in achieving the goal of economic growth. And in these situations it may also aid in attaining a better distribution of income. A severance tax or a yield tax, however, can be avoided by postponing the cut. The latter kind of tax does nothing to encourage utilization and may in fact inhibit use.

4. Since a yield tax has the same effect as a lowering of the stumpage price it also has the effect of making marginal stands inoperable.

## THE INCOME TAX

The year 1913 is a memorable date in United States tax history. In that year the Sixteenth Amendment to the Constitution was ratified, providing that "the Congress shall have power to lay and collect taxes on incomes, from whatever source derived, without apportionment among the several states, and without regard to any census or enumeration."[8] Since then, taxes on both corporate and personal income have steadily increased in relative importance until now they are the primary source of Federal tax revenue and are of increasing importance to state and even city governments as well. In 1970 the personal income tax provided 46.2 per cent of all Federal budget receipts, and corporation income taxes accounted for another 18.6 per cent (U. S. Department of Commerce, 1970). Thus, nearly two-thirds of Federal budget receipts came from the two

[7] But since in developing countries most legislators are property owners—often very large ones—the likelihood of securing passage of property taxation is not high.

[8] An income tax was levied during the Civil War and in 1872 was held to be an indirect tax and hence valid. But parts of the 1894 personal income tax were held to be direct, and therefore unconstitutional. The Sixteenth Amendment resulted, although the corporation income tax was enacted in 1909, four years ahead of the ratification of the Sixteenth Amendment.

income taxes. (Other receipts: social insurance trust funds, 22.5 per cent; excise taxes, 8 per cent; estate and gift taxes, 1.8 per cent; customs duties and miscellaneous receipts, 3.5 per cent). The role income taxation now plays in Federal and state financing, together with its absolute size (personal and corporation Federal income taxes totaled 129.2 billion dollars in 1970) makes the income tax of primary importance to corporations and to citizens of the United States. Since state income taxes vary greatly, attention here will be confined to the Federal income tax.

### Personal Income Tax

To discuss income taxation sensibly some agreement as to what constitutes income must be reached. The basic definition used here is the one now accepted by most fiscal theorists (Due, 1963, Musgrave, 1959). Income is equal to consumption during the given period, plus the increase in net worth. It is, as Musgrave says, "potential spending power for consumption." As a practical expedient, however, tax legislation in the United States interprets the concept of income as a realized flow of wealth during the specified period. While the difference may at first appear slight it is important. To illustrate: if stock I own increases in value from $1,000 to $2,000, my net worth has similarly increased, and quite clearly my "potential spending power for consumption" has risen by $1,000. But until I actually sell the stock—until the income is *realized*—it is not taxable. There is no recognized flow of wealth until an exchange transaction of some kind occurs.

This approach is a practical necessity simply because it would be virtually impossible to revalue all assets every tax period. It is not, however, a logical necessity, and the basic definition must be kept in mind when looking at income taxes from a policy and from an equity standpoint. And the practical approach does introduce inequities. For example, the rent I save by owning the house in which I live is tax free; the use value I receive is not counted as income. But a tax would have to be paid on any part of my salary used to *rent* a house. There are many other such inequalities—the value of services rendered by housewives, of home-produced food and clothing, of the values received from do-it-yourself projects, and others.

The fact that the tax is levied on income as received, along with the requirement that income determinations be made on a yearly basis, are two of the most important features of the United States income tax. The third is the progressive nature of the tax—the rate of tax increases as the income base increases. Many of the problems associated with this tax, in forestry as well as in other areas, stem from these features.

**Progressive rate structure.** The steeply progressive rate structure is usually given top billing as the most obvious, and to some, most painful, feature of personal income taxes. Rates, of course, change from year to year. For several years during the late 1950's and early 1960's the maximum marginal rate was set at 91 per cent. At this level many believed the consequent reduction in incentive was a real drag on economic growth. The incentive to work or to invest for additional income is certainly reduced if the worker or investor receives only 9 per cent of the marginal reward obtained by his efforts. In 1965 this maximum was reduced to 70 per cent—still a substantial level, but one having considerably less effect incentive-wise. It must be recognized that rates are *marginal,* applying only to that part of one's income *above* specified levels. Thus the tax is really much lower than the rate structure indicates. For example, in 1968 a married man with two children whose $10,000 salary was his entire income, who filed a joint return and took standard deductions, would have had a *taxable* income of $6,600. His total tax liability would have been $1,197.55—slightly under 12 per cent—but the *marginal* rate was 19 per cent. At a salary of $20,000 per year but with all other conditions unchanged, he would have paid 18.4 per cent of his total income in tax, but 28 per cent of the last dollar of his earnings. The maximum rate (70 per cent) applies only to amounts in excess of $200,000. (The Revenue Act of 1969 reduced the maximum marginal rate on *earned* income to 50 per cent, while leaving the rate structure otherwise essentially unchanged.)

**Agreement with taxation principles.** Personal income taxes are reasonably in accord with the three major principles stated earlier in the chapter. A tax on personal income does, of course, restrict individual freedom of choice, and quite probably has a substantial effect on the choice between work and leisure. Such a tax, however, is generally held to be relatively free from direct interference with allocation and use of resources; rewards for some kinds of labor are curtailed more than for others, and there is some influence on factor costs, but this interference is minor. The largest impact comes through its effect on income distribution. Because of the steeply progressive rate structure the income tax is generally held to have an unfavorable effect on the rate of capital formation. (Rich people pay heavier taxes, and they are the ones most likely to save.) This in turn adversely affects the rate of growth of the economy, and hence the employment level.

But there are offsetting favorable factors that also stem from the income redistribution effect. Poor people spend a higher percentage of their income on consumption than do richer people. Since more is available for spending on consumption (than would be if a flat rate tax were used to raise the same revenue) more goods must be manufactured, hence more jobs are created, etc. The evaluation of these conflicting effects is and must be a value judgment each citizen makes for himself.

In general, the personal income tax is regarded as the most equitable means of apportioning the costs of government among the population. The distribution of the tax load conforms with the income distribution pattern approved by society and helps attain this desired pattern. The real costs of collection seem within reason (less than one per cent) though one might hope for considerable reduction in the cost of compliance, which may be excessive. Because of the large amounts involved the personal income tax has become an important tool of fiscal policy in maintaining economic stability, especially since the Federal (and most state) tax on salaries and wages is collected at the source.

### Corporate Income Tax

A corporation is a legal entity, quite distinct from its owners. As such it receives income, but it does not necessarily pay this income out to the people who own the corporation. Corporate earnings not paid out as dividends cannot be reached by a personal income tax except to the extent they are reflected in taxes on capital gain. Without some sort of parallel tax treatment the corporation would become a major device allowing profits to be reinvested tax free. Hence, the corporation income tax. Yet plainly, the same rate structure could not apply to corporations as to individuals. Ability to pay has little meaning in a corporation context, while the use of progressive rates would be highly discriminatory against stockholders in large companies as contrasted to smaller companies.

**Rate structure.** As with the personal income tax, rates for the corporation income tax vary from year to year. For a considerable period during the 1950's and early 1960's the basic rate, applying to taxable profit (net revenue), was 30 per cent, with a surtax rate of an additional 22 per cent on all profit over $25,000. For firms making more than $25,000, then, the effective tax rate was 52 per cent. The rate progression was based on the idea that the tax might otherwise restrict growth of new businesses, since reinvestment of earnings is a major source of expansion funds. Since 1965 (and until at least 1970) the corresponding rates have been 26 per cent and 22 per cent, making the effective rate 48 per cent for firms in the over $25,000 profit group.

Both personal and corporate tax laws recognize that losses as well as gains can result from business activities. In this respect the corporation tax is somewhat more generous since it permits a carry-over of losses for a five-year period (and a carry-back for three years), a great advantage to any new enterprise or to one with a fluctuating income.

Three items require some further development before we can look into the effect of personal and corporation income taxes on forestry: capital gains, depreciation, and depletion.

**Capital gains.** Both corporate and personal income tax laws recognize two types of gain or loss: *ordinary* gains or losses, and *capital* gains or losses. Capital gains "consist of all increases in the sale value of property held, except merchandizing profits on goods regularly bought and sold as a business activity" (Due, 1963). For example, if one sells a forested area for a higher price than was originally paid, the difference is a *capital gain.* In the United States the major source of capital gains (in dollar terms) is common stock of corporations, with real estate a poor second. Population growth, urbanization, speculation, the reinvestment of corporate earnings, a rising general price level, or a decline in the market rate of interest—all these and many other factors produce capital gains.

There are some who argue strongly that capital gains should not be taxed. Canada views such gains more leniently than does the United States, and until 1962 Great Britain did not tax capital gains at all. Several other countries still regard such gains as non-taxable income. Part of the argument relates to the fact that an individual's income may be used to enhance future earnings rather than for consumption. Certainly, if one changes the definition of income from "potential spending power for consumption" to "potential earning power" a different type of tax might result, and especially if equity principles are revised or ignored. To illustrate: it was seen in Chapter 10 that a drop in interest rates would normally increase security prices. If interest rates were halved, for example, bond prices might approximately double; a "perpetual" bond yielding $8 annually that sold for $100 when interest rates were around 8 per cent would increase in value to around $200 when interest rates dropped to 4 per cent. From the standpoint of earning power there has been no gain in capital at all, for it now takes $200 to earn what $100 earned before the decline in interest rates. If one sold the bond for $200, however, the $100 gain would be just as available for spending as $100 earned as salary or wages. Naturally, the extent of any increase in price due to a lowering of interest rates depends on the time to bond maturity.

Similarly, if one buys a piece of forest property for $5,000 and sells it for $10,000, the $5,000 gain is as much income (potential spending power for consumption) to that person as any salary or consulting fee might be. To ignore this income would mean a failure to treat people receiving equal incomes in the same fashion. And since capital gains are much more important to high- than to low-income groups, failure to tax such income would violate the ability to pay criterion as well.

**Problems of capital gains.** The difficulty created by capital gains arises primarily from the previously mentioned characteristics of the income tax: steeply progressive rates, the tax levied on income as realized, and the fact that income is segregated into definite yearly periods. If capital gains could be taxed as they accrued, whether realized or not, most of the argument for special treatment would disappear. But this is an administrative impossibility, and the present treatment constitutes the compromise achieved. The alternative of permitting capital gains to go untaxed, however, would produce inequities that most people are apparently unwilling to accept (other than on gains accrued until death on assets bequeathed to heirs). Capital gains offer major problems to income taxation, and timber production is by no means the least troublesome area for such difficulties.

The situation is made more difficult by the fact that the *rate* applied to capital gains is much lower than that on ordinary income for either corporations or for individuals. In 1969 two alternatives were offered to the taxpayer: he could include one-half of all long period [9] capital gains as ordinary income with the remaining one-half being tax-free, or he might apply a 25 per cent rate to all such gains. The corporate rate on capital gains in 1969 was a flat 25 per cent, but increased to 30 per cent in 1970 (not including surtax). Clearly, for any individual, especially for those in upper income brackets, or for a corporation having more than a $25,000 profit, there is a real advantage to have gains classified under the capital gain alternative. For a corporation it meant in 1969, essentially, an increase of 50 per cent in the profit after taxes.

From the standpoint of the general economy the most serious criticism of present capital gains treatment stems from the so-called "locked in" problem. Investors are encouraged to hold on to securities or other properties on which gains have been made rather than selling them and reinvesting in some other venture. The problem becomes most severe in the case of older people, since capital gains on property held until death are not taxed at all. The true significance of this problem is difficult to assess: since the maximum rate of the tax is 25 per cent the rational investor would not be greatly influenced if the opportunity for greater gain was clear—but there is no denying the fact that the *margin* is altered.

**Depreciation and depletion.** Depreciation and depletion problems are also traceable to the time element, but arise from the nature of capital expenditures rather than gains or losses. When a firm buys land, or buildings, or equipment, or invests capital in road construction, bridges, a fire tower, etc., the firm is not allowed to "charge off" the entire sum as an expense in the year during which the purchase or expenditure was made as long as the asset has a life expectancy of over one year. Yet common sense demands that the expense be allowed some way, for clearly such costs are just as necessary to the conduct of a business as those incurred for gasoline, or coal, or wages. Such expenses are recovered by the deduction of an annual depreciation allowance, designed to permit the total recovery of the entire cost (less salvage value) over the life of the asset. Should it be evenly divided? Should the original cost or the replacement cost be the basis of the total amount recovered? How does one determine the life of a piece of equipment before it is

[9] Long term gains are those from assets held over 6 months. Short term gains are derived from the sale of assets held 6 months or less and are treated as ordinary income.

used; etc? To avoid confusion these questions have been somewhat rigidly answered by the Internal Revenue Code. Many businesses and individuals use the "straight line" method, allocating the purchase price over the expected life of the asset in equal annual amounts, although alternative methods can be used.[10]

A "depletion allowance" is very similar to that for depreciation, but is designed to permit recovery of the investment made in an exhaustible asset. The oil, gas, and mining industries of the United States are the recipients of very special favors in this respect—entirely because of the manner in which the allowance can be calculated. Firms in these particular industries are permitted to choose between a "cost" or a "percentage" depletion, and it is the latter method that usually gives the greatest advantage. Percentage depletion permits the deduction of a particular "specified percentage of *gross income* for tax purposes, year after year, without end" (Due, 1963). The extreme case is that for gas and oil, which in 1969 allowed a percentage deduction of 27½ per cent. Percentage depletion was criticized for years by almost all students of taxation, but the results were always the same; as Due remarked, "continued successes for the pressure groups benefitting, and failure for the Treasury and the taxpayers as a whole." In the Revenue Act of 1969, the maximum rates allowed were reduced to 22 per cent. To avoid any possibility of misunderstanding it should be emphasized that timber does not share in such favored treatment. The timber depletion allowance is much the same as the depreciation permitted other industries and makes possible the recovery of an initial investment.

## Forestry and Income Taxation

What has all this to do with the problems of timber production? Quite a lot. Because of the peculiarities of producing timber, forestry has been a particularly difficult production process to fit into our income tax structure. Timberland is often held for substantial periods of time before the timber itself is harvested—certainly it

[10] The two most used alternatives give an advantage under situations in which it is desirable to have a larger amount deducted in the first years. These are:

*Diminishing Balance Method:* A uniform rate, not to exceed twice the applicable straight line rate, is applied each year to the remaining cost or other value basis, without adjustment for salvage.

*Sum of years-digits method:* This is best explained by an example: Assume the asset has an estimated life of 5 years. The sum of years is then 5+4+3+2+1 = 15. For the first year, 5/15 of the cost less salvage is allowed; for the second year, 4/15, etc. (United States Treasury Department, 1969).

would not be unusual to have a period of over six months between the original purchase and the harvest cut. Should the resulting gain from such cutting be counted as a capital gain or as ordinary income? If I buy stumpage for some future cutting should I be allowed to recover my capital expenditure through some sort of depreciation? But what if the timber *grows* between the time I purchase the forest and the time I sell the timber? And what about planting costs? If I am operating on a clear cut and plant system of pulpwood production can I subtract the planting costs as a regular expense? These are a few of the questions that arise.

To provide complete discussion of income taxation as it applies to timber production today is far beyond the scope of this text, yet to ignore such an important determinant of profitability in an economics text would seem unrealistic. The discussion which follows is at best a compromise. The intention has been to concentrate on the principles that are involved. The actual working out of particular cases is left to others (see, for example, Williams, 1963).

**Capital gains on timber.** Prior to 1943, providing one had owned standing timber for six months or more, he could *sell the timber outright* and claim the difference between the original purchase cost and the final sale price as a capital gain. If the timber had been acquired before 1913 he could claim the difference between its estimated value in 1913 and the time of sale as a capital gain (which, incidentally, sometimes created problems of evaluation). If timber were bought, then sold, with six months or less intervening, any difference was treated as ordinary income. Clear enough! But what if a farmer cut his own timber and sold logs? Or what if a lumber company cut timber that it had held for 20 years and sawed it into lumber? Could the increase in value (or the loss, if any) then be treated as a capital gain? Before World War II the answer was a definite NO. To qualify as a capital gain the timber had to be sold outright and neither of the above cases constitutes a sale of *timber*. Then in 1944, Section 117(k) was added to the Internal Revenue Code, an addition now known, in the Internal Revenue Code of 1954, as Section 631(a) and (b). Section 631(a) states that the cutting of timber for sale or use by a taxpayer who owns or has a contract right to cut the timber *shall be considered as a sale or exchange* of the timber cut, providing he has owned the timber or held the cutting contract for over six months at the beginning of the tax year. Section 631(b) gives the same right to any owner who "retains an economic interest" in the timber, a step made necessary by the fact that stumpage payments are often made periodically as the cutting progresses and may take place a considerable time after the sale itself.

In reporting on this change the Senate Committee stated:

> Your committee is of the opinion that various timber owners are seriously handicapped under the Federal income and excess profit tax laws. The law discriminates against taxpayers who dispose of timber by cutting it as compared with those who sell timber outright. The income realized from the cutting of timber is now taxed as ordinary income at full income and excess profits tax rates and not at capital gain rates. In short, if the taxpayer cuts his own timber he loses the benefit of the capital gain rate which applies when he sells the same timber outright to another (Senate, 1943).

With this change it became possible for *any* timber owner to count as capital gains any change between the purchase price of his timber and its market value, provided he had held the stumpage (or retained an economic interest in the stumpage) for six months prior to the beginning of the tax year. Market value, for tax purposes, is calculated "as of the first day of the taxable year in which such timber is cut."

Two other difficulties of timber income taxation deserve mention: the necessity for special means of handling depletion allowances, and the difficult but necessary division between what is to be called a capital expenditure (which is recovered through depreciation) and operating expenses (which are deducted directly from gross income).

**Depletion.** When timberland is purchased, part of the purchase price goes for land and the rest for whatever timber exists at the time. For tax purposes the two must not only be kept separate, but the "timber account" will usually require three subaccounts: (1) merchantable timber, (2) young growth, and (3) tree planting or seeding. Of the original purchase cost, only that portion attributable to merchantable timber is recoverable through the depletion allowance. However, if any appreciable time elapses between purchase and harvest the original volume will have changed, and because of the dual nature of standing timber there is no practical means of separating the original from the newly added product.

To overcome this problem the Code specifies a procedure for calculating a "depletion unit," usually expressed in dollars per thousand board feet or cords. Included in this "depletion unit" are the various carrying charges and the capital expenditures—the latter encompassing tree planting expenses if any have been incurred. Successive cuttings require successive adjustments of the depletion unit. The procedure is somewhat involved, and there is always a variety of items that seem not to fit the standard case. Still, there are no really basic problems of principle involved in the calculation of the depletion allowance under current income tax legislation. The details are well covered in the excellent handbook prepared by Ellis T. Williams (1963).

**Capital expenditures.** A "capital expenditure" is an expenditure made for equipment, buildings, trucks, machinery, office furniture, copyrights, or patents or other property with a useful life of more than one year. Such expenditures can be recovered through depreciation over the useful life of the property. Operating expenses, on the other hand, can be deducted directly from gross income. Almost every business has expenditures of both kinds, and in nearly all operations there are "borderline" expenditures that are difficult to classify. Because of the unusually long time period involved in timber production, forestry enterprises tend to have special difficulties, most of which have been settled by administrative ruling.

**Effect on production decisions.** To see the effect of an income tax on production[11] decisions of a going forest enterprise the usual set of cost curves proves useful (Fig. 10.4).

Since the tax is on *net* income it will be levied on some vertical "slice" of the shaded area of the upper diagram—the area between the total cost and total revenue curve—since it is this that represents net revenue. To facilitate seeing the effect of the tax this net revenue is reproduced directly below the total cost–revenue diagram, with the output (horizontal) scale maintained in the same units, but with the dollar units converted to net revenue.

The income tax always represents a *percentage* reduction of the net revenue. With the corporate income tax, the tax *on production* amounts to 25 per cent annually, for all output will (under our present assumptions) qualify for capital gains treatment. The optimum output level will, as we well know now, be that output maxi-

[11] Here, as with the property tax, the "production" decisions meant are those involving *how much* to produce from a given forest property through changing the *inputs* or the intensity of management.

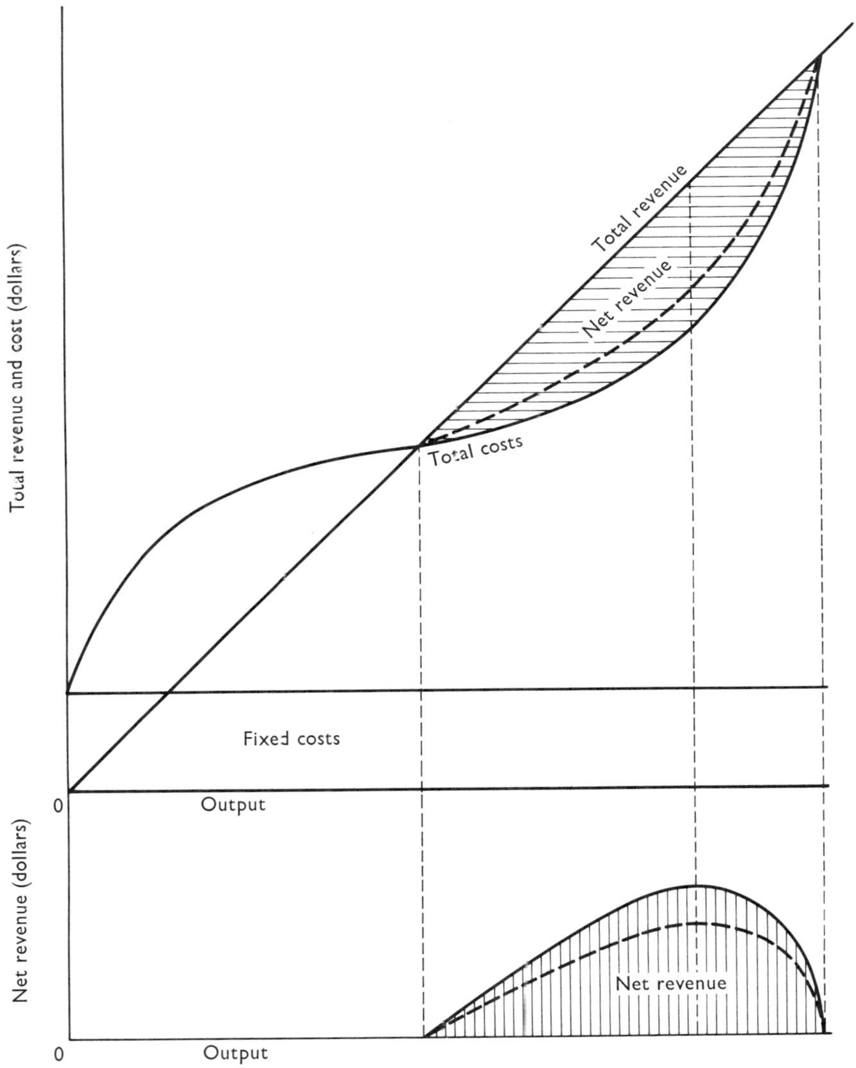


**Fig. 10.4** The effect of an income tax on usual production decisions.

mizing net revenue—where the slopes of the two curves (cost and revenue) are equal. And on the lower diagram this same point will be the one where the slope of the net revenue curve is zero, where it is perfectly horizontal. As long as the income tax is a percentage reduction of net revenue there will be no effect on production decisions. The point of optimum operation *before* application of the tax will be exactly the same as the optimum output *after* the tax. Hence, on decisions of this type, the income tax is *neutral.* The dotted line in the lower diagram is drawn to indicate a 25 per cent reduction, as is the similar line in the upper diagram.

The same reasoning will hold true for the personal income tax. It is not necessary that the individual's income be sufficiently large to make the 25 per cent tax rate applicable, for as long as the progression in tax rates is such that the *marginal* tax is less than 100 per cent (so that whatever the rate, after tax, income will increase if the total net revenue increases) the optimum output point will be unchanged by this tax.

**Effect on investment decisions.** Clearly, any tax affecting the absolute size of the net return will have a definite influence on investment decisions. Suppose, for example, we consider the problem facing the board of directors of a paper corporation when deciding whether or not to embark on a land acquisition and planting program. If the rate of return on the planned investment worked out at 12 per cent *before* tax, it would approach 6 per cent *after* tax under the ordinary corporation income levy but would be about 9 per cent if the increase could be treated as a capital gain. As a consequence the land acquisition and planting program would be "pushed up" along the curve of investment opportunities for the corporation (see Fig. 9.4 of the previous chapter); forestry is moved up on the capital budget of the firm. Offered two investment opportunities paying the same rate of return before taxes, if profit from only one of them can be treated as a capital gain, that is the one that should be selected (other things being equal).

When income taxes are paid on all net income under the same rate schedule there is usually little effect on resource allocation. But if income from one kind of investment is given preferred treatment over another, then investment is channeled toward the favored industry. Section 117(k) and its successor, Section 631(a) and (b), eliminated an undesirable situation, for prior to its passage there was a strong tax incentive to break up a forest property, selling the land and timber together, rather than to manage the stand, holding the land, and selling only the timber. Nor is there any doubt that Section 631(a) and (b) provide a decided encouragement to investment in forestry enterprises.[12] The rapid increase in investment in timber growing following World War II has been cited as proof of the efficacy of the new provisions (Briggs, 1962). How much of this increase is really due to the change in the Code, and how much to other factors is a moot question, but there is no doubt about the increase itself.

**Basic policy issues.** Given the need for amending the previously undesirable situation, and accepting as fact that forest investments are encouraged by the present tax treatment, we can move on to two somewhat more difficult—and disturbing—questions: "*Should* forestry be thus favored, and is this particular kind of treatment the best means of attaining the desired ends?"

A good many factors are, or have been, cited in support of the present treatment. Forestry yields many social benefits for which owners are not recompensed, and there are many reasons why a national government should encourage forest ownership. A tax advantage is one way of offering this encouragement. Moreover, stumpage production is a long, time consuming process that extends over many years. It would be "wrong" to tax all income from the growth of these many years in any one single year just because the harvest happened to come at that time. Besides, other industries are given even more generous treatment—for example, the petroleum and natural gas industries. Why should not forestry be the recipient of some favors? [13]

Offsetting such positive items is the knowledge that any interference or alteration in resource allocation by a tax is presumed undesirable unless the alteration can be proved to promote a closer attainment to the optimum. Is forestry more important to the nation than, say, automobile manufacture, or newspaper publishing, or educational institutions? The fact that some resources are given unwarranted benefits should be no real excuse for enlarging the area of mistake! An adequate system of averaging gains would take care of difficulties caused by harvesting the growth of several years. And finally, there is no direct connection between forest manage-

[12] In 1963 the United States Treasury's Department of Tax Analysis studied returns of 15 large wood-using corporations, all having assets of 25 million dollars or more. The effective tax rate paid by these corporations was 33 per cent as compared to the average effective rate of 48 per cent paid (both in 1959) by corporations as a whole.

[13] This last, almost childish, argument ("If Joe does it, why shouldn't I?") scarcely deserves mention in a serious text, but is frequently cited by serious people.

ment practice and benefits received from the tax. Under present legislation any owner, regardless of his management or lack of it, can claim the difference between purchase cost and market value at the beginning of the tax year as a capital gain. The purchaser of national forest stumpage can benefit just as can the owner who has carefully managed his stand for several years, or the owner who has *not* managed his stand at all.[14] In fact, it can be highly profitable to cut heavily or even to liquidate forest properties that have been built up over the years. Cash purchases of large tracts may enable a buyer to obtain the holding at well below the figure that can be established as the market value. Gain from product manufacture can then be shifted to the timber side and thus be given capital gain treatment.[15]

Most foresters, and almost all industrial operators, have expressed very strong opinions that Section 631(a) and (b) have proved to be in the past, and are still today, highly beneficial to forestry, and that therefore the general public is the gainer (United States Congress, 1963). The case for the first part of the statement is much stronger than for the latter part.

### Estate and Inheritance Taxes

Generally known as "death duties" or "death taxes," estate and inheritance taxes are imposed only once, at the death of the owner. They differ in that estate taxes are levied on the entire estate as a unit, while inheritance taxes are imposed on the share left to each heir. A few people have argued that since it represents an increase in wealth to the recipient an inheritance should be included in regular income rather than being given separate treatment. Most people believe that separate treatment is required since very frequently—as in the case of an inheritance of a widow or children—the receipt of the property does not represent any real increase in well-being. The beneficiaries have already had full use of the property before the death of the owner. An additional reason for separate treatment is the highly irregular nature of their occurrence. Estate and inheritance taxes most often represent a deliberate

[14] Presumably, any really careful manager would stand to benefit more than an owner who did nothing to his stand, since the values present in the forest should be higher in the former case.

[15] The gains through such tactics are for the most part limited to companies able to raise large cash sums—often through borrowing.

policy, aimed at avoiding excessive or undue concentration of wealth.

The United States Federal government imposes an estate tax on all estates in excess of $60,000. Rates are much lower than in some countries, but are quite steeply progressive: in 1969 they began at 3 per cent, and ranged upward to 77 per cent on the excess over 10 million dollars. The Federal estate tax is backed up by a gift tax which at least partially prevents avoidance of death duties by "giving away" the estate prior to death. State taxes vary considerably, most states using inheritance taxes although a few employ an estate tax. The 80 per cent tax credit allowed by the Federal government for state inheritance and estate taxes makes these latter reasonably uniform—the express purposes of the credit allowances.

**Possible effects on forestry.** The effect such taxes *can* have on forestry is most easily seen by an example. Suppose an uncle were to die, leaving you a forest property of 10,000 acres, which area was duly (and conservatively) appraised at $100 per acre. About the time you recovered from the shock of inheriting a million dollars you might be somewhat disconcerted to learn that (at 1969 rates) the Federal estate tax could reach as much as $500,000. You might be a little surprised by the lack of sympathy your friends would provide at your inheritance being cut to a half million, but your real problem will be "Where can I raise $500,000 to pay the tax?" Most probably you would find it necessary to sell off part of the estate, thus raising the required amount—and it may well prove difficult to find a buyer.

But suppose your uncle had been building up a real sustained yield forestry concern with a modern sawmill designed to cut and finish the annual output of the forest, an all-weather road system, fire suppression equipment, etc. If you still find it necessary to sell part of the forest, your property is no longer the well-balanced operation that your beloved uncle succeeded in constructing. Quite possibly you will decide it will be to your benefit to sell the whole property to a much larger corporation, exchanging it for shares of the company's stock at a favorable price. The tax could then be paid by selling whatever stock might be necessary, but keeping the forest property intact.

It may also be pointed out that this situation is made to order for anyone holding spare cash and looking for a way to make a fast dollar. Such a person might well offer you a *cash* price considerably below the market value of the property, which you might take in order to get your share of the estate. The new owner could then liquidate the entire operation, taking as capital gains the difference between his purchase price and the highest value he can establish for the stumpage. He could recover his cost through depreciation, sell off the mill, take his gain on the timber, and have the land left! It should also be noted that under the described circumstances the law would permit the estate tax to be paid in installments over a 10-year period—perhaps out of net yield!

Death taxes impose a decided disadvantage on individual forest management as compared to corporate forestry, for a corporation is essentially deathless. Such taxes also tend to stimulate mergers and thereby increase large forest ownerships. Since large ownerships tend on the average to be more efficient than small ones, the estate tax may act to produce a more efficient allocation of resources, although some might deem it less desirable from a social standpoint. The estate tax, then, can have a serious effect on forest properties by forcing dismemberment of an efficient operation. It could also act as a deterrent to investment if the potential investor decides to anticipate such a likelihood (Yoho, 1965). On the other hand, it might not have either such effect, and it could even be beneficial.

Death taxes in Great Britain offer an interesting contrast to such possibilities, through a sort of "double reverse" of the impact. English death taxes are large, and they have forced dismemberment of some estates. But standing timber is not subject to death taxes; that is, the value of standing timber is not included when evaluating an estate. Hence one finds wealthy estate owners devoting large areas to forestry as a hedge against death duties. The timber can be cut and sold after the death of the owner and used to pay whatever tax is levied on the estate!

Empirical evidence on the actual effect of the United States estate tax is neither abundant nor convincing. Few economists would agree with Greeley (1951), who (in his "Forests and Men") placed the estate tax at the top of his list of deterrents to United States private forestry. There is an undoubted effect on liquidity, probably the most serious factor so far as forestry is concerned. Because the tax is levied only once, and then at the very outset of the initiation of management by any single owner, the effect on management decisions would usually be slight. However, the decisions taken to *avoid* payment of estate and inheritance taxes can at times be very substantial, as witness the case in Great Britain.

From an equity standpoint the estate tax seems quite justified. The recipient of a bequest or inheritance has certainly been made better off and this increase can justifiably be taxed. The number of people receiving significant additions to their incomes from inheritances is so small that those who do should (from an equity standpoint) make some additional tax payment. Perhaps one of the best reasons for such taxes today is that they provide a means of reaching a large sum that otherwise escapes taxation altogether. Present legislation permits any capital gain *on property transferred at death* to escape taxation completely, and death duties place at least some tax on these large amounts.

## OTHER INSTITUTIONAL FACTORS

Interest, credit, taxes—all are important elements of the institutional framework within which any United States forestry enterprise must operate. There are, of course, a good many other institutional factors we have not mentioned. The legal system, for example, is one of the most significant components of the institutional setting: anti-trust laws have often played an important role in the affairs of American forest industries. Our market structure, our educational system, our entire attitude toward business and the code of business ethics, and literally hundreds of other factors—all wield an important influence on forestry in the United States. But most of these things produce an effect on forestry through indirect means. The factors treated here operate very directly, and some understanding of their effect is vital to all foresters who would term themselves "professional."

# 11

# Marginal Analysis and Input Decisions

Given the major peculiarities of timber growing and the general framework within which forestry must operate, we can get on with the job of applying economic analysis to timber production. In most respects forestry offers a relatively straightforward production problem. Were it not for the complication induced by time one might almost apply elementary production theory directly.

It is necessary, however, to expand the simple production framework a little further if the really interesting problems of timber production are to be treated. By restricting attention to problems of maximizing net revenue in the single-input–single-output situation the general analytical framework of production economics and price theory could be reviewed quickly, but this simple case is not suitable for very many forestry problems. Forestry frequently requires the balancing of several inputs, and may also involve the simultaneous production of several products.

## THE TWO-INPUT CASE

The problem of determining the optimum combination of productive factors will be approached by first examining the special instance of two inputs, for this has the great advantage of lending itself to graphic treatment. From the two-input case one can move easily to the general problem involving any number of productive services (the "n" input problem). The two-factor case can best be approached in two steps:

1. Given two productive services, what is the optimum combination of these services when producing any quantity of product?
2. What particular combination of inputs will assure maximization of net revenue?

### A Production Surface

To show graphically the effect on output of varying the combination of two inputs, three dimensions must be used (Fig. 11.1). A

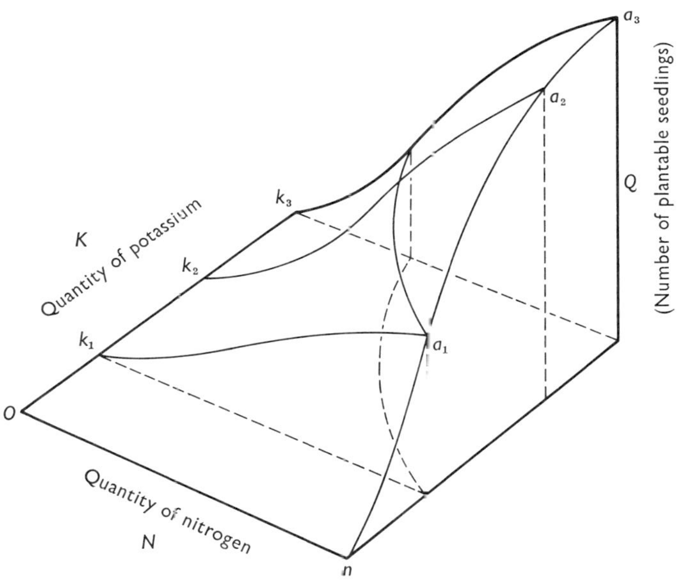


**Fig. 11.1** A production surface, generated by two input factors.

forest nursery will again serve as an illustration. The increasing inputs of nitrogenous and potassium fertilizer are indicated along the two sides of the base ($N$) and ($K$), while the vertical axis represents the output measured in terms of plantable pine seedlings. It is assumed that the two inputs are added to a fixed factor (a given amount of land) and that all other inputs such as water, seedlings, weeding treatments, etc., are held constant. If no potassium is used, additional nitrogen will have no effect whatever; no plantable seedlings will be produced, nor are any shown along the axis *On*. If the level of potassium is held constant at, say, $k_1$, the output response to changing levels of nitrogen will be shown by the curve $k_1a_1$. Similar total physical product curves are identified on the diagram for two other levels of potassium, $k_2$ and $k_3$. The same general pattern of output response characterizes changing levels of potassium if nitrogen inputs are held constant, although Fig. 11.1 shows only the curve $na_1a_2a_3$.

With a "production surface" such as the one shown here, any vertical cross section cut parallel to either the *N* or *K* axes will show the characteristic production function of a single input as described earlier, since such a cross section *does* show the response for a single input. A similar, but different, curve of output will be generated by cutting this surface vertically, but with a line passing through the origin, O. Any such "slice" through the origin shows the effect on output of changing *both* *N* and *K*, but always maintaining the two in some precise proportion: the ratio $N/K$ is a constant on any line drawn through the origin. This case of constant proportion will be met again.

### Isoquants

Foresters have had long familiarity with topographic mapping, so the technique of portraying the production surface on a two dimensional diagram will create no problem. Clearly, if lines of equal quantities of product are traced on the production surface they will form "contour" lines, just as in any two-dimensional representation of surface topography. Projected on the base of the diagram they can be labeled according to the quantity represented, forming a "map" of the production surface. These lines are called "isoquants," or lines connecting points indicating equal quantities of output, much as lines connecting points of equal temperature are called "isotherms," and lines connecting points indicating equal barometric pressure are called "isobars."

Transferred to a different drawing, this time a two-dimensional one, these isoquants show the various combinations of nitrogen and potassium that can be used to produce a given quantity of seedlings. As with contour lines, we can draw any number of them we wish, limited only by the detail needed on the "map." Almost any configuration is theoretically possible, the one restriction being that isoquants can never cross. In Fig. 11.2 the isoquants labeled 100 and 200 show outputs of 100 and 200 seedlings respectively and indicate all the different combinations of $N$ and $K$ that can be used

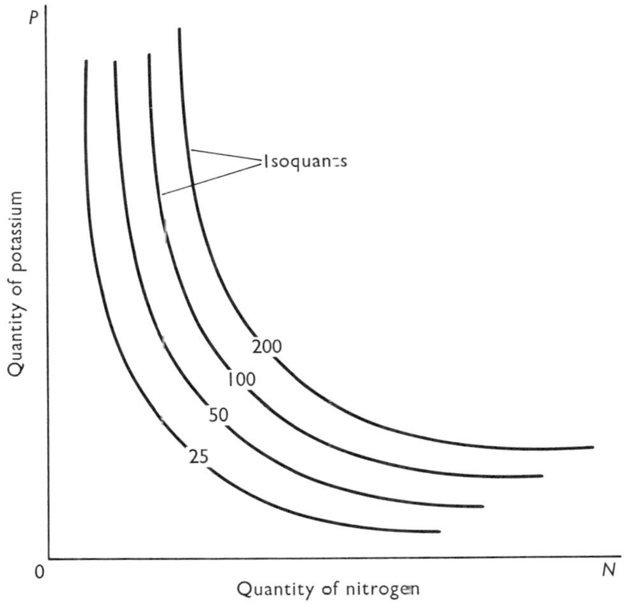


**Fig. 11.2** Two-dimensional representation of the production surface of Fig. 11.1. The four isoquants show outputs of 25, 50, 100, and 200 seedlings per unit area.

in combination with the particular set of fixed factors to produce these quantities. However, given the necessary information, one could readily draw other isoquants showing the combinations of nitrogen and potassium necessary to produce 150 seedlings, or 125, or (if we could draw them fine enough) lines showing the different $N$ and $K$ combinations for outputs of 100, 101, 102, 103, and on up to 199 and 200 seedlings.

## Isocosts

Now suppose one asks, "If I had a given sum of money for buying nitrogen and potassium, how much of each should I order?" At this point the problem takes on an economic slant, for one would want to purchase the combination of the two inputs that would produce the maximum number of seedlings for the money spent. As long as the purchased amounts of the productive services are not enough to influence their prices (pure competition in the purchase of the productive services is assumed), one can determine how much of either input might be bought for any given sum by dividing that sum by the price of the service. That is, given that the expenditure is to be $C$ dollars, and that $P_n$ and $P_k$ are, respectively, the prices of nitrogen and potassium fertilizers, then $C/P_n$ or $C/P_k$ represents the quantity of nitrogen or potassium fertilizer that can be bought for $C$ dollars (Fig. 11.3). Now, by connecting with a straight line

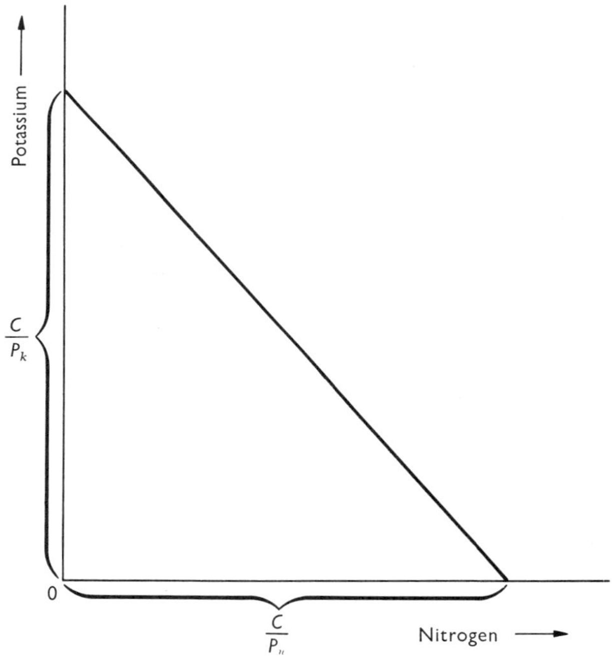


**Fig. 11.3** Graphic representation of the quantities of services that can be purchased for a given expenditure.

the two quantities so determined, all possible combinations of the two input factors that can be purchased for $C$ dollars will be identified. This line, a type of *isocost* curve (isocost = equal costs) is only one of a family of isocosts, since a different line will be generated for every different value selected for $C$. No matter how many of these are drawn, as long as factor prices remain unchanged all the lines will have exactly the same slope, and they will all be parallel. The slope of the line is independent from the expenditure level; the slope depends entirely on the relative prices of the two factors.[1]

## Least Cost Combinations

If the two diagrams (Figs. 11.2 and 11.3) are now combined into one—if we superimpose one on top of the other—then the first part of the problem posed at the outset of this chapter can be solved. The isocost curves of the figure (Fig. 11.4) show all combinations

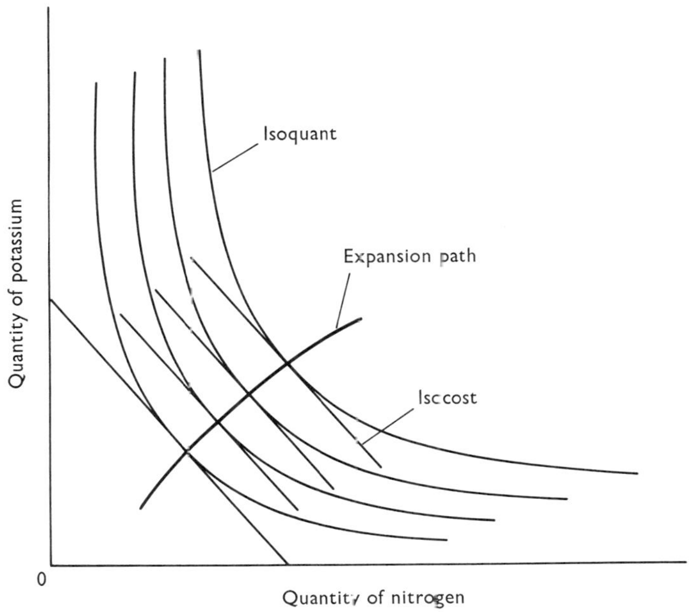


**Fig. 11.4** Solution to the first phase of the problem—optimum combinations of the two inputs that can be used under given conditions to produce any feasible level of output are indicated by the expansion path.

[1] The slope, or tangent, of the isocost line of Fig. 11.3 is:

$$\frac{\frac{C}{P_k}}{\frac{C}{P_n}} = \frac{\not{C}}{P_k} \cdot \frac{P_n}{\not{C}} = \frac{P_n}{P_k}$$

of $K$ and $N$ that can be purchased for a given level of expenditure, while the isoquants show how different combinations of $K$ and $N$ can be used to produce particular quantities of seedlings. The points of tangency of the two curves (of the isocosts and isoquants) identify the combinations of $K$ and $N$ that will permit any given production level at lowest possible cost. These points of tangency identify the *least cost combinations* of the two inputs. A curve drawn so as to connect all these points of tangency is called the *expansion path* or *scale line,* since it shows the least cost combinations of input factors for the whole range of operating scale. At any given production level the manager will always minimize production costs by using the combination of services indicated by the expansion path. The variable production costs, therefore, will always be those indicated by the isocost curves at their points of intersection with the expansion path. Output will similarly be indicated, except the quantities will be read from the isoquants. The relation of cost to output quantities can be read directly from the graph of Fig. 11.4 and, if desired, replotted in more conventional form as in Fig. 11.5. In each case the output indicated by the intersection of the expansion

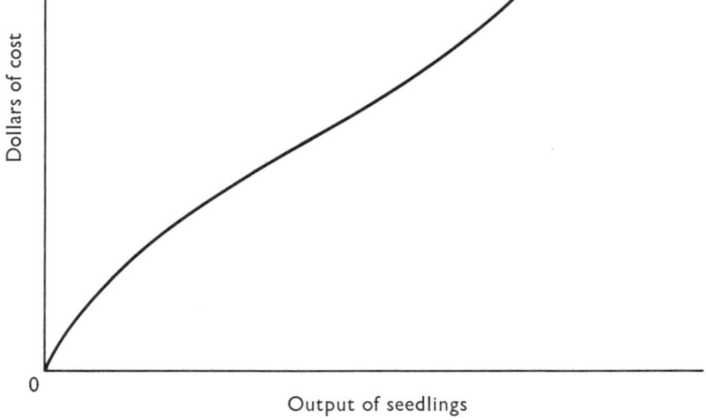


**Fig. 11.5** The total variable cost curve derived from Fig. 11.4.

path with an isoquant has been plotted at the cost level indicated by the corresponding (and also intersecting) isocost curve.

### Maximizing Net Revenue

By this procedure the maximization problem has been reduced to exactly the same form as the one already solved in the single input case reviewed earlier. And the same graphic procedure can be used. Fixed costs must be added, along with the total revenue curve, and the profit maximizing output will be found where marginal cost and marginal revenue are equal (Fig. 11.6), which is where the slopes of the total cost and total revenue curves are identical.

While it is now possible to answer the two questions initially posed, and to determine the particular combination of inputs that should be used to maximize net revenue, one must continually remember the assumptions built into the model. Note again the sig-

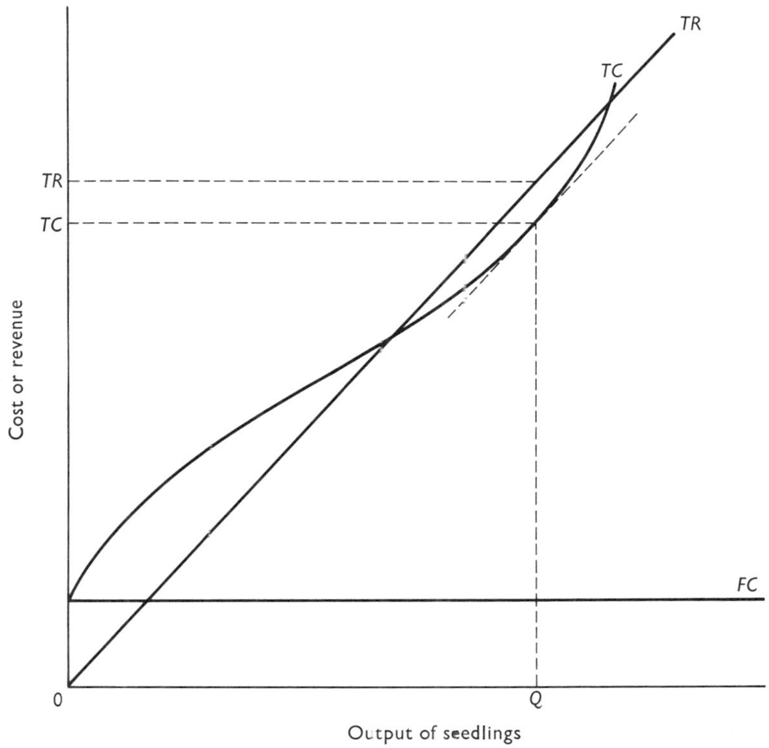


**Fig. 11.6** The solution of the maximization problem.

nificance of the production surface: any change in technology—and any error or uncertainty introduced by lack of knowledge of how seedlings might respond to the particular inputs being studied—will be reflected in a shift of the isoquants, which, in turn, will dictate a new combination of the productive factors. On the other hand, any change in relative factor prices would change the slope of the isocost curves and hence shift the location of the expansion path. This also would dictate a new combination of input factors.

The two-input model represents a substantial advance in the production framework; not because of the more complicated geometry, or even because the number of inputs has been doubled, but because by this still relatively simple approach one can intuitively see how any number of inputs can be handled. The relationship between output and the two inputs is reproduced in Fig. 11.7. Beginning at the extreme upper left side of the diagram with the isoquant labeled "Q," follow this isoquant along its entire path. In-

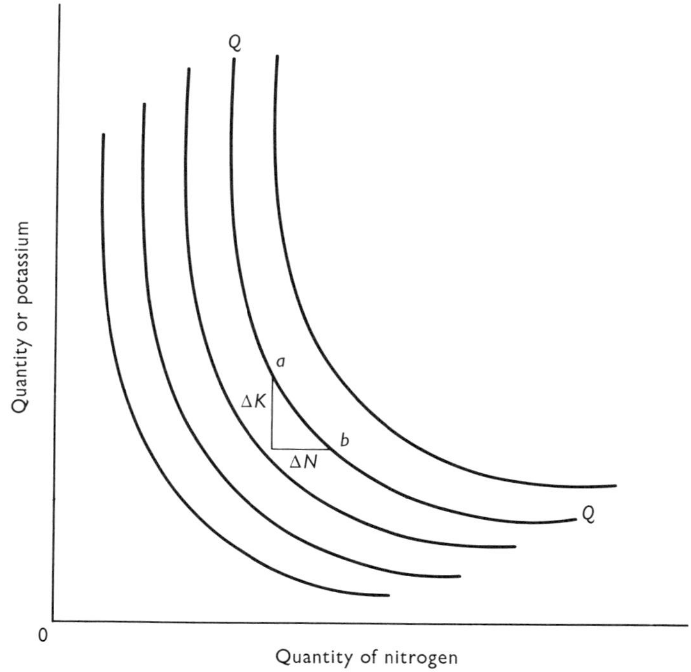


**Fig. 11.7** The slope of an isoquant (here $\Delta K/\Delta N$) is equal to the marginal rate of substitution.

itially one is using a large amount of $K$ and very little $N$, but as the isoquant is followed "down" the curve, more and more $N$ is substituted for the $K$.

## A GENERAL GUIDE

If we take any two points on this isoquant, say $a$ and $b$ in Fig. 11.7, then $\Delta K$ is the *change* in the amount of potassium that must be "balanced" by an increase in the amount of nitrogen ($\Delta N$) in order to hold output constant, and $\Delta K/\Delta N$ represents the ratio of these changes. At any point on the isoquant, then, $\Delta K/\Delta N$ is the *marginal rate of substitution* (while holding output constant).

Previously it was seen that the slope of the isocost curve was dependent on the relative prices of the two services, and was in fact $P_n/P_k$. At the point of tangency between the isocost and isoquant curves $\Delta K/\Delta N = P_n/P_k$, *or the marginal rate of substitution of the factors is equal to the inverse of the ratio of their prices.* Hence, if the price of nitrogen ($P_n$) rises relative to that of potassium ($P_k$), one should use relatively *less* nitrogen and *more* potassium. Expressed in very common-sense terms, if there is some measure of substitution between two factors of production being used to produce a product, when the price of one factor falls, increase the use of that factor. Substitute the cheaper material for the more expensive.

Intuitively one can follow this rule with any number of inputs. Substitute one input for any other input (or any other group of inputs) until their marginal rate of substitution is equal to the inverse of their price ratio. (Substitute cheap inputs for expensive ones as long as output can be kept constant!). The $n$ input case can be developed easily by using differential equations for isoquants (Carlson, 1937; Baumol, 1965).

### Degree of Factor Substitutability

A little reflection will reveal that the isoquant pattern depends directly upon the physical relationships involved. The degree to which one factor may be substituted for another varies tremendously. It varies, in fact, from perfect substitutability to no substitutability whatever and encompasses almost all degrees that lie between these two extremes. Perfect substitutability of productive factors yields an isoquant pattern similar to that of ($a$) in Fig.11.8,

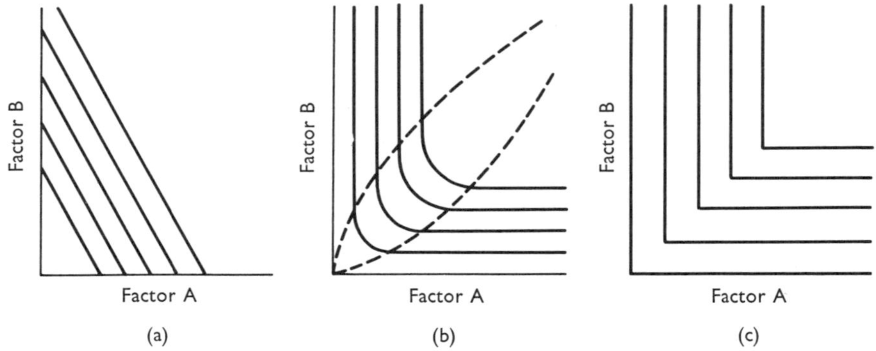


**Fig. 11.8** Isoquant patterns showing different degrees of substitutability; (a) depicts perfect substitutability; (b) shows intermediate substitutability (the two factors are substitutes over a limited range); (c) shows no substitutability.

while (*c*) gives the opposite case—that of no substitutability whatever.

If the situation is that of (*a*) there are three alternatives, with choice dependent on the relative prices of the two factors. One would use either all *K* and no *N*, or all *N* and no *K*, or (if the isocost curve had exactly the same slope as the isoquant) the choice of input combinations would be immaterial. In the situation shown by *C* (no substitutability), regardless of the relative prices the optimum combination is that indicated by the "corner" of the isoquant; there is no other possible point of tangency.

The more usual case is the intermediate one of Fig. 11.8(*b*). Here the two factors can be substituted for each other, but over the limited range indicated by the dotted lines. Outside this range there is no substitutability. No matter what the relative prices of the two factors are, some of each will be required, but there is an opportunity to substitute one for the other if relative prices change.

**Constant proportions.** As stated earlier, the case of constant proportions is of special interest. Given any isoquant map or diagram relating output to two productive services, any straight line through the origin connects input combinations of *constant proportions.* In many cases productive services can only be used in specific proportions. Tractors and trucks require a driver; power saws require an operator; etc. When factor proportions are absolutely fixed there can be no problem of finding the "optimum combination." Only one combination makes sense, so it can be treated as a single factor.

There are, however, situations where one or several combinations are possible. Fig. 11.9 provides a simple illustration of production

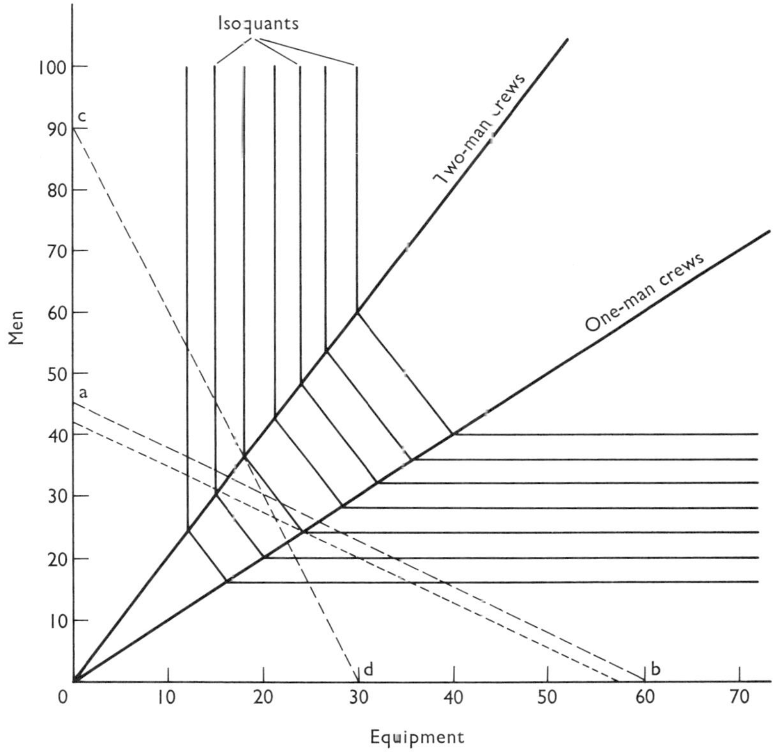


**Fig. 11.9** Constant proportions, where two combinations are possible.

under conditions of constant proportions where there are two possible ratios. For this example, suppose either one-man or two-man crews can be used for felling, limbing, and bucking timber, each *crew* being equipped with a power saw and an axe. If (1) these are the only combinations of men and equipment possible, (2) returns to scale are constant (each crew performs equally well and output per crew is the same regardless of the number of crews), and (3) production is as shown by the isoquant pattern of Fig. 11.9, then along either of the two "rays" from the origin the proportions of manpower to equipment are constant, and indicate the two possible combinations. Note that as drawn, production increases at a constant rate as one moves out along either ray.[2]

Costs of the two inputs can now be introduced. With equipment costs of $12 per day for each "package" (one power saw plus one axe), and with wages at two dollars per hour ($16 per day), the dashed line *ab* becomes an isocost, showing all combinations of men and equipment that can be purchased for $720 daily. Any desired number of isocosts can now be constructed, all paralleling line *ab*, and any number of isoquants can also be drawn if helpful. But clearly, the point of tangency between isocost and isoquant can only be at the lower "corner" of the isoquant (as shown by the dotted line parallel to *ab* indicating that under the stated conditions, one-man crews would be best for producing *any* output.

But suppose wages were only $4 per day while equipment costs remained at $12. The isocost pattern would now shift to that shown by *cd* and the upper "corner" of the isoquant would become the only possible tangency point. At these relative costs, then, the *two*-man crew would be best.

In this particular illustration a wage rate of $6 per day (with the same equipment costs) makes the choice between one- and two-man crews immaterial—the isocost curves parallel the isoquants along the significant part of the production pattern. This is simply another way of saying that if wages are less than $6 per day, the two-man crew should be chosen, while at any higher wage a one-man crew will be less costly (assuming equipment costs and productivity remain unchanged).[3]

## GENERAL APPLICABILITY

The problem of optimum input combinations is one only now beginning to emerge in forest production. Up to World War II nearly all timber was produced from what were, essentially, "natural" stands. To be sure, there were estate forests in England and

[2] The assumption of constant returns to scale is not necessary at this point. *Any* pattern of isoquants could be substituted so far as the *progression* of output is concerned. The assumption is, however, necessary if the problem is to be treated by linear programming.

[3] This illustration has many characteristics of a typical "linear programming" problem, a topic that will be further mentioned, but not developed, in the following chapter.

a few well known public forests in Germany, France, and Switzerland that had been under management for centuries. But for the most part these forests were managed from a biological rather than from an economic viewpoint. Today, plantation forestry is becoming commonplace in many parts of the world, with large acreages of highly productive man-made forests in the United States, Chile, South Africa, Kenya, New Zealand, England, Spain, Italy, and elsewhere. The problem of determining optimum input combinations is no longer an academic one. It has taken on very practical aspects. Efficient forest management requires far more knowledge of the response of trees to managerial inputs than we have been satisfied to have in the past. The foreseeable future, with ever-growing population pressures on land resources and an increasing variety of ways to use them, demands that forestry, like other productive ventures, gets the most out of its inputs.

The general framework given here can be applied directly to many forest production problems. Optimum combinations of pruning and thinning, of fire prevention and suppression, of different kinds of equipment or of crew composition, are examples of problems that can be solved by the simple techniques described in this chapter. More complex cases can be dealt with effectively by using the proportional relationship between marginal rates of substitution and factor prices, or through a simple mathematical extension (Baumol, 1965; Carlson, 1937).

The real value of the framework, however, does not lie in its literal application but in the fact that the approach is a "technique for thinking." It permits analyzing any problem that arises when there is a possibility of substituting one factor for another. One should always think of substituting the cheaper productive factor for the more expensive as long as output can be maintained, and of using more and more of a combination until the two factors are being used so that the last unit just repays its cost.

Economics, it has often been said, is a way of thinking!

# 12

# Marginal Analysis and Output Decisions

Marginal analysis has frequently been criticized by both businessmen and economists as an unrealistic, unworkable, and unwieldy guide to output decisions. Managers, it is said, almost never know their marginal costs of production and, in fact, could seldom define the term "marginal." There is much truth in such comments, even though really top-flight executives are surprisingly aware of such terms. Modern factory production methods are complex, often involving hundreds of processes requiring close integration. Here, any direct or precise application of marginal analysis is obviously difficult.

Yet forestry, and especially timber production, abounds with situations where marginal analysis can aid in making significantly better decisions, and it is to these particular situations that this chapter's subject matter is directed. Moreover, the marginal principle is almost universal in its applicability.

## "UNIVERSALITY" OF MARGINAL PRINCIPLE

The classic example of the "universality" was described more than 50 years ago. Wicksteed observed that man even applies marginality subconsciously, and illustrated his point by describing the actions of an individual who, though dreading any sort of high dive, would readily leap into a river from a five-foot bank to save a drowning stranger (Wicksteed, 1933). The same man, however, would be conscious of the conflict of the two forces—the costs and revenues—"as he took off from 8 feet." At a height of 10 feet he might screw up his courage and "take a header"; he would walk away "with some measure of self-contempt" if the bank were 12 feet high. But at a height of 20 feet he would feel no pangs of self reproach; the costs would have become too high!

Wicksteed then changes the revenue side of the picture. The same man would unhesitatingly dive in from 12 feet to save his friend, or from 20 feet to save his wife or child "though even in this case it would not occur to him to take off from 40 feet." The point is clear; one *does* act on a marginal principle, even on occasions when a money calculus is inappropriate.

The same is true today. Should you study another hour before retiring? Only if the expected gain is at least equal to the "cost" of the lost sleep (don't stop reading here). Should you take another year of collegiate work and obtain an M.S., providing Dad will foot the bill? Only if the benefits to be derived are more than the costs.[1] Should you . . . the list is endless. Full coverage of the applicable situations met in timber production is impossible, but the range of application may be illustrated to demonstrate the power and usefulness of this analytical tool.

Some cases need only be mentioned and the appropriateness of the marginal approach is immediately apparent. An operator watches a crew of men piling lumber in his yard. Will the addition of another crew member speed the work enough to warrant hiring the new man? If so—if marginal cost is equal to or exceeded by marginal revenue—the man should be added to the payroll. A sawmill owner looks critically at a piece of equipment in his mill. Will a new edger increase either quality or quantity enough to cover the cost incurred by its purchase and installation? Will *MC* be equalled or exceeded by *MR*? These cases also illustrate that in practice the changes tend to be incremental, for both productive factors and

[1] Here both benefits (revenues) and costs are more difficult to quantify, but the principle remains the same. And in this case one must also discount the benefits to the present, since most benefits are expected in a future time period.

product may be, practically speaking, indivisible. One adds a man to a crew, not part of a man; one replaces an edger, not part of the machine (although edgers of different speeds may be purchased).

## THE MARGINAL LOG OR TREE

Many cases are much less obvious. One of the most discussed economic problems of timber production, and one of the most simple in theory (though not in practice) is that of determining which logs or trees should go to the mill, and which should be left in the woods. There are many variations on this same theme.

### Marginal Sawlog Size

Consider first the case of a "gyppo" operator who has purchased cutting rights to a farm woodlot for a lump sum.[2] He receives $50 per MBF Doyle scale[3] for all logs he delivers to the sawmill. At the moment we meet him he is examining the four logs of a tree that has already been felled and bucked. The three lower logs are unquestionably worth taking, but he studies the top log, 14 inches in diameter. He has already paid for the tree, for its felling, and for the bucking and limbing. How should he decide whether to take or leave the log?

This is a clear marginal case. If taken to the mill the log will increase his total revenue by $100 \times \frac{50}{1{,}000}$ or $5.00. The only costs to be considered are the *additional* ones involved from this point forward. If the cost of skidding, loading, and delivering the log to the mill is less than $5, the log should be taken, but if these costs amount to more than $5, it should be left in the woods. If these costs are exactly $5 then it makes no difference to the operator if the log is taken or left, because the added cost is just covered by the added revenue (although it could make a difference to others).

[2] A purchase of timber which entitles the buyer, by payment of a given sum, to cut any (or all) timber he wishes from the area not specifically reserved by stated terms in the contract. Usually contrasted with a "purchase on scale," wherein payment is made according to the measured (scaled) volume removed.

[3] The Doyle rule for 16-foot logs is: Board Foot Content $= (D - 4)^2$ where $D$ is diameter inside bark at top of log, in inches. Other lengths are proportional. This rule is widely used in the southern United States.

## Digression on Sunk Costs

But what of the costs already incurred in felling and bucking? Hasn't the operator already invested money in this log? Is he simply to leave the log to rot and lose whatever he has already invested? In this case the answer is a very definite "Yes! Leave it to rot!" For at this point in time all past costs are irrelevant. They are "sunk" or "historic" costs and should have no effect whatever on the decision at hand. A short detour to clinch this point is believed well worthwhile.

Suppose you were to buy a portable electric calculator for $300. One week after your purchase you learn that a new and revolutionary electronic gadget has been placed on the market. For $10 you can now buy something that will outperform your $300 machine in every way. Should you buy the new one or not? If you decide you "can't afford to throw out the $300 calculator because there is so much invested in it" you are making a serious mistake. So far as the question of purchasing the new $10 machine is concerned, whether the old one cost $300, or $3,000, or $3 is completely unimportant. If the new $10 machine will be worth more than its cost, then buy it! If the additional (marginal) cost is covered by the additional (marginal) revenue created, the cost is not only warranted, it should be incurred. Past costs should never be permitted to influence present decisions except that, of course, one always hopes to learn from mistakes! [4]

This is why our operator could and should properly ignore all costs already incurred. At any point in a logging operation, or in any undertaking, the question should always be "Will the additional costs incurred because of the particular decision being made be matched by the resulting revenue?" But, one must bear in mind that this question may be answered differently by people in different circumstances.

## Marginal Pulpwood Size

Of somewhat greater complexity is the case of a pulpwood contractor with cutting rights on a 200-acre tract of loblolly pine. Want-

[4] The *present* value of the asset may enter into the decision, but only because it could reduce the marginal cost of the new item. (If you can unload the calculator for which you paid $300 to an unsuspecting, or unthinking, friend for $8, you could get the new machine for only $2.) The original cost, however, is unimportant.

ing to convert the old field stand to pasture, the owner has sold the timber for a lump sum ($8,000), with nothing specified as to cutting diameter limits, trees to be left, etc. The contractor must now decide how to cut the timber so as to maximize his *net* revenue.

Since it is seldom a problem when cutting pulpwood the quality aspect can be ignored. The major question becomes one of size: what minimum diameter should the contractor specify to his cutting crew? Very probably he will have developed a "rule of thumb" that will identify this diameter sufficiently closely for his purposes, but how might he go about the job of determining the minimum diameter with reasonable precision?

Again, as you should certainly suspect, the answer can be found by taking a marginal approach. In this instance the problem may easily be related to output, for output will increase directly as cutting diameter is decreased. If typical logging costs in this type of stand had been found to vary as shown in Table 12.1, then the marginal costs may be calculated by completing the three right-hand columns, as has been done in the table.

**TABLE 12.1 TOTAL COSTS OF LOGGING 200-ACRE TRACT TO DIFFERENT MINIMUM CUTTING DIAMETERS, EXCLUSIVE OF STUMPAGE PAYMENT.**

| Minimum cutting diameter (inches) | Total volume delivered to mill (cords) | Total cost (dollars) | Change in cost (dollars) | Change in volume (cords) | Marginal cost per cord (dollars) |
|---|---|---|---|---|---|
| 20 | 3,300 | 58,728 | | | |
| 18 | 4,200 | 63,408 | 4,680 | 900 | 5.20 |
| 16 | 5,000 | 68,208 | 4,800 | 800 | 6.00 |
| 14 | 5,700 | 73,178 | 4,970 | 700 | 7.10 |
| 12 | 6,320 | 78,510 | 5,332 | 620 | 8.60 |
| 10 | 6,920 | 84,750 | 6,240 | 600 | 10.40 |
| 8 | 7,500 | 92,000 | 7,450 | 580 | 12.50 |
| 6 | 8,000 | 99,750 | 7,500 | 500 | 15.00 |
| 4 | 8,400 | 106,700 | 7,200 | 400 | 18.00 |

The same data may be portrayed graphically with the relationship between diameter cutting limits and output shown by means of a double scale (Fig. 12.1). In this case the marginal cost *is* clearly the change in cost that accompanies a change in output,

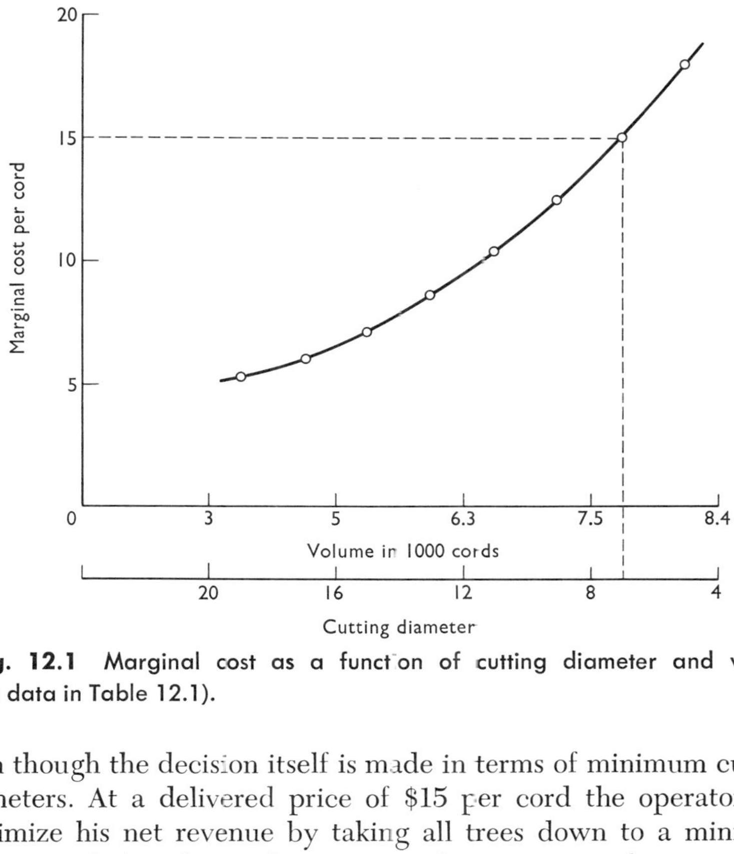


**Fig. 12.1** Marginal cost as a function of cutting diameter and volume (from data in Table 12.1).

even though the decision itself is made in terms of minimum cutting diameters. At a delivered price of $15 per cord the operator will maximize his net revenue by taking all trees down to a minimum diameter of 7 inches, where marginal cost is equal to marginal revenue (price). Recall also that marginal costs are plotted at the *midpoint* of the interval.

If price *rises* to, say, $18 per cord then clearly the cutting instructions should be changed. Under these new circumstances the timber should be taken down to a diameter of 5 inches. And if, instead of rising, price should *fall* to $12 per cord the minimum cutting diameter should be increased.

What about all the other costs—salaries of assistants, depreciation that must be charged to trucks, power saws and other equipment—all the overhead costs that are involved in any contracting business? These costs do not change under the impact of a change in cutting diameter and, hence, have no effect on the decision. They can and should be ignored. In fact, if overhead costs are apportioned to the trees by any means whatever there is great risk that their inclusion will lead to a wrong decision.

Before leaving this rather straightforward case it may be worth noting that the landowner would have a quite different viewpoint. Hoping to convert the area to a cattle ranch the landowner would realize that everything left on the ground will have to be removed at his own expense. He would want the smallest possible lower diameter specification. On the other hand, had he intended to stay in the timber business a cutting limit well above that of the contractor might conceivably be preferred.

## Effect of Scale Used

Now change the picture a little. Instead of a pulpwood contractor, visualize the problem faced by a gyppo operator who has paid cash for unrestricted cutting rights to a tract of pine which he intends to harvest, selling the logs to a local sawmill at $50 per thousand board feet, log scale. He pays all his laborers by the day while his assistant, his two foremen, and his bookkeeper are salaried.

The operator must answer the same question that faced the pulpwood producer. To what minimum diameter should the stand be cut in order to maximize his net revenues? The same basic approach should be taken and approximately the same data must be at hand, the data of costs and returns from trees of different diameters.[5] Because of the peculiarities of timber measurement, however, there are a few additional complications. The basic information is presented in Fig. 12.2, where the required cost data have been plotted over thousands of board feet using both the traditional Doyle scale[6] and the International ¼-inch rule. Clearly, the way in which the timber is measured will influence the minimum tree diameter to which the stand should be cut. Or, viewed strictly in price terms, one could say that a price quoted in board feet, Doyle scale, is a different price from the same dollar figure in terms of the International rule.

In this illustration it was specified that the contractor paid his workers at a *daily* rate. If instead, wages had been paid by log scale,

[5] It should be apparent that there is a great deal of carryover from one stand to another. Once this sort of analysis has been made, the application to other stands can be made with little new effort required. Costs must be adjusted to the new conditions, of course, but this seldom requires any time-consuming new study.

[6] The Doyle rule is defined on page 238, footnote 3.

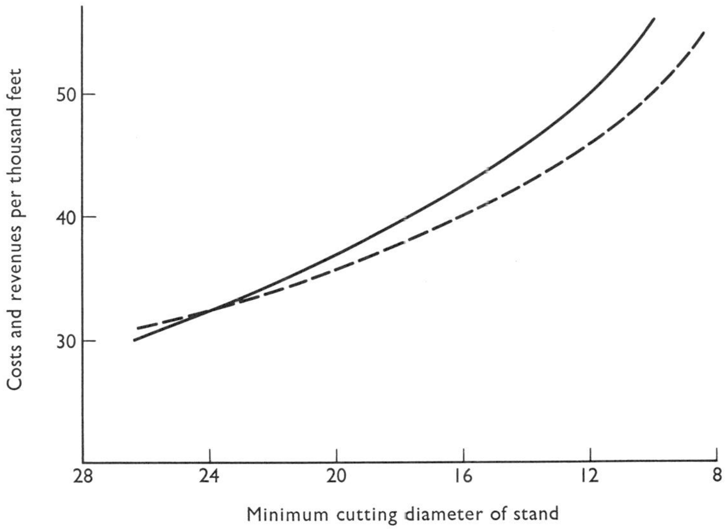


**Fig. 12.2** Marginal costs, per thousand board feet, of delivering logs to sawmill if stand is cut to various diameters. Solic line indicates Doyle scale; dotted line International $\frac{1}{4}$-inch scale.

the relationship between cost and log diameter would again change. And, like revenues, the costs would depend not only on tree diameter *per se*, but also on the method used in measuring volume. Similarly, if payment were to be on a piece-work basis (so much per tree or per log) the analysis would necessarily reflect the variation.

## Effect of Purchase Method

A change in the *method of purchase* by the contractor can also have a significant effect. If, for example, the contractor in this case had bought the timber according to scaled volume instead of for a lump sum, then for this reason alone he would tend to cut to a higher diameter limit. This is most easily seen from the total cost and revenue picture (Fig. 12.3). A double horizontal scale has been used, as in the example of Fig. 12.1, to show both volume produced from the 500 acres and the corresponding diameter to which the stand would necessarily be cut. The effect of stumpage price as well as the manner of payment can be seen from these graphs. With a lump-sum purchase, as in Fig. 12.3(a), stumpage cost is fixed

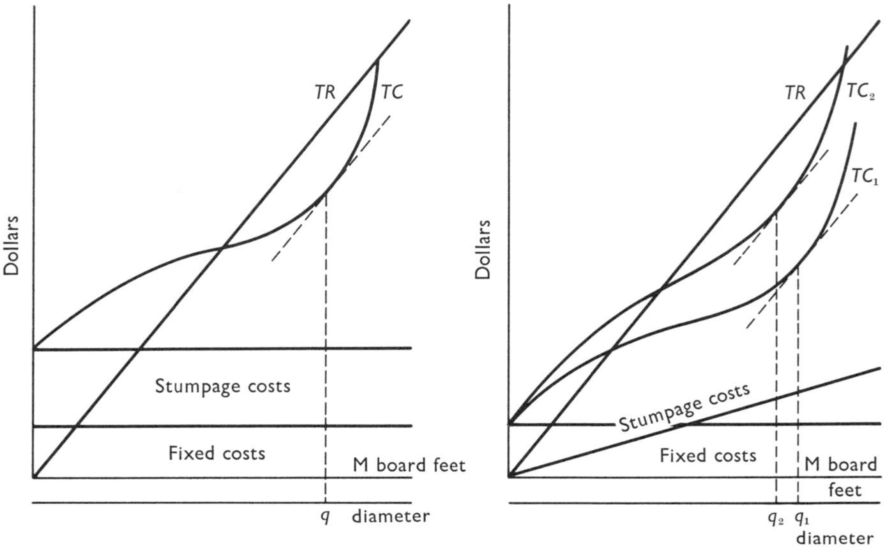


(a) Sale at lump sum (b) Sale by scale

**Fig. 12.3 The method of sale may also influence the optimum cutting diameter (see text).**

and, as with any fixed cost, its level will not influence the operating decision.[7] But in the present instance, Fig. 12.3(b), each additional 1,000 board feet of timber cut must be paid for at the agreed-upon price. In Fig. 12.3(b), where the output scale is regularly graduated, the cost of stumpage will appear as a straight line passing through the origin, its slope being equal to the price per thousand board feet of stumpage. Adding this cost to the variable cost curve will clearly "attenuate" the total cost curve, pushing it upward from $TC_1$ to $TC_2$. Since the slope of total cost will be increased, optimum output will be shifted to the left, toward lower outputs and, hence, toward higher diameters. In the present example production drops from $q_1$ to $q_2$. From the operator's standpoint there has also been a gain in flexibility. The choice of diameter is no longer so critical a decision, for the cost curve has been stretched out and become more "flat," increasing the zone of profitable operation.

[7] Of course, once the lump sum has been paid it is also an *historic* cost, so clearly can have no effect on the cutting diameter.

The same result will be attained if one regards the stumpage price simply as a reduction in total revenue, much as in the yield tax example of Chapter 10. Viewed in this light, the stumpage charge simply "swings" the total revenue curve downward, from $TR_1$ to $TR_2$ in Fig. 12.4, again indicating a reduced output and consequently a higher diameter limit for cutting. And clearly, even though the timber might be purchased under any one of many log rules, the effect will *always* be such as to make payment by log scale tend to reduce output, increase optimum cutting diameter (and increase the operator's cutting flexibility), as compared with a lump-sum purchase without specification of diameter limits.

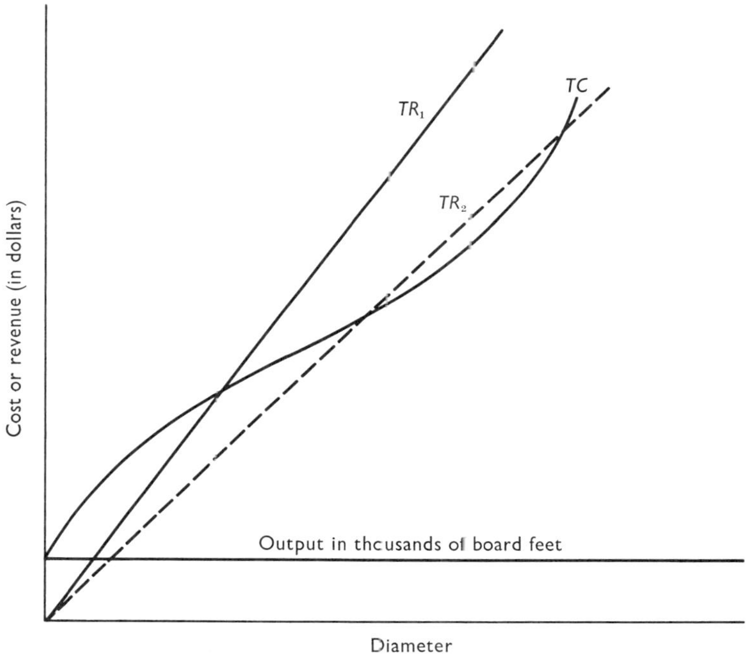


**Fig. 12.4** Stumpage price can also be treated on the revenue side.

This does *not* mean that the operator's net revenue from the tract will necessarily be reduced by reason of sale by log measurement. It may, in fact, be increased. Net revenue is the *difference* between total cost and total revenue. In the comparison being made here it depends on the level of the lump-sum payment on the one hand, and on the level of stumpage price on the other. In the case of a lump-sum purchase the *level* of the purchase price has no impact whatever on the optimum diameter limit for cutting, but it certainly does influence profit. In the case of sale by scaled measure, the higher the stumpage price the higher the diameter limit to which the stand should be cut, provided, of course, that the relationship between log diameter and costs is the usual one, similar to that shown in the preceding figures.

The variations on this same general theme of minimum cutting diameters are almost endless. Perhaps the most instructive case to think through is that of an integrated operation, where the corporation owning and operating the sawmill also has its own logging crews, its own trucks, etc. In such a case the cost aspects must be carried all the way through to the final sale of lumber, provided there are factors that cause costs to vary as diameter changes. Sawing costs, of course, also change with log diameter—and so does over-run,[8] but few of the other costs not associated with harvesting will vary with diameter. Marginal *revenue,* on the other hand, may vary quite sharply, for lumber quality and seasoning de-grade may both act to decrease the value of lumber cut from smaller diameters, while typically the over-run will increase the value of the smaller diameters by raising the yield.

### Effect of Quality

Variation in tree or log quality must be given some attention simply because it is so very important in attaining maximum net revenues from a lumber operation. Neither tree nor log grade have a very significant effect on cost, but the effect on revenue is pronounced. As long as the logging operator is paid on a flat per M basis (or on any other volume basis, such as cubic feet or cubic meters) the matter of log or tree grade will not enter into his calculations of profitability or influence in any way his minimum cutting diameter.[9] Log quality will, however, greatly influence the profitability of any sawmilling operation where the product is sold

[8] Over-run is the amount by which the actual amount of lumber cut exceeds the log scale. With the Doyle scale, for example, it is not unusual for a mill to get as much as 200 per cent over-run on 8-inch logs, and to cut slightly *less* than scale on logs over 25 inches in diameter. Doyle scale is seldom used on really large logs.

[9] If a minimum diameter or log quality is specified by the mill operator, the contractor will modify his cutting accordingly. But within this constraint, quality will have no bearing on *his* cutting.

by grade. Many mills could improve their profit position substantially by paying different prices for logs of different size and quality, providing logging contractors would accept the innovation—which would also help the logging contractors. If these marginal revenue curves were to be applied to cutting limits in field operations, decidedly different tree or log diameters should be specified for the different log or tree grades, as is illustrated in Fig. 12.5. Note that in this illustration only the revenue curves are shifted.

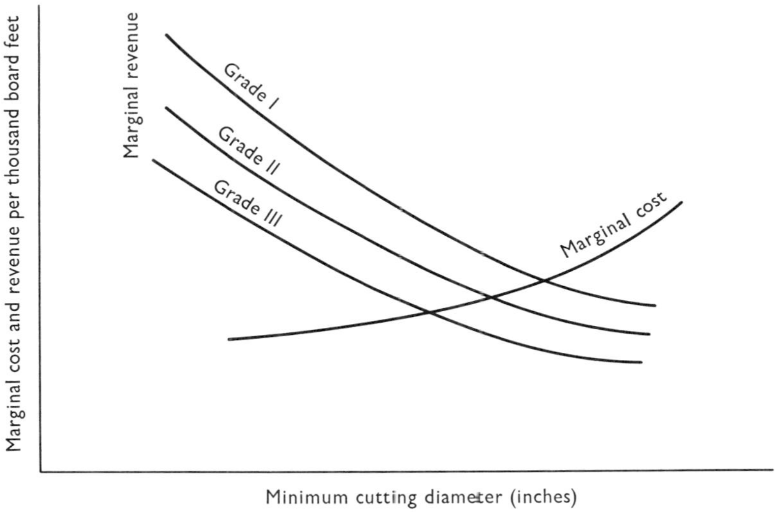


**Fig. 12.5** Effect of tree grade on minimum cutting diameter as long as revenues are made effective for logging operator.

The mill owner of earlier days faced a different kind of situation. In the Lake states of the 1880's, for example, tremendous stands of timber were waiting for anyone willing and able to take them. There was no shortage of timber, nor was there a shortage of loggers. There was a premium on getting the timber out quickly, for if one stayed very long in one place a rival logger would be cutting over the next section.[10]

[10] The situation is repeated today in many of the developing countries. True, most of these nations have strict legislation regarding timber removal, but these laws are seldom more stringent that those existing in the Lake states during the 1870's and 1880's. And, unfortunately, far too often they are no better enforced!

### Opportunity Cost

This brings us to the second detour of the chapter, one occasioned by another important and somewhat peculiar kind of cost called *opportunity cost.* In the most simple terms, an opportunity cost is *revenue foregone.* The early Lake states lumberman provides an excellent example. Why did he "cream the stand," that is, cut the best trees and leave many which could have been taken at a profit? Because of the *opportunity cost* involved in cutting the smaller, or less valuable, trees. The old time lumberman could tell his crews (translating from his much more colorful terms, and recognizing that really big trees were uncommon) "Get me my logs from the biggest and best trees you can find!" His instructions were valid, for the opportunity cost of cutting smaller, or less valuable, timber was very high. Every 1,000 feet of lumber cut from 20-inch timber would yield an average profit of, say, \$5, while 1,000 feet cut from 40-inch trees might yield \$20 in profit. The opportunity cost, then, of cutting 20-inch timber would, under these conditions, be \$15 per thousand.

The principle of opportunity cost has wide applicability. Consider the following examples. A hydroelectric development creates a sizable lake bordering a privately owned forest. Highly productive timberland suddenly becomes desirable for an outdoor recreation development. The owner is offered \$300 per acre for his forest property, but decides to maintain his ownership for timber production, reasoning that since he paid \$20 per acre for the tract he is earning 6 per cent annually. But from an economic standpoint conditions have changed. The \$20 per acre is no longer the relevant cost, for each acre of his forest property now has an opportunity cost of \$300. The return of \$1.20 per acre is not yielding 6 per cent annually, but instead a paltry 0.4 per cent.[11]

The minister of agriculture in a South American country recommends to the planning commission that the overgrazed brushland on the rolling high plain be gradually converted to improved pasture. Land prices are presently so low that the return from beef cattle

[11] This does not necessarily mean that the property should immediately be sold, or even converted to recreational use. For some people a return of 0.4 per cent from timber production may be preferable to sale for recreational purposes, which would stongly imply that such individuals may be receiving return (satisfaction) in addition to monetary income. This return, although non-monetary, affects the opportunity cost of holding the land in timber production.

could be as much as 10 per cent. But the chief of the forest service states flatly that *Pinus radiata* will yield at least 25 cubic meters per hectare per year at $2.40 per cubic meter—an annual return of $60 per hectare. To solve this controversy the planning commission utilizes the opportunity cost concept, calculating land values under each alternative use (its opportunity cost) and *also* product values under alternative uses. Using the area for producing beef cattle may force importation of newsprint and thereby impose a serious drain on the country's foreign exchange balance, an opportunity cost of beef production not foreseen by the Ministerio de Agricultura. On the other hand, conversion of the brushland to timber may mean a reduction in protein intake for the nation's people, an opportunity cost of timber production completely ignored by the Jefe de Servicio Forestal.

## THE MARGINAL PROPERTY

The marginal log and marginal tree are concepts quite readily accepted by most operating managers, even though an analytical approach to determining just which logs or trees are marginal may represent something quite outside their experience. The extension of this reasoning to determine marginal properties is less easily accomplished, for it involves more than one dimension—precisely the reason for redefining marginal cost as the change in cost under the impact of some particular decision. There are many such dimensions, only a few of which can be considered here.

Consider, for example, a sawmill operator looking for timber to supply his mill. How far from the mill can he afford to go? What represents the "marginal tract" to him? Clearly, distance is only *one* of the several factors involved: a 50-acre woodlot of scrubby second growth within a mile of his mill could well be submarginal, while a tract of high-quality old growth 150 miles distant but near the expressway that skirts the mill yard could be highly profitable. To the sawmill operator, then, the marginal property is one which returns just enough after conversion to lumber to cover all the expenses of acquisition, logging, transport, milling, seasoning, and sale. With timber of a given species of average size and quality, growing in an area with little change in topography, distance may be the limiting factor. In such instances one might plot the changing costs over distance (the marginal cost per thousand board feet of extending the logging radius from the mill) and, by equating this with revenue, determine the "marginal property" under that particular set of conditions.

For a *mill owner* buying timber for immediate use it is unnecessary to go into such detail. He need only determine whether the *additional* costs involved in purchasing and logging a specific piece of property will at least be covered by the *additional* revenue stemming from its acquisition. If he remembers to include any opportunity costs involved in his decision he can be certain it will be the correct one.

With a *timber grower,* however, a large group of additional factors will be involved, for everything affecting timber growth and its cost must be considered. Once a paper corporation has decided to build a new pulp mill, for example, their forester in charge of land acquisition must decide what land to buy to produce pulpwood for delivery to the mill site.[12]

## Effect of Site

Probably the most important variable to consider when buying land for timber production is that of soil productivity. No agriculturalist would overlook this factor, but perhaps because forestry has been regarded as an extensive land use there has been a tendency to disregard basic productivity when acquiring land for growing timber. Yet in some ways timber growing is more sensitive than agriculture to site changes. Yields from each are significantly affected both quantitatively and qualitatively, but the effect of site may be even more pronounced with timber because the timing of the yield can be changed. This will be considered directly in Chapter 14. It is enough at this point to refer to Fig. 12.6, showing the present net worth of *Pinus radiata* land, sites 1, 3, and 5, plotted as an aid to land acquisition in South America.[13] Clearly, a property consisting solely of site 5 land is "submarginal" for timber growing if its price is more than \$50 per hectare, while land of site 1 would be a good buy at anything under \$150 per hectare. Data of the same

[12] Essentially the same factors would apply to a lumber or a plywood corporation.

[13] Details of the calculations behind such curves will be developed in Chapter 14. These curves were based on a price of 8¢ per cubic foot (about \$7 per cord); planting costs were \$70 per hectare (a hectare is about 2.5 acres); annual costs were \$5 per hectare; discount rate was 9 per cent.

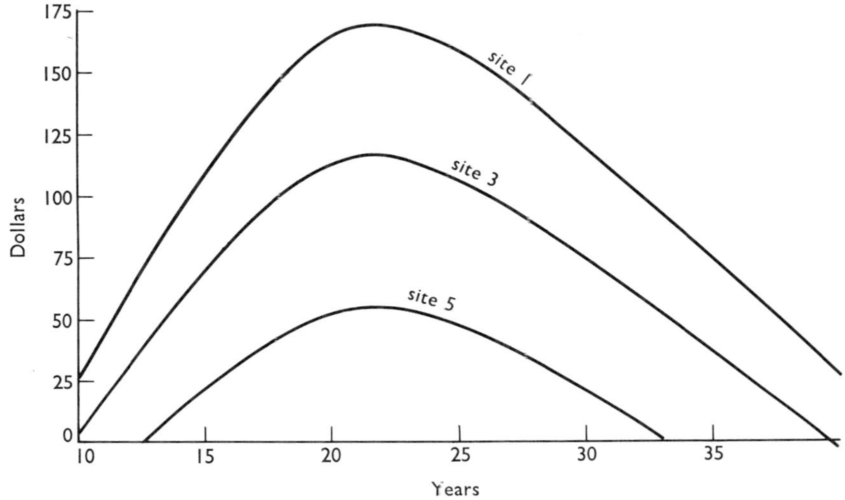


**Fig. 12.6** Present net worth of *Pinus radiata,* sites 1, 3, and 5 (per hectare basis).

general kind, but based on loblolly pine in the southern region of the United States, is given in Fig. 12.7.

## Effect of Transport Cost

Transportation cost is another very important factor in determining the marginal property for timber production. Other things being equal (which they seldom are), land values decrease as their distance from the mill site increases. For acquisition purposes, or for tentative evaluation of a region's potential for forest development, it is often helpful to map areas of timberland according to "cost contours" based on transport costs, thus establishing the "timbershed" of the mill and at the same time dividing it into cost "zones." Fig. 12.8 gives an example of this kind of map and illustrates also the effect of existing transportation routes.

If the mill is located in such a fashion that land might be purchased in different states (or even counties, in some cases), then tax levels must be taken into consideration. Similarly, items such as insurance, workmen's compensation, and minimum wage legislation are other factors that tend to vary with political boundaries.

The possible applications of marginal analysis to timber output

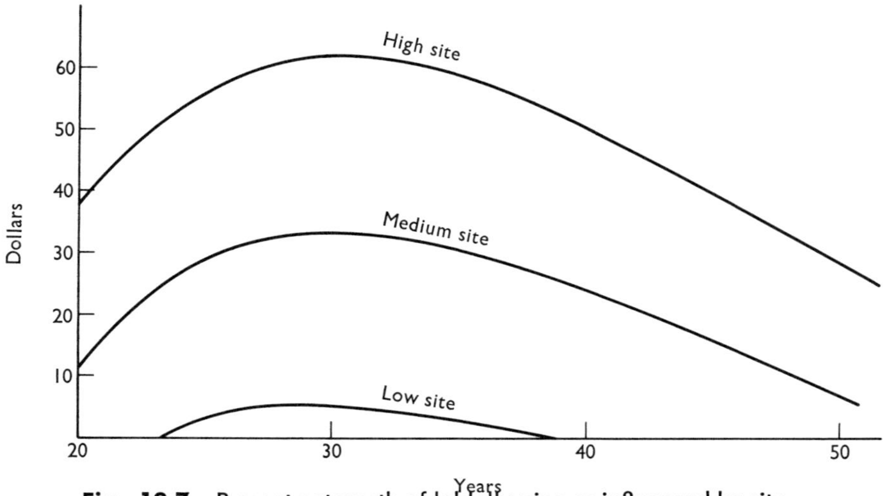


**Fig. 12.7** Present net worth of loblolly pine as influenced by site.

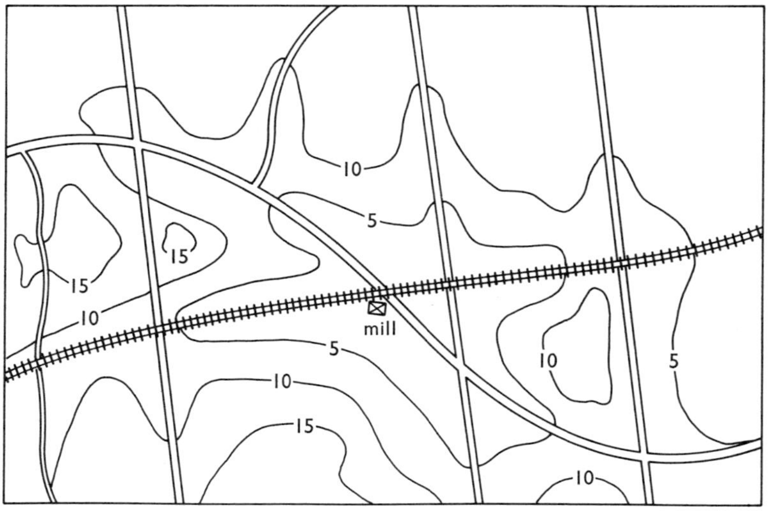


**Fig. 12.8** The timbershed of a mill. Note the influence of roads, railroads, etc. Cost contours in dollars of transport cost.

decisions are virtually inexhaustible. The objective here has been that of illustrating the variety of problems amenable to the approach rather than to present the details of any specific application. Every problem calls for a careful identification of the costs and returns that stem from the particular decision being made. Sometimes, as will be seen when non-timber products are discussed, it is this very problem of identification and measurement that makes for difficult decisions.

# 13

# Marginal Analysis and Multiple Products

From the simple one-product–one-input problem with which this section on timber production economics began, the framework was expanded to permit considering multiple input factors, then the marginal approach was applied to a variety of timber output decisions. Up to this point, however, the development has always been in terms of a single product. In reality, timber production almost always involves not only multiple inputs, but the production of several products as well. Even though a forest might be devoted to producing only sawtimber it still produces logs of different grades from which lumber of a variety of grades and sizes is manufactured. And in today's economy it is a rare instance indeed to find a forest producing sawtimber as its only product: at the very least pulpwood is almost always sold at some stage of the production process, and most often other products as well. Clearly, if economics is to be applied to modern forestry the production framework must be expanded to permit inclusion of this additional complication.

The essential features of the required expansion follow easily from the development of the two-input case. There, it will be recalled, the varying quantities of two productive factors were related to output by a system of "isoquants" which showed how the input combinations might be altered while still holding output constant

(Fig. 11.2). For the present problem of two products a similar diagram is most useful.

Consider, for example, the diagram in Fig. 13.1. With the quantities of two products, *X* and *Y*, measured along the axes of the dia-

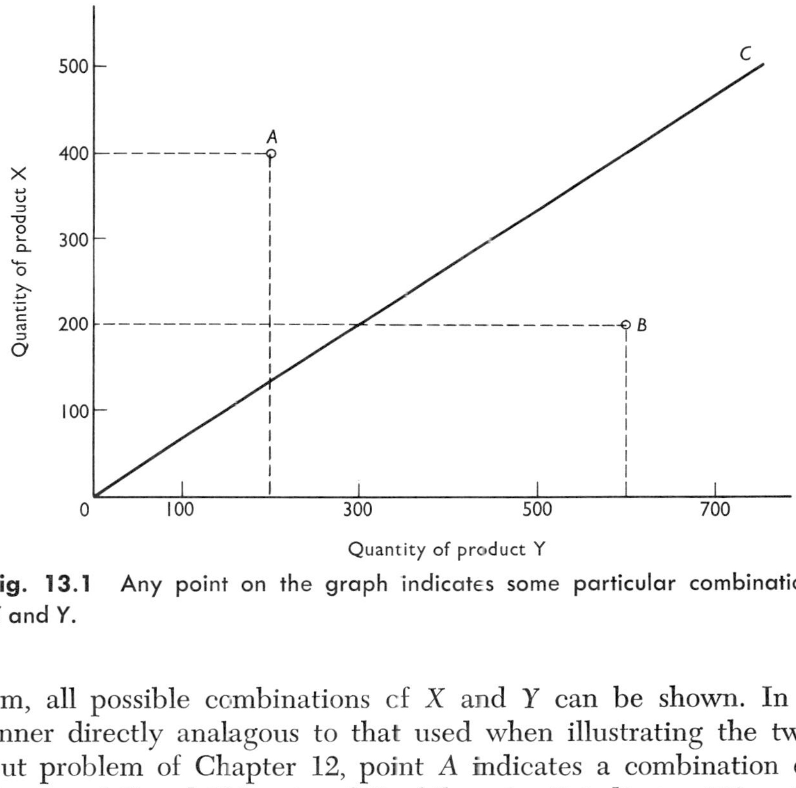


**Fig. 13.1** Any point on the graph indicates some particular combination of *X* and *Y*.

gram, all possible combinations of *X* and *Y* can be shown. In a manner directly analagous to that used when illustrating the two input problem of Chapter 12, point *A* indicates a combination of 400 units of *X* and 200 units of *Y* while point *B* indicates 200 units of *X* and 600 units of *Y*.

## JOINT PRODUCTION

As long as the two products are produced by entirely separate processes no special problems arise. The analysis can be made by the usual techniques, but using two separate production functions, two separate cost functions, etc. The interesting cases arise when the same production facility—the same factory—is used to produce two or more products. This is the case of "joint" production, so called because at least one of the factors, often the entire group of fixed factors, is "jointly" used.

Two types of joint production may be distinguished. In one the products are produced in "technically fixed proportions" (wheat and straw, beef and hides, mutton and wool, etc.) and in the other the product proportions are "technically variable."

### Technically Fixed Proportion

With technically fixed proportions, since all technically possible product combinations hold a constant ratio to each other, the product combinations always fall on a straight line passing through the origin when plotted on a diagram such as Fig. 13.1. The line $0C$ illustrates the particular combination of 2 units of $X$ to 3 of $Y$. Forestry offers few examples of truly fixed proportions, although some forest areas are handled as if the proportions were unchangeable. Naval stores and sawtimber, or in some cases pulpwood and sawtimber, might occasionally be so considered, although in actual fact these proportions can be drastically altered.

Joint production with technically fixed proportions can always be reduced to the single product case. If, for example, in studying a particular sawmill one should find that one ton of pulpwood chips can be produced with every 1,400 board feet of sawnwood, the problem of joint production can be solved by treating this as a "combination" product. The total cost and revenue picture might then appear as in Fig. 13.2, where single cost and revenue curves are used while the horizontal axis shows quantities of both lumber and chips at all relevant outputs. The profit maximizing solution is obtained exactly as before, and a full set of *total* cost curves can be derived. But *average costs cannot be derived.* An average cost for lumber or for pulpwood chips cannot be calculated. It does not exist. Try to derive it! Whenever two products are produced jointly there is simply no way of determining the average cost for either product. This is probably the major point that needs remembering so far as joint production with technically fixed proportions is concerned, and it is a point that holds good for *all* joint production.

In purely competitive situations (where prices of both products are given to the producer) no particular treatment of the firm's demand is necessary. But, from the standpoint of *industry,* market analysis, and pricing the two jointly produced products may be quite

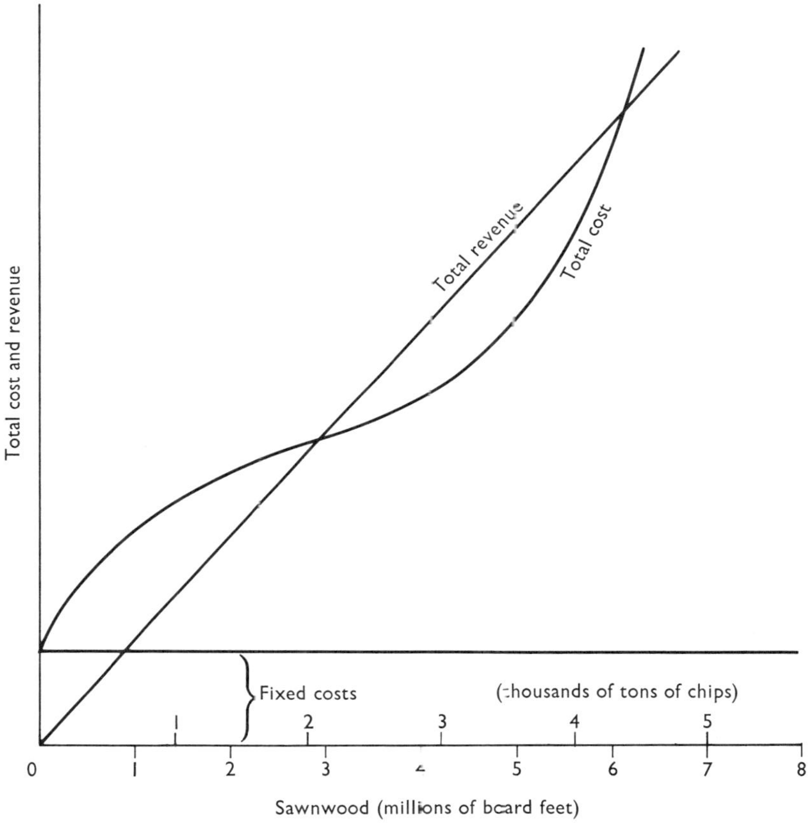


**Fig. 13.2** Joint production with fixed proportions means, essentially, a combination" product. It requires no change analytically.

different. Even though the demand curves may be completely independent, the supply curves are not, and price is a function of both demand and supply. In Kenya, for example, the derived demand for sawlogs—derived from the demand for lumber—might resemble $D_s$ of Fig. 13.3(a), and the demand for chipable waste material from these same logs (after Kenya has built a pulp mill) might resemble $D_c$ of Fig. 13.3(b). The industry demand for sawlogs *and* chipable waste could then be regarded as the vertical addition of the two demand curves, the curve labeled $D_{sc}$ in Fig. 13.3(b). If the supply curve for the combination product were that shown as

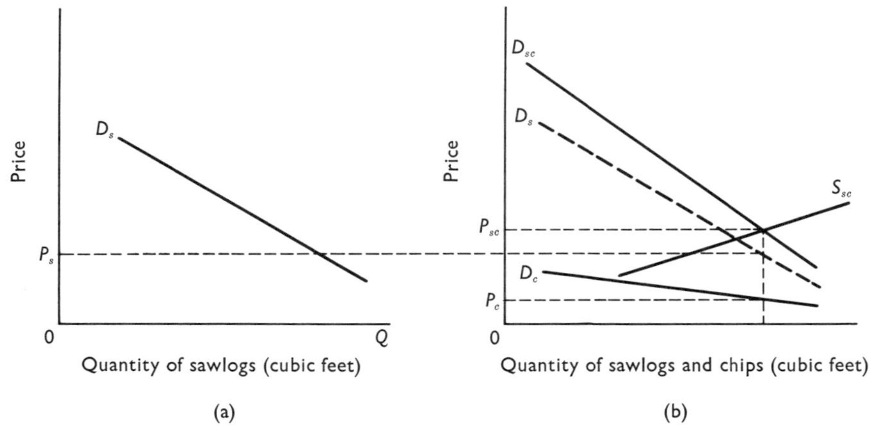


**Fig. 13.3** The demand curves for sawlogs and chippable waste would be added vertically to form the total demand ($D_{sc}$).

$S_{sc}$, the price of logs would eventually settle at $P_{sc}$.[1] Actually, in almost all cases *some* costs can be separated from the total and identified as applying only to the output of one or the other product. Product sales and transport costs are typical examples of such "separable" costs. These can be brought into the supply schedules (see footnote below) but serve at this time primarily to emphasize that, while average costs of production cannot be calculated for jointly produced products, it is often possible to calculate *marginal* costs for both items. These marginal costs, of course, include all the additional costs incurred by the production of the particular product, and it is the *marginal* costs that are relevant to the production decisions.

## Technically Variable Proportions

The second type of joint production—and by far the more interesting to forestry—is known as joint production with technically *variable* proportions. Such products have the particular characteristic that, although joint use of some productive factor or service is

[1] This represents a decided oversimplification. It is introduced here primarily to indicate the existence of problems on the demand as well as on the production side of joint operations. Full treatment of these problems, and especially of price determination, often involves the theory of discriminatory pricing and necessitates recognition of the co-existence of pure and non-pure competition in the product markets. While highly interesting, such treatment was believed beyond the scope of this text.

involved, the proportion in which the products can be produced may be varied. This kind of production may be illustrated by Fig. 13.1 if one assumes it is technically possible to produce either *A*, *B*, or any combination on the line 0*C*. More often one finds the range of product proportions limited.

Forestry abounds with examples. A given acre of forest can be made to yield sawtimber and pulpwood in varying proportions. The same holds true for a particular tree, while nearly every log can be cut to yield varying proportions of high- and low-grade lumber. And although used to illustrate production with fixed proportions, in practice lumber and chips provide a prime example of variable proportions. A different type of joint production is that involved in any mixed forest, where widely varying species combinations can be produced from the same acre or tract; in fact, a primary problem of management is that of determining the optimum species combination. Management of tropical forests provides the extreme case here, with so many product combinations that the silvicultural techniques have yet to be established in most types. And, as we will see in Chapter 18, even the much maligned term "multiple use" can be analyzed as a problem in joint production where the principal difficulty is that of determining the optimum product combination.

Since maximization of net revenue has been taken as the primary objective of management, both costs and revenues must be brought into the picture, which so far deals only with the technical aspects of product combinations. Costs and revenues may be introduced, at least, in a fashion again analogous to that used in determining input combinations, but with a few twists of its own.

**Production possibility curves.** The starting point can be that of Fig. 13.1, a "map" of the various combinations that can be produced with the given set of productive factors. *Costs* are introduced by connecting all those product combinations that might be produced for some specified amount. Take, as an example, the situation described by Fig. 13.4. Here the output of sawlogs in board feet is shown on the vertical axis and pulpwood output in cords on the horizontal axis. The curve labeled \$16 connects all combinations of the two products that can be produced (from each five acres) for a total cost of \$16, and similar curves are identified for other total cost outlays. Such curves are called "production possibility curves," or isocosts, here relating products instead of inputs. There would be

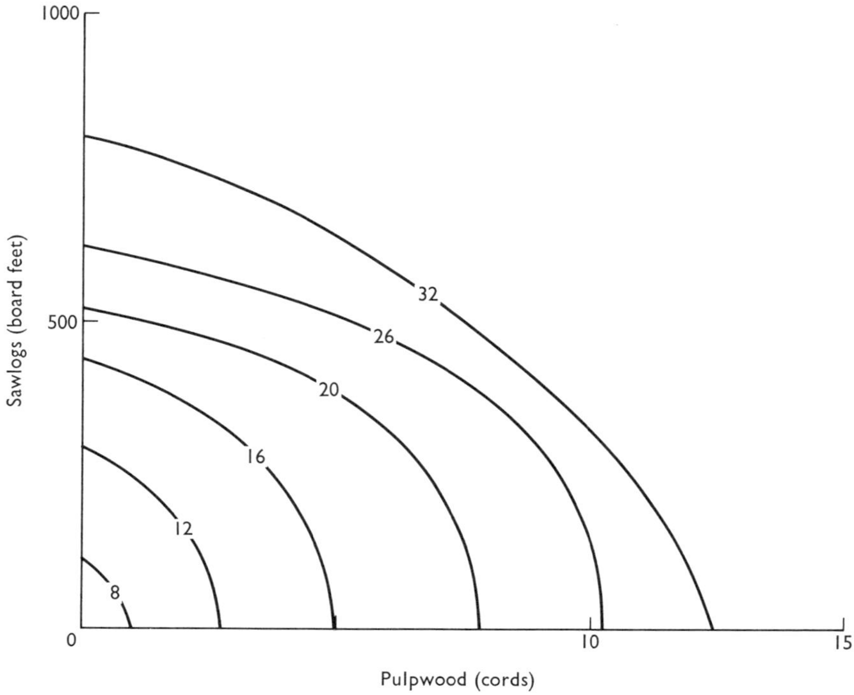


**Fig. 13.4** Production possibility curves, or isocost curves, for jointly producing sawlogs and pulpwood (five-acre basis).

an isocost for any total cost one wished to identify, and hence a "family" of isocosts.

**Isorevenue curves.** Since the objective of production is taken to be maximization of net revenues, it is still necessary to bring revenues into the picture. Revenues stem from product sales, so competitive conditions in the markets for the two products determine the shape of the isorevenue curves. There are, of course, many conditions that can and should be recognized, but for the moment we will deal only with the one most frequently found in forest enterprises. Typically, no one forest producer influences product price by his production decisions. Under such (purely competitive) conditions the family of isorevenues will be parallel straight lines. If, for example, the two products are priced in a 1:12 ratio (as they would be if the price of pulpwood were \$2.50 per cord and sawtimber \$30 per thousand board feet) then the isorevenues would be as shown in Fig. 13.5.

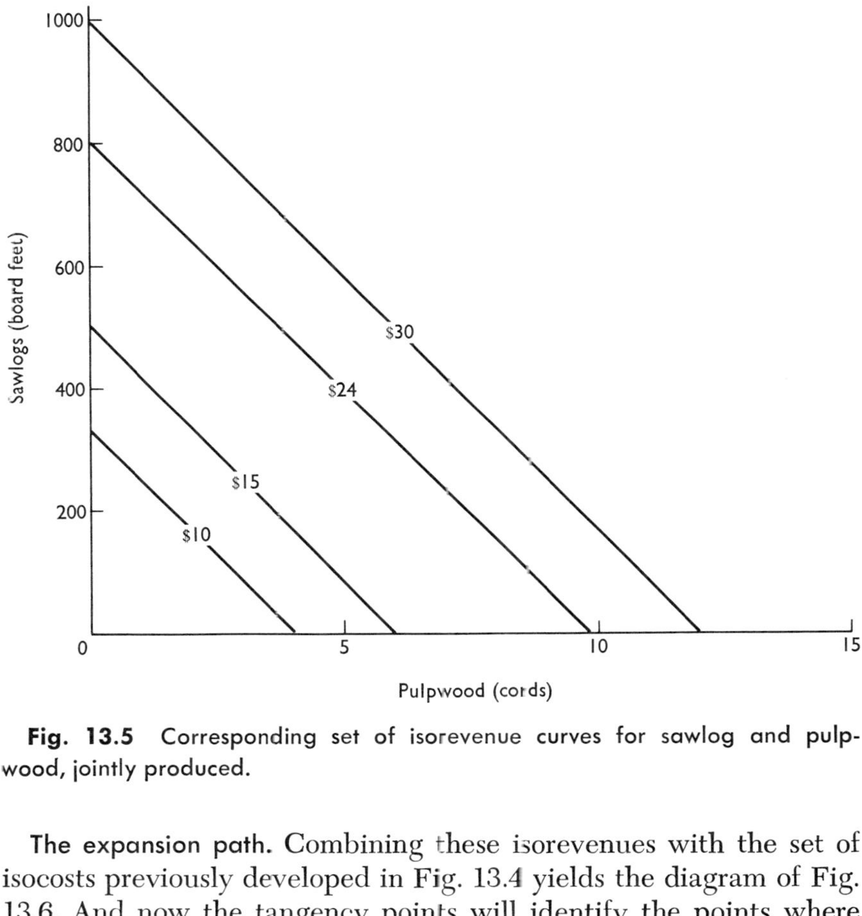


**Fig. 13.5** Corresponding set of isorevenue curves for sawlog and pulpwood, jointly produced.

**The expansion path.** Combining these isorevenues with the set of isocosts previously developed in Fig. 13.4 yields the diagram of Fig. 13.6. And now the tangency points will identify the points where revenues are at a maximum for any particular cost outlay. Hence, any rational producer would always operate in such a manner as to produce one of the product combinations identified by a line connecting all these points of tangency; that is, he would always operate at some point on the "expansion path."

**Maximizing net revenue.** The final step is identical to the single product case, differing only because two products are now involved in varying combinations. The complete solution can be shown as

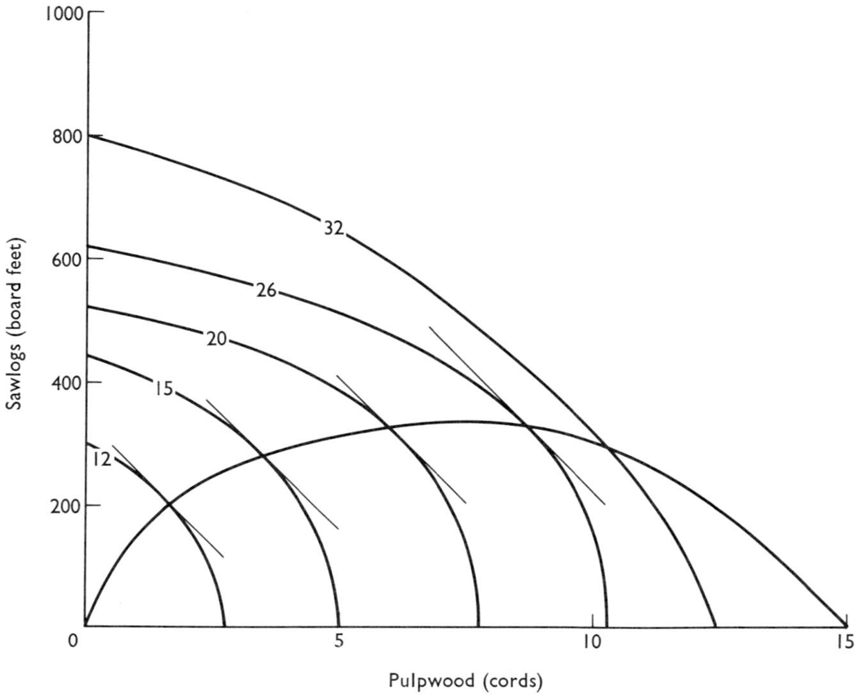


**Fig. 13.6** By combining the isocost and the isorevenue curves the expansion path is determined.

in Fig. 13.7—a combination of the previous figure (Fig. 13.6) with the resulting total cost/total revenue curves. The maximizing combination can be read from the point on the expansion path that falls directly below the point where the slopes of the total cost and total revenue curves are identical. In the particular case illustrated, this occurs with an output of 340 board feet of sawlogs and 8 cords of pulpwood.

In such cases no set of average costs can be shown for either product. An average cost of pulpwood production *or* sawlog production simply does not exist. Conceivably one could construct a psuedo-average cost diagram by employing a device similar to that of Fig. 13.7, but the idea of an average cost has been completely destroyed by the joint nature of the production process: average cost equals total cost divided by output, and one cannot

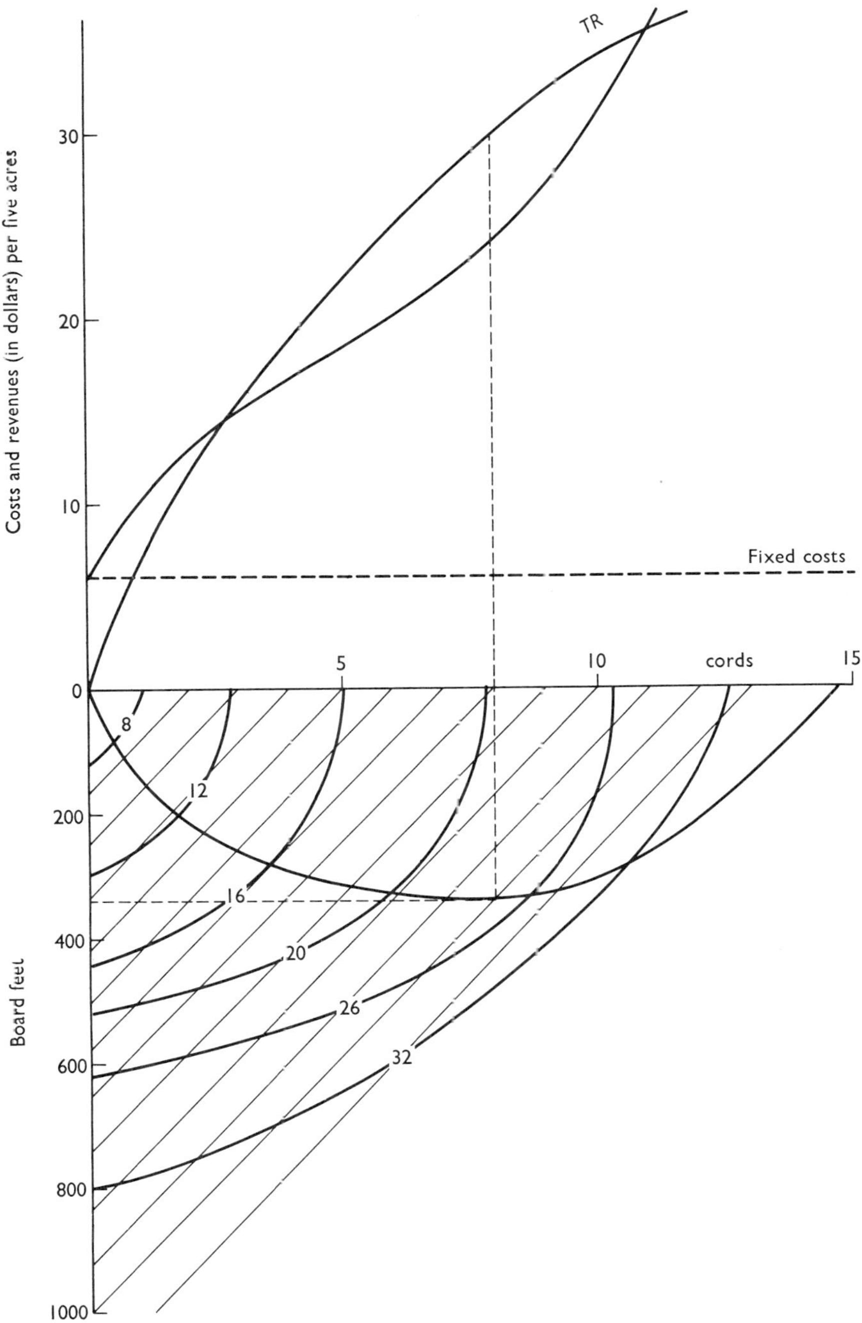


**Fig. 13.7** Determining the profit-maximizing combination of products.

divide by "output" when the makeup of that output is constantly changing. In joint production, average costs are meaningless.

But marginal costs do exist and can be determined for either product. Because the marginal cost equation for either product will contain the output of the jointly produced product as a variable, the two marginal cost curves must be solved simultaneously.[2] Joint production, therefore, always leads to making output decisions on *both* (or all) products simultaneously, and this is true whether or not it is recognized by the manager or decision maker. There is probably no more basic point to be made in forest economics than this, because so many products *are* produced jointly by a modern forest in today's economy. Decisions affecting one product affect the output of all other jointly produced products—by definition.

**Product relationships.** Although the principal problem of joint production has been "solved" (at least conceptually), there is still much to be gained from a consideration of some of the complications that might occur. For example, the shape of the isocosts reveals something of the nature of the relationship between the two products from a production standpoint. Consider, for example, (a) and (b) in Fig. 13.8. With the relationships indicated in (a) one might produce either 200 units of $X$ and a quantity of $Y$ ranging from 0 to 200 or, alternatively, and with the same total cost, an output of 200 units of $Y$ but with the quantity of $X$ ranging from 0 to 200. Isocosts such as these are characteristic of joint production with technically *fixed* proportions: the corners of the isocosts follow a straight line from the origin as shown in the earlier diagram, Fig. 13.1. Clearly, no matter what the price relationship between the two products might be, the isorevenues could only

[2] Given the total cost as a function of the output of two products, $X$ and $Y$,

$$TC = f_c(X,Y)$$

then the marginal costs of $X$ and $Y$ become, respectively,

$$MC_X = \frac{\partial(TC)}{\partial X} \quad \text{and} \quad MC_Y = \frac{\partial(TC)}{\partial Y}$$

Note that here the marginal costs are defined as *partial* derivatives. Hence $\frac{\partial(TC)}{\partial X}$ is the change in total cost that accompanies a change in the output of $X$, *when the output of $Y$ is held constant.* In any marginal cost equation involving joint production of two or more products the output of all products produced jointly will appear as variables.

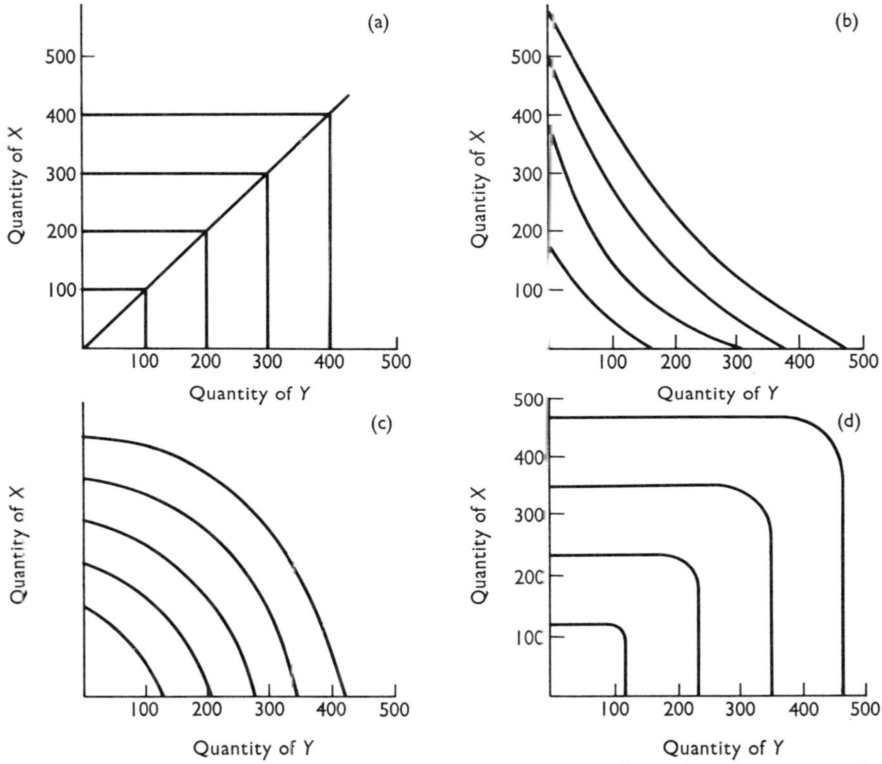


**Fig. 13.8** Four possible shapes of isocost relationships (production possibility curves).

be "tangent" at the corners of the isocosts, i.e., on the straight line passing through the origin.

Isocosts such as those pictured in (b), while quite possible (and, in fact, very common) are of genuine interest only when they appear after some other pattern has been experienced or expected. Whenever isocosts are convex to the origin, as in (b), then the two products should be produced separately—not jointly. At either end of the isocosts a small increase in yield of one product is possible only through decreasing the output of the other product by a large amount. Again, one need only imagine an isorevenue superimposed on these convex isocosts to realize immediately that the only possible rational combinations are along the two axes. This necessarily means a "combination" in which the output of one or the other is zero. The slope of the isorevenue will determine which of the products is to be eliminated from the enterprise.

Isocosts that are concave to the origin (are bowed outward) as are those of diagrams (c) and (d) indicate a process that *should* be joint: the two products can be produced *jointly* at a lower cost than would be possible if separate facilities were to be established. In contrast to the situation in (b), at either end of the isocosts in (c), a small decrease in yield of one product results in a large increase in the output of the other. The introduction of isorevenues in this case results in precisely the type of solution already outlined in Fig. 13.6.

In real life most products produced jointly are not completely variable in proportions. The isocosts (if one were to plot them) would be curved, but only for certain combinations—as in diagram (d). When isorevenues are superimposed on such isocosts one sees immediately that production will always be joint, and that the optimum product combination can only vary over a limited range.

**Form of isorevenue curves.** The shape of the isorevenues is controlled by the competitive market situation for the products. As long as the markets are purely competitive the isorevenues will be straight lines and will be parallel to each other. Their slope will always be the inverse of the product prices. Take, for example, the isorevenue of Fig. 13.9. To construct such an isorevenue the easiest approach is probably that of taking any value, $\$C$, and determining the quantity of each of the two products that can be sold for this amount. These amounts are then measured out on the respective axes and the two points identified connected with a straight line. The appropriate quantity of $X$ is thus $\$C/P_x$ and of $Y$ is $\$C/P_y$, where $P_x$ and $P_y$ are the prices of the two products.

The slope of the isorevenue is, therefore,

$$\frac{\dfrac{C}{P_x}}{\dfrac{C}{P_y}} = \frac{\cancel{C}}{P_x} \times \frac{P_y}{\cancel{C}} = \frac{P_y}{P_x}$$

and is clearly the inverse of the product prices. This being so, any change in the price of either product that results in a change in the price ratio will influence the optimum combination of the products. If the price of $Y$ rises relative to that of $X$, the optimum com-

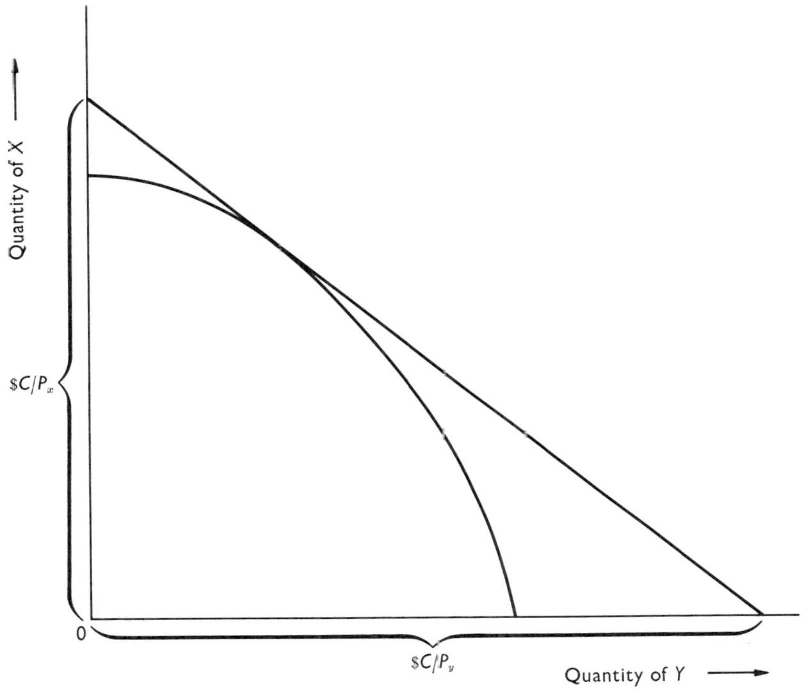


**Fig. 13.9** The slope of the isorevenue curve will be inversely proportional to the ratio of the two products.

bination will consist of a higher proportion of $Y$ than before the price change, and vice versa.

If the markets for the two products are *not* both purely competitive then the isorevenues will not necessarily be parallel, or even straight lines, but a few generalizations can be made. If increased production of product $X$, for example, led to a lower price for $X$ while the price of $Y$ remained constant (or if the situation were reversed), then the isorevenues would become concave toward the origin. But, if such increased production lowered the price, then production would no longer be purely competitive since the producer has at least some control over price. Under these circumstances the producer would cease sale of the product at the quantity producing the maximum revenue, and the excess quantity would be destroyed (wasted). In earlier years the mines of Boliden in Sweden, producing gold and copper as principal products, also yielded more arsenic than the entire world's consumption (Carlson, 1937). The disposal of arsenic actually became a major production problem. While no such clear-cut examples can be cited from the forestry field per se, much the same thing could be said of the production of vanillin from the black liquor of sulphate pulp, of the production of garden mulch from redwood bark, or of the production of fuelwood as a byproduct from sawmill waste in a good many mill towns.

In much of the world, including the United States, the forest owner most able to influence prices of sawtimber and pulpwood is the state or national forest service. This possibility stems entirely from the fact that they control by far the biggest block of stumpage in the nation (and indeed, in many nations, the entire forest area). In such instances the production policies favored by the public forest managers could have a decided influence on supplies, and hence on prices, of jointly produced products even though there was no deliberate effort in such directions. In certain of the African nations, for example, where early colonial foresters directed their efforts toward classical plantation management with long rotations aimed at producing high-quality sawlogs, there is a present inability to supply needed pulpwood, although the tremendous flexibility of marketing associated with timber production prevents this from becoming too serious. In the United States, where public forest management has favored sawlog over pulpwood rotations for many years, it is doubtful that there has been any effect on pulpwood prices. As long as pulpwood demand is met at prices no higher than production costs, additional pulpwood production (that might have resulted from a policy of shorter rotations) would serve as an indirect subsidy to the pulp industry. Quite probably the policy of long rotations has aided in keeping sawlog prices from showing an even more rapid increase than that exhibited in the 1950's and 1960's. In specific locations, however, there may well have been quite different results. Long rotations can also *restrict* sawlog supplies over the short term. Public forests of the West Coast have supplied many such examples during the decade of the 1960's.

It takes little thought to discover many opportunities for applying the conceptual tools outlined in this chapter to problems in everyday forest production. Quality, for example, is very frequently of prime importance when choosing between one or another managerial step, and quality—provided it can be measured—fits into this multiproduct framework with practically no special treatment. A particular product that has two quality levels, for example, can also be called two separate products and treated accordingly.

### Linear Programming

For multiproduct or multiquality situations a *linear programming* approach will usually be preferable, especially if computer facilities are available. Real life problems of this general type are characteristically subject to *constraints* of one type or another. Manpower, or machines, or storage space, or shipping space is limited; some given amount of one product must be produced under any circumstances; marketing agreements restrict production to certain levels, etc. Mathematical programming is a powerful tool for particular kinds of problems, especially those that can be treated under assumptions of *linearity* (constant proportions, constant returns to scale, etc.), and those subject to a series of constraints. Even an introductory presentation, however, is believed beyond the time requirements imposed on most introductory courses in forest economics, and consequently the techniques of programming will not be further developed in this text.[3]

[3] Anyone interested in gaining a good understanding of linear programming would do well to read Dorfman (1953) and Baumol (1965). The Baumol text is particularly good for developing the basic techniques.

# 14

# Financial Objectives for Timber Production

Up to this point our study of timber production has concentrated primarily on quantitative aspects—the *quantity* of product that leads to maximum net revenue, how *much* of each necessary input should be used, the relative *amounts* of different productive factors that make up the least cost combinations, or the proportionate *volumes* of the several timber products that might be produced. Occasionally the qualitative aspects have been mentioned, and in the last chapter quality was identified as a special case of the two-product problem. The time aspects—the *when* questions—remain to be considered, and among the time-related problems are some of the most troublesome issues of all production theory. In forestry, these problems involving time, perhaps because they are troublesome, have led to the development of traditional methods of solution and today they encompass what is undoubtedly the largest body of "doctrine" to be found in our profession. The enshrinement of these traditional methods by past and present generations of forest managers dictates that these approaches must be thoroughly understood by every student of forestry, regardless of how inappropriate, or even useless, they may be.

Nowhere are traditional approaches more apparent or controversial than in the various procedures advanced to determine the optimum rotation for timber production. As a plain matter of fact, rotation length per se is seldom critical from either a theoretical or a practical standpoint. To be sure, it is much easier to organize a forest property if the rotation age for each stand is known precisely, yet rotation ages not only can but always will be changed to suit the changing objectives of management.

The problem of rotation determination, however, may be regarded as the forestry expression of a basic economic issue. Fundamentally, this is a problem in capital theory. Forest growing stock is forest capital, and the question of how much capital should be invested for how long a time is surely an important one to timber production economics. This in turn necessarily involves an equally basic issue of economic theory: what, if anything, does the firm attempt to maximize over time? These two issues form the central theme of the present chapter—investment, and the financial objective of the firm.

## TIME AS A FACTOR IN TIMBER PRODUCTION

As a start, time will be treated as an explicit productive input—as if it were simply another type of fertilizer singled out for special attention as in Chapter 3. The necessary assumptions are much the same as in the previous one-product—one-input example: a given area of forest land; a particular species, density, type of management, etc. As a specific illustration, all physical inputs required for a one-acre plantation of *Pinus radiata* are assumed to be absolutely fixed, then increasing amounts of *time* (the variable service) are added. The result is shown in Fig. 14.1. It is clearly of the same general family as most biological production functions.

What might come as a mild surprise is the fact that such curves have long been part of most forester's training. Only the terminology has been altered. Change output to "yield," substitute "age" for time, replace total physical product with "total yield," and the diagram could be substituted for the traditional "yield curve" found in almost every forest management text published over the past century. And the parallel continues for the two curves derived from the total yield relationship. Divide the total physical product by the corresponding input, and one derives the average physical product (*APP*)—in forestry terminology called "mean annual increment" or *MAI*. The *change* in total physical product

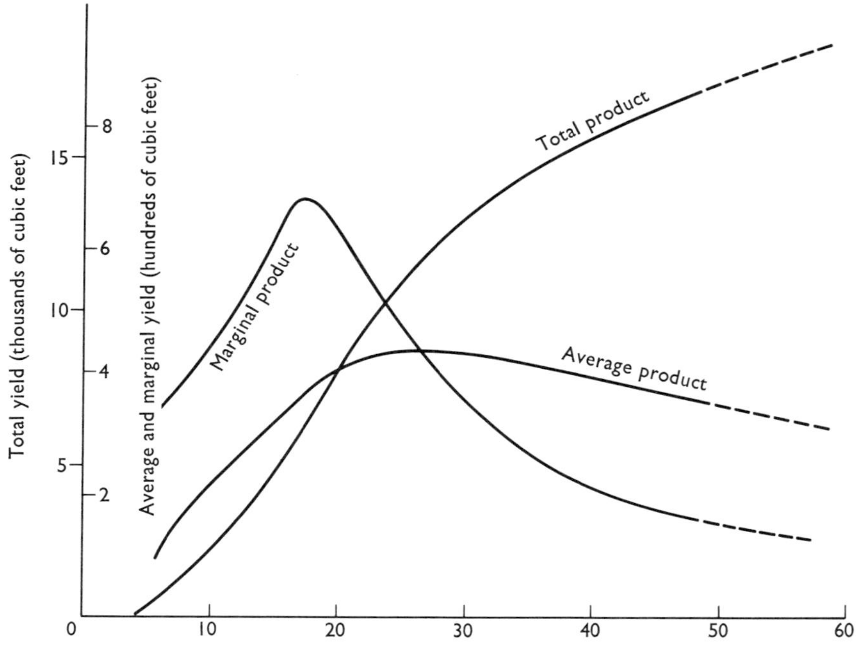


**Fig. 14.1** Production function for *Pinus radiata* (total product). Average product and marginal product also shown (note different scales).

that accompanies a *change* in input is called marginal physical product (*MPP*) by economists, but when time is the input and is kept in units of one year, foresters call this "current annual increment," or *CAI*. These curves have exactly the same relationship to each other as those given in Chapter 3. The set of traditional forestry curves is not, in fact, a parallel to the physical product curves of production economics. When time is considered the input they are identical!

Having established this identity it would seem sensible to calculate costs and revenues and determine the time input that would maximize net returns from the operation just as was done with the previous production examples. The problem is not new, although in many cases the answer itself is of little importance. Determining the time input under the stated conditions amounts to determining the optimum rotation for the area under consideration. The procedure used previously required transformation of the physical product and inputs into dollar terms—into revenues and costs. In the present problem there is no difficulty posed by the revenues; as long as the product price is available, total revenue can be obtained for any output level by direct multiplication. But what costs should be attached to the input side?

In most production analyses of economics the time cost is ignored—partly for convenience, but even more because it is of minor significance. With timber production, time is one of the chief inputs, and to overlook time in a timber production analysis would be as serious an oversight as overlooking labor in an analysis of automobile production. But what is the cost of time? Does time itself have a cost measurable in dollars?

Regardless of the philosophical overtones these questions create, as long as one's view is restricted to production per se, time itself has no specific value. To be sure, individuals and institutions attach great significance to the timing of incomes and expenditures, but the problem is not really one of attaching a value to time itself. Rather, it is one of precisely comparing revenues and costs that occur at different points in time, of determining the effect of time on incomes and expenditures, of determining the influence of time on decisions involving comparisons between incomes or expenditures that can be made at any one of several different points in time. It is perhaps a subtle distinction, but an important one. The basic task is one of discovering how time influences the effect on timber production of interest rates, growth rates, taxes, uncertainty, and other such factors. Then, and only then, can the "when" problems of timber production be answered. Now, back to the example.

By scaling the horizontal axis of Fig. 14.1 in time units each output has been dated. With the vertical axis scaled in dollars, every input can be similarly identified time-wise and both revenues and accumulated costs plotted for any chosen point in time. This procedure not only facilitates comparisons, but since it also permits one to accumulate costs and revenues as time proceeds, the relative changes of costs and revenues over time can be better examined.

### Revenues

The Monterey pine data already used in Fig. 14.1 will provide the specific means for illustrating time effects. From these physical data the revenue at any point in time is obtained by multiplying output by price ($TR = P \cdot Q$). For simplicity, a single product (pulpwood) will be assumed and the value for stumpage will be held at four cents per cubic foot (about \$3.20 per cord). The resulting revenue curve shows the value per acre for ages ranging from zero through 45 years for *Pinus radiata* grown under the specified conditions (Fig. 14.2). Calculations are shown in Table 14.1.

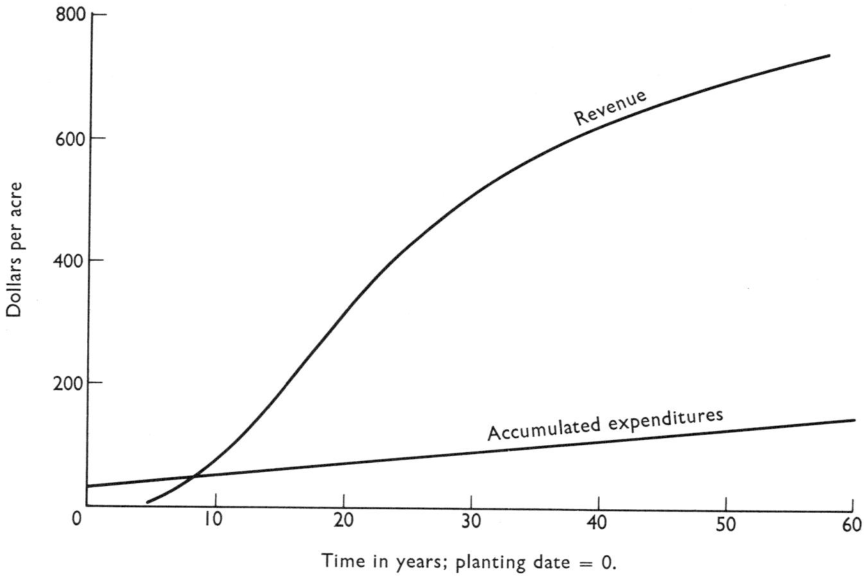


**Fig. 14.2** Revenues and accumulated expenditures per acre for *Pinus radiata.*

## Costs

Costs will also be kept as simple as is compatible with reality. Holding the question of land cost in abeyance for the moment, it is assumed that planting costs are \$28 per acre (including all costs of planting, soil preparation, seedling production, etc.) and are incurred at the beginning of the planting period.[1] Two other costs are identified: a weeding cost of \$4 per acre incurred one year after planting, and the annual maintenance costs of \$2 per acre for taxes,

[1] The costs, also, are very roughly those reported for Chile.

TABLE 14.1 CALCULATION OF PRESENT NET WORTH, SOIL EXPECTATION VALUE, INTERNAL RATE OF RETURN, AND FOREST RENT. (INTEREST RATE AT 6 PER CENT, PRICE AT 4¢ PER CUBIC FOOT, ANNUAL COST OF $2.00)*

| Year | Yield | Total revenue | Marginal revenue | Total costs | Marginal cost | Net revenue | Forest rent | Se value | Present net worth | Internal rate of return |
|---|---|---|---|---|---|---|---|---|---|---|
| 5 | 270 | 10.80 | | 53.80 | | −43.00 | −6.24 | −127.14 | −32.13 | — |
| 6 | 597 | 23.88 | 13.08 | 59.03 | 5.23 | −35.15 | −3.35 | −83.99 | −24.78 | — |
| 7 | 946 | 37.84 | 13.96 | 64.57 | 5.54 | −26.73 | −1.16 | −53.08 | −17.77 | — |
| 8 | 1,320 | 52.80 | 14.96 | 70.44 | 5.87 | −17.64 | 0.59 | −29.71 | −11.07 | 1.3 |
| 9 | 1,721 | 68.84 | 16.04 | 76.67 | 6.23 | −7.83 | 2.09 | −11.36 | −4.63 | 4.3 |
| 10 | 2,150 | 86.00 | 17.16 | 83.27 | 6.60 | 2.73 | 3.39 | 3.44 | 1.52 | 6.3 |
| 11 | 2,608 | 104.32 | 18.32 | 90.26 | 6.99 | 14.06 | 4.57 | 15.64 | 7.40 | 7.5 |
| 12 | 3,096 | 123.84 | 19.52 | 97.68 | 7.42 | 26.16 | 5.65 | 25.84 | 12.99 | 8.4 |
| 13 | 3,616 | 144.64 | 20.80 | 105.54 | 7.86 | 39.10 | 6.66 | 34.50 | 18.32 | 9.0 |
| 14 | 4,171 | 166.84 | 22.20 | 113.87 | 8.33 | 52.97 | 7.63 | 42.00 | 23.42 | 9.5 |
| 15 | 4,766 | 190.64 | 23.80 | 122.70 | 8.83 | 67.94 | 8.57 | 48.64 | 28.34 | 9.7 |
| 16 | 5,403 | 216.12 | 25.48 | 132.07 | 9.37 | 84.05 | 9.50 | 54.56 | 33.08 | 9.9 |
| 17 | 6,078 | 243.12 | 27.00 | 141.99 | 9.92 | 101.13 | 10.41 | 59.73 | 37.55 | 9.9 |
| 18 | 6,763 | 270.52 | 27.40 | 152.51 | 10.52 | 118.01 | 11.25 | 63.63 | 41.34 | 9.9 |
| 19 | 7,440 | 297.00 | 27.08 | 163.66 | 11.15 | 133.94 | 11.97 | 66.12 | 44.26 | 9.9 |
| 20 | 8,095 | 323.80 | 26.20 | 175.48 | 11.82 | 148.32 | 12.59 | 67.10 | 46.24 | 9.9 |
| 21 | 8,712 | 348.48 | 24.68 | 188.01 | 12.53 | 160.47 | 13.07 | 66.87 | 47.20 | 9.7 |
| 22 | 9,295 | 371.80 | 23.32 | 201.29 | 13.28 | 170.51 | 13.44 | 65.49 | 47.31 | 9.5 |
| 23 | 9,845 | 393.80 | 22.00 | 215.36 | 14.07 | 178.44 | 13.73 | 63.27 | 46.71 | 9.2 |
| 24 | 10,363 | 414.52 | 20.72 | 230.28 | 14.92 | 184.24 | 13.93 | 60.42 | 45.50 | 9.0 |
| 25 | 10,850 | 434.00 | 19.48 | 246.10 | 15.82 | 187.90 | 14.07 | 57.07 | 43.77 | 8.8 |
| 26 | 11,310 | 452.40 | 18.40 | 262.87 | 16.77 | 189.53 | 14.16 | 53.39 | 41.66 | 8.5 |
| 27 | 11,745 | 469.80 | 17.40 | 280.64 | 17.77 | 189.16 | 14.21 | 49.48 | 39.22 | 8.3 |
| 28 | 12,156 | 486.24 | 16.44 | 299.48 | 18.84 | 186.76 | 14.22 | 45.42 | 36.53 | 8.1 |
| 29 | 12,545 | 501.80 | 15.56 | 319.44 | 19.96 | 182.36 | 14.20 | 41.27 | 33.65 | 7.9 |

* Basic yield data adopted from that of Lota, Chile. Planting cost assumed at $28 per acre, and a single weeding made at end of first year with cost of $4 per acre.

fire protection, supervision, etc. These costs are also shown in Fig. 14.2. Of the three costs, only that for maintenance varies with time, increasing $2 every year.

Part of the "time problem" is now apparent. The revenue is always to be received at some future date, while some of the costs are incurred at the very outset of the venture. In Chapter 10 the particular "rate at which the value of future income is discounted to make it comparable to present income" was defined as *individual time preference*. If the time preference appropriate to the owner of this particular stand were known, all potential revenue could be discounted to the present, thus making revenue directly comparable (in time) to the planting costs. And if all the other costs involved were similarly discounted, then all discounted costs could be summed and subtracted from the discounted revenues, thus obtaining a present net worth, or discounted net value, of the stand for any chosen age. In many ways this is an easy and altogether natural approach.

The more usual procedure in forestry, however, is to treat all costs as if they were met by borrowing the required funds at some specified interest rate.[2] The costs, together with the accumulating interest charges, are then carried to any desired date and compared directly with the revenues at this same point in time. A net revenue appropriate to this date can also be calculated. If this *net* revenue is discounted to the proper point in time, and if a single interest rate has been used throughout, the result will be identical to that obtained by subtracting discounted costs from discounted revenues.

**Cost of interest.** Figure 14.3 depicts the same basic cost and return picture as that of Fig. 14.2, but includes the interest charges (the shaded portion of the graph) as they accumulate at a compound rate of 6 per cent. Note how the "cost of interest" builds up as the time between planting and harvesting (the rotation) is increased.

## POSSIBLE FINANCIAL OBJECTIVES

The influence of time, and of interest charges, on net revenue is readily apparent. But because of time, a simple maximization of

[2] Or one could argue that, since the funds used for planting, weeding, and meeting the yearly costs *could* have been invested, and would yield some particular rate of return, the interest charge is an opportunity cost.

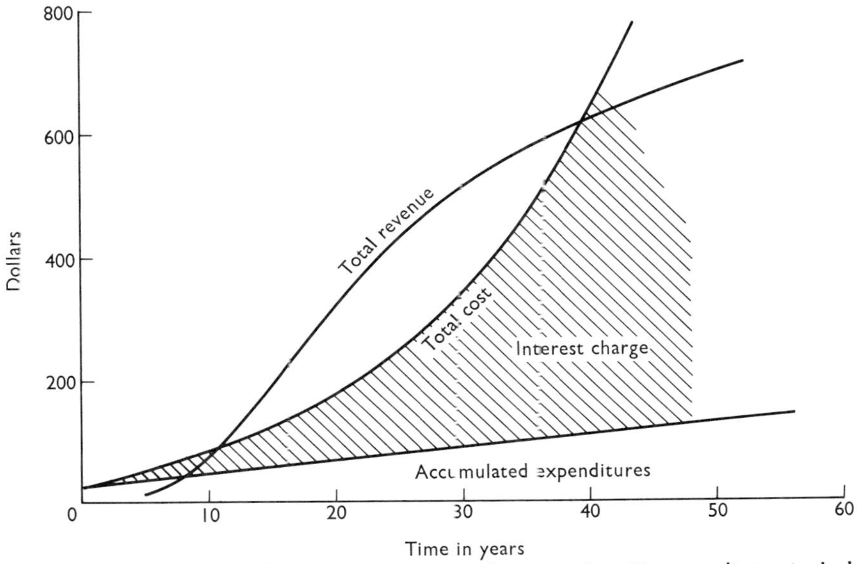


**Fig. 14.3** Costs and revenues per acre for growing *Pinus radiata*, including interest charge of 6 per cent.

this net revenue is not enough, for an income received in the future is worth less than an equal amount received today, and the longer the income is deferred the less valuable it becomes to most individuals living in the present. For example, 26 years after planting the net income indicated for this operation is \$189.52, while the net income expected at 20 years is \$148.31. Which is preferable? If money can be borrowed at 6 per cent, then either rotation is highly "profitable." But which one is best? Which rotation maximizes . . . *maximizes what?*

There are two parts to this latter question, and the more fundamental of the two determines the answer to the second part. Once the quantity to be maximized has been agreed upon there is little difficulty in determining how long the timber investment should be allowed to earn dividends before cashing in and reinvesting—before harvesting and starting a new stand. But what should be maximized? What is the logical financial objective in managing a forest? Is there some quantity that corresponds to the net revenue concept used in earlier chapters? Time is, indeed, a most complicating factor!

Numerous economists have puzzled over this problem but have not always agreed on the answer. [See, for example, Allen (1933), Boulding (1948), and Lutz (1957).] Foresters have often chosen to disregard the financial aspects, preferring to look at the problem from a purely biological viewpoint. While a thoroughgoing discussion of the "what to maximize" question is beyond the scope of this text,[3] it is necessary to review at least some of the suggested solutions if there is to be a possibility of evaluating the impact of time on production decisions.

Purely biological approaches of maximizing total yield or average yield in physical units might conceivably be acceptable in a world where land, capital, and skills were all costless and the effect of time unimportant to forest owners and managers. Once its boundaries had been established, Adam might have chosen to manage the Garden of Eden forest properties under such guiding principles (although events indicate he and Eve were early believers in multiple use). But for many years land, capital, and technological skills have not been free, and most people regard time as of some importance. Purely biological answers to the time input problem are easily calculated but are simply not relevant in a modern economy.

The economic approaches will be discussed under four headings, each identifying the quantity selected for maximization. The first approach constitutes the general model, and here the quantity to be maximized is the *present net worth* of the forest, now often called "discounted cash flow." Included in this discussion is a special case called *financial maturity*. The other three approaches maximize, respectively, the *soil expectation value*, the *internal rate of return*, and the *forest rent*. The last of these three, forest rent, should not be termed an economic approach; it is, in fact, a most *un*economic approach. It is included here primarily because forest rent is an important part of the forest management mystique.

Following a somewhat simplified presentation of the four approaches, major reasons for differences will be identified through a direct comparison. The more important (and more interesting) business of identifying appropriate financial objectives for timber management can then be tackled.

[3] Probably the most thorough discussion of this problem is that of Gaffney (1960). A more available review, and one highly recommended, is that of Bentley and Teeguarden (1965).

## Present Net Worth

Present net worth (PNW or discounted cash flow) calculations have become such a routine operation in business today that formulas to fit almost any need are available. Timber production is no exception; expressions specifically including any (or every) conceivable revenue source and cost item can be constructed. For purposes of this chapter the following will be sufficient:

$$PNW = \frac{\sum_{t=0}^{r} R_t(1+i)^{r-t} - \sum_{t=0}^{r} C_t(1+i)^{r-t}}{(1+i)^r}$$

where $R_t$ is a revenue received at time $t$, $C_t$ is a cost incurred at time $t$, $r$ is the rotation age, and $i$ is the interest rate, with the present being defined as $t = 0$.

To see the changing relationship between cost and revenue over time, however, a graphic model is preferable. For this kind of presentation the calculation of present net worth can be thought of as taking place in two stages, or steps. The first stage consists of calculating the *net worth*, or net value, of whatever is being considered—an individual tree, a stand, a forest, or a whole property. In the example used here this stage involves determining the anticipated net revenue for a plantation under various rotations. The second step consists of discounting this net revenue to ascertain its present value, the present net worth of the plantation.

Figure 14.4(a) is identical to Fig. 14.3 and shows the estimated costs and revenues for *Pinus radiata* under the conditions stated earlier. Since no value for land is included it is the net revenue of the *stand* that is represented by the distance between the cost and revenue curves. This net revenue is shown in (b) of Fig. 14.4 as a separate quantity, permitting the vertical scale to be expanded. Net revenue is seen to be maximized under a 26-year rotation—an input of 26 years.

In (c) of Fig. 14.4 the net revenue that could be produced at each age has been discounted for the indicated number of years, using the selected rate of interest (here, 6 per cent). Given this pattern of net revenue and a discount rate of 6 per cent, the present net worth of this one acre of *Pinus radiata* plantation at time zero (just before planting), excluding land values, would be maximized with a rotation age of 22 years and would have a present net worth of $47.31.

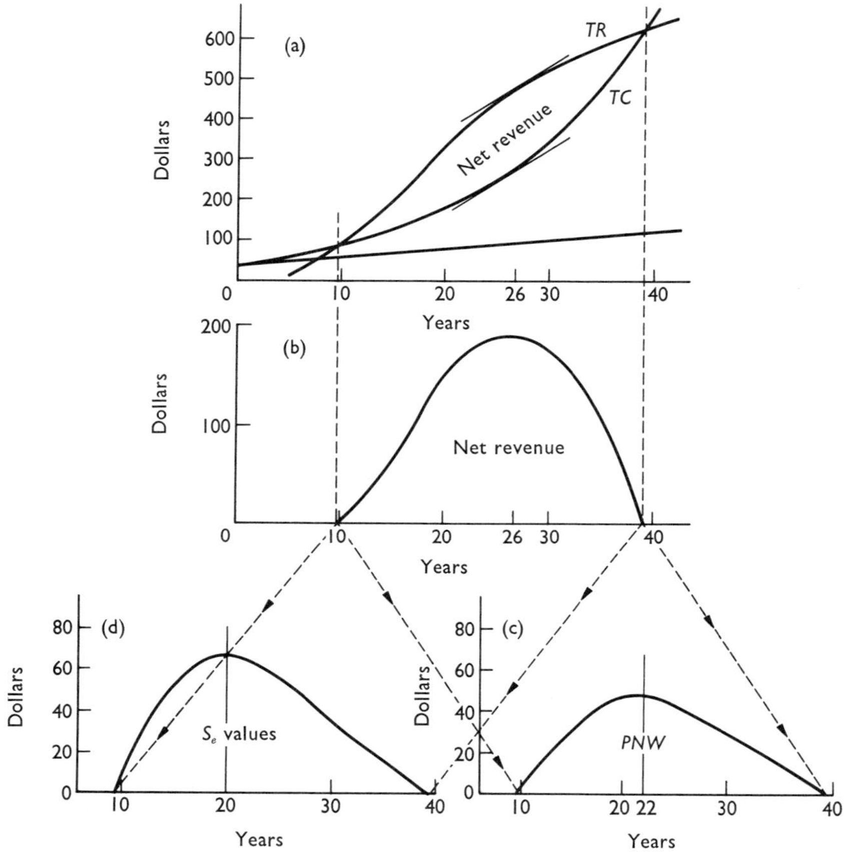


**Fig. 14.4** Derivation of present net worth and the determination of the rotation of maximum *PNW* is shown in (a), (b), and (c). Derivation of soil expectation values and determination of the rotation of maximum $S_e$ value is shown in (a), (b), and (d).

**Inclusion of land value.** But there is no necessity whatever for excluding land values from this model. The effect of land values is seen in Fig. 14.5. To make the illustration specific it has been assumed that land was purchased for $50 per acre at the outset of the venture, and will be sold for the same price when the timber is harvested at the end of the rotation.

Under this simplifying assumption of constant land value the total revenue curve is shifted directly upward by (in this case) $50;

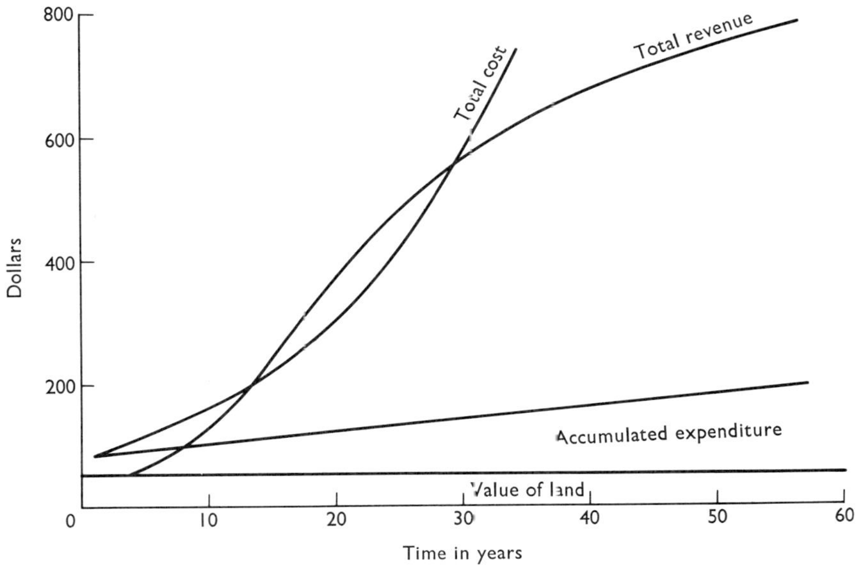


**Fig. 14.5** The effect of including land value.

sale of the land will bring in $50 at any selected rotation. The initial starting point of the total cost curve will also be shifted upward by the $50 land cost. The effect on the total cost curve, however, is not constant, for the base upon which the interest charge is levied has been enlarged. As a consequence, the total cost curve now increases at a more rapid rate than before. The extent of this effect (the increase in the slope of the total cost curve caused by the land value) will depend entirely upon the land value itself; the higher the land value, the more the slope of the total cost curve will be increased under any given level of interest. Clearly, at some value the net revenue would be reduced to zero, and at this point the plantation would just be earning the specified rate of return on all invested capital, including that paid for land.

Nor is it necessary to start with bare land. Values of any existing growing stock can also be entered in the model. The procedure is identical to that followed with land values. The starting point of the revenue curve will be increased by whatever growing stock values exist at time zero, and the shape of the total revenue curve will also change (presumably, the yields would be quite different).

The total cost curve would also change, for whatever growing stock values are added would necessarily increase the interest charge. Establishment costs would probably be less, and in the case of a fully established stand would be zero. Taxes, protection charges, residual land value, etc., would also be changed, but there is no conceptual difficulty involved. As a matter of fact, the present net worth model is easily adopted to the problem of determining the growing stock and cutting cycle of an all aged forest, as will be seen in the following chapter. For the remainder of this chapter, however, we will stick to the somewhat less complicated case, that of bare land.

### PNW in Marginal Terms

Return now to the diagram of Fig. 14.4, the calculation of present net worth *excluding* land values. That this calculation is a direct application of marginal analysis may be seen from Fig. 14.6(a) and (b). In (a), the annual *change* in revenue (marginal revenue) and in cost (marginal cost) is plotted over time. The intersection of the two curves at 26 years corresponds exactly to the maximization of net revenue shown in Fig. 14.4(b). And similarly, when the *marginal discounted revenue* and *marginal discounted cost* curves (calculated from the *discounted* total revenue and total cost data) are plotted over time as in Fig. 14.6(b), the point of maximum present net worth is established by the intersection of the two discounted marginal curves. In the data used here, the maximization point is at 22 years, just as in Fig. 14.4(c).

**Inclusion of land value.** If land values were included there would also be a change in the marginal cost curve of Fig. 14.6. Marginal revenue will (under the assumption of constant land values) remain unchanged; the total revenue curve shifts directly upward but its slope remains unchanged. The marginal cost curve, however, will shift upward (the rate of increase of the total cost curve has been increased) and its slope will be increased (the absolute differences in cost will also be larger). The same statements necessarily apply to the discounted marginal curves of Fig. 14.6(b). The discounted marginal revenue curve would remain unchanged, while the discounted marginal cost curve shifted upward. The net result would be one of shortening the rotation, and the shortening effect will increase as the land value is increased.

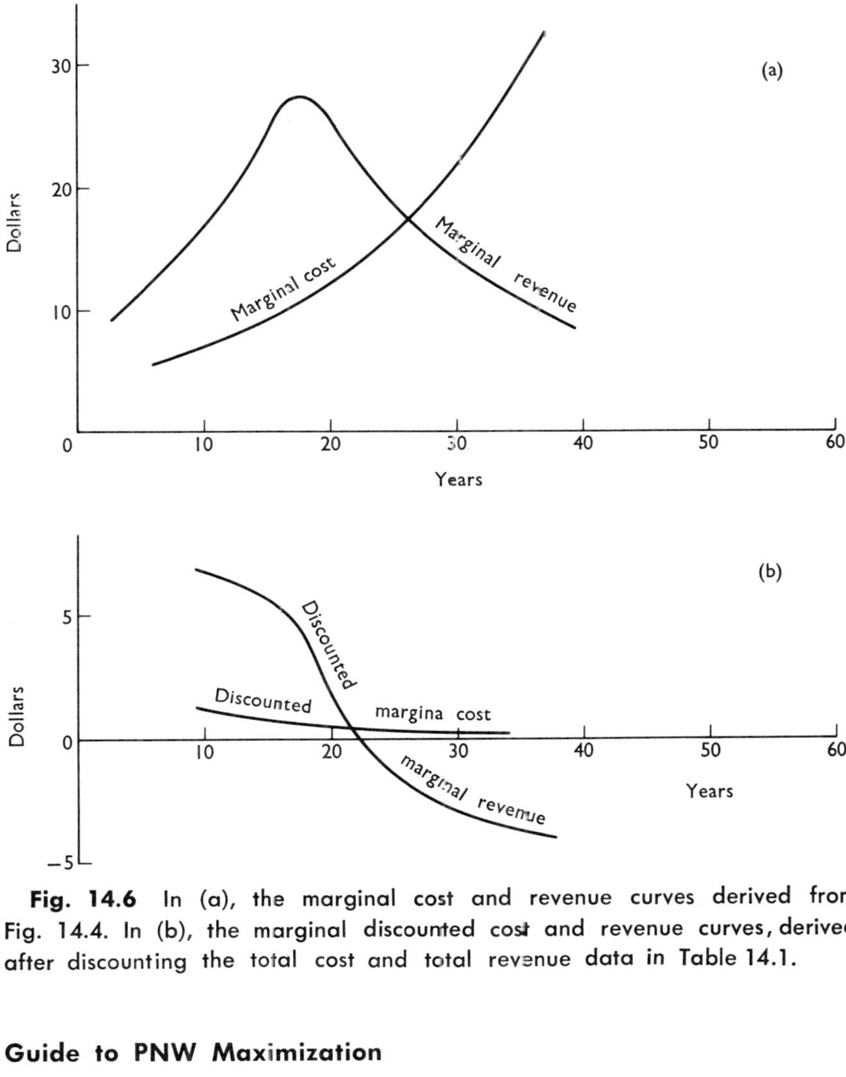


**Fig. 14.6** In (a), the marginal cost and revenue curves derived from Fig. 14.4. In (b), the marginal discounted cost and revenue curves, derived after discounting the total cost and total revenue data in Table 14.1.

## Guide to PNW Maximization

Calculating discounted marginal costs and revenues used to be a tedious job, and still is unless one has ready access to computer facilities. Fortunately there is another guide to maximizing present net worth, easily derived from the same set of curves used to illustrate the reciprocal nature of compounding and discounting in Chapter 9. It will be recalled that two factors influence the slope and position of a curve tracing the growth of capital over time—the interest rate, and the amount to which the rate is applied. In Fig. 14.7 the curves originating at *a* and *b* show the growth at 6

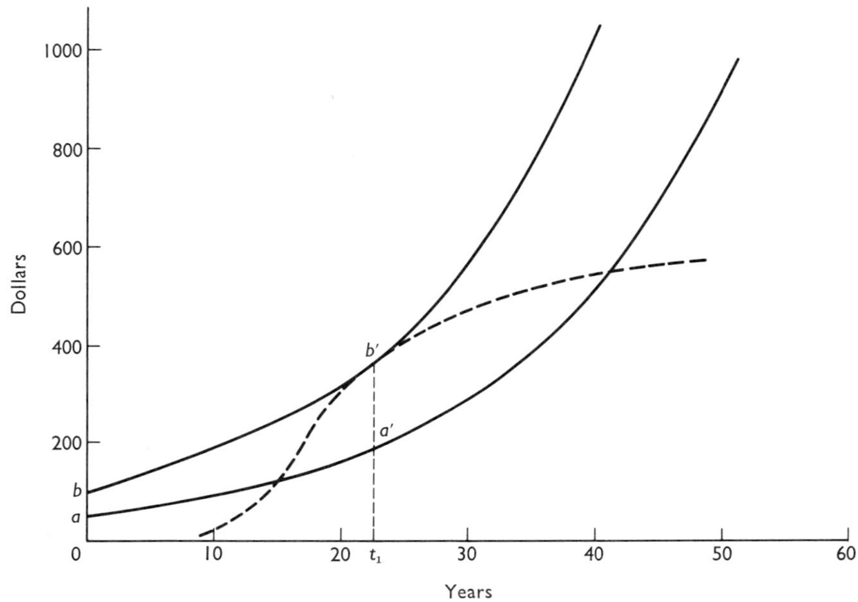


**Fig. 14.7** The guide to maximization of present net worth.

per cent of the two amounts, 0*a* and 0*b*. Because discounting is the reverse of compounding, 0*a* and 0*b* represent, respectively, the present worth of the values indicated by the two curves at any point in time (see Fig. 9.1).

Now suppose the quantity 0*a* were the total cost involved in establishing the forest, and that no other costs were involved. Under this special condition the annual interest charge would be the marginal cost of holding the stand for another year, for *there is no other factor causing any cost change*. Total revenue (the dotted curve) is increasing at exactly the same rate as total cost at point *b′*—where *bb″* is tangent to the total revenue curve—for *aa′* and *bb′* are both increasing at 6 per cent. Clearly, *a′b′* is the net revenue, and its present net worth is *ab*. Note that the *change in cost* does not equal the *change in revenue* at age $t_1$. It is the *rate* of change that is important in this instance. The guide to maximization of present net worth is that *the per cent change in revenue must equal the per cent change in costs*. And under the special condition that all costs are incurred at the beginning of the rotation one can also say that present net worth will be maximized when total revenue is increasing at the preselected rate of interest.

## Financial Maturity

In a general sense the procedures discussed in this chapter are all designed to determine when timber is "financially mature." The term, however, has had a special connotation in forestry since Professor W. A. Duerr and his coworkers introduced the above guide to forest managers under the name "Financial Maturity" (Duerr et al., 1956). One significant change in the guide suggested by Duerr et al was that the rate of return from the stand should be equated to the available alternative rate of return on investment capital (instead of to the per cent change in costs). The stand is then said to be "financially mature."

To calculate the rate of return from a forest stand the expected change in market value during the selected period is divided by the market value at the beginning of the period, then normally reduced to the corresponding one-year rate. Using the Monterey pine data again, at 21 years the rate of return is $\frac{23.32}{348.48} = 6.69\%$, and at 22 years is $\frac{22.00}{371.80} = 5.92\%$. Fig. 14.8 shows the rate of return, based on the same data used throughout this chapter, plotted over time.

**Alternative rate of return.** The alternative rate of return is the opportunity cost of capital to the owner of a timber stand. If the best of the alternatives open to him—bank deposits, United States savings bonds, municipal bonds, house mortgages, common stock, or investment in a different timber growing opportunity—should happen to pay, for example, 6 per cent, and if the alternative is of equal risk, etc., then 6 per cent becomes the opportunity cost of investing in timber growth. When the return from timber production on the property in question drops below 6 per cent of its market value the stand should be harvested.

Actual costs of timber production are not specifically brought into the analysis. In effect the principle of sunk costs is employed: past costs of growing the stand are unimportant because, once made,

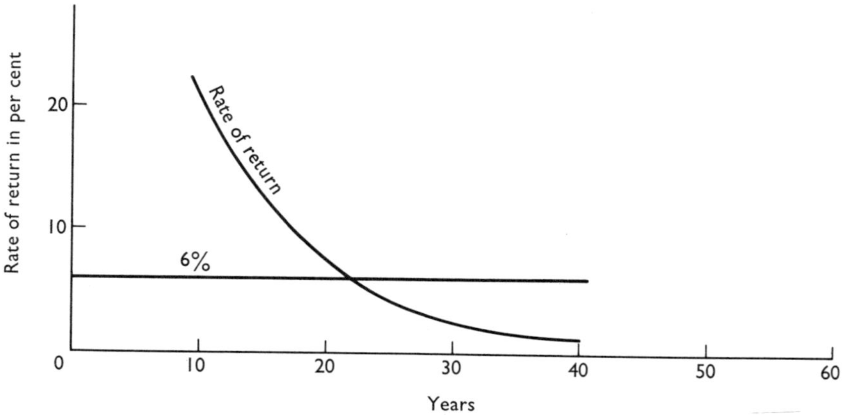


**Fig. 14.8** The rate of return (in per cent) compared with an alternative rate of 6 per cent.

they cannot be changed. As long as the timber stand value is growing at a satisfactory rate—"satisfactory" being defined as the fastest rate at which the capital can grow elsewhere—then it should be left to grow. If the rate of value growth falls below the best available alternative, capital should be taken out of timber and put to the alternative use. But bear in mind that the best alternative may be that of starting another rotation.

## Soil (or Land) Expectation Value[4]

The approach leading to soil expectation values was first presented in 1849 by the German forester, Martin Faustmann, as a solution to the problem of evaluating forest land for tax purposes. Accepted in 1858 by Pressler and Judeich, Professors at the famous Forest Academy at Tharandt, Saxony, these soil expectation values (or Se values) have been a cornerstone of forest management since that time. The procedure has also been a major source of irritation and frustration to practically all succeeding generations of forestry students.

Herr Faustmann must have reasoned along lines somewhat as follows: If I were to start planning a forest from scratch, how much

[4] As Davis points out "land" is a more accurate translation of the German term "boden." Davis therefore uses the term "*land* expectation value" rather than the older "*soil*," which was introduced by Schlich (Davis, 1966).

could I afford to pay for bare land? In calculating the value for agricultural land one first determines the net income the particular field will produce, then determines how much must be invested to obtain an income stream of this same amount. If the going interest rates were 6 per cent, for example, then one should be willing to pay $100 for land that will *net* six dollars or more every year, that is, after all costs including those of management have been deducted from the sale value of the crop. The value of agricultural land is equal to the cash amount that, when placed at interest, would return every year the same net income as the land. The land value is the "capitalized" value of the expected annual net income.

The transition from crop to forest land posed only minor difficulties to Herr Faustmann who would have reasoned:

> Foresters say trees are a crop! OK! How much can I afford to pay for bare land if I intend to use it for *timber* production? To be consistent I'll have to treat it just the same as any other crop, but because of the time interval I'll have to include interest charges. And because my income won't be "incoming" until harvest, I'll have to figure its value today, which means I'll have to *discount* the net revenue. But that isn't all; there's one more step. Since I'll be operating on a sustained yield basis I'll be getting this income every so many years. Therefore, I'll have to figure the cash amount that would return this same net income once every rotation. It's obvious! *The value of bare forest land has to be the capitalized value of the infinitely long series of expected periodic incomes from the timber.*

Figure 14.4(a), (b), and (d) shows the derivation of Se values from the same *Pinus radiata* data used in calculating present net worth. The first stage is exactly the same, Fig. 14.4(a); net revenues must first be calculated for each rotation by subtracting all costs and carrying charges from the total revenue. It is only the second step which differs. Each net revenue is capitalized as if it were the first payment in an infinitely long series of identical periodic payments.[5] The result of this capitilization is shown in Fig. 14.4(d). For each rotation the Se value represents the cash amount

[5] There are many formulas for calculating Se values, varying primarily in the degree to which costs and revenues of different types and timings are explicitly recognized. A basic formula is:

$$S_e = \frac{\sum_{t=0}^{r} R_t(1+i)^{r-t} - \sum_{t=0}^{r} C_t(1+i)^{r-t}}{(1+i)^r - 1}$$

where $R_t$ is a revenue received at time $t$, $C_t$ is a cost incurred at time $t$, $r$ is the rotation age, $i$ is the interest rate.

that would have to be invested today at the given (6 per cent) interest rate in order to yield, at intervals equal to the indicated rotation, periodic incomes just equal to the net revenue expected from the forest. The answer to Faustmann's question of "How much could one afford to pay for bare land?" would be the highest value calculated. Under the conditions here illustrated the *maximum* Se value ($67.19) will be attained with a time input of 20 years.

### Internal Rate of Return

In a timber growing enterprise the internal rate of return (IRR) may be defined as the average rate earned on all costs made prior to the time of timber harvest. If only one cost is incurred, and if this occurs at the beginning of the period, the maximum internal rate of return may be found by the graphic procedure illustrated in Fig. 14.9[6]. Costs and revenues are plotted on a logarithmic (vertical) scale and time on an arithmetic (horizontal) scale, so any quantity increasing at a constant rate will appear as a straight line. Under these conditions the line *AB* identifies the maximum average rate that can be earned on the initial cost, 0*A*, when total revenue is shown by the curve *CBD*. Therefore, 0*M* becomes the investment period that will maximize the internal rate of return.

Because annual costs are an important component of total costs in almost every forestry operation, the graphic approach of Fig. 14.9 (which assumes all costs occur at the outset of the period) can seldom be used. One can, however, readily solve for the investment period yielding the maximum IRR by trial and error.[7] In the *Pinus radiata* example the maximum rate is 10.1 per cent and will

[6] Under the very simplified conditions of no intermediate costs or revenues, if $i$ is internal rate of return and $r$ the rotation,

$$\text{Cost }(1 + i)^r = \text{Revenue}$$

[7] A useful and simple formula for such calculations is:

$$\sum_{t=0}^{r} C_t(1 + i)^{r-t} = \sum_{t=0}^{r} R_t(1 + i)^{r-t}$$

where $C$ = cost; $R$ = revenue; $i$ = internal rate of return; $r$ = rotation; $t$ = time, or year in which cost or revenue is incurred. After a first approximation of the rotation and rate in this formula, successive substitution of different rates and ages, using the usual tables of $(1+i)^n$, permits identification of the two unknowns, $i$ and $r$.

Programs for computing internal rates of return are available at almost any computing center. Some have been specifically designed for forestry problems and are described in various journals. Among these programs are those by Row (1963), Hall (1962), and Schweitzer, Lundgren, and Wambach (1967).

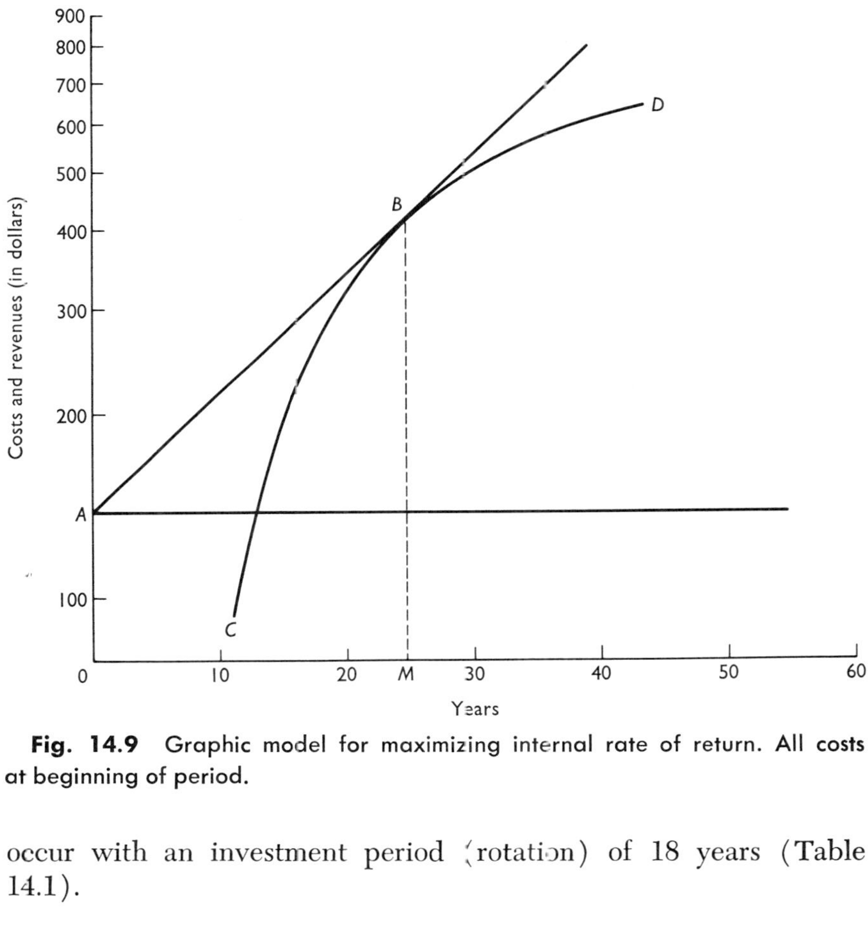


**Fig. 14.9** Graphic model for maximizing internal rate of return. All costs at beginning of period.

occur with an investment period (rotation) of 18 years (Table 14.1).

### Forest Rent

Forest rent is the term given to the *average annual net revenue* when net revenue is defined as total (gross) revenue less total cash outlay (or total revenue minus all costs *except interest charges*). Figure 14.10(a) is simply a repetition of Fig. 14.2, showing the estimated revenue and costs for *Pinus radiata* under the conditions stated earlier. The net revenue, defined as above, is shaded for identification.

Figure 14.10(b) shows the net revenue itself (the shaded area) plotted over the time axis. Dividing this net revenue by the corres-

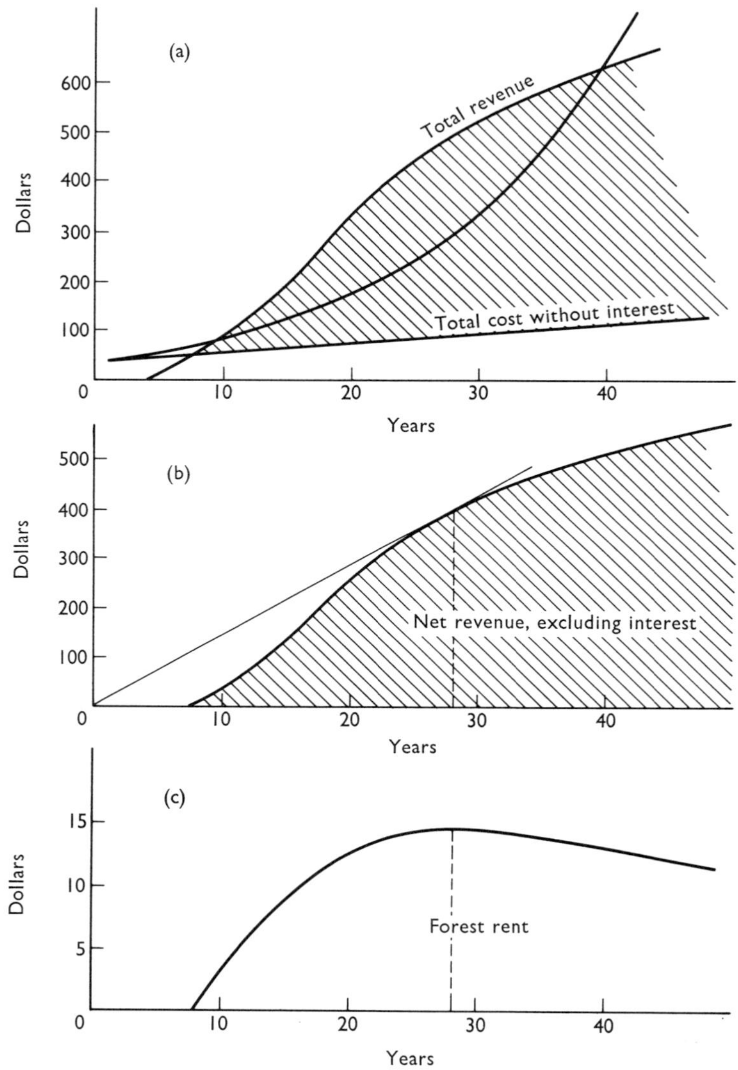


**Fig. 14.10** Derivation of forest rent.

ponding age yields the average annual net revenue or forest rent, as shown in Fig. 14.10(c). In the example used here the maximum forest rent is returned at 28 years, when the average annual net revenue is $14.22.

Viewed geometrically, the average annual net revenue at any age is the tangent of the angle formed by connecting the origin (0) with the net revenue at that same age, since

$$\text{Average net revenue} = \frac{\text{Net revenue}}{\text{Age}}$$

The rotation of maximum forest rent, therefore, may be found by determining the point of tangency between a line drawn from the origin to the net revenue curve, Fig. 14.10 (b). If forest production were entirely costless, and if timber prices were constant, then the rotation of maximum mean annual increment (the culmination of *MAI*) would coincide with that of maximum forest rent. But timber prices tend to rise with tree size (age), and forest production is seldom costless!

## COMPARING THE OBJECTIVES

That these various financial objectives are closely related is intuitively apparent. The purpose of the following discussion is to clarify these relationships.

### PNW and Se Values

Differences between present net worth and Se values can be reconciled in several ways,[8] the most direct probably being to observe that the Se value is simply the present net worth of an infinite series of periodic incomes. As usually calculated, present net worth calculations consider income from only one rotation, while Se values are derived from incomes extending over an infinitely long succession of identical rotations. In our example (see the computations in Table 14.1) the present value of a 20-year rotation is less than that of a 22-year rotation ($46.24 vs. $47.31). But a periodic payment of $148.31 received every 20 years is worth more today at 6 per cent than one of $170.50 received every 22 years, as is indicated by the Se values in Table 14.1.

It sometimes clarifies the relationship between Se and present net worth to calculate the present net worth of a forested property (the timber value plus land value, as presented earlier in Fig. 14.5), using the land value that maximizes Se under the selected interest rate. This is illustrated in Fig. 14.11, where the land value is shown

[8] For a somewhat different and more detailed comparison, see the article by Bentley and Teeguarden (1965).

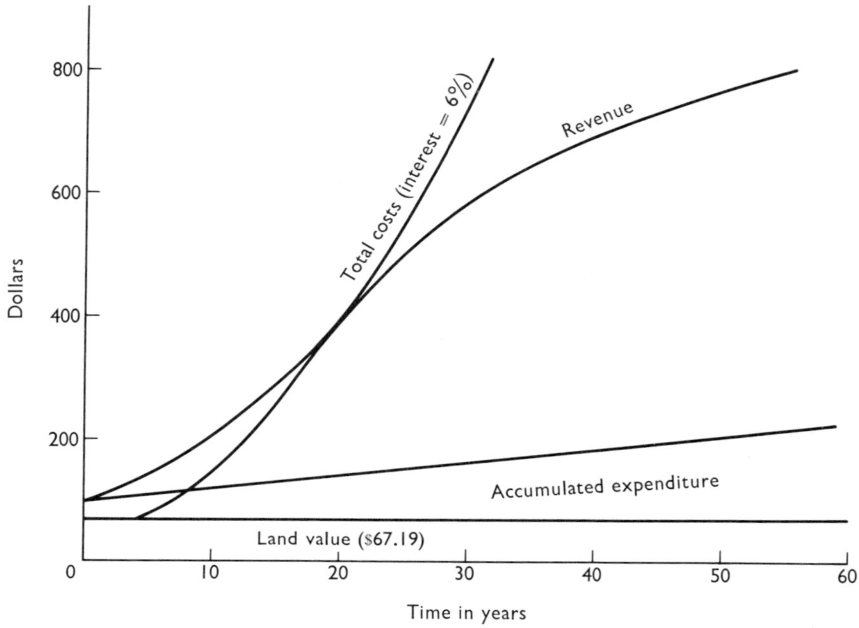


**Fig. 14.11** Total costs and revenue, with land value of $67.19—the soil expectation value.

with other production costs and thereby raises the initial (time = zero) value of both total cost and revenue curves by $67.19. Since the total cost curve now just touches the revenue curve at 20 years, the rotation indicating a maximum present net worth is identical to that maximizing the Se value. Or, put another way, if a 6 per cent return is to be earned on all costs including this land value of $67.19, then the present net worth of the property is exactly zero. The Se value is the highest amount one could possibly attribute to the bare land and still earn the specified rate of return (6 per cent) on all investments for the forest operation—just as Herr Faustmann had calculated!

## Se Value and Financial Maturity

In discussing the financial maturity approach it was observed that costs were not specifically brought into the picture. Omission of two cost items leads to a difference between rotations determined by Se and financial maturity approaches; annual costs, and the value of successive rotations. Include these two and the result will be the same as that of Se calculations. Include the first (the annual costs) and the results will be identical to that obtained by a present net worth evaluation under an assumption of zero land value. Both omissions were recognized by Duerr et al. (1956), but the gain in simplicity was believed worth the theoretical error thereby introduced. Gaffney has observed that the omission can lead to substantial differences in the calculated optimum investment period (though these differences may still be unimportant).[9]

### Se Value and IRR

The internal rate of return rule can also be brought into agreement with the Faustmann approach, and thereby with present net worth and financial maturity calculations, by including all costs and treating the land value as an initial cost item. A glance at Fig. 14.11 shows that when land is accorded its full Se value (here, \$67.19) the maximum internal rate of return on all costs is exactly 6 per cent, and this rate will be returned only on a 20-year rotation.

### Forest Rent

The easiest and most direct comparison between forest rent and present net worth or soil expectation calculations is simply to point out that a forest rent calculation yields the same result as an Se calculation employing a zero rate of interest. This can be seen graphically by comparing Fig. 14.10 with Fig. 14.4; a mathematical proof has also been given (Bentley & Fight, 1966).

Such a demonstration, however, ignores the heart of the difficulty, for forest rent is a "different breed of cats." It has been argued by some foresters that one should view a forest property as a single entity organized for continuous production, and not as a series of independent age class segments. The argument runs as follows: Suppose one were to approach the time problem from the standpoint of a "going concern" in which a fully regulated forest is being operated on a specific rotation. Each year the area harvested, planted, etc., would be $1/r$ (total area). For example, if the rotation

[9] A substantial difference in rotation length as calculated by the Faustmann formula and financial maturity guides does not necessarily indicate a serious deficiency. If discounted net revenue is essentially flat over a considerable time span the maximizing point may shift from one extreme to the other, but profitability (or present net worth) may be virtually unchanged.

were 25 years, then each year 1/25 of the area would be harvested, 1/25 planted, 1/25 given a first weeding, 1/25 given a first thinning, etc. Taxes and protection charges would be paid on the total area every year and would be covered by the income received this same year. The forest firm (so the argument runs) interested in maximizing its net revenue can ignore the time problem and the related interest payments. Each year there will be a group of costs, but also each year brings income from harvesting, thinning, and perhaps from other revenue-producing actions. Because such a firm must look at the entire property it should maximize the average net revenue from the whole forest, not just from the particular acres being cut. This is equivalent to maximizing the average net revenue *per acre* from the forest property, and this, of course, is maximizing forest rent (defined as average net revenue). Dr. Eugen Ostwald, an early and unconvinced student of Judeich and Pressler at the Tharandt Forest Academy, was the early champion of this approach (see Markus, 1967).

Much of the idea's popularity may be traced to its simplicity and to its appeal to elementary reasoning. A forest should be viewed in its entirety and not as a series of isolated tracts that are planted, thinned, and eventually harvested. True rent is the return from the forest—forest rent. Interest need not be calculated or paid, since each *cost* is taken from yearly *receipts*. If a cost is covered by recepts during the same year, why should one calculate any interest payment?

The idea is, admittedly, appealing in its simplicity. Unfortunately, it is bad economics, bad business, and bad forestry. In the modern world, capital should not be invested by any rational individual, corporation, or agency unless the desired return from the investment is anticipated.[10] If one plants timber, every seedling represents an investment and must be expected to earn the desired rate of return. A "going concern" must earn a return on each investment —or it will be gone, not going! A large, fully regulated forest represents a major capital investment, and the largest single cost in operating such a forest will usually be the revenue foregone—the opportunity cost incurred by the decision to maintain growing stock at the specified level. A forest owner who can earn, say, 6 per cent

[10] This is not the same as saying "unless the alternative interest rate is earned." The return from investment in timber may often be in a nonmonetary form, such as control of supply, or control of raw material price.

on capital in alternative investments of less or equal risk than forestry should not be content with any smaller return from capital tied up in growing stock.

Finally, if interest is paid on the discounted value of each age class in a fully regulated forest, the rotation maximizing annual net revenue, or forest rent, will be exactly the same as the rotation determined by maximizing present net worth under the same interest rate. Forest rent would be an important subject to anyone studying the history of thought in foresty. It deserves no serious consideration in courses concerned with forest economics or forest management. It has been treated here solely because of the position this "doctrine" has occupied in the literature of forest management.

## FINANCIAL OBJECTIVES FOR TIMBER PRODUCTION

We can now return to the central theme of this chapter and face the question raised at the outset. What is the logical financial objective in managing a forest? Four different quantities have been identified, defined, discussed and compared.[11] Of these, forest rent was found to be an objective completely inappropriate to modern forestry. Maximization of soil expectation value, of present net worth, or of the internal rate of return still remain possible objectives. And to determine, even tentatively, logical objectives for today's managers requires a second look at the owners of forest land, for the goals of forest management must be consistent with the goals of the forest land owners, and there is no a priori reason for expecting uniformity among either owners or objectives. As previously, it is helpful to recognize at least three broad ownership categories: small private owners, corporate owners, and the public.

### Small Private Owners

**Farm owners.** In some countries such as Germany in the 1850's, other parts of Europe, and even some areas of the United States today, farms have been owned and operated by the same family for generation after generation. The goals of such owners will generally include the objective of handing down to succeeding generations a farm at least as productive as the one inherited. In fact, this objective will often rank very high on the list of goals held by such

[11] Financial maturity was also discussed but is basically a guide to maximizing present net worth.

owners. Thus a 70-year old Swiss farmer may readily undertake the establishment of a plantation of Scotch pine, not because he will ever see the end product of his efforts, but because he is secure in his knowledge that his children's children will benefit, and will undoubtedly recall that the Scotch pine were planted by *grandpère*. The financial counterpart of this objective is, quite clearly, maximization of the *capitalized* land value, or, in forestry terminology, management to maximize Se values.

Generally speaking, the situation in the United States today is very different. Here the family farm has virtually ceased to be a major factor in agricultural production. In most ways life on the farm has already become almost indistinguishable from life in the city, and urbanization is proceeding at an increasing rate. Capital requirements for farming are high. By 1962 the average capital investment per farm worker was substantially higher than for industrial workers ($22,000 vs. $16,000). Land itself is not a limiting factor; it is available at market prices almost everywhere.[12] And farm ownership is not blessed by continuity.

A modern farmer's investment decisions will be made within the framework of his farm enterprise as a whole rather than from the standpoint of any single part. As the owner–operator of a small business he might logically take maximization of the present value of the income stream to be generated by his farm over the next 10, 20, or even 30 years as his guiding financial objective. If timber were the only product of the woodlot, then maximizing its present net worth within this same time horizon would be the logical financial objective. And since the value of his woodlot is its opportunity cost to him as owner of a going farm business, this opportunity cost (often its market value) would be the appropriate starting value for management planning, with the interest rate his alternative rate of return.[13] Probable interest rate levels for farm woodlot investments will be discussed in the following chapter.

**Non-farm owners.** Non-farm owners have a host of possible objectives for woodland ownership, ranging from pure speculation with no inputs of any sort anticipated, through recreational use with or without additional investment, to long term management

[12] This does not mean that a 20-acre woodlot can be expanded to 60 acres by easy purchase. It means that farms of different sizes and characteristics can be purchased in most places.

[13] A short discussion of interest rate levels in forestry will be found in Chapter 15.

for timber production. If timber production is involved, maximization of present net worth would usually be consistent with other objectives of such an owner. For planning purposes land would be valued at its opportunity cost—normally the market value of the forest property.

## Corporate Holdings

**Small or medium-sized corporations.** Little more need be said about holdings of small- or medium-sized corporations. While theoretically corporations may be considered deathless, in practice they seem to operate with rather short time horizons. Their objectives appear similar to those of any small- or medium-sized business concern and subject to much the same set of influences. Maximization of present net worth, with land value established by its opportunity cost to the concern in question, would seem the appropriate financial objective for reasons similar to those already given for the farm enterprise.

**Large corporations.** The large corporation, however, poses a much different and more interesting problem. Galbraith has argued powerfully that the large corporation has brought fundamental changes to economics, even though few economists, he says, are willing to acknowledge the fact (Galbraith, 1967). The following paragraphs outline some of the changes Galbraith believes can be seen, changes that are especially pertinent to a discussion of managerial and corporate objectives.

When a corporation has many stockholders it becomes difficult to rationalize profit maximization as a serious managerial objective—the owners are not even known to the managers. Very large corporations are owned by thousands of stockholders, none of whom can exercise any real voice in management by virtue of his stock ownership. Instead, management is exercized, says Galbraith, by the "technostructure," an inner group of executives, designers, engineers, financial experts, and other highly trained and extremely competent specialists who together have the composite technical knowledge required to make the tremendously complex decisions that are characteristically needed by large corporations. No other group could possibly control the corporation, for there is no other group sufficiently knowledgeable.

Members of this technostructure are not likely to be motivated by hope for pecuniary gain, but instead tend either to adapt the objectives of the corporation to their own goals (adaptation), to accept its goals as their own (identification), or to work out some combination of these two possibilities. The goals of the corporation and of the technostructure thus become almost inseparable.

The first, and most basic, goal of the technostructure—as with almost any organism—is *survival,* and the second objective is *corporate growth.* To meet these goals the corporation must earn a "satisfactory" rate of return. But it cannot price its product to *maximize* profit, for this would usually reduce its rate of growth, which is measured in terms of sales. Growth, however, is substantially within control of the large corporation; it establishes its own pricing policy, and to such firms *capital has ceased to be a limiting factor* of production. The large corporation generates its own investment capital through retained earnings and is virtually free from money markets.

Because of the complexity of modern production the corporation necessarily plans its output and its purchases of raw materials far in advance. This ability to make plans, and the power to enforce them, are the two most significant corporate facts of life today. Very large financial commitments are involved over substantial time periods. The modern large corporation simply cannot risk substantial losses from either a price collapse for its product or from a drastic cost increase in its raw materials. But such risk is unnecessary, for the large corporation is not operating in a purely competitive market structure. To a very large extent it controls the quantity of products produced, the prices at which these products are sold, and the prices at which raw materials are purchased. Through advertising, consumers are induced to conform to the carefully structured plans of the corporation. On the whole, the result is the most efficient production system ever developed. The thoughtful student may question the worth of this efficiency if the cost is that of individualism, but the production efficiency per se cannot be seriously doubted.[14]

Like the competitive models of economic theory which are so often criticized, Galbraith's picture of the large corporation is undoubtedly an oversimplification. But if this fact is kept in mind there is much to be gained from accepting his general approach.

[14] The system may not be efficient if viewed from other than production aspects. It is to this latter point that Galbraith addresses his most critical, and caustic, comments.

The large forest product industries could well serve as prototypes for Galbraith's argument. These corporations experience no problems in obtaining capital for growth,[15] for their investment needs are met by funds generated as a normal part of their regular activity. While one might argue that large lumber, plywood, and paper corporations do not control their product prices, any person would indeed be naïve who maintained these corporations are without influence in this direction, or that corporations have no effect on raw material prices. It does not seem far-fetched to attribute to large wood-using corporations the same overall objectives as those held by other large corporations—maintenance of a satisfactory rate of dividend payment, steady corporate growth (as measured by sales), and the avoidance of any foreseeable risk of product price collapse or raw material price rise. Planning for growth and the reduction of uncertainty should be a major preoccupation of these large corporations, and almost surely is. A new paper machine will return 18 to 25 per cent on its investment cost, but only if pulp of the proper quality is available in adequate quantity at suitable costs. Pulp-making facilities, either long-established or new, will earn at their anticipated rates only if pulpwood is available in the required quantities at sufficiently low prices. To assure these desirable outcomes it is not necessary that the corporation own land sufficient to produce *all* the anticipated raw material, but it would certainly be desirable to control a sufficiently large proportion of the supply to insure that raw material prices cannot rise disastrously. A tremendous additional advantage may be gained through capital gains treatment of all stumpage produced (see Chapter 10), and a large amount of good will and public approval is achieved through corporate holdings of forest land made available for public use. There is also a very high advertisting value in the ownership of forest lands, as a glance through almost any issue of *Time, Life, Fortune, Newsweek,* or *Look* will demonstrate.

Does this rule out the likelihood of a rigorous financial objective for forest management decisions on the part of large corporations?

[15] For various reasons information on sources of investment funds is not always published and consequently no uniform data can be cited. However, five of the largest wood-using corporations in this country (International Paper, Minnesota Mining and Manufacturing, United States Plywood and Champion Paper, Owens-Illinois, and Weyerhaeuser) gave these data in their 1966 and 1967 annual reports. During this two year period these five corporations secured, on the average, over 70 per cent of their funds from internal sources.

Not completely, but it certainly circumscribes the area within which these objectives are operative. Maximizing the present net worth of a forest holding may still be a sound *working* objective, but the forest manager must be prepared to have this objective overridden upon occasion. It is quite clear that protection of future supplies, and to a somewhat lesser extent the building of corporate good will, is more important to the large wood-using corporation than is making a return of, say, 6 per cent on the *forest* invested capital. Such corporations have been making forest investments that return no more than 4 to 6 per cent when calculated in the usual fashion, yet the opportunity cost of their invested capital has been no less than 15 per cent over the decade of the 1960's (see Chapter 15). In this connection it is worth noting that the largest of the forest-land-holding corporations is presently (1970) sharply reducing the period for liquidating its old growth.

In summary, large corporations probably do not manage their forest properties with financial objectives that are strictly profit maximizing. Since neither land nor capital are limiting factors in any absolute sense, one should not anticipate a maximization of returns to either. Hence, neither Se values, internal rates of return, nor present net worth are entirely suitable as financial objectives. The primary goals of corporate security (achieved by protecting raw material supplies and prices) and corporate growth must be met first. The timber-growing operation functions within this framework. Corporate security and growth serve as the constraints within which are set the working objectives of the forest manager. Given these constraints, maximization of the property's present net worth can be a useful guide, and especially if land values are viewed realistically. But such calculations must be regarded as guides, not as rules.

### Public Holdings

Public holdings pose a much different kind of "what to mazimize" problem, and in certain respects the problem is more complex. Most public operations have been given over to the public sector because society believes, or suspects, that the private market mechanism would either fail to produce the wanted result, or that even if the wanted result were to be produced there would be undesired side effects.

Because some three-fourths of this country's commercial forest land *is* in private hands and the bulk of the timber *is* produced by the private sector, it is difficult to argue that timber production per se should be guided by non-market objectives. Recognition that timber production is not the sole, or even, in many cases, the primary purpose of public forest land resolves part of the dilemma, but only part.

Over the past two decades public investment problems have increasingly been analyzed through so-called "benefit/cost" procedures, a subject believed most appropriate to the next section.[16]

Yet there are a few things to be said here about timber production on public lands, for this production requires the investment of capital. Growing stock is capital whether it grows on public or on private land, and the question of how much capital should be invested for how long a time is still relevant. Once benefits have been measured, maximization of *net* benefits over time is the objective by means of which the optimum size of public investment is determined. For timber sold to private operators as a part of their raw material requirements market price may be used to determine product values, and hence benefits. Efficient production then requires that each productive factor should be used so that the last unit used (the marginal unit) will produce just enough product to cover the cost of the productive factor (marginal value product must equal marginal factor cost).

It may be said, then, that insofar as stumpage prices reflect public timber values, any one of the three financial objectives could conceivably be appropriate. If the limiting factor for timber production is indeed believed to be land, then Se values would be appropriate. This might well be the case in India or in other nations where rapidly expanding populations press on the available land, but it would be very hard to maintain such a viewpoint in the United States today.

In developing countries the limiting factor is far more likely to be capital, and in such cases the internal rate of return is often taken as a useful guide. But in a country such as the United States one

[16] Benefit/cost procedures, as such, are also used by private industry but to a far lesser extent. The words *"benefit"* and *"cost"* have long been used in public and private investment problems, but with the publication of the Report of the Inter-Agency River Basin Commission (Subcommittee, 1950) the term "benefit/cost analysis" took on a special meaning.

would certainly question any suggestion that capital is a limiting factor in public forestry.

All things considered, maximization of present net worth is believed to be the most flexible and useful financial guide to public, as it is to private, timber management. This does not preclude prior establishment of even strictly biological goals; one might, for example, accept an administrative decision making high-quality sawtimber the primary product of public forestry. The public might even be willing to accept such a decision, but it should insist that this sawtimber be produced efficiently. The forest manager must decide on the amounts of land, labor, capital, and management he will devote to sawtimber production, but if this is done efficiently his decisions will lead to maximization of present net worth under the constraint that sawtimber rotations will be utilized (that investment periods will be set without reference to economic principles). Land value would be included as a matter of course, taking comparative market value as the guide (which values might be derived from pulpwood rotations). Interest rates used in the decision making process would be those charged for other governmental investments. And there is no reason to restrict the limiting constraints to sawlog production, for in the eyes of some groups other biological, social, or political constraints would be equally valid.

It should be emphasized that the preceding paragraph included a statement to the effect that ". . . one *might* accept . . ." an administrative decision of the kind cited. This is scarcely the same as saying one *should* do so, and an economist would necessarily raise the question of whether land *should* be devoted to long rotations required for large sawlog production if pulpwood, or recreation, or any product requiring a shorter investment period will yield higher rates of return on the necessary investment. If the answer is a definite "Yes" it means that other values are involved. From almost any viewpoint one can imagine it then becomes desirable, almost necessary, to identify these values even if they cannot be accurately measured.

### Big Government and Big Business

It is believed a fitting conclusion to this chapter to point out the growing similarity between big government and big business, for there is considerable evidence that the basic objectives of these two economic sectors are becoming almost indistinguishable. The discussion of private management strongly suggested that the purely financial objective of profit maximization was no longer adequate to explain the actions of large corporations; that their primary objective was corporate survival, achieved through maintenance of satisfactory earnings, growth, and protection of product and raw material prices. Almost everyone agrees that purely financial objectives are not suited to public forest management, but there is less agreement on just what the objectives either are or should be.

It seems time to recognize that with government agencies, too, the most basic goal is survival. Moreover, there is considerable evidence to show that agency growth is another highly important objective, and one that may often take precedence over economic efficiency. And the parallel does not stop at this point, for the means of achieving agency survival show strong similarities to those employed by corporations. A public agency does not measure its earnings entirely in dollars, but in terms of political support. To maximize such returns from any one segment would create strong opposition from others, just as maximization of profits might reduce growth in sales. As a consequence, land management agencies may only strive to keep a wide base of support, making sure there is no violent opposition to their land management activities. More on this score will be ventured in Chapter 18. Here it seems necessary only to point out that a "satisfactory" level of political support must be maintained, and this will often require compromising any objectives of maximization.

The desire for continued growth by government agencies has been a topic of political discussion for many years. There are many ways of securing such growth and it is not difficult to find evidence suggesting most of these ways are well known to public land management agencies. Actions which might increase timber production, for example, but which might shift funds away from the concerned agency tend to be viewed with a very jaundiced eye, while one would expect proposals for silvicultural measures which increase manpower requirements and district budgets to be favorably received—and not necessarily under conditions such that *all* other things are equal.

The third corporate objective, protection of product and raw material prices, is not easily adapted to public forest agencies unless one argues that interagency struggles to control the various activities and products of public forest lands offer a close parallel. But to the author, the more important (and in the long term, far more dangerous) point is the obvious pressure for ever closer cooperation between big industry and big government. When federal agencies control much of the raw material upon which industry depends, the mutually shared objective of survival will assure that the two will begin to work closely together. It is not entirely coincidental that cooperation between federal forest agencies and the wood-using industries, especially in the western states, dates from the early 1950's. Nor is it surprising that this cooperation has the strong approval of virtually all professional foresters, for it works to their own separate *and* mutual advantage.

The economist, however, must occasionally recall that in its earlier days his basic science was called "political economy." Sometime in the not-too-distant future political decisions must be made concerning the ever closer cooperation of big business and big government and the fundamental need to maintain basic democratic controls. The issues involved here are clearly far beyond that of forestry and forest economics, yet the forester must be careful to maintain his perspective as a citizen while acting as a forest manager. The all too obvious effect on the present generation of college students of the military-industrial complex has finally created an awareness of at least some citizens to the inherent dangers of such combinations. That the military-industrial combination is only one aspect of the big-government–big-business problem should be obvious to any thinking individual.

# 15

# Time, Production Decisions, and Interest

A general framework for analyzing the economic significance of time in timber production problems was developed in the preceding chapter. Now, using this framework, a direct approach can be taken to many of the important problems involving time. Not all the "when" questions will be treated, nor will any single example be developed completely, but enough cases will be discussed to show how the analytical machinery works, and to develop a few tentative answers.

The specific problems to be treated are those of growth factors (such as site, spacing, species, etc.), product prices, interest rates, uncertainty, and taxes. In all of these the central difficulty can be analyzed by determining the impact of these factors on the optimum period of investment, optimum always being defined in financial terms. The chapter will conclude with a discussion of the related and important (but often ignored) question of "what interest rate is appropriate to forest investment analysis?"

## TIME AND PRODUCTION DECISIONS

### Impact of Growth Factors

If one is to see how growth factors influence timber production economics he must first determine the extent to which physical production is affected. The revenue and cost curves of the previous chapter provide a convenient starting point, as seen in Fig. 15.1(a), (b), and (c). *TR* and *TC* in (a) are the total revenue and total cost curves, with net revenue (*NR*) shown in (b). Directly below, in (c), is the resulting discounted net revenue (*PNW*, or discounted cash flow) changing in amount as the investment period is altered over time.[1]

Now suppose a new, genetically superior strain becomes available for planting. With price and cost conditions held exactly as before, new total revenue, net revenue, and present net worth curves will result (*TR′*, *NR′*, and *PNW′*). The (hypothetical) new strain illustrated in the figure not only grows faster in its early years but holds its rapid growth rate for some time. Under these particular conditions the investment period should increase. Present net worth is maximized at a longer rotation.

The significant thing to note here, however, is not the degree or the direction of the shift in present net worth but the fact that it *does* shift. A little experimentation with a sketch pad will reveal that either a lengthening or shortening of the optimum investment period can result from introducing a faster growing variety, or, for that matter, a faster growing species.

Using the above procedure leads us, eventually, to the proper result, but it is somewhat cumbersome. The relationship between total cost and total revenue over time is not of itself a very useful guide to investment decisions. Net revenue must first be calculated, then discounted, and the comparison over time made in terms of present net worth. The approach was particularly useful in Chapter 14, when comparison of the present net worth, soil expectation value, and financial maturity concepts was a major objective, but it is not always the best means for comparing investment alternatives.

Early in Chapter 14 it was observed that an alternative calcula-

[1] In this figure, and throughout the remainder of the chapter, maximization of present net worth will be taken as the financial objective. Somewhat different numerical answers would be obtained if Se values or internal rates of return had been used, but the general conclusions would remain unaltered.

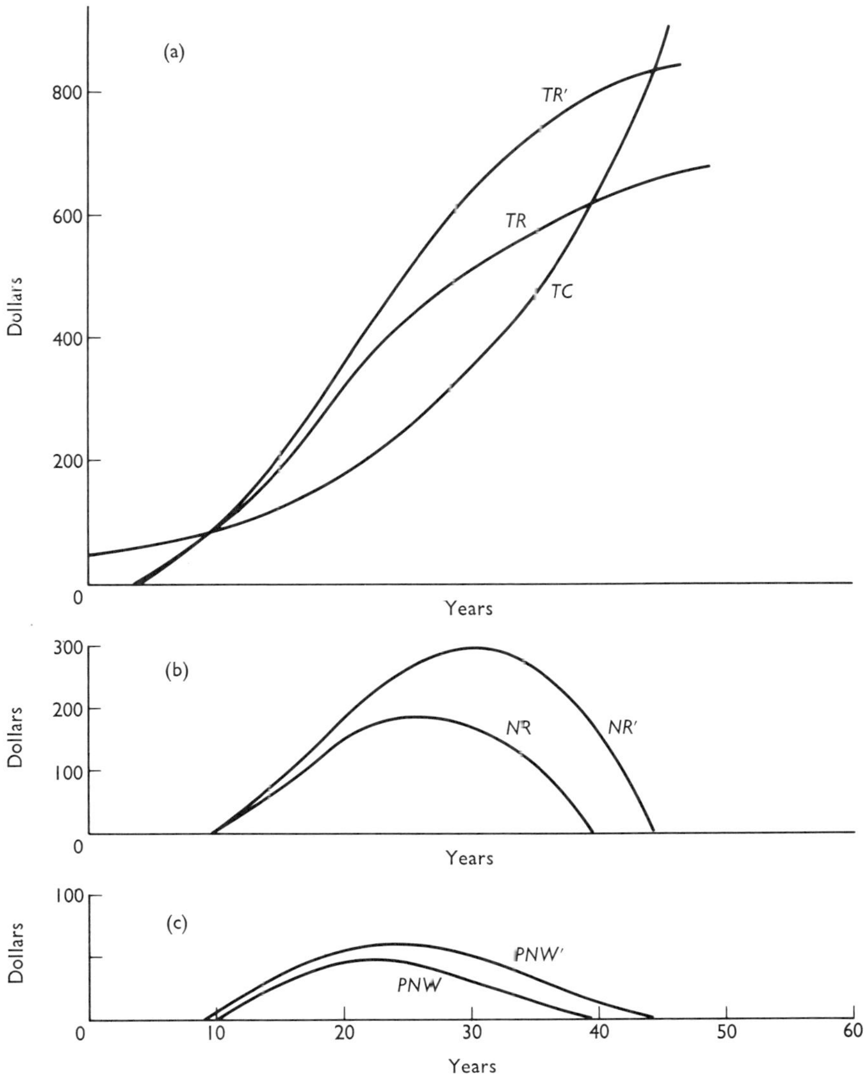


**Fig. 15.1** One possible effect on optimum investment time of substituting a faster growing species or strain.

tion was equally feasible. Total costs and total revenues can be discounted and the point of maximum present net worth determined by subtraction or by determining when the slopes of the two curves are equal—where marginal discounted cost and marginal discounted revenue are equal. The behavior of these discounted cost and revenue curves over time can be quite revealing. From the total costs and revenues of Fig. 15.1 are derived the corresponding discounted cost and discounted revenue curves of Fig. 15.2,

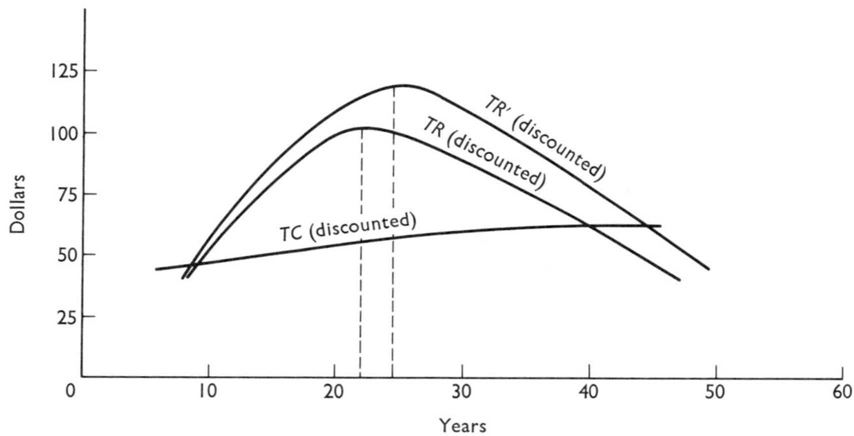


**Fig. 15.2** Discounted total cost and total revenue curves and the point of maximum present net worth.

with, of course, the same point of maximum present net worth. This latter approach—the use of discounted total cost and revenue curves—will frequently be used in the rest of this chapter.

## Effect of Changing Product Prices

An increase in product price has an effect very similar to that of increased growth rates: the total revenue curve is shifted upward. Perhaps the most significant effect of price changes, however, comes from its possible effect on the shape of the total revenue curve. Consider, for example, the fact that with trees the value per unit of volume usually increases as tree diameter increases; trees grow from pulping size to poles, to sawtimber of low quality, to high-quality sawtimber, to veneer logs. With each progression the value per cubic foot tends to increase, sometimes very substantially.[2] The result is a constant upward "stretching" of the total revenue curve, illustrated in Fig. 15.3(a). Under any given cost situation each unit value increase tends to lengthen the desirable investment period, as is seen from the discounted curves of Fig. 15.3(b).

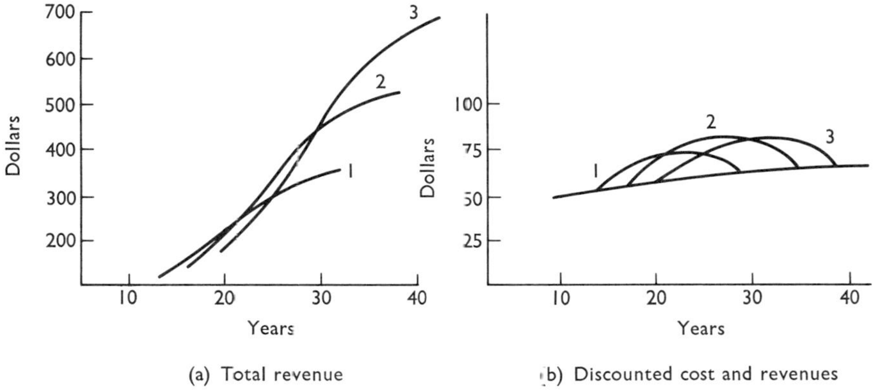


**Fig. 15.3** The effect of increasing value over time. The three total revenue curves of (a) are the result of increasing unit values. The corresponding discounted total revenue curves, together with the discounted total cost, are shown in (b) (6 per cent discount rate).

Values can also decrease as diameter increases. With many western species, for example, lumber yields from 10- to 12-inch trees may be in tight-knotted upper grades, but yields from 18- to 20-inch trees may be in loose-knotted lower grades. As growth continues, clear and select grades may be produced and value per unit of volume can rise substantially.

In either situation (changing products, or shifting grade yields) the price changes could result in two or even more maxima. Conceivably, two of these could be equal and a choice would be required, probably easily made on non-economic grounds. A double maxima situation, usually of unequal values, is most likely under conditions such that the manager must make an "either—or" selection: either grow all pulpwood or all sawlogs, for example, as is indicated by the two solid line curves of Fig. 15.4. In most cases

[2] Logging costs can, and often do, decrease with increasing log diameters. This, however, is primarily a function of equipment design, not of tree growth.

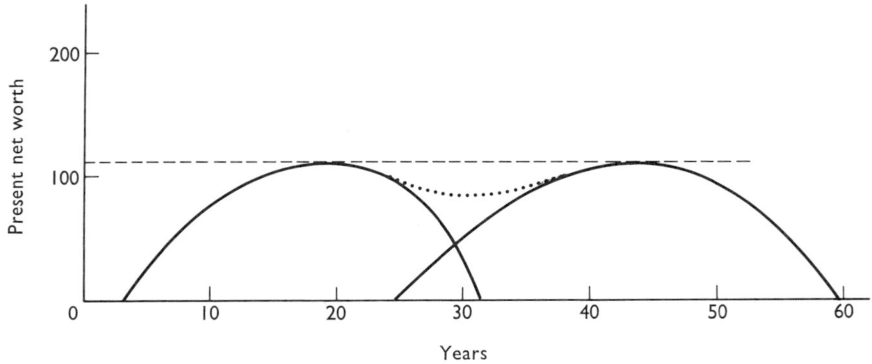


**Fig. 15.4** Changing products, or changing product values, could result in a double maximum.

the choice is really a multiple one in that some of each product can be produced. In such cases the curves are really joined, as indicated by the dotted line in the figure. As shown in Chapter 12, with any product combination the optimum investment period is found just as with a single product, using the costs and revenues of producing the combined products, but a double maximum is always a possibility.

Ideally, the total revenue curve should reflect the pattern of value changes anticipated for the stand. Under some situations a price increasing at some constant percentage may result in a closer approximation to reality than discrete price changes made at particular age-size classes.[3] A constant price increase yields a smooth revenue curve, and such a curve is probably more realistic than one resulting from discrete price changes since not all trees in a stand are identical. The revenue curve is "stretched up" again in exactly the same way that it is affected by an increase in the growth rate,

[3] The technique is particularly useful with computers, but a word of caution is necessary. For example, if at 20 years pulpwood is worth 4¢ per cubic foot (about $3 per cord) and at 60 years sawlog stumpage is priced around $60 per MBM (about 42¢ per cubic foot) then an annual price increase of 6 per cent, starting with 4¢ at year 20, yields a price close to $60 per M after 40 years. If the stand values follow the implied total revenue curve this kind of price adjustment may be very useful, but if the value changes are concentrated in any short period, acceptance of the results could lead to wrong conclusions.

The technique is also useful if, as is often true, a constantly rising price for timber and timber products is indicated for the planning period under consideration.

and if costs are unchanged will always lengthen the investment period, just as an increasing yield. This very fact—that from an economic viewpoint price increases are equivalent in effect to yield increases—sometimes provides a needed perspective on biological and economic forces. While direct comparisons are not strictly valid because costs can also be affected, a 4 per cent *annual* price increase has an effect not unlike a 4 per cent increase in *annual* growth, and a 20 per cent decrease in price may have an even greater financial effect than a 20 per cent loss of the year's cut through fire.

The effect of site differences on length of investment period is very similar to that produced by a percentage increase in price. Regardless of how site is measured, yield curves by site classes typically resemble those of Fig. 15.5(a), and if product price is constant the same curves can represent total revenue. Then, as long as the yield from higher sites is increasing faster than that from lower sites (as is usually the case, at least as indicated by harmonized yield curves) and providing costs do not vary with site, one can state that improved site usually means longer investment periods, even though mean annual increment often occurs earlier on high than on low sites. The discounted revenue curves, derived from the yield curves of Fig. 15.5(a), are shown in Fig. 15.5(b) with a

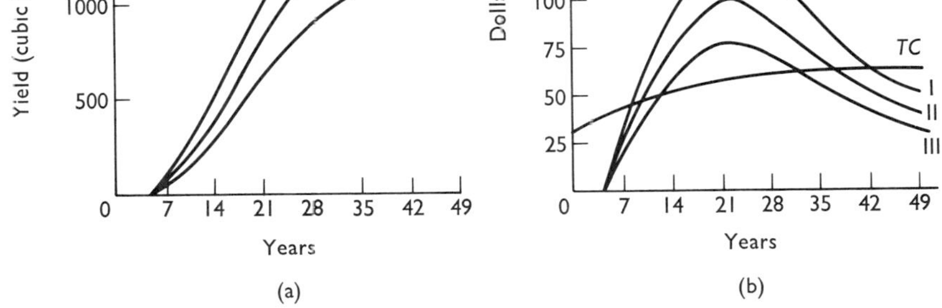


**Fig. 15.5** Yield by site classes is shown in (a), and the corresponding discounted revenue and costs by site classes is shown in (b). (Monterey pine)

single discounted cost curve. The resemblance to the increasing price situation and the tendency toward longer rotations is confirmed, although the discounting process reduces the latter.

But there is a difference between the two cases. Improved sites bring an opportunity for increased silvicultural activity which, if carried out, means higher costs.[4] And with increased costs one can no longer be sure investment periods should lengthen on better sites, for (other things being equal) increasing costs lead to *shorter* investment periods. Only after analyzing the specific relationship between costs and revenues can this particular question be answered.

## Silvicultural Inputs

Many silvicultural measures may be recommended to the timber manager, and each measure can affect either the amount or the timing of revenues and costs. Viewed as a separate operation, a direct marginal approach is the easiest means of evaluating any particular silvicultural treatment; does the change in discounted revenue produced by the treatment more than offset the change in discounted cost? In *planning* a total forest operation, however, each anticipated silvical measure should be considered part of the normal production process. It becomes, in fact, simply another element of the production function. And, since any such operation can affect either costs or revenues, and is likely to affect both, the length of the investment period may also be affected.

Tracing through the effect of a silvicultural input on cost and revenue relations over time may reveal more than just the effect on investment period. Thinning is perhaps the most common type of silvical prescription and will serve as an example.[5] In Fig. 15.6 the dashed line, *TR*, shows the yield and revenue from a plantation under assumptions that no action will be taken until harvest (revenue scale in dollars on left, cubic foot volume on right). Costs accumulate in the expected fashion (the dotted line, *TC*) and net revenue is readily determined as the area between the two curves.

Now suppose the manager, noting the dropoff in growth at about

[4] Sometimes this becomes a necessity; hardwood control, for example, may be needed on good sites but not on poor ones.

[5] The starting curves of yield, cost, and revenue are those of the *Pinus radiata* example used in the preceding chapter.

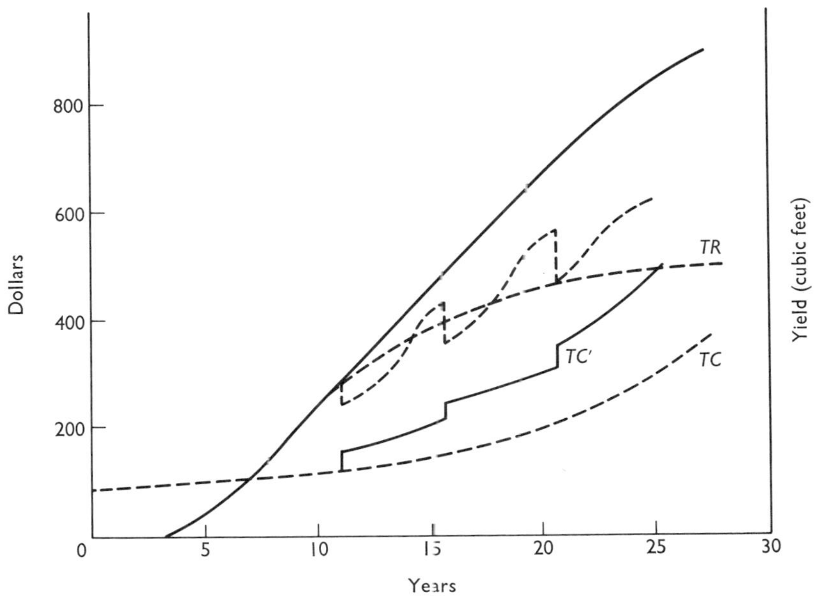


**Fig. 15.6** Effect of thinning on cost and revenue functions.

20 years, reasons that by making the proper type of thinning he can prolong the high annual growth—in fact, hold it at the same level for two or even three 8-year periods. To make the illustration more interesting and instructive we will also assume that returns from the first thinning will just offset costs, while the next two can be made at a profit. Each thinning costs 40 dollars, and the revenues from the three operations are expected to be \$40, \$80, and \$100 respectively.

The effect on costs is one of imposing a series of upward steps on the total cost curve, as seen in the figure ($TC'$). The revenue side is more complex. The dotted line shows the value of the growing stock—after every thinning it drops down by the value of the timber removed—but the *revenue* curve continues unbroken. In fact, it may actually increase its slope, for now the revenue obtained from the thinning operation is also "growing."

The effect on net revenue is seen to be one of spreading out or lengthening the time period of positive returns. Yet because of the discounting process the lengthening effect on optimum investment period is relatively small, as is shown in Fig. 15.7, where the discounted cost and revenue curves under the thinning program are given. The princpal effect of this kind of thinning measure is one of making the determination of rotation length far less critical.

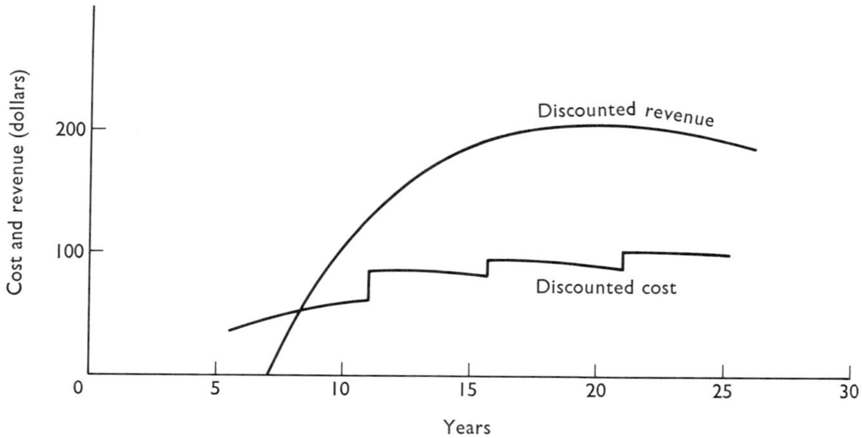


**Fig. 15.7** The effect of thinning on financial aspects of the operation is to spread out the period of positive return. Rotation determination is even less critical.

One can, to be sure, imagine cultural operations that would have a more pronounced effect, and in some instances the influences on present value can be quite dramatic. Thinning in stands that stagnate easily is one such case, and there are others. More often, however, the effect of thinning is the one noted above; the choice of investment period is made less critical. That this is no small gain will become clearer when we look again at uncertainty.

## Changing Interest Rates

The effect of increasing interest rates is best seen from the curves of discounted cost and revenue. Fig. 15.8 gives two sets of curves, derived by using interest rates of 3 per cent and 6 per cent. Total costs rise faster with increasing interest rates, and hence call for shorter investment periods. At the same time the higher rates also move the discounted revenue curves to the left. The end result is the one familiar to all students of forest management: high interest rates mean short rotations.

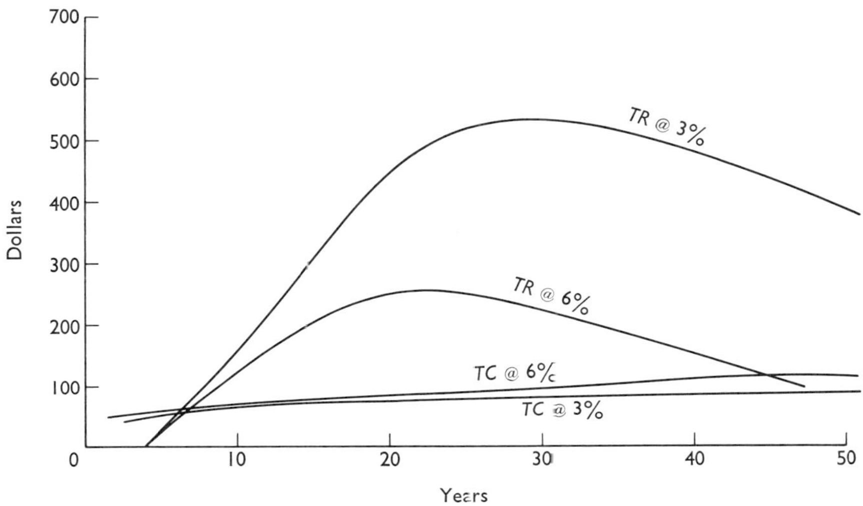


**Fig. 15.8** The effect of increasing interest rates. Identical revenue and cost data, but discounted at the two indicated rates.

## Time and Taxes

The "time bias" of the property tax has already been pointed out in Chapter 10, but with the framework available now the effect can be much more readily appreciated. Other taxes, too, may influence production decisions differently as the time period of the analysis is varied, and changing tax rates or kinds can directly affect the investment period.

The *property tax*, under almost any kind of application, will act to *shorten* investment periods as compared with no property tax, and the effect will be more pronounced as the tax rate is increased. The property tax requires annual payment in almost all instances. Under any positive interest rate this means that the total cost curve will rise more rapidly with an increase in the tax level. Fig. 15.9 illustrates this phenomenon, again by using discounted cost and revenue functions. Clearly, as the tax rate goes up the cost curves rise, even though the tax base per se is held constant. The resulting shift of the discounted cost curve tends always toward shorter rotations. If the property tax were to be levied on the actual value of the property (on the market value as indicated by the total revenue curve) the costs would rise at an even faster rate and the leftward

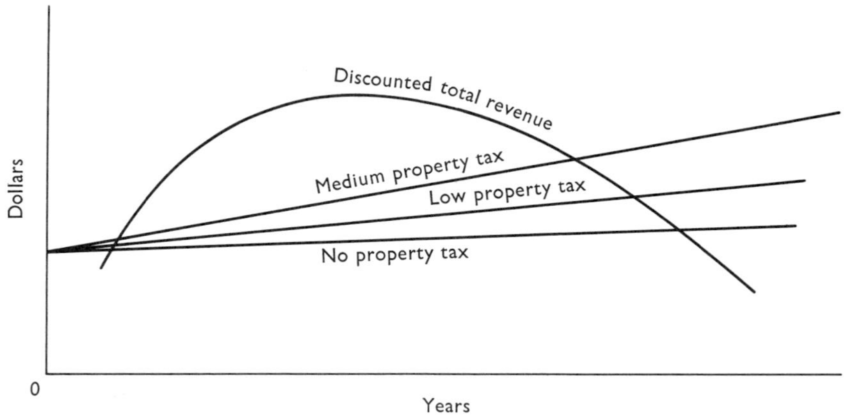


**Fig. 15.9** **The property tax acts to reduce the optimum time input (the period of investment).**

shift would become even more pronounced. The curve of discounted revenue is unaffected, and since the tax effect is always to increase the slope of the costs, rotations will always be shortened by a property tax increase. The same result will attend any increase in annual costs.

A *yield tax* usually reduces total revenue by some constant percentage (providing the rate does not differ by products). Since the slope of the revenue curve is thereby reduced while costs are unchanged, yield taxes, like property taxes, tend to shorten the optimum investment period if viewed in isolation. If such taxes are substitutes for the property tax, however, the effect on rotation length will usually be negligible; removal of the property tax would tend to lengthen rotations by lowering costs, this lengthening being offset to some degree by the imposition of the yield tax.

*Severance taxes* are often assessed at a flat rate per unit of volume produced (per ton of coal, per barrel of oil, per thousand board feet of timber, etc.). If so assessed the rate may in reality vary by product. With sawlogs, for example, a tax of one dollar per thousand board feet of logs harvested results in a higher rate per dollar of revenue from low-quality logs (low value per thousand) than from high-quality logs. The overall effect would still be to shorten the optimum investment period, but to shorten it more for low quality timber than for that of high quality.

*Capital gains taxes* (the most likely form of income tax on stumpage) are levied on *net* revenue increases. Under most situations the effect on investment period would be slight, although if net revenue increased gradually over several years the apparent effect on rotation could be substantial. In such cases, however, the difference in present value would be small and, hence, the determination of optimum investment periods (within this time span) would become a relatively unimportant decision.[6] The more important effect of an income tax probably arises because of uncertainty as to its probable future level, a matter to be discussed under the general aspects of uncertainty.

The effect of death and inheritance taxes depends upon property values (given the tax rate), and since forest values normally increase with age, the tendency would appear to be for death or inheritance taxes to favor shorter investment periods; the larger the growing stock at the death of the owner, the greater the tax. Here, however, the peculiarities of the particular tax laws must be considered. In this country the actual death or inheritance tax would have this effect, but it must be observed that capital gains are not taxed upon the death of the owner (see Chapter 10). The effect is even more pronounced in the United Kingdom, where standing timber, unlike other real property, is not subject to the inheritance tax. Hence owners of large estates have come to regard their forests as a means of holding wealth in a non-taxable form, yet keeping it available if it should be needed to pay the tax on other parts of the estate. Under such conditions the effect on rotation length can be very large. Conceivably the growing stock level might be determined by the level of tax payment anticipated at the owner's death.[7]

## Effect of Uncertainty

While the interrelations of time and changing interest rates, prices, growth rates, taxes, etc., on production decisions can be demonstrated by examining their effect on the optimum investment period, the most important effect of time itself on such decisions undoubt-

[6] In a recent study the effect of four alternative forms of capital gains taxation on optimum investment period was analyzed (Shirley, 1969). The results indicated that the effect on choice of investment period is very slight.

[7] More precisely, the anticipated tax would set either an upper or lower limit on growing stock levels depending on the growth characteristics of the stand, the attitude of the owner, his age, and his "trade-off" between wanting income today for his personal expenditure and wanting to preserve the estate for his descendants.

edly comes through the uncertainty that is almost invariably associated with future happenings. The accuracy with which optimum rotations may be determined is strongly dependent upon the accuracy with which future costs and prices and yields can be forecast. Here, as always, time acts as a veil, a mist that increasingly impedes our view as we try to examine more distant events.

The result of this veil is partially demonstrated in Fig. 15.10. Given a yield curve similar to that used earlier (the solid line of the figure),

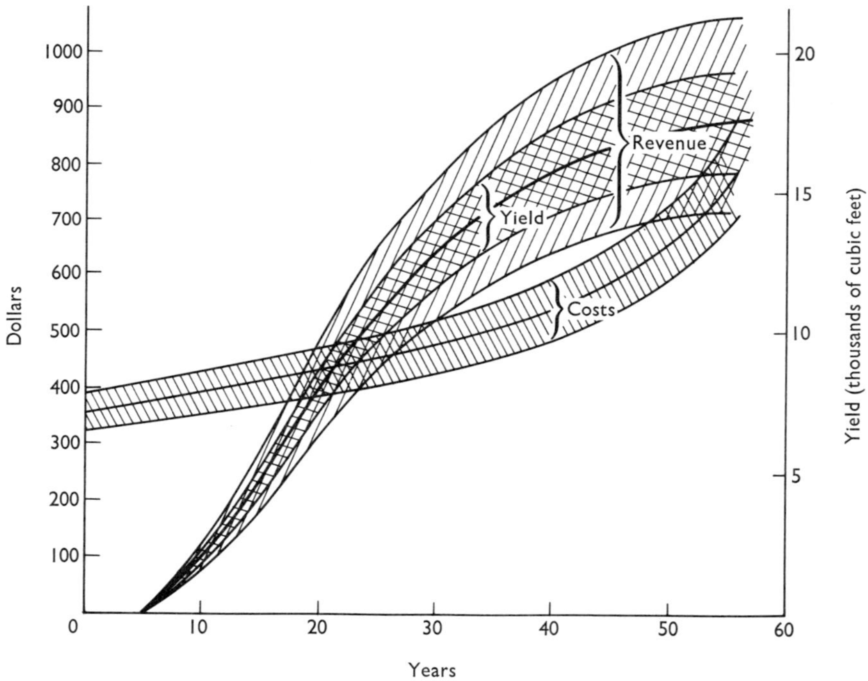


**Fig. 15.10** The effect of uncertainty.

suppose one were to grant that volumes could be predicted with a $\pm 10$ per cent accuracy, an accuracy, incidentally, which few silviculturists would be willing to guarantee. This would result in the "band" of possible yields shown in the figure.

This band of possible volumes must be transformed to expected revenues from the anticipated products, which in turn requires the estimation of future prices. A study of past stumpage price behavior, even after deflating, is not at all reasuring, but for illustrative purposes one might asume that price also can be predicted within a $\pm 10$ per cent range. The spread in total revenue that results from the uncertainty attached to both yields and product price is indicated in Fig. 15.10 by the still wider shaded band.

Costs would probably show at least equal variation, although judging from past behavior it would seem that labor or tax costs are less likely to go down than to rise. Technological changes, however, might well serve to reduce costs. On balance, then, perhaps a $\pm 10$ per cent in cost expectations would seem reasonable, or even conservative, over the 50 year period covered by the diagram. The resulting range in costs is also shown as a band in the figure. And in passing it might be noted that this cost range is predicated on use of a single interest rate, which might in reality also vary.

All kinds of possible cost/revenue combinations are now seen to be possible. Under the circumstances illustrated in Fig. 15.10, net revenues (before discounting) range from a low of \$50 to a possible high of around \$470, while the range of possible "break-even" points in rotation length is from 17 years to 30 years on the short side, and from 48 years (the minimum) to well over 60 years on the long side.

Into this already clouded picture the uncertainty of future income tax levels should still be introduced. Here it must be recalled that income taxes are levied on *net* income, so as far as length of investment period is concerned this particular uncertainty can only serve to cut down on the extremes. The greatest effect of uncertain income tax levels, however, is not apparent from a diagram such as this. If tax levels are expected to rise in the future, one can partially avoid the consequences by cutting today. The reverse of this—shifting harvest cuts to future years to take advantage of anticipated lower taxes—is also possible but considerably less attractive, for future gains are worth less than present ones. Besides, income is needed today even though taxes are high, and some people might even prefer to have incomes increased during high tax periods so they would be able to maintain accustomed high-expenditure patterns.

After these quite conservative allowances for uncertainty, a somewhat different shade of importance is taken on by problems of optimum rotation determination. The rotation maximizing present net worth of the property (not illustrated here) would apparently lie somewhere between 25 and 40 years, but there is no a priori reason for choosing any particular rotation within these limits as being "best."

Before leaving this topic it should be pointed out that up to this time it has been assumed the degree of uncertainty did not vary over time. Yet if time really acts as a "veil," then even the degree of uncertainty should be increased as the event shifts further into the future; if visibility is limited to one mile, an object two miles away is 100 per cent unseen. Under this kind of assumption, therefore, the width of the uncertainty band is not a constant percentage but an increasing one, and optimum investment periods would normally be shortened. On the other hand, the uncertainty attached to some events might be entirely independent of time, and in such cases the costs and revenues related to that particular event are the only ones involved.

## INTEREST RATES FOR TIMBER INVESTMENTS

A glance at Fig. 15.8, or for that matter, at almost any of the diagrams used to illustrate time effects in the previous chapter, should suffice to demonstrate that the level of interest rates *could* play an important role in most timber production problems where time was a factor. General principles underlying market rates of interest were discussed in Chapter 9; now a closer look is needed to see if there is some way of determining the rate that might be applicable to particular problems and owners.

Asked to suggest an interest rate for forestry enterprises, most businessmen would probably recommend taking the "market rate." But *which* market rate? At any given time there is no one market rate, but rather a whole host of them. And even if one were to specify the borrower, his purpose, the size of the loan, its duration, and the method of repayment, the interest rate to be charged would vary with when and where the loan was to be made, for the supply and demand of investment funds changes with time and place. True, there is nothing in the time model we have used that would prevent us from using different interest rates—one for planting, a different one for annual charges, or even one that changes from year to year. But *what* rates should be used?

It would seem that past forestry textbook practice has been to select a rate that is (1) sufficiently low to assure demonstrating that "good" forestry will pay, and (2) sufficiently high not to be rejected immediately by students primarily familiar with interest rates paid on savings accounts. But is 2, 4, or 6 per cent a realistic rate to be charged on forest investments? Or, much more important, is 2, 4, or 6 per cent a sensible rate to use when discounting financial yields from these investments?

## Effect on Investment Period

The theoretical question of "how do rotations change under the impact of interest rates" can be answered by glancing at any of several diagrams presented earlier. Any increase in interest rates clearly tends to shorten rotations, either by increasing the slope of the cost curve or by decreasing the slope of the revenue curve, or both. Now we need to look at a more fundamental question. Interest rates really provide a basis for deciding between different forest investments, and between forest investments and those possibilities in other (non-forest) areas. What rates are appropriate for timber production?

In discussing this question it is necessary to recognize that ownership necessarily plays an important role. Public and private investment opportunities should certainly be separated, and at least the small farm woodland owner from the large corporation. In the discussion which follows, keep in mind that these are only the extremes in ownership classes, isolated for the purpose of illustrating the range of applicable interest rates and the consequences of their use. Not all farm woodland owners, or all corporations, or even all public agencies, behave alike. The answers are frequently intertwined with the market structure of the organization or agency in question.

## The Small Woodlot Owner

A direct approach to interest rate levels applicable to particular borrowers is that of determining the rate regularly paid by that group. Loans are sought by farmers for a variety of purposes, one of the most frequent being to finance new equipment. Farm equipment loans, like those made for automobiles, are often handled directly by the manufacturer through a subsidiary corporation. Very commonly a 6 per cent loan covering two-thirds of the purchase price is made, with the interest being charged *on the total amount borrowed throughout the entire loan period.* Repayment is by equal monthly installments extending over two or three years. The net effect of paying on the total amount borrowed over the entire period is to double the nominal interest rate. A borrower under this sort of system, if paying the nominal 6 per cent is actually paying the equivalent of 12 per cent on a mortgage-type loan. Bank loans can, of course, be arranged, with rates depending on the individual, his credit rating, and on the collateral put up for the loan (most often the equipment itself). Bank rates for equipment loans may be as low as 7 per cent (1968) but are usually higher.

The alternative rate of return to a farmer buying equipment under these conditions should certainly be no less than the borrowing rate, and one would normally expect it to be considerably higher. One could scarcely expect a rational individual to borrow funds at 12 per cent if he held capital earning only 5 or 6 per cent. A farmer should be expected to charge a rather high rate of interest on any cash outlay made, including those made for timber production purposes.

What about the discount rate? Here individual time preference plays an important role. Time preference rates are generally correlated inversely with incomes—low-income groups have high time preferences. There are factors associated with agriculture tending to offset this generalization, yet by and large it would seem that farmers have a high time preference (and also probably a wide range of rates).

For any steps requiring cash expenditures, then, or for discount rates applicable to future timber values, one should expect farm owners to use relatively high interest rates, from perhaps 8 to at least 15 per cent. Even these high rates would not necessarily rule out timber production. Both the rate and the amount of the cost are important in time problems, and a considerable part of the necessary investment in farm woodlot operations is in terms of labor and non-specialized farm equipment. It is the opportunity cost of labor and equipment that must be calculated, and in many instances this cost will be low.

Because of the high discount rate one would expect to find few farm owners interested in long term forestry endeavors. Woodlots presently understocked or having a high proportion of low grade timber species (the typical farm woodlot) are unlikely to be greatly improved. It would be unusual to find a farmer consciously planning to grow large sawtimber unless his stand had already been established. On the other hand, suggestions to establish plantations of fast-growing pulpwood that would interfere little with other uses of farm labor and equipment might find ready acceptance by any moderately ambitious farm operator, and especially so when such activities are subsidized through free planting stock, soil conservation payments, etc.

### The Large Corporation

As with any investor, alternative rates of return are significant to the large corporation. Because the capital requirements of large corporations are met to a considerable extent from retained earnings, the guiding alternatives are those available within the corporation itself. The desire for corporate growth takes second place only to the desire for long range security, and most large corporations today are pushing expansion plans on several fronts simultaneously. In a few cases their alternatives lie within the range offered by a single product type—paper, or chemicals, or automobiles—but most of the truly large corporations produce an amazing array of products. This, together with the long-continued expansion of the United States economy, has resulted in average rates of return to industry over the past decade of at least 10 to 12 per cent, *after taxes*. The significance of the range of investment opportunities to the interest rate appropriate to large timber-based industries is direct. If, for example, a new plywood plant or particle board plant can earn, say, 20 per cent on the required investment, or a new paper machine could earn a similar rate of return, investments yielding lesser rates will be taken only if other reasons account for the difference.

Timber-based industries are not an exception to the general run of United States corporations. In 1966 the average rate of return on invested capital of the 15 largest forest products firms in the United States (all those in the 200 largest United States industrial corporations for that year) was 11.13 per cent.[8] Four of these firms[9] were heavily involved in producing products other than those derived from wood, whereas the remaining 11 [10] were primarily

[8] From *The Fortune Directory 1967: The 500 Largest U.S. Industrial Corporations.* The percentages quoted above are taken from the net income after taxes, expressed as a percentage of the invested capital. Invested capital is the net worth—the sum of corporate stock, surplus, and retained earnings at year's end.

[9] American Can, Continental Can, Olin Mathieson Chemical, and Owens-Illinois.

[10] Boise-Cascade, Container Corporation of America, Crown-Zellerbach, Georgia-Pacific, International Paper, Kimberly-Clark, Mead, Saint Regis, Scott Paper, United States Plywood and Champion Paper, and Weyerhaeuser.

forest-product industries. The average rate of return on invested capital for these last 11 was 10.87 per cent. An indication of the importance of these large corporations to forestry is given by the fact that together they own, lease, or control cutting rights to at least 35 million acres of forest land in the United States and Canada.[11]

From the above it would seem that alternative rates of return, after taxes, for large timber producing corporations in the United States must be no less than 10 per cent, leading to a conservative estimate of returns on invested capital of at least 15 per cent *before* taxes. If this strikes the reader as being too high he should contrast these rates with those expected from investment in developing countries. Here the *minimum* anticipated return must normally be well over 20 per cent if the investment is to be given serious consideration.

The number of opportunities for timber growing in the United States that will return 15 per cent on invested capital is, at best, somewhat limited under current cost and price conditions, especially if returns are calculated according to usual textbook examples. A 15 per cent interest charge on costs would make most plantations decidedly unprofitable, yet United States corporations have made large investments in timber lands, in plantations, in various forest management activities, and are heavily committed to major timber investment programs. The obvious discrepancy between the nominal rate of return from forest plantations and an alternative rate of return on capital of 15 per cent needs some explanation.

The preceding chapter discussed modern corporate objectives at some length. It was suggested there that the three prime objectives of management in large corporations are:

1. maintenance of a satisfactory rate of dividend payment,
2. steady corporate growth as measured by sales, and
3. the avoidance of foreseeable risk of either a product price collapse or a raw material price rise.

Clearly, paper machines could not earn from 15 to 20 per cent on their investment costs (and therefore dividend payments could not be made at a satisfactory rate) unless fed with pulp. Steady corpo-

[11] According to statements made in published reports to stockholders.

rate growth cannot be anticipated unless new supplies of raw material can be foreseen. The surest means of preventing a disastrous price rise in wood supplies comes through owning or otherwise controlling a sufficiently high proportion of the required wood to prevent any combination of producers from restricting the needed supply.

What rate of return does a modern corporation really earn from its forest? The rate earned depends upon the value these objectives are accorded by management. Their attainment may not be assured by forest ownership, but it is certainly enhanced. If investments are made in forest plantations when the alternative rate of return on available capital is 12 to 15 per cent, one can be sure that management believes the forest investment will pay no less!

### Public Agencies

As brought out in the previous chapter, public agencies seldom have purely financial objectives for timber production. Nevertheless, interest rates, like other factor prices, are an important element in the allocative mechanism of our economy. Baumol obviously regards the choice of interest rate in evaluating public expenditures as crucial, stating "At stake in the choice of an acceptable discount rate is no less than the allocation of resources between the private and public sectors of the economy" (Baumol, 1969). The substantial proportion of benefit/cost literature discussing choice of interest rates indicates the significance of this allocative function to public investment decisions. As demands for public funds—especially federal funds—increases, it becomes more and more necessary to compare as directly as possible the alternatives, a fact well recognized by governmental officials.

While most of the discussion of benefit/cost analysis will be delayed until Chapter 20, the usually accepted view today is that the appropriate interest rate for public investment is the opportunity cost of capital taken from private investors by the tax system. This cost tends to be slightly below the market rate for private capital, and during the 1967–70 period was probably around 7 to 8 per cent.

It is, of course, all too easy to point to examples of low rates of return on Federal forest investments without taking cognizance of the severe restrictions imposed on the use of particular funds. It would seem desirable, however, to question policies which lead to investing public funds in growing sawtimber on 80- or 100-year rotations when analysis indicates rates of return of, at the very best, 2 or 3 per cent.[12] Such investments imply either that there is little call for public funds, that sawtimber prices are to be very much higher in the future, or that something else is expected to boost the public values of big timber. The first is obviously not the case; benefit/cost calculations for other Federal investments often employ rates of 8 to 10 per cent.[13] It is certainly possible that future sawtimber prices will be substantially higher than at present. It could also be that the Forest Service is acting in the belief that the long term needs of this nation's citizens require ready availability of large-sized timber to provide a safety margin for the future, or to assure that citizens-to-be will have a wider choice of materials.

The question of long term interest rates applicable to Federal investments will undoubtedly be debated more vigorously in future years than in the past. Forestry investments are, or should be, as much a subject for economic analysis as are those for river-basin improvements, irrigation, urban renewal, space exploration, or any other Federal activity. The fact that there is no neatly identifiable rate of return marked *"For Forest Investment Only"* does not mean present practices are above discussion, or even criticism.

[12] According to the Timber Trend Study, growth rates in Pacific Coast forests averaged only 0.44 per cent in 1962 (U.S.D.A., 1965).

[13] Standard rates for sensitivity analysis used by the Federal Highway Administration are generally 7.5 per cent, 10 per cent, and 12.5 per cent (United States Congress, 1968).

# 16

# Timber Demand, Supply, and Price

Production of stumpage is the traditional objective of forestry, and to this day the sale of timber remains the major source of revenue to most forest land owners. Stumpage price is consequently a key factor in the successful practice of forestry, and the study of timber price formation takes on a most practical aspect to all foresters.

In this price-forming process all the physical and economic factors of timber production and management and of wood product consumption are necessarily brought together within the political and economic framework of the nation. Small wonder, then, that this subject has a strong fascination for anyone concerned with forest economics.

The approach to stumpage price formation will parallel that taken earlier to timber product prices. Demand will first be examined, then supply, and finally the two components will be integrated in a discussion of price formation.

## DEMAND

Chapter 6 was devoted to a discussion of the nation's lumber markets, with lumber demand viewed primarily within an historic context. In Chapter 8 lumber was treated as a producer's good, with its demand viewed as being *derived*, primarily from the demand for housing and for other kinds of construction.

One may similarly regard the demand for *stumpage* as derived from the demand for other wood products. Stumpage is required if lumber, pulp, plywood, or any wood product is to be manufactured. People demand housing for which lumber is needed; lumber requires sawlogs, and sawlogs must come from stumpage. A similar chain stretches from paper to pulp to pulpwood to stumpage, and if an understanding of stumpage price formation is sought none of these steps can be bypassed. Since lumber is still the most important single wood product, the development of stumpage demand will be in terms of sawlog stumpage, although pulpwood, plywood, or any other wood product would have served equally well.

### Derived Demand for Sawlogs

Because there is virtually no substitute for sawlogs in manufacturing lumber, some writers have concluded that the demand for sawlogs must be inelastic.[1] The fact that log demands may be derived from paper and plywood as well as from lumber does not alter this conclusion as long as one approaches sawlog demand as if there were a national sawlog market. But national markets for sawlogs do not exist! As soon as one begins thinking about stumpage, questions of "What stumpage?" and "Where?" come to mind. Because of the immobility of standing timber, the *location* and *accessibility* of stumpage are highly important aspects of its demand. Furthermore, we know that *species* and *quality* are important factors in the demand for any specific timber stand—and only specific areas are sold. Finally, because the location of the tract might limit either the number of possible buyers or the alternative sources of stumpage supply to the buyer, or both, the *structure of the stumpage market* itself might prove to be a significant factor. Even though usual price theory treatment assumes these factors to be constant, a theory of stumpage price must permit taking these factors into account if it is to be at all realistic.

**Sawlog demand for a single mill.** The starting point for our analysis will be the demand of a single mill for sawlogs. As a specific ex-

[1] See especially, Mead (1966), page 50. Unfortunately, Mead did not distinguish between demand for logs and demand for stumpage.

ample, assume the mill to be an efficient-sized southern pine sawmill producing a variety of grades and sizes, and selling on the national market. With annual sales of about ten million board feet, its output accounts for only 0.2 per cent of total southern pine sales, which in turn must compete with other species from other regions of the country.

One would expect the demand curve for this firm's product to be quite elastic within the usual range of output, with departures from perfect elasticity, probably insignificant in most cases, arising primarily because of customer loyalty.

To derive the sawlog demand curve for this mill requires not only knowledge of its product demand, but also of the supply curve for all productive factors other than logs used by the mill. Even though substantial variation can be expected in an industry containing some 30,000 units, some useful generalizations are possible.

If "normal capacity" is defined as the annual output resulting from a single shift working a 5-day week, 8 hours a day, then output can be reduced by (1) instituting a shorter work week (fewer hours per day, or a 4-day week, for example), or (2) closing down the mill for longer periods. The possibility of reducing output by such means is decreased during periods of abundant alternative employment opportunities for unskilled labor, or by any employment of specialized labor.[2] Output can be increased by adding overtime or by increasing the average log diameter, the latter a possibility that will be ignored for now. Within a single shift operation, then, the non-timber marginal costs of this sawmill should be nearly constant up to the normal capacity output, and would then increase quite steeply, due to increased costs of overtime labor ($S_1S_1$ in Fig. 16.1).

Really significant increases in output could be achieved by operating on a two- or a three-shift schedule. The second shift would have somewhat higher costs ($S_2S_2$), for the new labor would presumably be a little less efficient and the needed supervisory staff would be less skilled. Both factors would tend to be more pronounced for a third shift, thus raising the supply curve for non-log factors even higher ($S_3S_3$).

The relation between the product (lumber) demand for this one mill and the three supply curves for non-log factors is given in Fig. 16.1. The area between the relevant segments of the supply curves

[2] Such means of obtaining production flexibility are often opposed by organized labor.

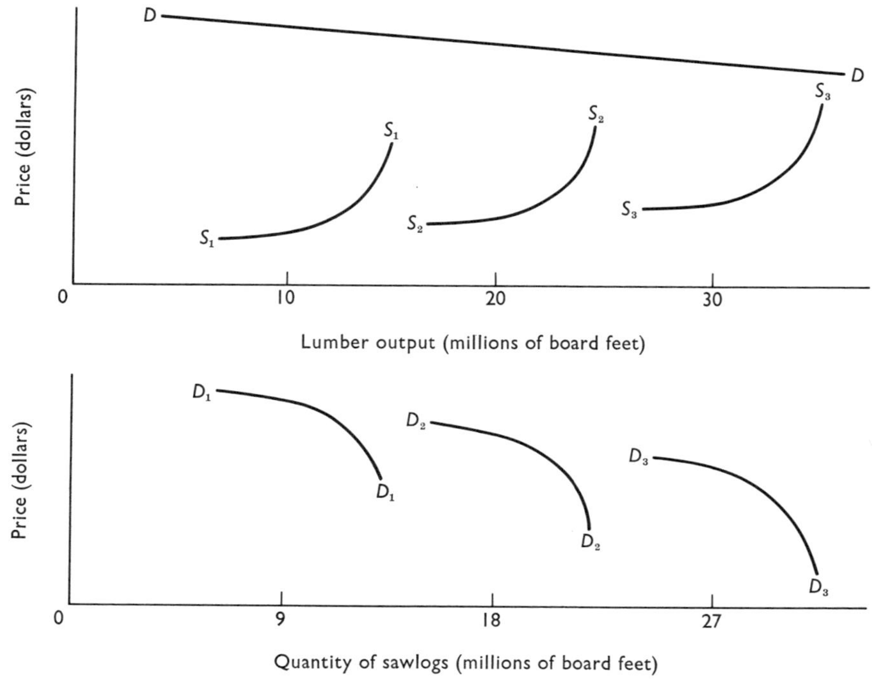


**Fig. 16.1** Derivation of sawlog demand for a single mill.

and the firm's marginal revenue curve (here assumed to be the demand curve) is plotted directly below, yielding the three derived demand curves for sawlogs.[3] The two quantity scales differ because of the characteristic "overrun" of most sawmills. As shown here the demand is elastic over part of each shift and inelastic over other portions, but conditions leading to different results could easily be envisaged, especially under nonpure competition.

**Sawlog demand for a group of mills.** According to usual price theory, the *market* demand for sawlogs would be determined by aggregating the sawlog demand curves of all mills in the sawlog market area. A summation of the supply curves for all non-log productive factors

[3] Practically speaking, a second or third shift is not normally instituted unless the mill's demand curve has "shifted" outward far enough to assure coverage of these added costs. Such shifts do occur, sometimes as a result of increased construction activity in the locality where the mill's output is normally sold, or because of decreased availability of lumber from other areas due to inclement weather, shipping difficulties, or a variety of other reasons.

is thereby implied, but would be acceptable only if the sawmills were independent in their demands for all productive factors. The traditional assumption of independence between cost curves of individual firms, however, is clearly lacking here.

The lumber demand curve ($D_1$), together with the supply curve for all non-log productive factors ($SS$) is shown for a group of mills in Fig. 16.2. Product demand could still be regarded as being per-

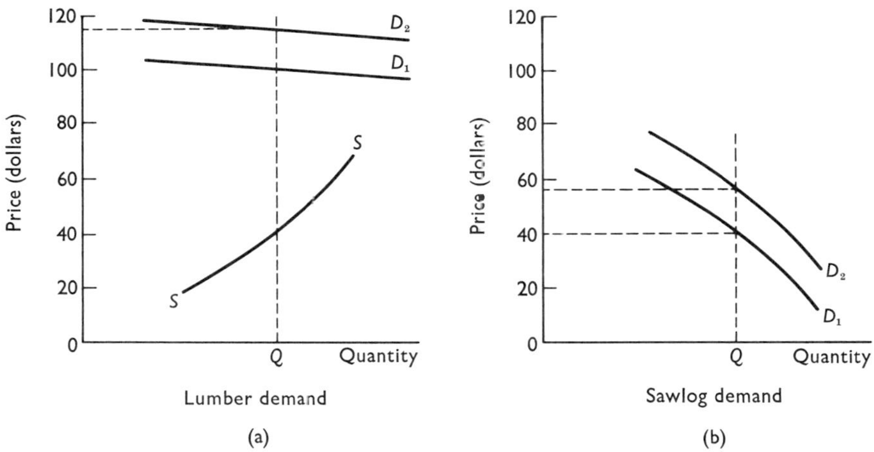


**Fig. 16.2** Derivation of sawlog demand for a group of mills.

fectly elastic in most instances. The supply curve for other factors ($SS$) would probably tend to be *in*elastic, primarily because a general expansion of the labor force would be needed if a group of mills were to increase output, and such a general expansion would be difficult. The discontinuities noticed in the illustration of the single mill (Fig. 16.1) would be less pronounced since each mill in the group would have a different production schedule for adding shifts and for hiring new men, or for going on overtime. As a consequence of these two factors, the derived demand curve for sawlogs would tend to be smoother, and probably considerably less elastic, than that for a single mill.

If the demand for lumber were to shift sufficiently to cause a price increase of $15 (from $D_1$ to $D_2$), and if the supply of all other productive factors remained unchanged, then the demand for sawlogs would also shift up by \$15. The effect on sawlog prices could be quite dramatic; at the indicated quantity of $Q$, a \$15 increase in lumber price (15 per cent) produces a 37½ per cent increase in the price of sawlogs *with no change in the output whatever.* It will be necessary to examine this point further, but for now we will only note this apparent effect of changed product demand.

By aggregating all such sawlog demand curves for a given species over an entire region (say, for Douglas fir sawlogs in the Pacific Northwest) a regional demand curve for sawlogs might be constructed. Then, by aggregating these regional curves for each species, one might similarly obtain a "national" demand curve for sawlogs. This latter, however, would involve the same problems encountered in obtaining a demand curve for pets: one must add cats and dogs, canaries, turtles, goldfish, etc. The result would have no more, and no less, significance.

Three further practical complications must be mentioned. The one concerning species has already been alluded to; each species must be treated separately since the end product (and, therefore, the sawlog price) does vary by species. The second concerns quality; high-quality sawlogs are those yielding increased proportions of lumber in the more valuable grades. Because the demand curve for such upper grade lumber is at a higher level than that for lower grade lumber, the demand for high-quality sawlogs is "pulled up." Sawlog demand might be regarded as a "family" of curves, each particular curve being appropriate to a given sawlog grade or quality range.

The third complication concerns time. Lumber stocks in retail yards absorb sudden changes in construction activity just as mill stocks help smooth out shipments and production after orders have been received at the mill. But plans and contracts for log delivery need to be made well before product orders are received. The effective demand for logs is really derived from an anticipated demand for lumber, and this is surely influenced by the mill owners' expectations of construction activity. Since expectations are difficult to quantify this influence will reduce the accuracy of the estimated demand, and because expectations are generally more volatile than actual occurrences one should anticipate substantial fluctuations in log demand—perhaps even more than would occur in the lumber market itself.

## SUPPLY

Earlier chapters have emphasized that an industry supply curve is an aggregation of marginal cost curves, and that it serves as a producer response curve by showing the quantitative response made by producers when the price offered for their product is changed. The supply curve for sawlogs must similarly reflect the marginal costs of sawlogs, and must also show how producers will respond quantitatively to changes in sawlog prices. Construction of the required sawlog supply curve will be aided by considering the cost of delivered sawlogs as being composed of two parts. The cost of transforming standing timber into logs on the mill deck will be called the *cost of availability*. The second part is the cost of standing timber, or stumpage, which will be related to (but is not identical to) its *reservation price*. Both cost of availability and reservation price require explaining.

### Cost of Availability

With no topographic variation and a homogeneous timber stand, log production costs of felling, limbing, bucking, skidding, and loading can be regarded as constant per thousand board feet, except for the costs of getting men and equipment to the logging site. For any given area, then, total log production costs will increase proportionately to volume removed (Fig. 16.3), and costs of getting men to and from work—(a) in Fig. 16.3—will vary with location.[4]

The other significant cost variable is that of log transportation. With any given method, hauling costs per thousand board feet tend to be linear when expressed as a function of distance, with the unit cost directly dependent on the method used. Generally speaking, the more elaborate, the larger, or the faster the means of transportation, the higher the fixed charges and the smaller the cost per additional mile. The general pattern of transport cost curves consequently tends to resemble that of Fig. 16.4. Only four types of equipment are illustrated, but if one considers the possible variations and combinations it becomes realistic to regard the transport function as an envelope curve of the "family" of linear relationships, much as

[4] If logging crews report for work at the logging site, this cost may not be borne by the logging company. The cost curve would then begin at the origin. Such situations are not uncommon in the South, the Lake states, or in many nations other than the United States.

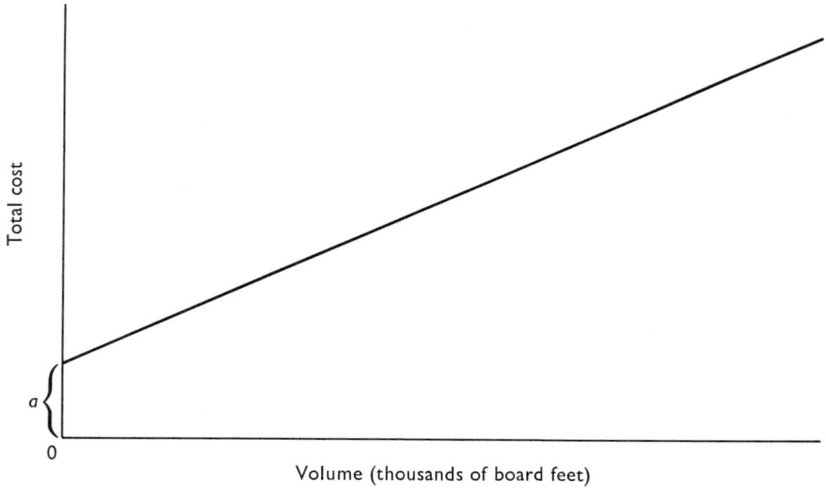


**Fig. 16.3** Total log production costs as a function of volume.

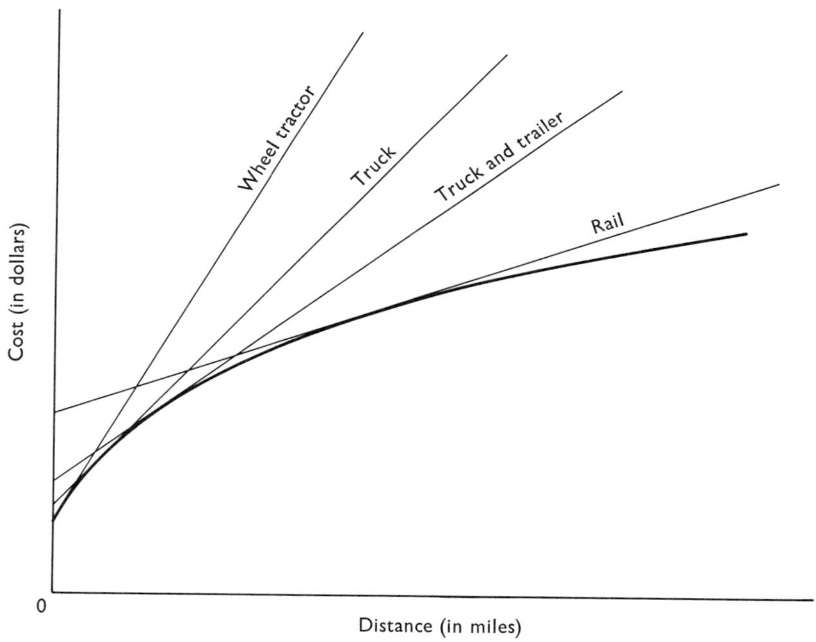


**Fig. 16.4** The usual form of a transport function (including transfer costs).

shown in the figure, with costs of loading and unloading indicated by the intercept on the vertical axis. When used to estimate transport costs, such a curve implies that for any particular distance the most efficient equipment is used.

If all other costs of log production (costs of felling, bucking, skidding, etc.) are added to the transportation cost the transport function will be shifted directly upward, as in Fig. 16.5.[5] And finally, this logging cost curve ($LC$) should be increased by the rate of re-

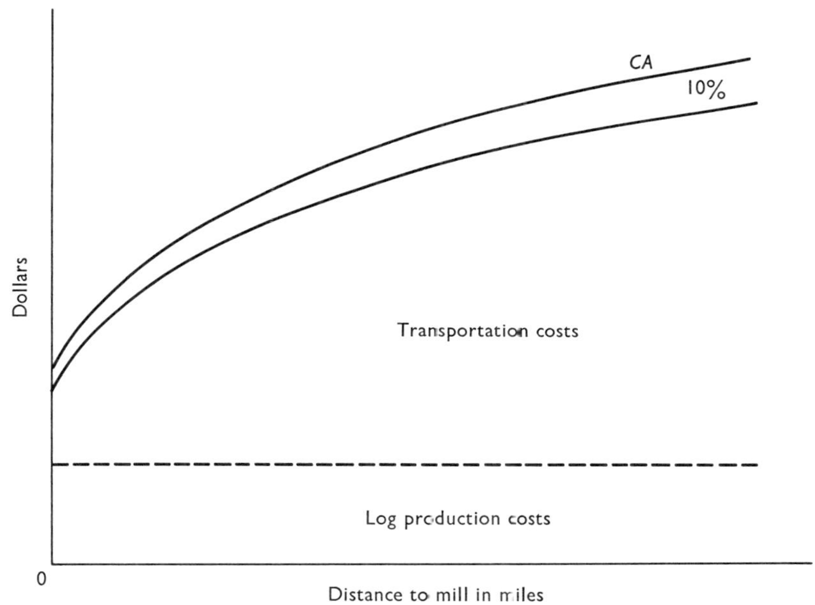


**Fig. 16.5** Derivation of the cost of availability.

turn regarded as normal by logging contractors. In Fig. 16.5 this rate is 10 per cent, yielding the *cost of availability* curve ($CA$).[6]

From this curve the cost of availability for any stand within the flat plain being used in the present illustration can be determined.

[5] The cost of getting crews to the logging site also increases with distance. This cost is not shown in Fig. 16.5, but its effect is readily seen.

[6] Ordinarily the return on investment would be treated no differently from any other cost; it is simply one of the costs of logging. For convenience at a later point, however, the separation is made here. It will also be noted that in this diagram the rate of return has been assessed against all costs, both fixed and variable. This may be debated, but is believed more realistic than would be a confining of the return to fixed investment. Moreover, it is in accord with modern concepts of cash flow.

Forests, however, are rarely on flat plains, and both topography and the established transportation facilities distort the cost of availability curve. Cost of availability should directly reflect not only distance, but all other factors influencing transportation and log production costs. At some points within this chapter it will become useful to make the cost of availability curve more generally applicable, and in these instances the term *cost distance* will be used—the distance that corresponds to a cost of $X$ dollars on the cost of availability curve.

Several important aspects of the $CA$ curve should be noted from Fig. 16.5:

1. Any reduction in log production costs—use of better power saws, of labor training, etc.—may increase the distance over which logging can be conducted profitably.
2. A lowering of transport costs will also increase the profitable logging radius. Here, however, the shape of the curve as well as its position might alter. For example, if the slope of the transport function were to be decreased (from $CA_1$ to $CA_2$ in Fig. 16.6), perhaps by use of larger or faster hauling equipment, the result

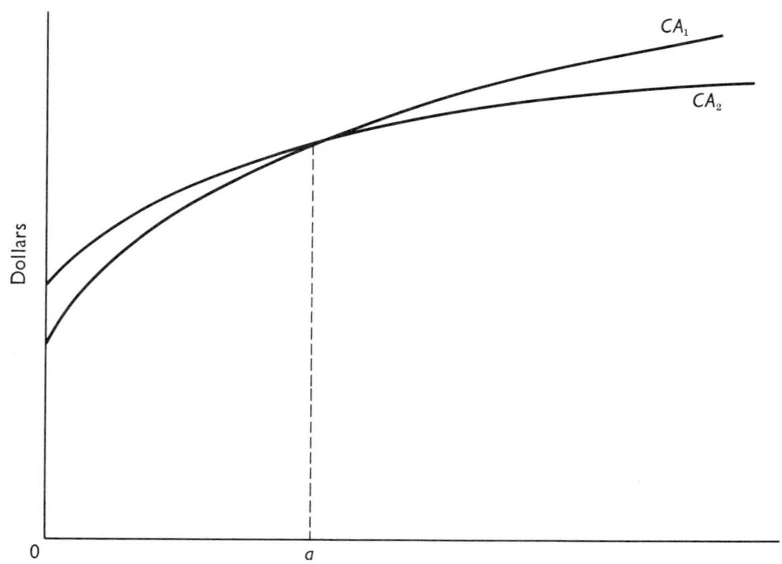


**Fig. 16.6** Effect on cost of availability of changing equipment used for transportation.

could be an upward shift of the $CA$ curve near the mills (between $0$ and $a$ in Fig. 16.6) but a downward shift at greater distances (beyond $a$ in the figure).

3. The logging radius has a non-linear relationship to the area that can be logged. For example: a cost reduction that would permit increasing the profitable logging radius from 30 to 40 miles would increase the potential timber supply area by almost 80 per cent.
4. A reduction in the rate of return on invested logging capital could, for similar reasons, produce a substantial increase in the potential timber supply area.

## Reservation Price

In any region with a reasonably dispersed forest land ownership pattern some owners will be actively seeking to sell all or part of their timber, others would be willing to sell if the price were sufficiently attractive, and there may be a few who are simply "not in the market." This variability has a variety of reasons; it is in part explained by differences in price expectations, in part by differences in time preferences, and in part by the many purposes for which forest land and the associated timber is held.

All three of these factors are straightforward in their effects. Owners who expect stumpage prices to rise in the near future will tend to hold their stumpage today. People with low time preference (those who value future incomes highly in comparison with income received today) will also tend to keep their timber, preferring to have their income in the future. And individuals who purchase or hold forest land for recreation, wildlife, or other non-timber purposes will be less interested in selling stumpage if its sale impedes realization of their major objective. The length of time a stand has been under management may also influence an owner's time preference.

These factors, and others, can be grouped under the phrase *reservation price.*[7] The reservation price for a particular volume or area of timber is that price necessary to persuade the owner to sell that volume or area. It is the owner's estimate of his opportunity cost of selling now, the highest present value of any future expected timber price—the lowest price at which he would be willing to sell. At any price below this reservation price the owner will "reserve" his timber.

[7] Many of the ideas presented in this chapter have grown out of discussions with students and other colleagues. The general concept of reservation price came from Dr. Henry Vaux of the University of California.

Individuals have many reasons for reserving timber, and not all are necessarily financial in origin. Sentimental, political, social, or almost any kind of reason may be sufficient to keep an owner from selling. Corporations wishing to protect future supplies, or a society wishing to preserve flexibility in its resource base and acting through state or national governments, may also reserve timber from present sale. Most private forest owners, however, have some price at which they would be willing to sell. And occasionally the reservation price may even be negative; should an owner want to clear land for pasture, for example, he might be willing to pay to have the timber removed. Colonization schemes in developing countries sometimes produce similar results.

Whatever the reason, it is the price fixed in the owner's mind that is important. Since only he has the final word on whether to sell or not to sell at whatever price is offered, it follows that one of the keys to timber supply is its ownership. Detailed information on United States timber ownership was given in Chapter 9; here we need only the reminder that so far as timber production and sale are concerned, the United States is a "mixed economy." Part of our standing timber is owned by private industry operating under competitive market forces, and part is publicly owned and managed. And since few people care whether the wood in the houses they own or the paper they read was sold by the National Forest Service or by the Joe Blow Tree Farm, both private and public timber must be brought together in the supply picture.

The short term supply function is the logical place for this marriage but both bride and groom must first be identified. In the pages which follow the short term supply curve for sawlogs from privately owned areas will be developed first, after which the supply curve for sawlogs from public timberland will be discussed, and to some extent compared with that for the private sector. The two will then be combined to produce a sawlog supply model which, together with the model of sawlog demand just presented, will permit examining the factors influencing sawlog prices. From this point it is a short step to the development of a "theory of stumpage price."

## Sawlog Supply from Private Timber

Any timber sale is necessarily made from existing forest growing stock. One might sell all the growing stock of a stand, the trees above (or below) some stated diameter or quality, individual trees as marked, or make almost any other kind of full or partial sale. It will aid initial development of the sawlog supply curve, however, to assume that a stand can only be sold outright.

**Elements of supply curve.** Suppose now that the reservation price had been determined for every stand in a given market area. If to the *reservation price* (per M) of each stand one now adds its *cost of availability* (per M), the resulting total will be the minimum price any buyer must pay if sawlogs are to be put on his log deck from the specific stand. This total is called the (per M) *reservation cost* (not *price*) of the stand. If all stands are now *ranked* according to their increasing reservation cost, and these increasing costs plotted over the accumulating total volume of sawlogs, the resulting cost curve will be the short term sawlog supply curve for logs delivered to the mill.

**A numerical example.** The approach is most easily demonstrated by constructing two related tables such as Tables 16.1 and 16.2. With

**TABLE 16.1 CONCEPTUAL PROCEDURE FOR CONSTRUCTING A SAW-LOG SUPPLY CURVE (THE DATA).**

| 1<br>Reservation price (dollars per M) | 2<br>Cost of availability (dollars per M) | 3<br>Reservation cost (dollars per M) | 4<br>Volume (millions of board feet) | 5<br>Rank (no.) |
|---|---|---|---|---|
| 42.00 | 56.00 | 98.00 | 1.200 | 10 |
| 34.00 | 43.00 | 77.00 | 1.600 | 3 |
| 60.00 | 24.00 | 84.00 | .800 | 7 |
| 20.00 | 62.00 | 82.00 | .400 | 6 |
| 52.00 | 36.00 | 88.00 | .400 | 8 |
| 72.00 | 24.00 | 96.00 | 1.200 | 9 |
| 44.00 | 29.00 | 73.00 | 1.000 | 1 |
| 46.00 | 34.00 | 80.00 | 2.200 | 5 |
| 30.00 | 45.00 | 75.00 | .400 | 2 |
| 60.00 | 18.00 | 78.00 | .800 | 4 |

this as a format, one could (at least conceptually) enter the data from each timber stand in the area—the stand volume, the owner's reservation price, and its cost of availability—then derive the saw-log supply curve.

To illustrate: in Table 16.1 the reservation price (column 1) and the cost of availability (column 2) are added to produce the reserva-

**TABLE 16.2 CONCEPTUAL PROCEDURE FOR CONSTRUCTING A SAWLOG SUPPLY CURVE (THE DATA AFTER RANKING, AND WITH ACCUMULATED VOLUMES).**

| 6<br>Rank (no.) | 7<br>Volume (millions of board feet) | 8<br>Reservation cost (dollars per M) | 9<br>Accumulated volume (millions of board feet) |
|---|---|---|---|
| 1 | 1.000 | 73 | 1.000 |
| 2 | .400 | 75 | 1.400 |
| 3 | 1.600 | 77 | 3.000 |
| 4 | .800 | 78 | 3.800 |
| 5 | 2.200 | 80 | 6.000 |
| 6 | .400 | 82 | 6.400 |
| 7 | .800 | 84 | 7.200 |
| 8 | 1.600 | 88 | 8.800 |
| 9 | 1.200 | 96 | 10.000 |
| 10 | 1.200 | 98 | 11.200 |

tion cost (column 3). Stand volume is shown in column 4, and in column 5 the stands are assigned a rank according to their reservation cost, with the stand of lowest cost being ranked first. In Table 16.2 the data have been regrouped according to their ranked order (columns 6, 7, and 8). The accumulated volume (column 9) has been calculated by adding the volumes stand by stand as they are listed in column 7.

The supply curve for sawlogs (*SS* in Fig. 16.7) is now easily constructed by plotting the reservation cost (column 8) over the accumulated volume (column 9). As with all marginal cost curves plotted from discrete data, one should use the midpoint of the added output for plotting. With the data of Table 16.2 the result is the short term sawlog supply curve of Fig. 16.7. In most market areas of this country, of course, there would be many stands and a smooth curve would normally result. For ease in illustration, only ten stands are utilized here.

Each stand in the market area has now been entered in the supply curve at its sawlog reservation cost per thousand board feet of sawlogs, so the sawlog supply curve shows how much an additional 1,000 board feet of sawlogs would cost the buyer at any possible quantitative level of production. This reservation cost curve of sawlogs is an aggregate marginal cost curve, just like any other short

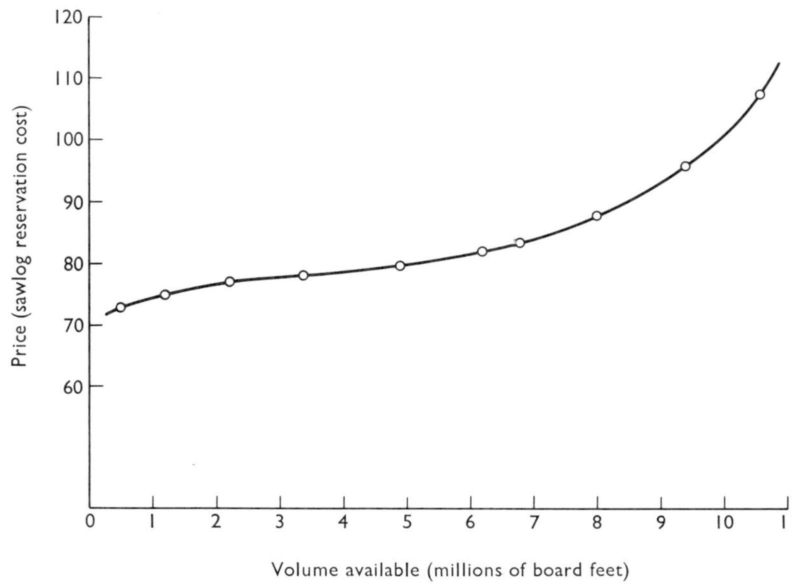


**Fig. 16.7** A short run sawlog supply curve.

term supply curve, and shows the response of sawlog producers to changing sawlog prices.

**Aggregation over distance.** The sawlog supply curve requires that volumes be "aggregated over distance." Fig. 16.8 may help to clarify this aspect. Here, 49 separate stands have been ranked according to their reservation cost and the supply curve, SS, plotted for the market area. The volume of each stand is indicated by the length of the interval occupied on the volume scale at the figure's base (at a zero price level). This same aggregate of volumes is "disaggregated" on the lower part of the diagram, where the volume of each stand is plotted at its distance from the market, with the distance being measured in cost miles. When the volumes are all added together the total volume on the quantity axis is completely accounted for. Since the stands were ranked according to their sawlog reservation cost, the curve formed is necessarily a rising one. Buyers look at every stand, regardless of its location, from the viewpoint of "What will the logs cost me at my mill site?" For this reason the supply curve for a market area must include all timber within the timbershed.

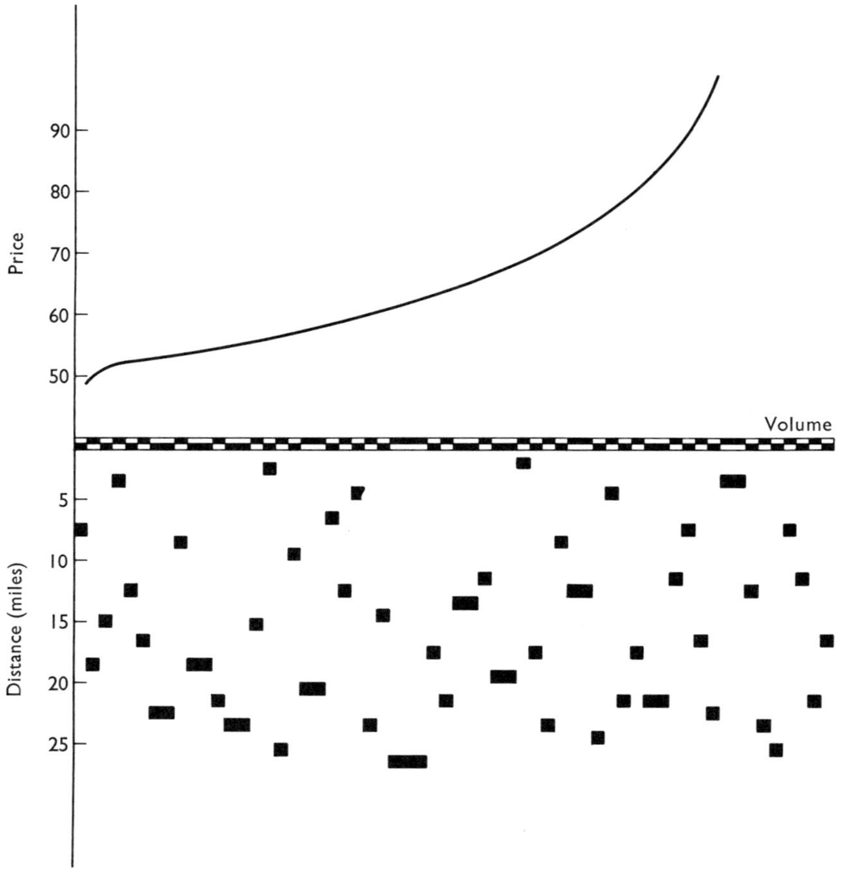


**Fig. 16.8** The sawlog supply curve requires that volumes be "aggregated over distance." The buyer is not interested in the stand's location, per se, but rather in the cost of logs delivered to his mill site. Stands are therefore ranked according to their reservation cost, regardless of location.

There could be cases where reservation price would not be the same for all possible sale volumes from the same stand—where different volumes could be removed from the stand at different reservation prices. In such cases the particular stand being considered must be represented by a *curve* of reservation cost over volume rather than as a single point. All such curves must then be added together to produce the short term sawlog supply curve, the added volumes accruing at each increase in reservation cost. The process is identical to that of aggregating the marginal cost curves of firms in an industry. The difference is not in the process of aggregation, but in the fact that each such curve has its origin at its own (ranked) reservation cost.

## Sawlog Supply from Public Timber

A thorough discussion of public timber supply would require a complete review of the policies of multiple use, of sustained-yield–even-flow, of appraisal, of bidding, and perhaps most important, of the procedures and policies which underlie the determination of the so-called "allowable cut" from forest service management units.[8] Fortunately, the background of Federal timber policy need not be reviewed to develop the supply relationships, although this particular subject area provides an excellent illustration of how economics and policy are, or should be, related. Very briefly, the significant facts of public land timber supply are these:

### Institutional and administrative factors.

1. Under the Organic Administration Act of 1897, national forests were established for watershed protection and ". . . to furnish a continuous supply of timber for the use and necessities of citizens of the United States." This firmly established the sustained yield policy, which was reaffirmed by the Multiple Use Act of 1960.
2. In accord with these principles and directives the annual "allowable cut" (the quantity of timber to be removed from a particular management unit in a year) is calculated by various means. For even-aged stands it is usually determined through some variation of Hanzlik's formula:

$$\text{A.C.} = \frac{\text{Volume of mature and overmature timber}}{\text{Rotation age}} + \text{Annual growth on immature stands}$$

   or for mixed-aged forests under some variant of the Austrian formula:

$$\text{A.C.} = \frac{\text{Actual volume} - \text{desired volume}}{\text{Conversion period}} + \text{Average annual growth}$$

3. In both cases one of the key components is the rotation period which is determined by the time required to produce timber of

[8] The discussion here is entirely in terms of the United States Forest Service, since 86 per cent of all public commercial forest land is under the administration of this agency.

the size and quality suited to the product in mind. On Federal lands the product objective has traditionally been sawtimber, and rotations are those necessary to produce prime sawlogs.

4. The timber is advertised for sale at "not less than the appraised value." Bidding may be either by submission of sealed bids or by oral bidding at a scheduled auction.
5. The contract for harvesting is awarded to the highest qualified bidder. (Qualified in that the bidder has demonstrated his ability and willingness to carry out the contract terms.)

From the standpoint of short term supply curve formulation the most important step is the determination of the allowable cut. The actual volume of mature and overmature timber is predetermined, although the measurement of these volumes may pose problems and can also produce variation in results. The rotation age is fixed by the size of tree desired and the expected growth rate, and growth on immature stands is determined by site factors and to a lesser extent by the expected level of management. Under present procedures, then, there is no possibility of the quantity of public timber offered for sale being influenced directly by market price. Once the timber is sold there is some flexibility in the time of cutting. Usually the amount of timber sold but uncut is around 2 to 2½ times the annual allowable cut, but due to the required road building, etc., there is less flexibility here than might at first thought be expected.

**The appraisal procedure.** The appraisal process will be encountered again toward the end of this chapter. For the present it is sufficient to know that (1) timber appraisal usually involves a residual calculation; (2) for sawtimber the starting point is lumber value, from which all costs of production other than stumpage are subtracted; and (3) this "conversion surplus" is then divided (not necessarily equally) into two parts, (a) an allowance for "profit and risk," and (b) a remainder for "stumpage return." Within the framework of sawlog supply already established in this chapter one can regard the appraisal procedure as requiring, for each stand, an estimation of the cost of availability which is then subtracted from the sawlog value.

**Constructing the short term supply curve.** The short term sawlog supply curve for public timber is constructed in the same manner as that for private timber—by adding together, for each stand, its cost of availability and its reservation price, with the latter now being the *appraised* price. Unlike private timber, however, there is a positive, administratively fixed minimum. If it is assumed that all appraisal estimations and calculations have been perfectly performed, the sawlog reservation cost for each stand in the present example will be identical.[9] The sawlog supply curve for timber from Federal lands will therefore be perfectly elastic up to the amount of the allowable cut, then become perfectly inelastic. Such a supply curve is illustrated in Fig. 16.9, with $Q_a$ being the amount of the allowable cut.

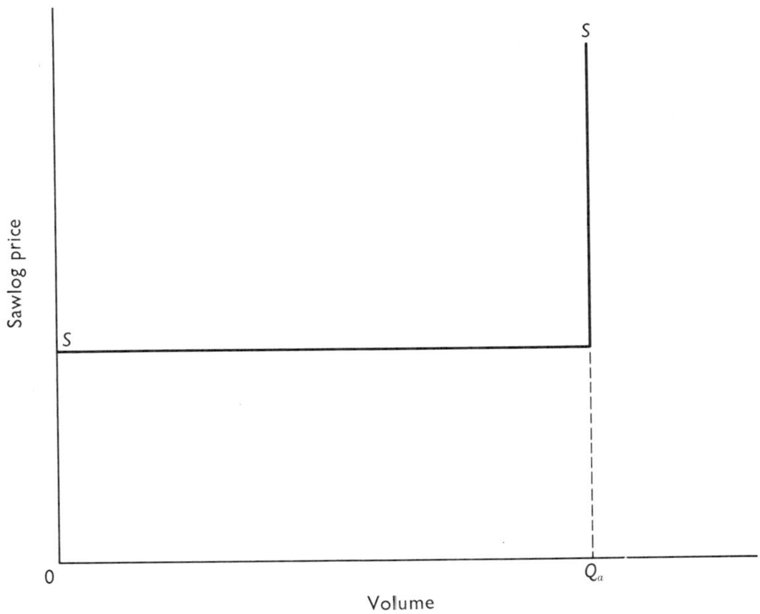


**Fig. 16.9** **Sawlog supply curve for timber from public lands.**

While both public and private short term sawlog supply curves are based on sawlog reservation cost, there are three significant differences between the two curves. The first is obvious: the calculation of an allowable cut makes the sawlog supply curve for public timber perfectly inelastic at the allowable cut quantity, but perfectly elastic

[9] This is necessarily true because of the assumption of (1) a perfectly homogeneous forest on (2) a flat plain, and (3) a single group of mills located at its center. Hence sawlog value less cost of availability equals appraised price.

for all quantities up to this amount. The second difference is only apparent if one attempts to quantify the sawlog supply curve for a particular area. Publication of the appraised prices of stands gives one the sawlog reservation costs directly, while the task of determining these costs for private timber can be difficult. The third difference deals with the stability of the curves. The sawlog reservation cost of private timber varies with the owner's expectations, while that for public timber is not influenced by expectations.

### Sawlog Supply from Mixed Ownerships

If the land ownership pattern in the area is either all public or all private then the sawlog supply curves shown in Figs. 16.8 and 16.9 are representative of what might be expected in typical situa-

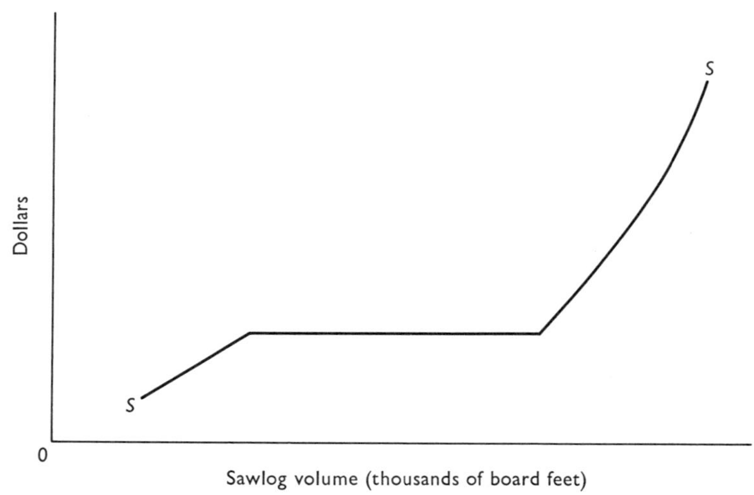


**Fig. 16.10** Sawlog supply curve for timber from mixed ownerships.

tions. In many parts of the United States, however, forest land ownership is decidedly mixed,[10] especially in parts of the West, and in such instances the two supply curves can simply be added together horizontally. Fig. 16.10 gives the result of such an addition, using the two curves previously developed as an illustration. The "break"

[10] The extreme in this regard is that found in some western states, where the public domain was used to encourage railroad construction by granting alternate sections to certain corporations, thus creating a checkerboard of public and private forest land ownership.

in the curve caused by the inflexibility of allowable cut calculations on public lands is readily apparent, as is the perfectly elastic portion.

## PRICE FORMATION—SAWLOGS

Given the short term demand and supply curves already developed there is no particular problem in price formation for sawlogs from either private or public timberland. With these two functions, and assuming a competitive market structure, the intersection of the two curves will indicate the equilibrium price and quantity. Fig. 16.11(a) and (b) illustrates for both cases how an equilibrium

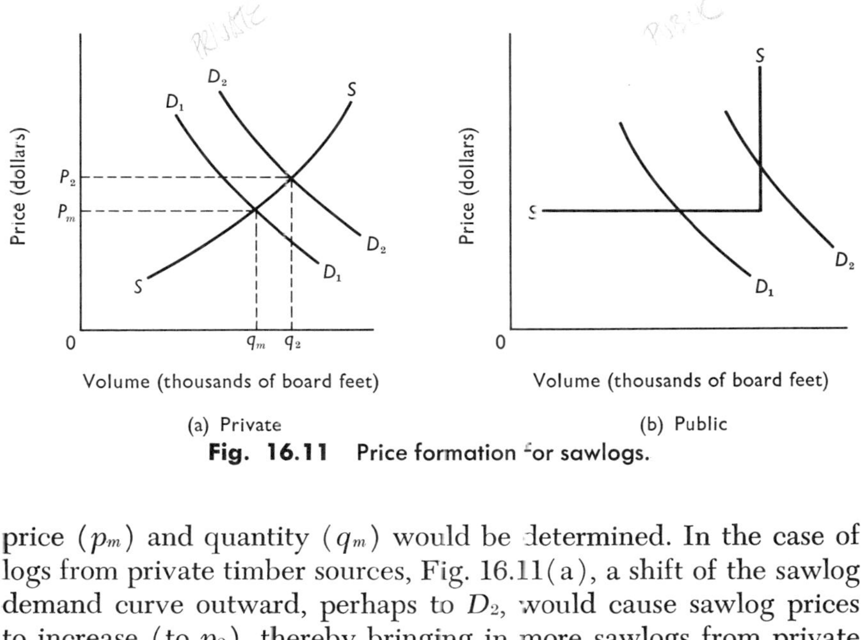


**Fig. 16.11** Price formation for sawlogs.

price ($p_m$) and quantity ($q_m$) would be determined. In the case of logs from private timber sources, Fig. 16.11(a), a shift of the sawlog demand curve outward, perhaps to $D_2$, would cause sawlog prices to increase (to $p_2$), thereby bringing in more sawlogs from private lands. With logs from only public sources, however, Fig. 16.11(b), there are two possible outcomes if demand increases:

1. If the demand curve intersects the elastic portion of the curve (as does $D_1$) any place short of the "corner" then the full allowable cut will not be sold.
2. Only if the intersection occurs at or above this corner—if it intersects the inelastic portion of the supply curve—will the full allow-

able cut be sold. The price effect will be transmitted backward to stumpage prices, which will be examined shortly.

The most interesting case is undoubtedly that of Fig. 16.12, wherein sawlogs are available from either (or both) public or private lands. Here the three possible situations are shown by the three demand curves, $D_1$, $D_2$, and $D_3$. The results in terms of sawlog prices and quantities, and in terms of who sells and who does not, are sufficiently clear to require no further comment.

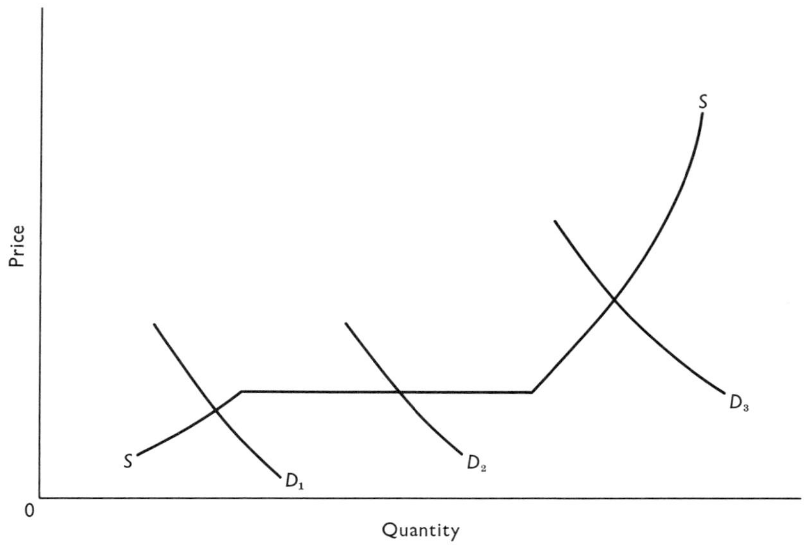


**Fig. 16.12** Sawlog price determination when sawlogs may be obtained from either public or private sources.

## PRICE FORMATION—STUMPAGE

In a perfectly competitive market the sawlog stumpage price at any specific location will equal the sawlog price at the mill minus the cost of availability. But forming stumpage demand and supply curves presents a most peculiar problem because the demand for stumpage originates at one point in space and the supply in another. The buyer (mill operator) views each possible stand from *his* market point, while the seller (the timber owner–producer) views it from his own very different market point. Both buyer and seller, of course, are aware of the difference, but this effectively prevents stumpage supply and demand curves, at least of the usual variety, from being constructed.

## Significance of Location

The significance of location, and the way it distorts *stumpage* demand and supply, may be clarified by reexamining a supply curve for sawlogs. Fig. 16.13 is the same sawlog supply curve shown in Fig. 16.7, constructed from the data of Table 16.2. Suppose the sawlog market price is at \$92.00, delivered at mill site—that is, assume the demand for sawlogs is elastic at a level of \$92.00. Now from the

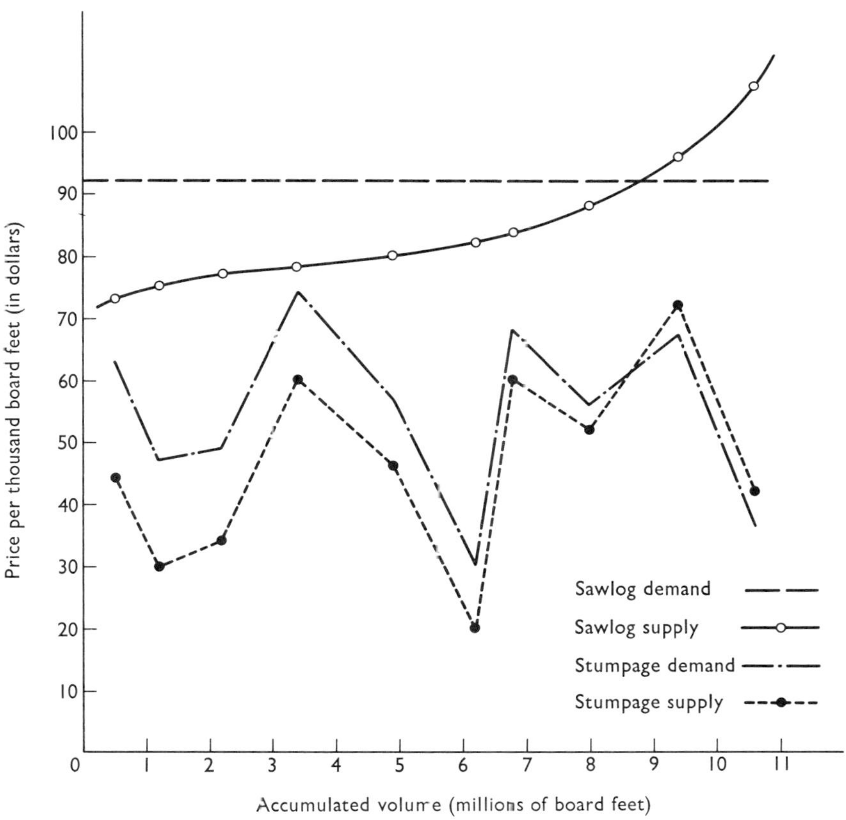


**Fig. 16.13** A "demand" and "supply" curve for stumpage—derived from the demand and supply curves for sawlogs.

given *supply* curve, we can identify exactly which stands will be purchased and can also pinpoint their cost of availability, for this cost is part of the data used to construct the supply curve. This cost of availability is the information needed to derive the stumpage demand curve, for if we subtract this cost from the market price the result will be the derived demand curve for stumpage, derived from a perfectly elastic demand curve for sawlogs. (There is no real need to assume perfect elasticity but the illustration encompasses only 10 stands, hardly enough to make up the total supply of an area.)

**Price for individual stands.** The result of this subtraction is shown by the dashed line in Fig. 16.13 and represents the stumpage price that will be paid for each stand. This very irregular line is in fact, a "perfectly elastic" demand curve for stumpage. The reservation price is also shown in the figure, connected by the dotted line, *SS*. In a very real sense this is the "supply curve" of stumpage (although not really continuous), while the demand curve is that of *DD*, the dashed line. The intersection of the stumpage demand and supply curves identifies the total volume of stumpage that will be purchased, which of the stands will be taken, and their prices. But the intersection does not indicate *the* stumpage price, for this price differs with every stand. The dashed line of stumpage price lies above the reservation price (supply curve) for all stands taken and below for all remaining stands, but the price level of the intersection has no general meaning.

**A more general model.** For most purposes it is believed preferable to use a somewhat simpler model that stops short of determining specific demand and supply curves for stumpage. In Fig. 16.14(b) the sawlog demand and supply situation (supply from private lands only) is shown as establishing the sawlog price, $P_m$, and the quantity, $Q_m$, at the selected market point. Part (a) of the diagram shows the same market price ($P_m$) that is established in part (b), but plotted over a horizontal axis scaled in distance instead of volume units. The cost of availability at each distance has been subtracted from $P_m$, and the result plotted over the appropriate point of the lower (distance) scale. In effect the cost of availability curve is reproduced, but now inverted, and with the origin transposed to $P_m$. The new curve, declining with distance, traces out the stumpage price as stand location is varied from the marketing point.

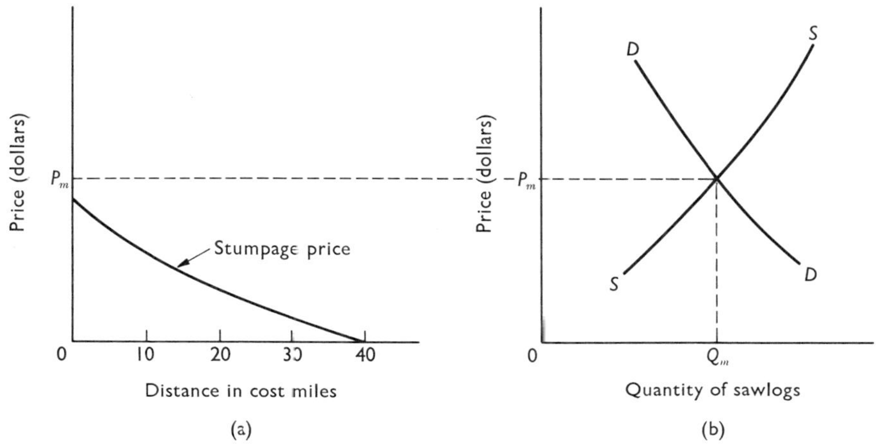


**Fig. 16.14** The formation of stumpage price.

**A model for mechanics.** For those who are somewhat mechanically minded the model of Fig. 16.15 may prove helpful. Suppose the cost of availability curve were traced on a piece of material such as Plexiglas, then cut out with a horizontally projecting arm (as in Fig. 16.15) and attached to a vertical rod or pin so it might

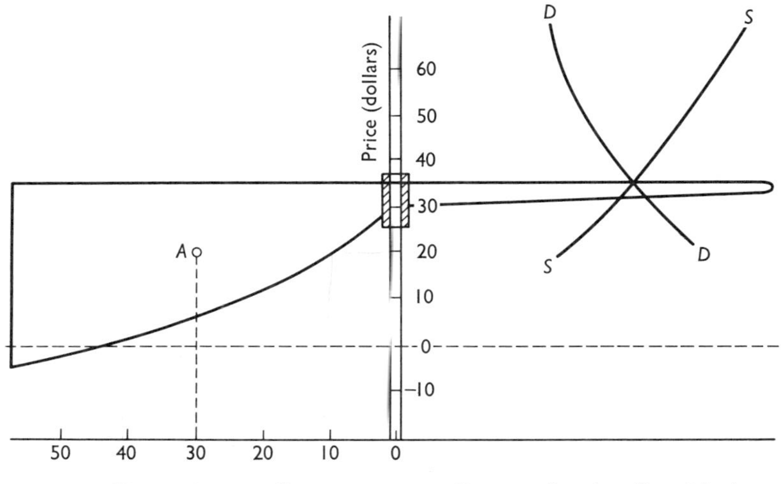


**Fig. 16.15** Stumpage price determination.

be slipped up or down as desired. The rod itself is scaled in dollar units. The right side of the model has as its components the sawlog supply and demand curves, with the vertical rod serving as the price scale and the horizontal axis scaled in timber volume units (board feet). A shifting of either the supply or demand curves raises or lowers the intersection, and hence the height, of the Plexiglas indicator, thus keeping the top of the indicator always at the market price of sawlogs, while the lower (curved) portion indicates stumpage price at any cost distance from the market—on the left side of the diagram. For example, in Fig. 16.15, with the demand and supply curves determining a price of $35 for sawlogs at the mill site, a timber stand located 10 miles from the market will have a stumpage price of $20, while at 45 cost miles the stumpage price would be zero. Note carefully the role of reservation price. A stand with a reservation price of $20 that is located 30 cost miles from the mill (*A* in Fig. 16.15) will not be sold under the prevailing conditions. Only those stands with a reservation price equal to or below the indicated stumpage price will be sold. But the sawlog supply curve is itself made up by a rank ordering of the reservation cost (reservation price plus the cost of availability) for all stands in the market, and exactly enough timber will be available at the indicated set of stumpage prices to supply the mills with the (equilibrium) quantity of logs they demand.

The mechanical approach can be carried considerably further, and there are advantages to discussing at least one additional step. Suppose a map of the entire timbershed—the flat plain of homogeneous timber—was available, drawn to a scale identical to that used in making the stumpage price curve of Fig. 16.15. Suppose further that the vertical rod, together with the attached indicator carrying the inverted cost of availability curve—the whole apparatus of Fig. 16.15—was then pinned into the map at the location of the group of mills. As the indicator, set at the market price for sawlogs and carrying the stumpage price curve, is swung about the rod it will describe a pointed, parasol-like surface over the entire timbershed. The height of this surface directly over any location on the map will indicate the stumpage price for that location. Fig. 16.16 conveys at least an idea of how the surface might appear under the hypothesized conditions.

The significance of this almost physical model of stumpage pricing is not its applicability per se, but rather the fact that it

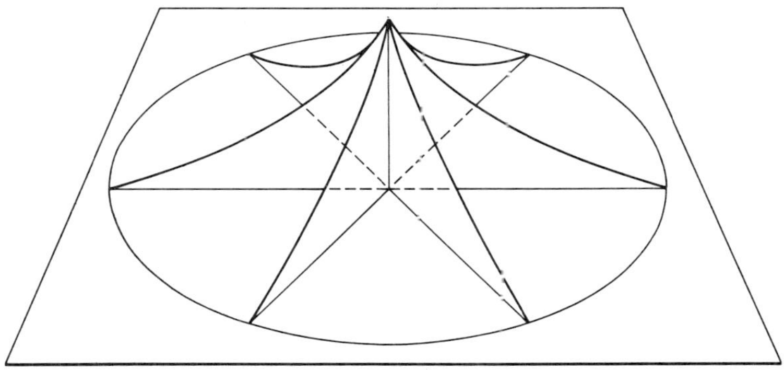

**Fig. 16.16** A hypothetical price "surface" extending over a timbershed, centered on the buying mill.

aids in seeing how the surface must change if it is to represent the real world rather than a flat, homogeneous forest. Instead of the perfectly symmetrical surface of the figure, the stumpage price surface over a forest area (and using real miles instead of cost miles) will be full of indentations and convolutions. Rough topography increases logging costs, so the stumpage price surface would dip down over any such area; a high-speed highway lowers transport costs, so here the surface would be further above the map. But the basic elements would still remain. Under competitive conditions the stumpage price for any stand in the forest will be the price of sawlogs at the delivery point *less* the cost of availability, and unless the owner's reservation price is at or below this price the stand will remain uncut.

It should also be clear that mills with different cost structures, or at different locations, could have different offering prices for any particular stand. But as each mill's stumpage price curve swings over its timbershed—whether in competition with other mills or not—it will "sweep up" just enough stumpage to meet its demand for sawlogs at its own equilibrium price. And note also that it is in no way necessary that every mill will have the same equilibrium price for sawlogs.

Short term reservation prices are particularly prone to change, and anything affecting future price expectations tends to shift the sawlog supply curve, changing timber as well as sawlog prices. Though there is no clear-cut line for distinguishing between short and long term factors in sawlog or stumpage pricing, and to some extent the division is an artificial one, it will be necessary to look at the more important factors that serve to shift stumpage prices under particular conditions and to see the effect of relaxing at least some of the more stringent assumptions. And it will be especially important to examine the impact of a few of the important institutional factors. Before proceeding to this analysis (Chapter 17) it seems best to review other approaches that have been suggested as pertinent to stumpage price formation.

### The Residual Approach to Stumpage Price

To many, the determination of stumpage price seems quite simple. Stumpage is a raw material. If it has a value at all it is because some firm or some individual believes this raw material can be fashioned into a finished or semifinished product that can be sold profitably. It follows that the starting point for determining stumpage price should be the selling price of the product or products that are to be produced from the raw material. By subtracting all costs from the product's final sale price—from felling costs through costs of manufacturing—and including whatever amount is necessary to repay the entrepreneur for his particular contribution,[11] one derives a residual value. This approach is the time-honored technique of timber appraisal, well developed in texts of forest management and forest valuation.

### Stumpage Price and Rent Theory

Approached as a topic in rent theory one might begin the study of sawtimber stumpage price determination with a diagram such as that of Fig. 16.17(a) and (b). Average total costs of lumber production ($ATC_1$) and the corresponding marginal cost ($MC$) are shown in their usual form. The average total cost curve shows all production costs, including whatever return on investment is necessary to keep the lumber producer content in his business. At a price of $P_1$ the firm will be in equilibrium, just covering all production costs, including the indicated stumpage payment, $s_1$. If one were to aggregate all the marginal cost curves of all the firms pro-

[11] In stumpage appraisal terms this is usually called "profit and risk." But in the usual economic sense profit is the reward for risk bearing—the return to the prime entrepreneurial function.

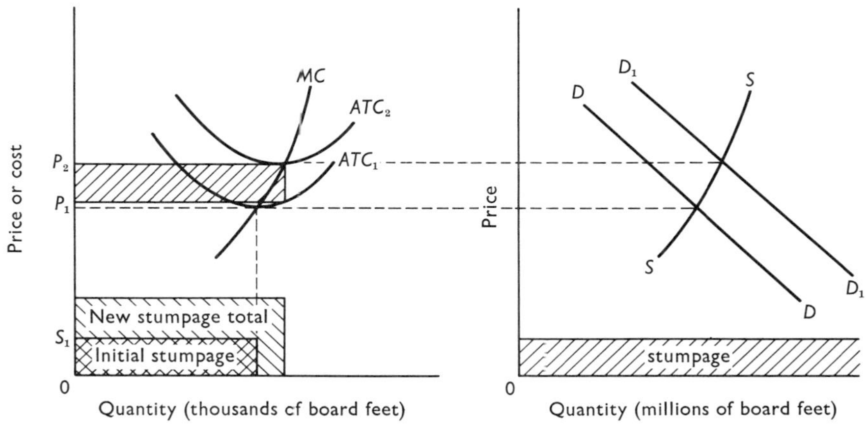


(a) (b)

**Fig. 16.17** Stumpage as a rent.

ducing lumber for this market the resulting supply curve would be that of *SS* in Fig. 16.17(b), which intersects the demand curve for lumber (*DD*) at a price of $P_1$.

Now suppose something happens to shift the lumber demand curve out to the new position, $D_1D_1$. As a result prices will rise to $P_2$, and all producers in the industry will increase their output until once again their marginal cost will be equated to the new price, $P_2$. At this price the firm will make a substantial profit, the size of which is indicated by the shaded rectangle of the figure and which will average $(P_2 - P_1)$ per thousand board feet.

But assume the timber from which the lumber is being produced is being purchased from another party who controls the supply. Previous to the lumber price increase the stumpage owner could charge only $S_1$ for the timber. Now, however, he can make an additional charge of exactly $P_2 - P_1$ per thousand board feet. At the new stumpage price the lumber producer will be just as well off as he was at the outset—he is no worse off than before the demand curve shifted. His average total cost curve will simply "slide up" the marginal cost curve, eventually reaching the position shown in Fig. 16.17(a). This is the classical case of "Ricardian Increasing Rent" referred to in Chapter 4, and it fits almost exactly the residual approach to stumpage pricing upon which the appraisal process rests; $P_2$ equals the selling price of the lumber, and $P_1 - S_1$ equals the average cost of production *plus the margin for profit and risk* which is normally included in the total cost curve.

Note that rent theory is not "opposed" to the supply–demand model. The differences stem from the assumptions underlying the two approaches. Practically speaking, one can say that the rent approach to stumpage price determination proceeds entirely from the manufacturer's side. All the elements of the analysis concern production of the wood product—lumber, pulp, or plywood, as the case might be.[12] On the demand side the differences in models are relatively slight. The derived demand approach also begins with an end product such as lumber, but by requiring knowledge of supply conditions for all other productive factors it necessarily involves a much wider spectrum of productive factors. The greatest difference lies in the treatment of supply. Neither rent theory nor the related appraisal procedure involve consideration of stumpage supply conditions at any one point, particularly the possibility of a reservation price on the part of forest owners. It assumes that the supply of all productive factors other than timber is perfectly elastic, that the supply of timber offered is perfectly inelastic with respect to price, and that the forest owner is in a position to extract the rent (which may not be the case if he is one of many small landholders selling to a few large timber buyers). As a consequence of these assumptions the rent-based appraisal approach to stumpage pricing provides no possibility for analyzing the effect of changing timber supplies on either stumpage prices or on product prices. Nor can one hope to assess the effect on stumpage price of changing stumpage production costs or the impact of important institutional factors; one cannot examine possible influences of specific stumpage market structures on timber and wood-product prices. In a stable, short term period the appraisal approach, if accurately applied, necessarily yields a valid estimate of the amount available for stumpage, profit, and risk. Under such conditions it is a perfectly straightforward accounting technique permitting one to find a point on the stumpage price surface. The purpose of this text,

[12] This itself has posed a severe conceptual problem to residual theory ever since pulpwood production became important to timber producers. Should one begin with the value of pulp, or of paper? What is a fair profit margin on pulp or paper manufacture? And how accurate, or even significant, is the calculation of a residual that amounts to less than 10 per cent—and often as little as 2 per cent—of the initial sum?

however, is to develop an understanding of how forestry fits into an economy, and how some of the important economic forces of the nation influence the possible and the probable returns to timber production and ownership. For this task the demand–supply model seems to be the only presently available alternative.

# 17

# Timber Prices in a Market Economy

The interplay of product and factor prices with consumer and producer actions was stressed when reviewing the role of prices in a free enterprise economy (Chapter 4). In the final chapter of this timber economics section, attention will be focussed on specific relationships of stumpage price to product price, institutional factors, production costs, and land use. One of the most significant of these is that of timber price to the price of final products. And since discussion of the rent-appraisal procedure provides an excellent introductory background, the relation of product prices to timber prices is a convenient starting point.

## RELATIONS BETWEEN TIMBER AND PRODUCT PRICES

As a practical example, suppose bank lending rates were suddenly reduced. With money available at lower interest rates, an outward shift in the demand for housing could result as seen in Fig. 17.1(a). Previously, it was seen that increased construction activity produces an outward shift in the demand for lumber, as shown in Fig. 17.1(b) by the two demand curves, *DD* and *D′D′*. The resulting increase in lumber price would be reflected, after some time lag, by a corresponding outward and upward shift in the derived de-

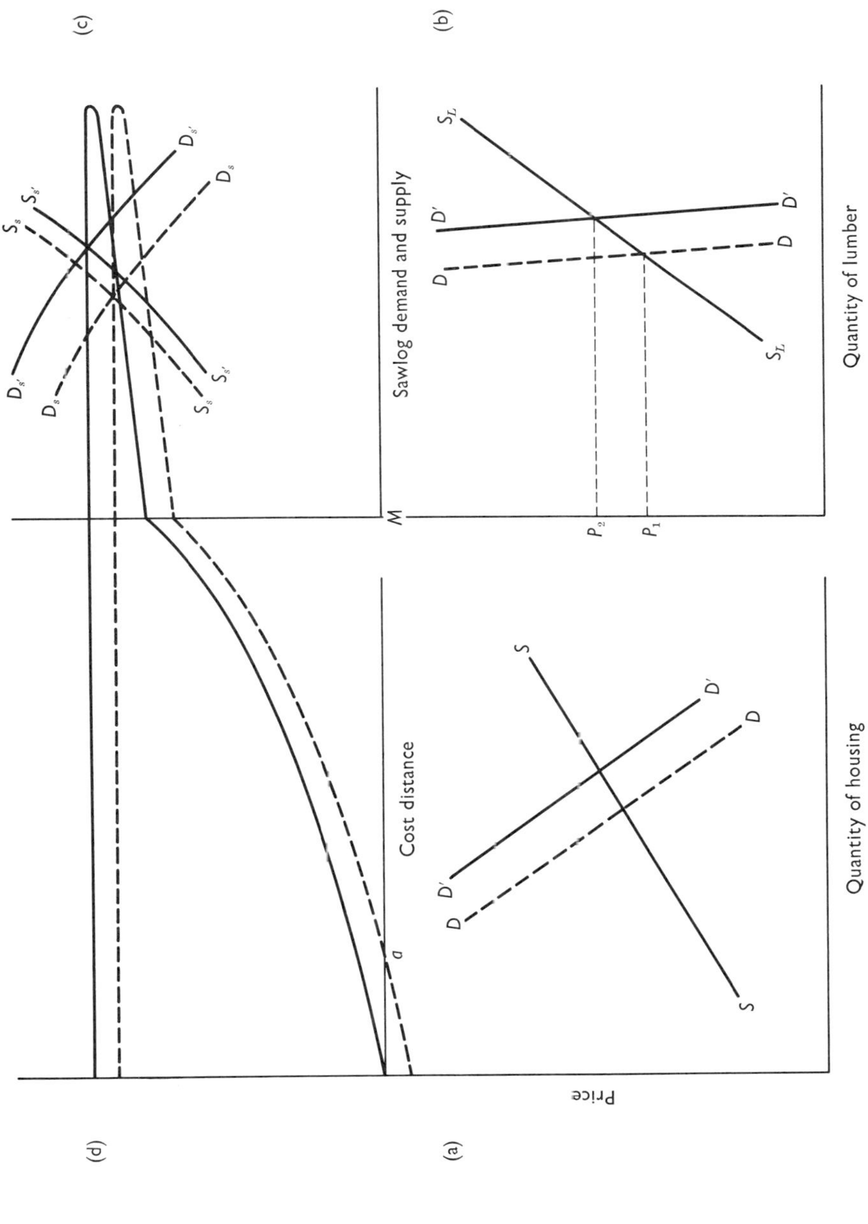


**Fig. 17.1 The relation between lumber and stumpage price (short term).**

mand curve for sawlogs, Fig. 17.1(c), at the mills being supplied by the forest area under study. This increased demand would, in turn, raise the entire schedule of stumpage prices to the new position shown in Fig. 17.1(d).

### Short Term Effects on Private Stumpage

Such an increase in prices would have two short term effects; within the zone of previous positive stumpage values (from point *a* to *M*) a number of stands would now be sold that were not previously available because of the high reservation prices held by their owners. Logs from these newly available, higher priced stands would tend to shift the supply curve for sawlogs outward at least slightly, perhaps enough to somewhat lower both sawlog and stumpage prices from the first high point. But this new stumpage price schedule would also increase the distance over which logs might profitably be transported to the mills, thus making new timber areas available as supply sources. With these new supplies the sawlog supply curve would again shift outward, resulting in a further drop in sawlog and stumpage prices. Eventually the equilibrium price of timber would settle at a somewhat higher than original level, with some new areas sending timber into the mills. If owner's stumpage price expectations rise as a result of these shifts, a further round of adjustments would take place with the end result depending on the relative shifts of the demand and supply curves. But note that no particular increase in growth or in sustained yield capacity is necessarily needed. An increase in timber price will normally (under conditions of private land ownership) result in more timber being sold over the short term.

### Effect on Stumpage if Ownership Is Mixed

The same increase in residential construction activity could have a quite different effect on sawlog and stumpage prices where sawlog stumpage was being obtained from both public and private sources. The sawlog demand and supply curves for public and private timber can be simultaneously treated by using "back-to-back" supply and demand diagrams, as in Fig. 17.2. Unfortunately such an illustration does not lend itself to illustrating directly the stumpage price curve, but by now this extension should be readily envisioned. With supply from private sources on the right side and from publicly managed areas on the left, the quantity taken by

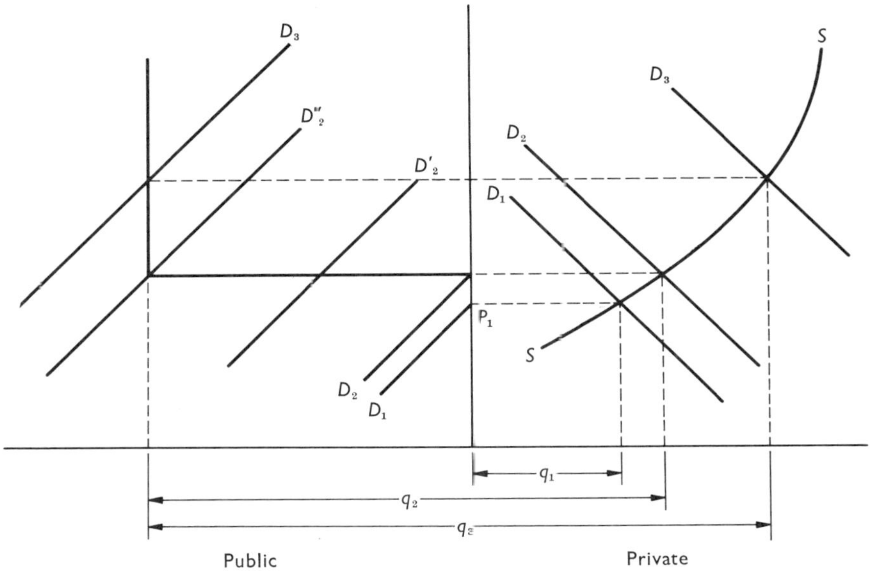


**Fig. 17.2** Effect of shifting demand on sources of supply.

lumber manufacturers is represented by the distance between the points of intersection of the demand curve(s). (The fact that two demand curves are used does not mean there are two separate sources of demand; the same demand draws from the two supply sources.) With demand at $D_1$, for example, sawlog price will be at $P_1$, and $q_1$ will represent the quantity of sawlogs delivered to the mills. At this level of demand all logs will be supplied by private timberlands; the reservation price of public timber (the appraised price) reserves it from sale. Under current conditions (1970) such a situation is unlikely, since use of market prices in the appraisal process would usually lower the appraised price of public timber to the $P_1$ level. During the early part of the 1930's, however, when lumber prices were extremely depressed and public timber was withheld from the market as an aid to private timber owners, the situation was not unlike the one shown in the diagram.

As demand shifts outward to $D_2$ (other things being equal), supplies from public lands will begin to become economically available, and under the conditions assumed in Fig. 17.2 the demand for sawlogs can continue to shift—through $D_2'$ as far as $D_2''$—until the full allowable cut is being purchased before any further rise in price will take place. Normally such a large increase would be of a long term nature. New mills would be required, and expansion of the old mills would undoubtedly take place. In the meantime the supply curves for both public and private timber would undergo considerable change.

But suppose demand for sawlogs expanded beyond that indicated by $D_2''$, say, for example, to $D_3$ (on both sides of the diagram, of course). With this level of demand, stumpage prices will increase *for both public and private timber.* So far as public timber sale is concerned, two events can be predicted with confidence: (1) bids for public timber will begin to exceed the minimum appraised price, first by small amounts, and then, as sawlog demand shifts further, the bidding may far exceed appraised prices; and (2) lumber manufacturers will start agitating for an increase in the allowable cut from public forests. And in part they will be correct, for if the allowable cut is increased the supply curve will shift outward and, if the shift is large enough, stumpage prices may begin to come down.

If future demands do not decline, however, increasing the allowable cut now may force a greater reduction at some future date. Such a reduction is especially likely if the increased cut reduces the growing stock (provided the increased incomes from larger cuts do not lead to intensified management offsetting this reduction). Under the impact of competitive bidding for the limited short term supply, and if the area concerned competes with mills that are in a better supply situation, some units may be forced out of business. A similar result may come about if the allowable cut now comes almost entirely from old growth timber, as is the case with many of our western forests. In this case, too, increasing the allowable cut may merely hasten the day when the old growth will be exhausted and the dependent mills forced to close down.

## Discriminatory Pricing

By separating public and private timber supplies the possibility of *discriminatory pricing* can be explored. Suppose, for example, the sawlog supply and demand situation is as shown in Fig. 17.3. The entire allowable cut from public lands is being taken, as indicated by the position of the demand curve, $D_1D_1$, and overbidding of the appraised price is usual.

Now suppose foreign timber manufacturers enter the market,

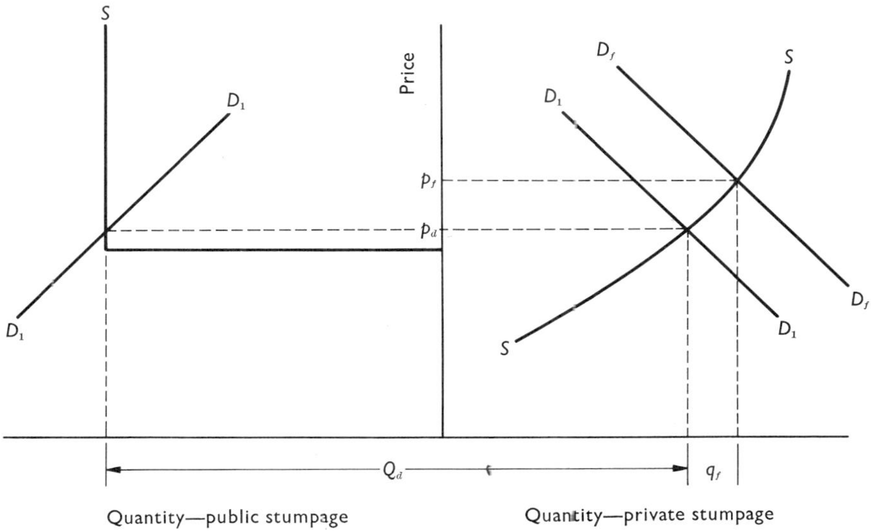


**Fig. 17.3** Discriminatory pricing of export logs.

seeking to buy timber at whatever price is possible. Any purchase of new quantities under the conditions illustrated here will raise sawlog prices and either force domestic producers to find supplies elsewhere, meet the competition by paying the price required to bring new supplies into the market, or go out of business. But there is another possibility. If local timber buyers can persuade government officials that it is "wrong" to sell public timber to foreign manufacturers when United States mills are themselves seeking new supplies (foreign competition is forcing American laborers out of jobs by closing United States sawmills and plywood factories, etc.), then an embargo might be declared on the sale of public timber to foreign buyers. If the embargo is enforced it is possible (but not probable) that domestic buyers can continue to purchase both public and private timber in the same quantity and price ($p_d$) while a few private timber owners sell the desired amount ($q_f$) to foreign manufacturers at the higher price of $p_f$—and reap the corresponding profits (subject to capital gains).

Such price discrimination is not at all rare in international sales, but it is usually possible to maintain price differentials of this type only if the market is firmly controlled. In the above situation, if demand is really active the new sales will, very shortly, raise the price of all stumpage, including that from public lands, to the new, higher price level. For under these circumstances the new supplies will almost surely be taken from timber formerly reserved for later harvest. This tends to raise price expectations, shifting the supply curve for private timber upward and raising the price to all buyers, domestic and foreign.[1]

### Expectations

So far the discussion has been in terms of the effect that shifting lumber demands might have on stumpage prices. But what happens if sawtimber owners, observing past increases in both lumber and stumpage prices, and reading of the boom in house construction, anticipate a decided rise in lumber and stumpage prices; that is, suppose expectations of stumpage owners rise. This change in expectations would quickly raise the reservation price of owner's timber, making sawlogs more costly by shifting the sawlog supply curve upward and, if the increase is at all widespread in the industry, cutting back production. Because short term lumber demand is quite inelastic, and because under specified conditions lumber producers are in a seller's market, lumber prices will be increased as stocks are diminished. Such an action would reinforce the expectations of timber owners and again raise their reservation price.[2] A mutually reinforcing spiral may thus be set in motion. Increased timber prices raise lumber prices; builders hurry to lay in stocks to protect their commitments; this permits lumber producers to further raise lumber prices, which encourages timber owners to boost stumpage prices again, which raises lumber prices, etc. The spiral is aided by the appraisal process, for with higher lumber prices the appraised price of public timber is necessarily increased. Present systems of allowable cut determination and pricing on public timberland have a decided escalator effect on timber prices.

If for some reason the demand curve for lumber shifts *inward* and lumber prices decline, then opposite forces are set in motion, but with one very important difference. A decline in lumber price,

[1] The log export market to Japan in 1968 and 1969 provided an excellent example of the situation described theoretically in these paragraphs.

[2] In this connection the more advanced student may wish to check the concept of "elasticity of expectations," the ratio of the per cent increase in price to the per cent increase in expected price (Hicks, 1946).

unless very prolonged, is unlikely to have a seriously depressing effect on price expectations of timber owners. The history of lumber and stumpage prices since the turn of the century has been one of almost constant advance. Declines have occurred, of course, but for nearly four decades each upward jump has been enough higher than the succeeding decline to result in a net advance of substantial proportions.

The interaction between lumber and stumpage price is of particular importance during periods of sustained price disequilibrium. In the years immediately following World War II, for example, lumber manufacturers could almost be described as being unconcerned about stumpage prices since any resulting cost increase was willingly absorbed by lumber retailers (and in turn by consumers). As a consequence, public land sawtimber prices often exceeded the "appraised" value by very large percentages, a phenomenon difficult to explain under any strict interpretation of residual theory. The same difficulty (though to a much lesser degree) has occurred in reverse during periods of declining lumber prices. Confronted with falling lumber prices but committed to buy stumpage at prices derived from the previously high lumber price level, mill owners have been placed in very difficult economic positions. The situation with respect to Federal timber sales was greatly eased when the Forest Service instituted a "review" procedure. Stumpage purchase contracts are now subject to renegotiation after price changes of given relative amounts. Procedures are carefully spelled out for determining the amount of the price adjustment, which may be either up or down.

## MARKET STRUCTURE AND STUMPAGE PRICES

A factor of great importance in shaping stumpage price is the competitive structure of the market in which the timber is sold. The point of departure for theoretical development is that of pure competition, the situation assumed in developing the supply–demand model of the previous chapter. It is doubtful, however, if timber markets with many sellers and many buyers should be called characteristic of United States timber markets. Real life conditions probably range from the extreme of pure competition on the one hand, all the way to bilateral monopoly (a single seller and a single buyer) on the other, and almost any conceivable combination of buyer and seller conditions could be found someplace in the United States.

Some degree of timber market imperfection is almost inevitable, simply because forests occupy extensive land areas, and transportation costs per unit of volume are high. Since no two processing plants can occupy the same point in space, *some* locational advantage will necessarily accrue to *some* firm in most timber buying situations. The question is one of degree; how much price protection is afforded depends on the cost of moving logs to or from the nearest alternative. Spatial monopoly is almost always present to some extent, but it can be of great significance in forestry.

The four combinations to be reviewed here are those involving (1) many sellers, or (2) a single seller of timber, with (3) oligopsony, and (4) monopsony, on the buyer's side of the picture. The review will also be confined to short term effects; the extension to long term effects on both price and quantity aspects of timber sales should be made by the student.

## Many Stumpage Sellers and a Single Buyer

The major departures from purely competitive conditions brought about by having a single buyer (monopsony) can be seen from the data in Table 17.1 and the corresponding figure (Fig. 17.4). Viewing a schedule such as that of the table, the monopsonistic buyer

**TABLE 17.1 SCHEDULE OF RESERVATION COSTS FOR PULPWOOD.**

| Quantity available (cords) | Price (dollars) |
|---|---|
| 150,000 | 19.00 |
| 175,000 | 19.50 |
| 200,000 | 20.00 |
| 225,000 | 20.50 |
| 250,000 | 21.00 |
| 275,000 | 21.50 |

would reason as follows: "I can buy 200,000 cords at an effective price of $20 per cord, or I can buy 225,000 cords at an effective price of $20.50 per cord. The additional 25,000 cords will, therefore, actually cost me $612,500, or $24.50 per cord, since in getting the

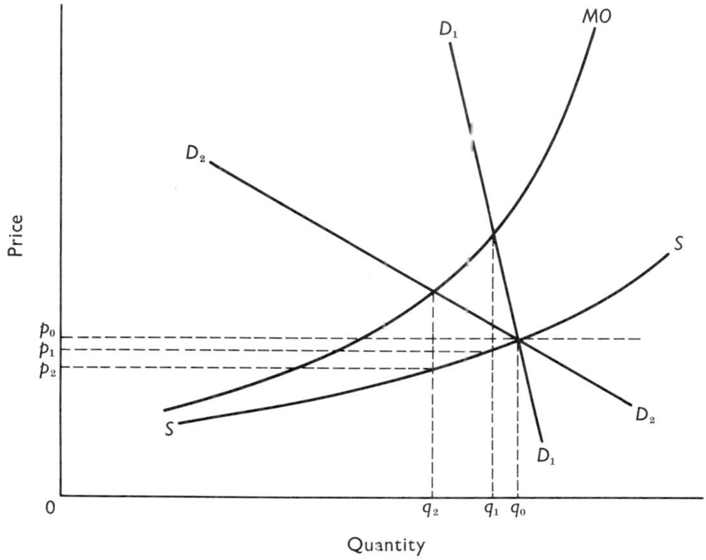


**Fig. 17.4** Price determination under conditions of monopsonistic buying from many owners.

25,000 additional cords I must raise the price of pulpwood by 50¢."[3]

Such a buyer will not consider the supply curve of pulpwood as such, but rather the *marginal outlay* required by his purchase. The relevant curve is his *marginal outlay* curve (*MO* in Fig. 17.4) showing the additional expenditure caused by the purchase of each additional unit. It is the curve marginal to the reservation cost schedule. A rising pulpwood supply curve will act to "push up" the marginal cost curve of any monopsonistic buyer and will also make it "steeper," since each additional unit purchased adds to his production cost.

To aid in evaluating the possible effect of monopsony on, for example, sawlog price and quantity relationships, two possible demand curves are shown in Fig. 17.4. One ($D_1D_1$) represents a very inelastic demand on the part of a monopsonistic buyer, and the other ($D_2D_2$) the relatively elastic demand that might be generated

[3] In this particular example: 225,000 cords @ \$20.50 = \$4,612,500
200,000 cords @ \$20.00 = —4,000,000
Cost of additional 25,000 cords \$ 612,500

by a single large sawmill selling on a national and highly competitive market. In appraising the effect of this market structure on timber price and quantity, the shape and elasticity of both the sawlog demand and supply curves must be considered. If the supply curve is perfectly elastic, monopsony will have no effect whatever on either price or quantity. The marginal outlay curve will be identical to the supply curve and will be horizontal. But if the supply curve is rising (as illustrated in Fig. 17.4, which is the more probable situation under reservation cost conditions), the effect of monopsony will be to reduce the quantity of sawlogs purchased and to lower the purchase price. The more *elastic* the demand curve, and the more *inelastic* the supply curve, the greater will be the effective reduction of both price and quantity. In the figure, for example, purely competitive conditions on both sides would indicate a price of $p_0$ and a quantity of $q_0$ under either of the demand situations shown. But the monopsonist with the demand curve $D_1D_1$ would purchase the quantity $q_1$ at the price $p_1$, while the monopsonist with the demand curve $D_2D_2$ would purchase $q_2$ at a price of $p_2$. From the very broad, national viewpoint one can say that a monopsonistic buyer purchasing in a purely competitive market would always result in fewer resources being devoted to producing and using sawlogs (and hence sawlog stumpage) and would usually mean (1) lower returns to timber owners, (2) higher profits for the buying industry, and, perhaps (3) higher prices for consumers, this latter depending on the market conditions under which the final product is sold.

In attempting to evaluate the actual effect of monopsony on stumpage markets one must bear in mind an important practical qualification. An elastic demand curve for timber ordinarily implies, because of its derived nature, an even more elastic demand for the final product. If product demand is elastic, then competitive pressure will normally force firms to construct highly specialized plants, which tend to have a distinctly limited output range, especially with processes such as chemical pulp production. This leads to an inelastic demand for raw materials such as pulpwood, which also limits the degree to which price and quantity can be affected by a buyer.

Studies of national forest sawlog stumpage prices have indicated that sales made to single bidders often bring lower prices than those involving several bidders, indicating the possibility of monopsonistic price influence. Usually other factors are involved, and careful study is needed before concluding that monopsony has had any serious effect on United States timber prices.[4] Yet common sense would tell one that whenever timber must be sold to a single buyer if it is to be sold at all, then prices would at least tend to be lower than if the sale were made under competitive pressure. This seems to be the situation in many localities, although no studies of a conclusive nature can be cited. In pure monopsony, with many sellers, one must also recognize the probability of discriminatory pricing.

### Many Timber Sellers and Few Buyers

If there are only a few buyers in an area with many timber sellers the theoretical effect on timber price can vary greatly. At one extreme oligopsony could have the same effect as that already described under monopsony, for the possibility of either collusive oligopsony or development of price leadership (where one buyer sets a low price with all other buyers following) must be recognized. This possibility is tempered by three facts:

1. The actual effect on price will depend upon the elasticity of both the demand and supply curves for the product in question. With a perfectly elastic supply curve there could be no effect whatever on stumpage prices.
2. Any oligopsonistic buyer, when considering a buying price reduction, would be limited by the knowledge that other buyers, even though operating under similar circumstances, might promptly "bid up" the price to limit his output or to undermine confidence in those from whom he is attempting to buy.
3. Since buying price agreements are illegal there is little hope of enforcing them, even if they are made

Whether oligopsonistic buying from competitive sellers has in fact depressed timber prices below those that might have obtained under competitive conditions cannot be claimed without careful study. In the single such example known to the author (a pulpwood case), it seemed clear that prices were actually higher, not lower, than the level to be expected under purely competitive conditions, apparently because buyers wanted to maintain good supply relations.

[4] Public timber sales have other constraints which limit the effect of monopsony. See page 372.

## Single Seller and Single Buyer

The possible influence of bilateral monopoly on both selling price and volume of standing timber sold can be assesed through Fig. 17.5. As before, a rising sawlog supply curve has been used, based

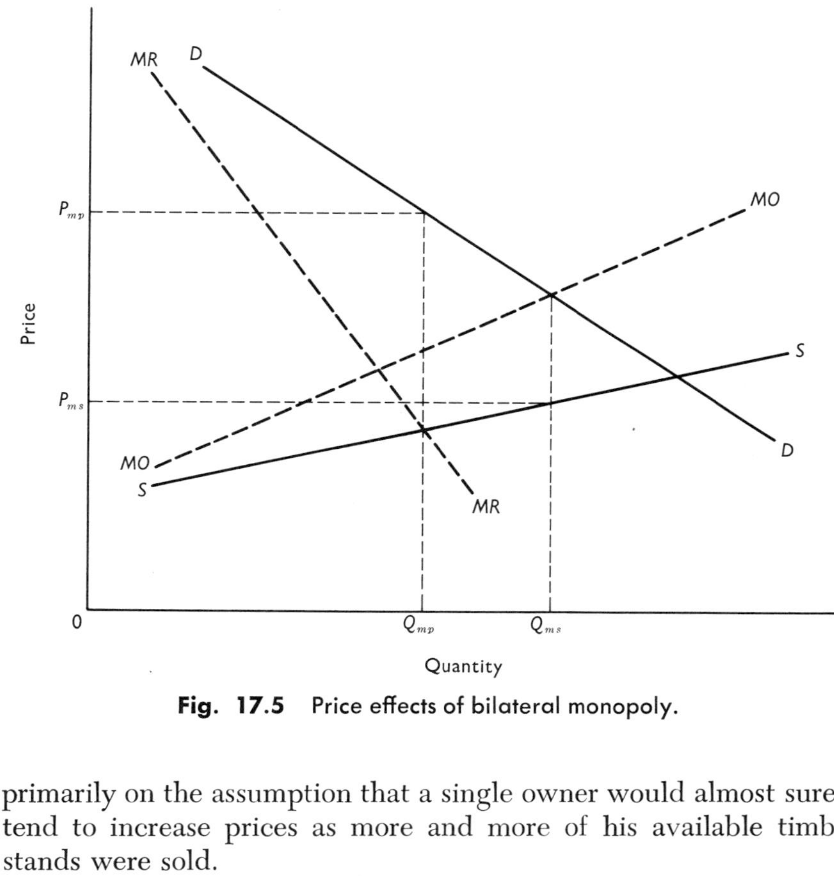


**Fig. 17.5** Price effects of bilateral monopoly.

primarily on the assumption that a single owner would almost surely tend to increase prices as more and more of his available timber stands were sold.

A monopsonistic buyer, observing his marginal outlay curve ($MO$) and with the demand situation as shown ($DD$) would want to purchase $Q_{ms}$ stumpage at a price of $P_{ms}$. Compared with the outcome under pure competition, this implies on the buyer's part a reduction in quantity purchased and in the price paid for sawlogs—and timber. The extent of the price reduction, and in fact, the degree to which the monopsonistic power can be effective, will depend primarily on the elasticity of the supply curve.

But with bilateral monopoly the buyer is not the only source of market imperfection. To a monopolistic owner the demand curve is his average revenue curve, and to take full advantage of his position he would try to equate *marginal* revenue to his supply curve. In the case illustrated this means selling $Q_{mp}$ at a price of $P_{mp}$, leading to sawlog (and stumpage) prices well above levels in a competitive market and reducing the quantity of wood products available to consumers. Because of the interaction between raw material and final product prices, the latter would also tend to increase if the amount produced under these conditions constituted a significant portion of the industry output.[5] The elasticity of demand for sawlogs in the monopolized area will be significant in determining possible price effects. With a given supply curve, the more inelastic the demand the greater the possible impact on price and resource allocation.

The outcome of bilateral monopoly is indeterminate. Any one of several results is possible, since the monopsonist buyer will strive for a low purchase price and a restricted output, while the monopolist seller will want a high price and also a restricted (but different) output. The final result could fall at either extreme if one party happens to hold a dominating bargaining position, or it could fall anyplace in between. If it happens that the bargaining power is equally divided, then price will probably fall somewhere near the middle and output will approach the competitive level. It is also theoretically possible for the two parties to get together, maximize their combined return, and divide the loot.

### Single Stumpage Seller and Few Buyers

The illustration used for bilateral monopoly (Fig. 17.5) is also appropriate for the monopoly–oligopsony situation, since the limiting case (where the oligopsonistic group unites) is that of bilateral monopoly. The possible price effects of the buying group are limited, however, by the elasticity of the supply curve (which is assumed to be rising when a monopoly exists), by legal constraints, and by the individual buyer's knowledge that his own actions may influence other buyers in their price offers.

The practical importance of monopoly on United States timber prices is difficult to assess. There are areas in this country where

[5] If the product is sold in a perfectly competitive market, then product prices could not be affected. Here, however, we are discussing market *imperfections*.

virtually all standing timber is owned or controlled by a single individual, firm, or agency. Most large private ownerships are held by wood manufacturing corporations, and in such cases the monopoly of timber supply is more probably a byproduct of the raw material security such ownership affords. Price effects could arise only if other manufacturers attempted to buy in the monopolized area, a situation very possible, but not believed to be of great importance in the United States. The use of private monopolistic ownership of timber is more likely to be directed toward restricting entry into the monopolist's market area.

The prime example of standing timber monopoly in this country is that of public land holdings. The United States Forest Service and the Bureau of Land Management provide examples of public timber monopoly, especially in the west, where large forest areas were withdrawn from the public domain and made part of the national forest system or where railroad land grants were returned to public ownership. In 1969 fully two-thirds of all sawlog stumpage was under public management.

**Constraints on public timber sales.** Timber sales from public agencies are made under policy constraints that would seem to prevent any direct economic exploitation of their monopoly position. With the United States Forest Service, two such constraints are particularly significant.[6] First: the allowable cut is determined almost entirely by biological rather than market criteria; and second, minimum prices are established by appraisal, using costs and returns from "operators of average efficiency." The first constraint means that the inelastic part of the supply curve from public lands (Fig. 17.6) is established at an output ($a$) that is unrelated to the economic monopoly position. The second means that the level of the completely elastic part of the curve is independent from, and probably decidedly lower than, the level that would be determined under monopolistic pricing.

[6] In addition to the two constraints cited above, Bentley gives a third—that "the demand for stumpage is exogenous." Technically, of course, this is necessary, but with public stumpage this does not seem to be very significant. The discussion should be referred to by anyone interested in details of forest service pricing policy. Four general criteria "that *should* guide selection of timber sale procedure" are cited: economic efficiency, economic progress, economic stability, and equity. These criteria are the same ones most often given for tax policy, for public investment policy, etc. (Bentley, 1968).

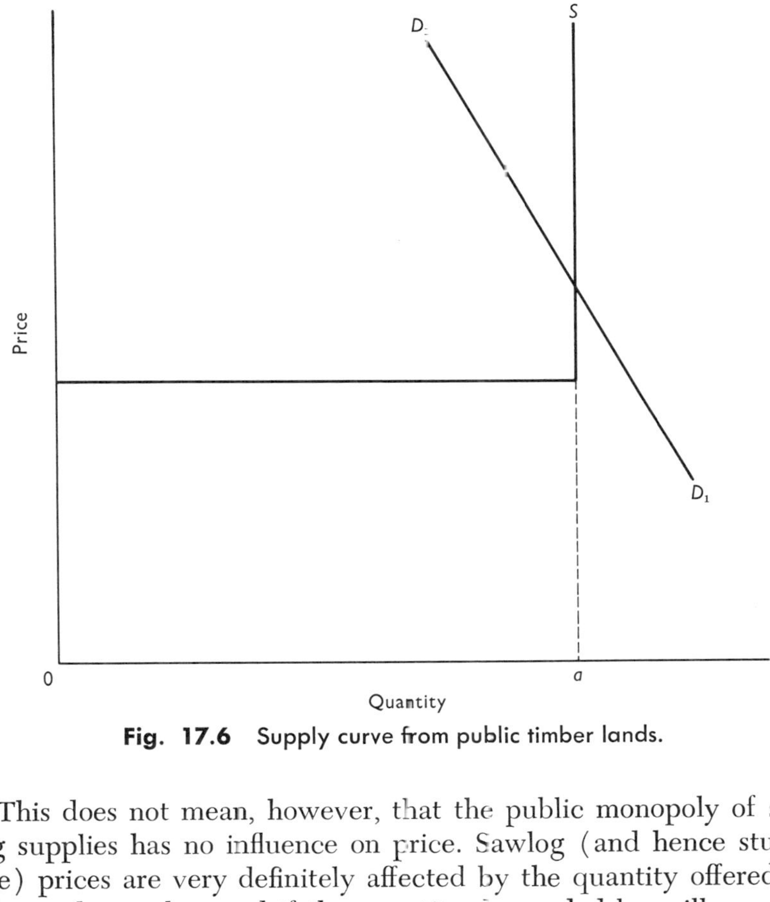


**Fig. 17.6** Supply curve from public timber lands.

This does not mean, however, that the public monopoly of sawlog supplies has no influence on price. Sawlog (and hence stumpage) prices are very definitely affected by the quantity offered for sale on the market, and if the quantity demanded by mills exceeds that established by the allowable cut policy—if the demand curve for sawlogs intersects the supply curve in its inelastic portion as does $D_1D_1$ in Fig. 17.6—then the appraised price is primarily useful as a starting point for the auction. It has little relevance to the final sale price. The *objective* of the restricted output level may not be that of acquiring monopoly gains, but the *result* so far as price and the availability of wood products to consumers is concerned may be very similar. If one further considers the interaction between stumpage and lumber prices over time, especially when the escalating effect of appraisal is included, there can be little doubt that the public monopoly of sawtimber has and does influence the price of stumpage, of sawlogs, and of lumber.[7] Whether the influence has been beneficial or otherwise is a quite different question and depends very considerably upon the criteria chosen. There is no reason to demand that the criteria be economic in nature; hence, the discussion of this point will be deferred. At this time we are primarily interested in establishing the economic framework and in exploring some of its implications as examples of how it should aid analysis.

## STUMPAGE PRICES IN DEVELOPING NATIONS

In developing nations the landholding pattern sometimes leads to monopolistic timber pricing even under conditions of strict bilateral monopoly. In Mexico, for example, large areas of forest land have been used to establish new *ejidos,* a form of communal land ownership.[8] Lumber and pulp producers, even though holding an exclusive right to purchase any timber sold in a specified area including one or more *ejidos,* are sometimes forced to close down because of excessive stumpage prices set by the *ejidatarios.*

The more usual case in developing countries is probably that of monopsony. To aid the industrialization process governments often grant timber "concessions" (exclusive rights to harvest timber over a specified area of publicly owned forest). Such concessions are made to individual concerns, either domestic or foreign. If private holdings are intermixed with public lands the only market for the private timber may well be that controlled by the concession holder, and monopsonistic pricing may result. Stumpage pricing, combined with sales procedures, can be a powerful tool to promote efficient timber utilization and to further the desired pattern of industrial

[7] Several analyses of forest service sales procedure have been published within the past decade and some include reference lists. Among the more significant analyses see Worrell, Lockwood, Lauridsen (1963); Bentley (1965); Davis et al. (1962); and Fedkiw (1965).

[8] As an agricultural system in a land-hungry nation the *ejido* pattern has much in its favor. Of ancient origin, *ejidos* are, and have been, of great political importance in Mexico, especially since the agrarian revolution. Over the past decade, because most of the land suited to agriculture has already been granted, an increasing amount of forest land has been taken to establish new *ejidos* for the rapidly expanding rural population. Unfortunately for Mexican forestry, for Mexico's wood-using industries, and for the nation, this has resulted in rapid destruction of forests, with erosion and desert the foreseeable end. Mexico simply cannot afford to waste its resources in this fashion, and a major change in policy must come about if real disaster is to be avoided.

development. The subject of concession pricing alone offers interesting opportunities for research.

## EFFECTS OF TAX STRUCTURE

The impact of property taxes on timber production was noted in Chapters 10 and 15. Here it seems necessary only to mention that any tax paid by a stumpage owner necessarily reduces his income, hence tends to lessen his financial interest in timber production, and in the long run must act to increase prices of wood products. One can scarcely claim that the property tax "discourages" timber production, for timber will grow whether taxed or not. But anything that reduces income from timber production tends to reduce the owner's interest in forest management.

Very often the indirect effect of a tax is of greater interest than the direct effect. Any tax levied on wood products, for example, or any tax tending to shift the supply curve of other productive factors upward (such as a tax on gasoline that would increase transportation cost) will either depress timber prices, raise lumber prices, or do both. Such a tax is almost always "shifted." Although levied on transportation and paid directly by the logging industry, it is ultimately passed along and paid by stumpage owners or by lumber consumers, or both. And conceivably, it could depress the price of gasoline as well.

### Capital Gains

Probably of more concern from the standpoint of stumpage price formation is the possible effect of the present capital gains treatment of timber. This is a somewhat delicate subject in corporate circles, and data to prove or disprove various hypotheses is understandably difficult to obtain. Section 631(a) provides (see Chapter 10) that the cutting of timber for sale or use by a taxpayer who owns or has a contract right to cut the timber shall be considered as a sale or exchange—providing the timber or cutting contract has been owned for six months at the start of the tax year. Section 631(b) gives the same right to any owner who "retains an economic interest" in the timber. While it is true that an owner who sells timber outright (or who sells land and timber together) can also treat the resulting net revenue as a capital gain, it is the effect of these two special provisions that are of most interest with reference to stumpage prices.

There are two possible influences of these provisions on stumpage price. As an example, suppose lumber company A had acquired large forest holdings years ago so that it now owns enough to supply all of its needed timber. Lumber from its sawmill sells at $100 per thousand board feet, with all non-timber expenses amounting to $50 per M. If the company could establish a price (value) for its own timber of $40 per M it could make a profit of $10 per M (a 10 per cent return on the selling price) and would pay the full corporate income tax on this profit. They would also show a positive return, and be taxed, on the *increased* value of the timber they cut—on the difference between the original cost of the timber and the present value of $40, less any capital charges—but on this increase in value they would be taxed at the capital gains rate of, roughly, one-half the regular corporate income tax.

Company B, located alongside Company A, owns no timber. It buys from any seller, but normally from public timber holdings as they are offered for sale. Like A, Company B buys stumpage at $40 per M, sells its product for $100 per M, and has operating costs of $50 per M. It also makes a profit of $10 per M upon which it must pay corporate income tax at the full rate. Company B makes no profit from its timber holding since it owns none (its purchased stumpage is always cut within six months of acquisition) and so of course pays no capital gains tax on timber. The only difference between the two companies, then, is that Company A had the foresight (or good luck) to acquire forest land when it was readily available. It earns a capital gain on this investment and on this gain pays a tax at capital gains rates. Fair enough!

But suppose Company A were able to establish that the price of its own stumpage should be $50 per M. Then the sawmill operation would be without profit—no corporate income tax would be paid at all. There would of course be a tax on the $10 increase in the value of the timber being cut, an increase that is certainly income to the corporation. The income, however, will now be treated as a capital gain with a tax (in 1969) of 25 per cent. Instead of paying $5 per M in regular corporate income taxes and having an after-tax income of $5 per M from its sawmill operation, the company will now pay only $2.50 per M as a tax on the capital gain. The corporation's after-tax income has increased $2.50 per M of output by having stumpage value raised from $40 to $50 per M. Clearly, any timber-owning wood-based corporation should try to have as large a part of its profits as possible coming from its timber production rather than from its sawmilling.

What about Company B? Suppose they decide to change their buying policy so they hold all timber purchased at least six months prior to the start of each tax year. They still buy stumpage at $40 per M, but now follow the lead of Company A and attempt to show that the timber they bought has increased in value to $50 per M. If successful they will reap the same benefit as Company A, and Company B's after-tax income will increase by 50 per cent—from $5 to $7.50 per M. By switching their profit from sawmilling to timber holding, *and by proving a stumpage price increase,* after-tax profits have been raised 50 per cent. The example here is extreme and the effect one of short term duration, but it is a valid illustration of this particular aspect of taxation.

The key to the questions raised here lies in determining how value can be placed on corporation-held timber, whether the timber has been owned for long periods of time or for periods as short as six months. Various methods are used, but the most usual involves transactions evidence. If the company can show that similar timber sold for $50 per M the court may accept the price.[9] It will, therefore, pay such a corporation to "bid up" the price on a few sales—enough to establish the higher price if called upon to do so —since for every dollar-per-thousand increase in timber value established, their profit per thousand will increase by 50 cents. There are severe problems in using evidence from a few purchases to establish a general price rise in stumpage values, especially if those purchases have been made by the company immediately concerned. There is no question, however, that there is a short term incentive for timber owning corporations or for those holding cutting contracts for six months or more, to "bid up" timber prices.

There is also an important secondary effect. Any general increase in stumpage price must eventually be reflected in the price of the final product. And with the usual public land appraisal techniques, higher product prices mean that a larger "residual" is available for

[9] The usual procedure, as with any income tax dispute, would be first to make the return using the $50 timber value. The Internal Revenue Service may accept, or it may dispute, the price. If the Internal Revenue Service insists that $50 is too high the case will eventually be brought to court, with the court deciding the issue. It is here that "transactions evidence" becomes important.

stumpage, which tends to raise timber prices further. From this standpoint Sections 631(a) and (b) tend to act as built-in escalators of stumpage price, although it must be remembered that the appraised price is primarily important as a floor whenever the demand for sawlogs is at a level such that the intersection with the supply curve from public timber falls in the inelastic portion.

That stumpage prices have in fact been pushed up by the corporate income tax structure, and especially by Sections 631(a) and (b), has not been demonstrated. In fact, Mead's study of the Douglas fir industry showed quite the opposite (Mead, 1966). Data to prove that bidding practices have tended to increase sales prices on either public or private timber would be very hard to obtain, and would certainly not be in the interest of any company to divulge. About all one can be sure of is that executives of large timberland owning corporations are not naive. Obvious opportunities for increasing the profitability of an operation will not remain unexploited for long. In the long run, increased stumpage prices will tend to force up lumber prices, inevitably reducing the quantity of lumber sold and, at least until the excess production capacity is eliminated, reducing the opportunity for profit. Whether this would most affect large corporations, who are better able to take advantage of capital gains and of scale economies, or smaller companies, that normally benefit little from capital gains provisions, is a point that has received no careful analysis.

## EFFECTS OF OTHER FOREST LAND USES

A full treatment of the demand for forest land, only in part derived from the demand for timber-based products, will not be developed here. At the same time it must be recognized that as demands for non-timber products increase, fewer acres will be available for stumpage production. The effect of this reduction on stumpage price could be considerable, but perhaps even more important is the possible effect on forest management for timber production.[10]

In general, as wilderness areas, national parks, roadside scenic

[10] The conclusion reached by Amidon and Gould in their study of three California national forests—that there has been little interference with timber output as a result of roadside reservations, etc.—only shows that for the area and time period studied such effects had not been important as far as reducing output is concerned. It could still have an important effect on price, either now, or in the future (Amidon & Gould, 1962).

strips, and other non-timber uses remove timberland from the market, reservation prices for timber will shift upward, thus tending to raise stumpage price. It also is possible that reservation prices will move up as a result of owner's anticipations of these other demands. A long term upward shift in reservation prices, however, would probably be limited by the knowledge that intensified management practices such as use of improved, high-yielding strains, the application of fertilization, etc., and particularly the widespread establishment of man-made forests (either plantations or newly produced, well-stocked natural stands) can bring about large increases in timber production per unit of area.

### Natural and Planted Forests

Figure 17.7 illustrates the probable short term relationship of natural to man-made forests. The demand curve (*DD*), is an aggre-

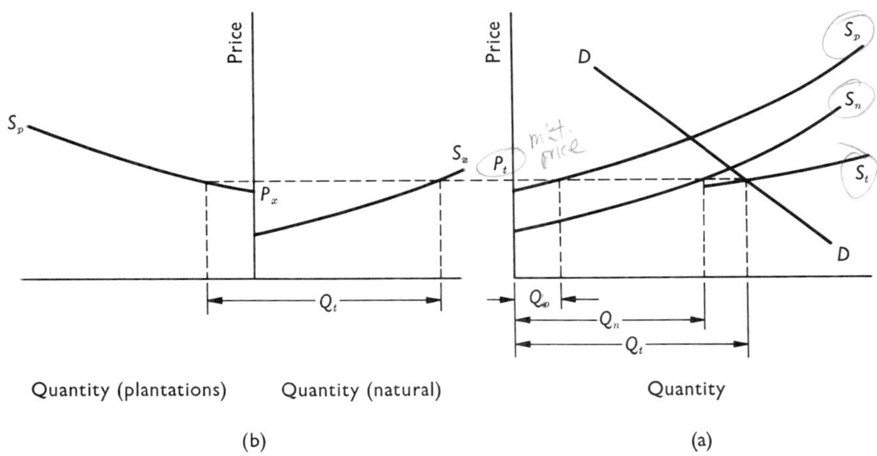


**Fig. 17.7** Short term relationships between natural and man-made forests.

gate of demand for sawlogs, pulpwood, veneer bolts, etc.[11] Two supply curves are shown, $S_n$ representing the supply curve for material from natural forests, and $S_p$ the supply curve for material from

[11] Demands for raw materials such as pulpwood, sawlogs, etc., may be aggregated in the same fashion as the derived demands for lumber used in shipping were aggregated in Chapter 8, page 152. It is, of course, necessary that all the demand pertains to the same kind of material in the same market.

plantations. Both curves are reservation cost schedules. $S_p$ includes the reservation prices of all plantation owners, primarily influenced by their expectations for *timber* prices, and $S_n$ includes the reservation prices of owners of all natural forests in the market area. These latter will be influenced primarily by expectations of demand for forest land to be used for camping, wilderness areas, watershed protection, esthetics, summer home sites, fishing camps, etc., and also by timber price expectations. Both supply curves are necessarily rising, and may be added together to form the single supply curve, $S_t$, shown in part (a). The discontinuity in this latter curve arises because man-made forests are not likely to be established unless price expectations are above some minimal level, while at least some natural stands will be available to supply timber at relatively low prices.

The intersection of $S_t$ with *DD* indicates the total quantity of timber that will be taken at the market price, $P_t$. At this price the amount $Q_n$ will be produced from natural stands and $Q_p$ from plantations. To facilitate the analysis that follows, the same two supply curves are shown in Fig. 17.7(b) as a set of "back-to-back" supply curves. Here, at any price above $P_x$, the total quantity of timber provided will be indicated by the horizontal distance between the two supply curves.

Figure 17.8 is an extension of this same diagram, constructed by incorporating time as the third dimension. At the right (a) of the figure are two conventional short term demand and supply curves for logs, the lower one representing the situation as it might appear today (essentially the same as that of the previous figure) and the upper showing the supply–demand picture at some future period, say, for the year, $T_1$. The supply curves of the two-dimensional figures on the right have been produced by adding together the two timber supply curves—the reservation cost schedules—of both natural and man-made forest areas in a given market. The differing levels and shapes of the two sets of curves indicate possible shifts that might occur as price expectations rise over time.

The same two supply curves are shown individually on the front and back of a three-dimensional diagram (b), this time using "back-to-back" supply curves so that each part can be examined. The addition of the time axis permits visualizing possible shifts of both supply curves over time. In the specific case illustrated the shifting curves produce the two supply surfaces, related to the de-

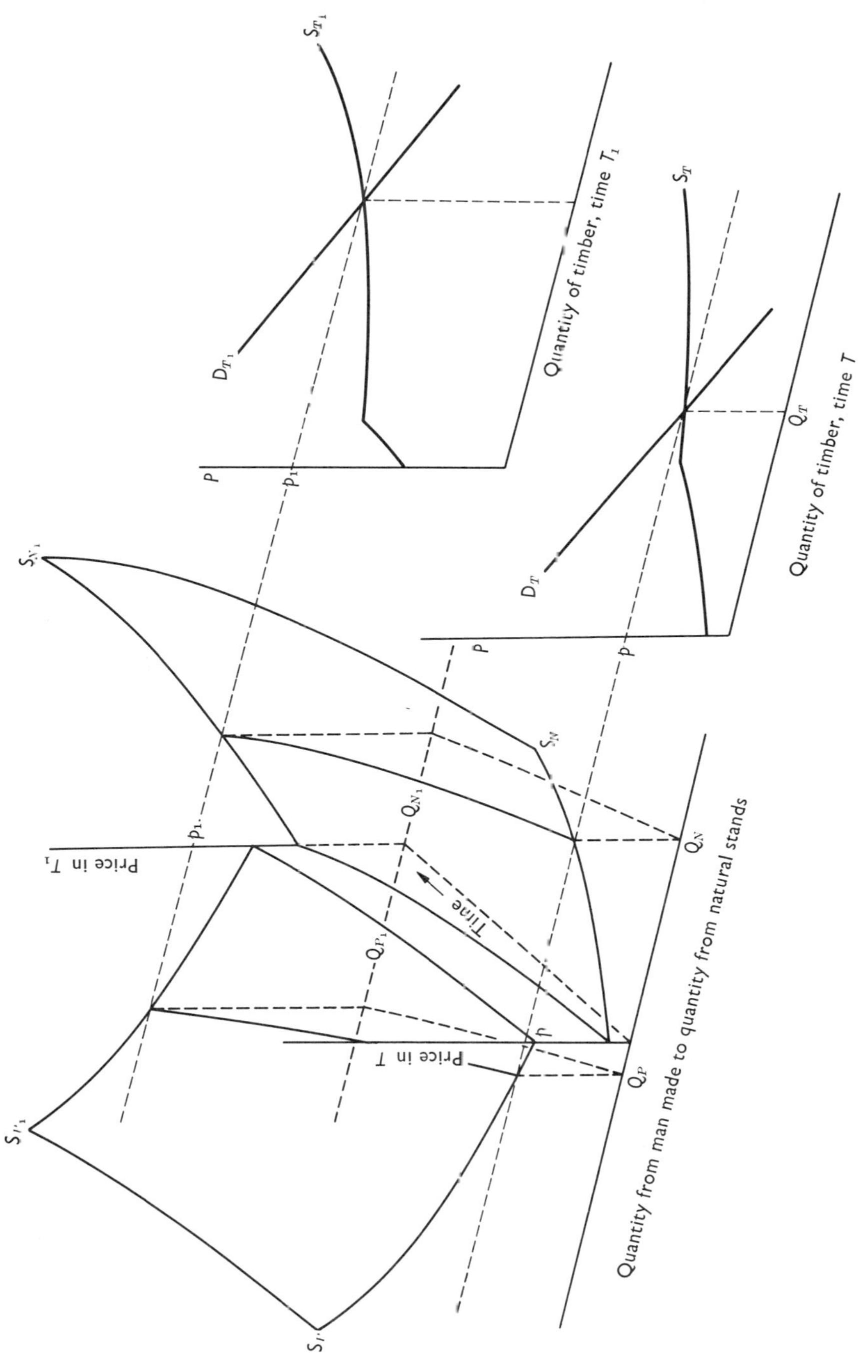


**Fig. 17.8** Three-dimensional representation of the changing relationships over time between natural and man-made forests.

mand aspects through price. The total quantity of timber available at any given price level is now the sum of the horizontal distances between the price axis and each of the supply surfaces. If price is sufficiently high, then plantation wood will be available and the total quantity will be the horizontal distance between the two surfaces.

## Rising Price Expectations

As shown in this particular case, the demand is shifting outward. Timber demand is increasing over time as more and more wood is consumed—by increasing populations, by newly discovered processes or products, etc. Available supplies of forest land, however, and consequently supplies of the timber this forest land can produce, are expected to decrease over time. If no other response was generated the quantity of timber available for wood products would decrease very substantially, to $Q_{N_1}$. But there may well be another response. Under the spur of higher price expectations, more capital and other inputs will be devoted to plantation establishment and intensive forest management. Because of the large increase in productivity that is possible from such measures, and because it seems likely that demand for such intensively managed areas by non-timber uses would be relatively low, the reservation cost schedule for these stands should rise less sharply than that for the remaining natural stands. The end result could be as illustrated—a large increase in production from man-made forests, perhaps more than enough to compensate for the loss of production from natural stands. Timber output in time $T_1$ is shown to be substantially increased over that at time $T$, with well over half the output coming from plantation sources. At the same time, prices are seen to be rising sharply, from $p$ to $p_1$.

The interaction between the two supply surfaces is probably the most intriguing part of this model. Price expectations of forest land owners result from many factors not connected with timber markets per se. Any possible use for forest land probably has some influence on price expectations of forest land owners, not only for the land, but also for the timber on its surface. Yet the actual price of timber over the long term is subject to a very real constraint. If price expectations rise rapidly, man-made forests can and will be established, and their production should serve to hold down timber prices. For if the quantity of stumpage available in the not-too-distant future can be substantially increased through intensive management, *higher cuts can be made on present stands.* And here again the institutional factor of public land ownership takes on far-reaching significance. If allowable cut policies on public timberlands are determined by applying biological formulas that utilize a 90- or 100-year rotation; if these formulas take no consideration of changing economic conditions, such as shifting demands; if no cognizance is taken of the possible increased output under intensified management; then this constraint to rising prices may be prevented from having any effect. Under such conditions both stumpage and lumber and veneer prices may rise drastically.

At the same time it must be emphasized that no claim for quantitative exactness can be made, or for that matter, for correctness in the relative importance of production from natural and man-made forests for any particular time. The direction of the change, and the influence of the two possible kinds of forest production, however, are clearly indicated. In fact, one could argue that the model is appropriate to forestry in both the United States and Western Europe today. In both areas there is a clear tendency toward reliance on plantation forestry for timber production, and toward diversion of the natural forests to non-timber uses.

# PART IV

# NON-TIMBER PRODUCTS ECONOMICS

# 18

# Multiple Use

## FORESTRY AND LAND MANAGEMENT

Earlier chapters have stressed difficulties of forest economics caused by the variety of products into which wood is manufactured, difficulties particularly noticeable on the demand side. For years the forest manager has been unable to say with certainty whether the timber he has produced will be used as an integral part of a frame house, as a member of the scaffolding temporarily erected during the construction of a highrise office building, as material for making grocery bags, or for all three—or for any of hundreds of other products. And the complexity of timber products is only part of the story. Today's forester may find himself managing a watershed protection unit, a game refuge, or a public or private hunting facility, and timber production may not even be a secondary consideration. He might need to devise a cutting method to provide scenic overviews from a high speed throughway, or to save old growth forest on a national park for as long as possible, or serve as manager of a wilderness area. In any one of these, whether the objective is aesthetic responses or sawlogs, overviews or veneer bolts, the forester is manipulating a land-based ecological system to produce a product or combination of products desired by man. This fundamental focus on land management and ecological relations rather than on the product itself has traditionally characterized forestry, although timber production tended to dominate forestry during the first two post-World War II decades in the United States.

### The Environmental Crises

A new factor became evident toward the end of the 1960's when the nation was swept with a wave of concern about environmental deterioration, particularly that attributable to pollution of soil, air, and water. Forestry per se will probably play a relatively minor part in pollution abatement, although some wood-based industries are major contributors to both air and water pollution loads and forests play a large, though poorly understood, role in air purification. But simply because forests occupy one-third of the nation's surface area, forestry and foresters in the United States are necessarily involved in the "environmental crisis."

The new awareness of environment has resulted in the elevation of ecology to academic respectability and has created an upsurge of interest in environmental problems by the so-called "hard science disciplines." That both were necessary and overdue is readily granted. What has not yet been generally recognized is that many of the environmental problems are not amenable to solution by these "hard sciences." The really basic difficulties lie as much, or more, in our political, social, legal, and economic institutions. Public concern with environmental problems now goes far beyond pesticides and pollution. Rachel Carson's *Silent Spring* did much to awaken the public to the environmental dangers a short-sighted science might create when coupled to a highly efficient production and distribution system (Carson, 1962). Of at least equal significance was the fact that this new awareness coincided with the rise of activism throughout the nation, generated by problems of civil rights, integration, military involvement, and a hard, harsh reexamination of our value system for modern relevance. The result has been a realization on the part of many that increased *quantities* of material things might cease to raise the *quality* of living, and might in fact have precisely the opposite effect. Applied to environmental problems this reexamination has brought on an awareness of deteriorating environmental quality. Serious questions are being raised about the diminishing availability of "wild space," the reduction in aesthetic variety, the stresses of urbanization stemming from man's increasing separation from nature and the serious results that could come if this separation is not somehow bridged, and the need for both mental and physical re-creation. The changing system of values that divides our society today is having a profound influence on forestry—on the ways that forests are managed, and especially on the purposes for which forests "should" be managed. Conservation has again been redefined. The early preservationist attitude was replaced by a concern for "wise use," and eventually defined in economic terms as the problem of securing an optimum allocation of resource use over time. Attention has again shifted, and now centers on problems of environmental quality. What end result this new concern will ultimately bring about is unclear, but it will have a decided influence on forestry, and especially on the relative emphasis which foresters will accord the non-timber products.

### The Challenge to Forestry

Public forest land managers are already under considerable pressure to devote more effort to recreation and aesthetics—to practice "people forestry instead of timber forestry." As a profession, forestry is being challenged to broaden its view, to prove it can satisfy the needs of a modern economic society which wants more and better houses, more forest recreation areas, bigger highways, and better hunting and fishing places. It may not be possible to satisfy all these demands within the economic and political framework that presently exists, but when treating non-timber products of forests we are inevitably involved with these kinds of problems.

A good many of these difficulties arise because forests may be regarded as, and also are a source of, "public goods," goods that for various reasons are produced in the public sector. Without public production, some of these goods might not be produced at all, and in other cases the goods are such that no one can be prevented from enjoying them once they have been produced. Public goods are often characterized by peculiar problems of measurement and pricing. Because problems of allocating public funds, and of determining budgets, programs, and projects are not always appropriately solved by the methods developed by private industry, a rather special kind of economic analysis called "benefit/cost analysis" has come into being over the last few decades. Public goods, and benefit/cost analysis, will necessarily be discussed in this section.

But public goods range from space programs and national defense to wilderness areas, fishing sites, and education. Some non-timber forestry products produced in the public sector are also produced privately, and these non-timber products of American forests are far too numerous to be dealt with individually. Moreover, any one of such major products—outdoor recreation, water, fish and wildlife, aesthetics, forage, or even air purification—is sufficiently broad, different, and important to require a separate volume, or volumes, to treat it in depth. The primary objective of this last section cannot be treatment in depth, but is rather that of presenting some of the current economic thinking, along with specific analytical tools and concepts, that may help present and future students to understand better, and perhaps in a few instances to resolve, these non-timber problems.

Being an ecosystem, a forest almost inevitably produces a variety of products. As stated in the Timber Report of the Public Land Law Review Commission (P.L.L.R.C., 1969):

> Rain falls, and water comes from the forest as an almost inevitable and completely natural sequence of events. Timber is grown, wildlife reproduces, and beauty is created without any conscious effort by man. But through either positive management and protection, or through negative neglect and carelessness, man can and does alter the amounts of the various products in both relative and absolute terms.

The term "multiple use" has been used for years to indicate recognition of this product multiplicity. Now more than recognition is being demanded. Methods must be devised to assure that when man alters the amounts of the various products, his lot is improved. The essential conceptual economic framework for such management has already been developed in Chapter 13 under the general topic of "joint production." This last section will begin by placing multiple use within the joint production framework, then proceed to a general discussion of public production, welfare economics, and benefit/cost analysis.

The three major non-timber products—forest recreation, water, and wildlife—will then be examined in some detail. Forage will be omitted, not because it is unimportant, but because methods for dealing with problems of timber demand, production, supply, and pricing can be adapted to forage with little difficulty.[1] Other non-timber products pose distinctly different problems than those met in timber and forage. Moreover, forest recreation, watershed pro-

[1] There are, however, policy problems of forage production, particularly on public forest lands, that differ sharply from those of timber. Most of these deal with vested rights to grazing that have come about over time.

tection, and wildlife are rapidly increasing in importance to forestry in this country as in many other areas throughout the world. Technical aspects of producing these products will necessarily be very lightly touched, and the economic aspects will be given a less detailed treatment than when timber was the focus. In part this will be because the general framework of production, supply, demand, and price has already been covered; it should not require as detailed a presentation the second, third or fourth time as the first. The larger reason for the less intensive treatment, however, is because the information on most non-timber products is too sparse to permit quantitative presentation. It should also be pointed out that in two of these areas—forest recreation and water—there is a growing body of literature. Foresters specializing in recreation or in watershed management should familiarize themselves with the appropriate works.

The final chapter does not really belong in this last section, yet someplace in a modern presentation of forest economics recognition must be given the fact that there are problems and opportunities created by the changing international picture of which each nation is a part, but only a part. These problems do not usually stem from non-timber products and services, but from forces so far outside traditional forestry concerns that it seems best to treat them as a separate entity.

## MULTIPLE USE IN FORESTRY

Just where the term originated is of little importance today, but the idea of "multiple use" is deeply embedded in forest history. Early German estate foresters were often more concerned with producing good deer hunting than with timber production, and some of the oldest European town forests were established as much for protecting municipal water sources as for furnishing fuel and other wood products to the local citizenry. In this country the multiple use idea was clearly embodied in Wilson's famous letter to the first Chief Forester (written by Pinchot himself) officially transferring the "Forest Reserves" from the Department of the Interior to the Department of Agriculture in 1905 (Pinchot, 1947).

> In the administration of the forest reserves it must be clearly borne in mind that all land is to be devoted to its most productive use for the permanent

good of the whole people and not for the temporary benefit of individuals or companies. All the resources of forest reserves are for use . . .

You will see to it that the water, wood, and forage of the reserves are conserved and wisely used. . . . The continued prosperity of the agricultural, lumbering, mining, and live-stock interests is directly dependent upon a permanent and accessible supply of water, wood, and forage . . .

### The Multiple Use–Sustained Yield Act

More than a half-century later, multiple use was finally given a statutory basis by the United States Congress when it passed, on June 12, 1960, the "Multiple Use–Sustained Yield Act" (United States Code 16, 1960). Listing the five most frequently acknowledged major uses in alphabetical order, this legislation specified that national forests would be administered for outdoor recreation, range, timber, watershed, and wildlife and fish purposes. Multiple use was defined (in Section 4) as follows:

"Multiple Use" means the management of all the various renewable surface resources of the national forests so that they are utilized in the combination that will best meet the needs of the American people, making the most judicious use of the land for some or all of these resources or related services over areas large enough to provide sufficient latitude for periodic adjustments in use to conform to changing needs and conditions; that some land will be used for less than all of the resources; and harmonious and coordinated management of the various resources, each with the other, without impairment of the productivity of the land, with consideration being given to the relative values of the various resources, and not necessarily the combination of uses that will give the greatest dollar return or the greatest unit output.

Readers should not be entirely surprised that this definition has fallen somewhat short of completely clearing up all questions regarding multiple use of forest lands.

The act itself was largely a restatement of existing statutes and procedures which the Forest Service had followed for many years. Perhaps its principal objective was that of giving support to officials who were under pressure from special interests to use the national forests for some particular purpose rather than for multiple purposes (United States Congressional Record, 1960). In 1964 legislation similar to the Multiple Use–Sustained Yield Act officially extended (on an interim basis only) multiple use to lands supervised by the Bureau of Land Management.

For many years *multiple use* was a major rallying point for those who favored public control of forest land. Today the larger timber-owning corporations are moving toward a multiple use view, and for a variety of reasons. Forests can and do produce many goods and services. If these products are in demand one should expect to find private operators willing to provide them wherever and whenever they can be provided at a profit.

## MULTIPLE USE AND JOINT PRODUCTION

The basic economic approach to multiple use was developed in Chapter 13 where the joint production of timber products was discussed. From a conceptual standpoint it makes little difference whether the products being jointly produced are as basically similar as sawlogs and pulpwood, or are as dissimilar as poplar pulpwood and white tailed deer, or scenic beauty and Douglas fir peelers. The same conceptual model can be used; it is in the application that differences (and difficulties) arise.

### An Economic Model

Joint production, it will be recalled, occurs whenever the same production facility is used to produce two or more products. In the model particularly appropriate to multiple use problems, two products may be produced in "technically variable" proportions (Chapter 13). The production of timber and forage provides a straightforward example.

With per acre production of timber measured in cubic feet and of forage in animal unit months (AUM's), Fig. 18.1 permits any possible combination of the two products to be shown.[2] A set of "production possibility curves" can be constructed by connecting all timber and forage combinations that can be produced from the area at a given cost. Since cost is constant on such curves they are also called "isocost curves."

Assuming that the amounts being produced are not large enough to influence the price of either product, the isorevenue curves (curves connecting all points yielding equal total revenue) will be straight lines, with the slope of the lines inversely proportional to the price ratio of the two products. In the example of Fig. 18.1, with timber at 25¢ per cubic foot, and AUM's at $1 each, either 100 cubic feet of timber or 25 AUM's will yield a total revenue of $25.

[2] An animal unit month (AUM) is the amount of forage required to support one cow, or five sheep, for one month. It is the forage unit commonly used in western forests.

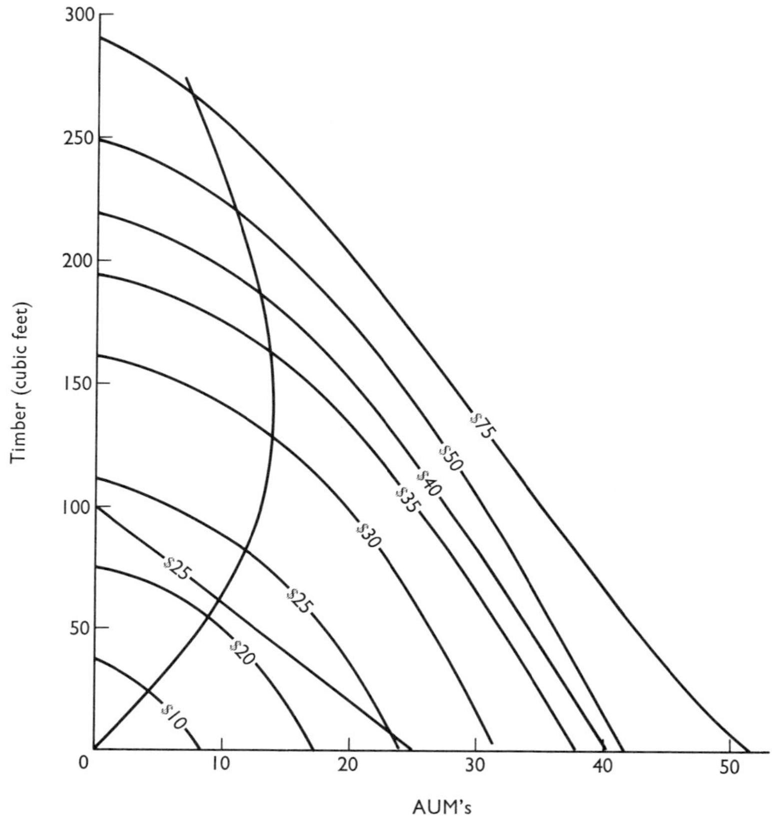


**Fig. 18.1** Production possibility curves (isocosts) and isorevenues for joint production of timber and forage.

And every combination of timber and forage indicated on the line connecting these two points will yield this same return. As long as prices remain unchanged, all isorevenue curves will be parallel.

The points of tangency between the two families of curves (isocosts and isorevenues) identify the various combinations of products that will produce the indicated revenue for the lowest possible cost (economic efficiency). Connecting these points of tangency yields the "expansion path," or scale line, the path of product combinations along which production would logically be expanded. The problem now becomes one of determining which point on the expansion path maximizes net revenue, a problem readily solved by the usual graphic or mathematical methods. The entire process can be combined into a single diagram such as that of Fig. 18.2, where the product combinations and expansion path are shown below, and the usual total cost and revenue curves directly above. Maximum net revenue is attained (in the example) at an output of 178 cubic feet of timber and 13.2 AUM's. With this general analytical framework as a basis we can examine the two most common interpretations of multiple use forest management.

### Traditional Interpretations

One group declares that multiple use must be interpreted primarily, or even solely, in terms of large areas—a ranger district, working circle, or perhaps an entire forest holding, either public or private. The interpretation may be visualized from Fig. 18.3. The total area mapped in the figure produces several products, but any specific area is devoted to a single (or at least primary) use. Under this approach the land manager settles upon a *primary* use for each subdivision of the total area. *Secondary* uses are permitted, even welcomed, as long as they do not interfere with the *primary* purpose of that area's management. Timber production, for example, might not be prohibited on a designated "recreation" area, but timber production would not be allowed to interfere in any way with recreation, the primary use of the particular area.

The alternative approach makes no area subdivision and declares no uses to be either primary or secondary. Instead, the primary objective of management is assumed to be maximization of net returns. Hence, every acre of the forest should be made to produce that particular combination of products that will maximize net returns to the owners, either public or private, and whether measured in monetary or other units.[3]

### An Economic Comparison

Now suppose one attempts to apply the "primary use" approach to three different areas, each having the production possibilities

[3] The list of authors writing on multiple use is far too long to attempt complete citation. Probably the first direct application of joint production theory to forestry problems was that of Ciriacy-Wantrup (1938). It was applied to multiple use by Gregory (1955). The two approaches cited above were called, respectively, the "Pearson approach" and the "Dana-McArdle" approach, after articles by these authors had, in essence, outlined the interpretations.

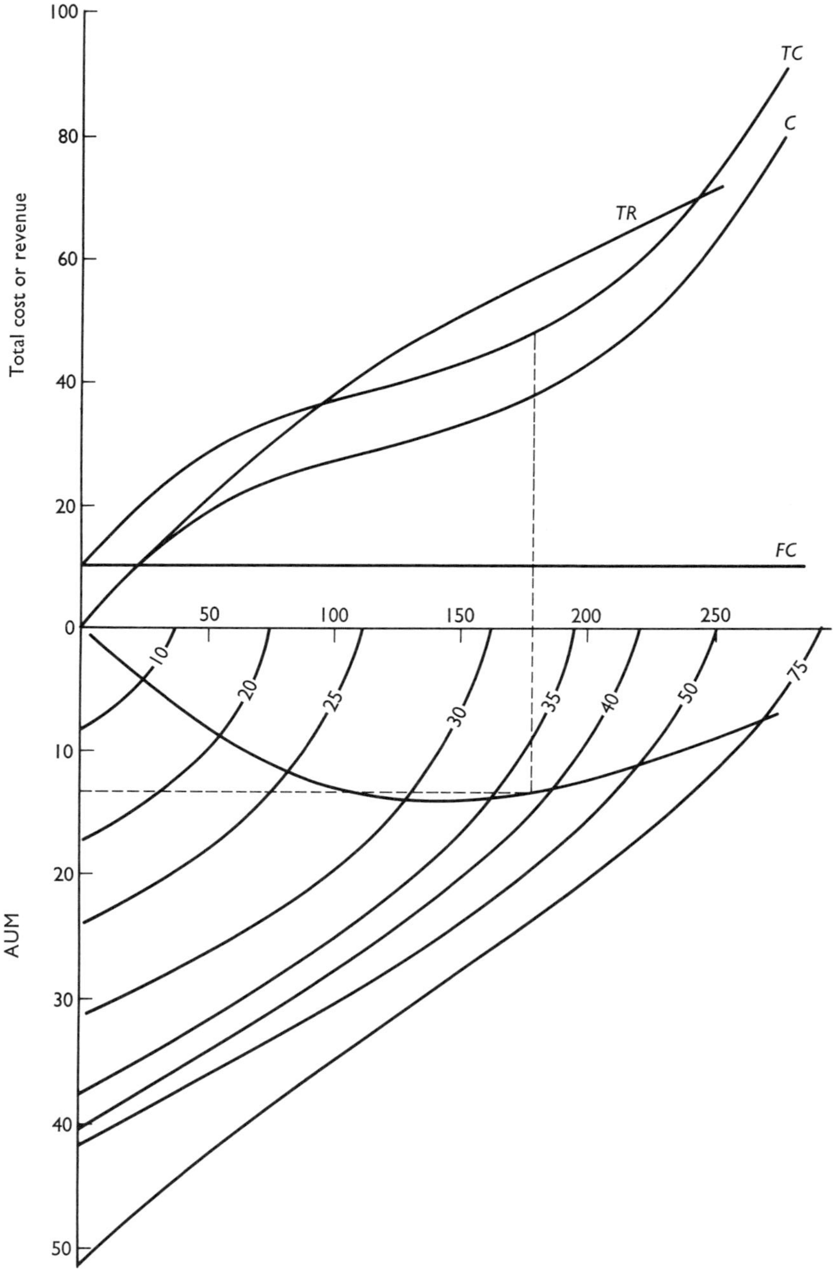


**Fig. 18.2** The expansion path and the point of maximum net revenue, indicating the optimum combination of timber and forage under the assumed conditions.

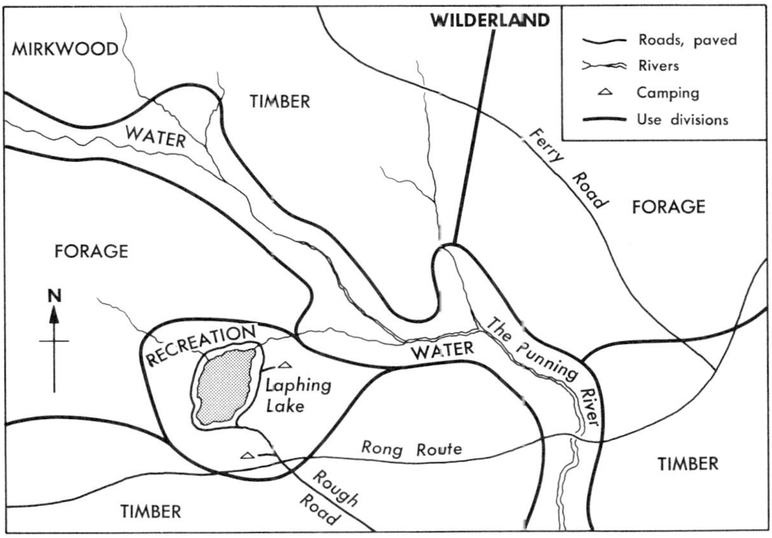


**Fig. 18.3** Map of forest area designating primary uses. Multiple use on total area, but single use on any specific area.

graphed in one of the three diagrams of Fig. 18.4(a), (b), and (c). In each diagram, production of one product (say, timber) is shown on the vertical axis, and of some non-timber product (here, recreation) on the other.

Of particular interest here is the shape of the production possibility curves that identify all combinations of the two products that could be produced for a given level of expenditure. The production possibility curves of (a) indicate that the two products are competitive (the word "substitute" is sometimes used) throughout their entire output range, but competitive in a rather special sense. Output of either timber, or recreation, or of both timber *and* recreation, could be increased by spending more money on production. But at any given level of expenditure the output of more timber will require a decrease in the visitor days of recreation being jointly provided. In fact, when producing any combination of the two products, regardless of the level of expenditure, output of timber can always be increased without any increase in cost simply by reducing the visitor days of recreation, until only timber is being produced

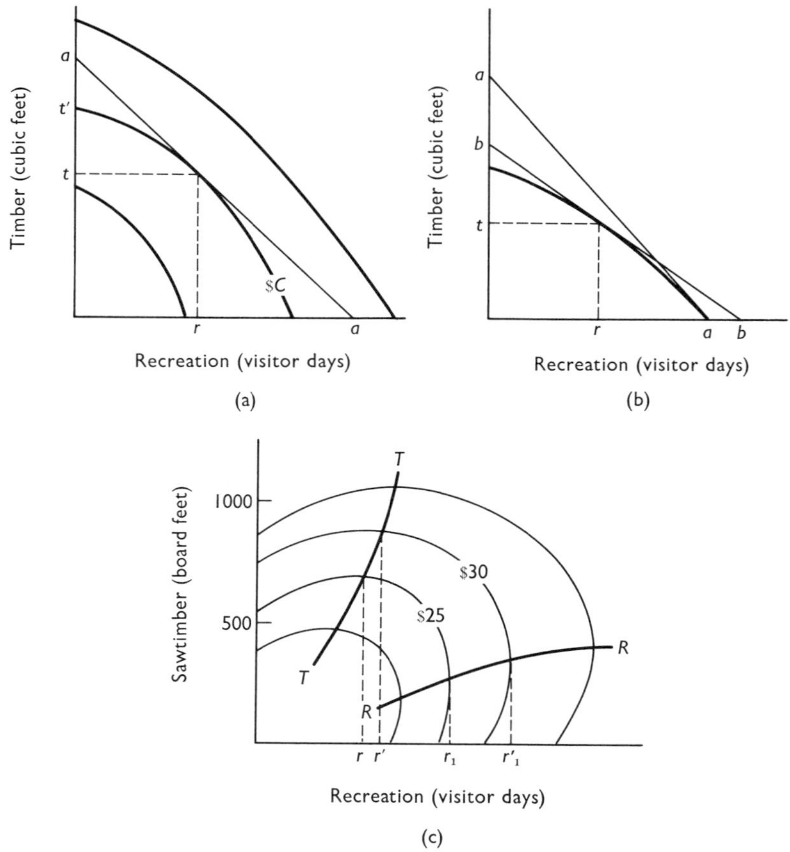


**Fig. 18.4** Production possibility curves for three different kinds of forest areas.

from the area. Clearly, if this area were to be designated as a "timber key" area—one where timber was declared the primary product—then only timber would be produced. At any expenditure level and at any output level, production of the secondary product reduces timber output.

Now suppose the relative prices of the two products were such that the isorevenue curve was that of *a–a* in Fig. 18.4(a). Given these relative values, if an expenditure of *C* dollars is to be made for managing the area, the optimum combination of the two products would be *t* cubic feet of timber and *r* visitor days of recreation. In this particular instance, making timber the primary product would cost the owner the equivalent, in value terms, of $a$–$t'$ cubic feet of timber, for with timber the primary product, $t'$ cubic feet would be produced. The reduction in total revenue is the opportunity cost of making timber the primary product, for costs are identical under either of the production combinations.

In Fig. 18.4(b) the isocost and isorevenue pattern is quite different. With the relative prices of the two products such that the isorevenue curves take the slope indicated here by $a$–$a$, only recreation should be produced if the production expenditure is that shown by the isocost curve. Here, it would seem, is a clear case for declaring the area primarily suited to the production of recreation. In fact, under these conditions one might seem justified in making recreation the *exclusive* use of the area so far as timber and recreation are concerned. But what if values change? Suppose, for example, timber prices rise while the value of recreation drops off until the isorevenue curves take on the slope of $b$–$b$. Now, unless timber production is permitted, net revenue will be considerably less than is possible; the optimum combination under the new conditions becomes $t$ cubic feet of timber and $r$ visitor days of recreation.

The situation shown in Fig. 18.4(c) is probably the most typical, and the most interesting. Here the two products are complementary in cost through part of the output range, but competitive throughout the remainder. As sawtimber output per acre (in this example) increases from the 670-board-foot maximum on the $25 isocost curve to the 875-board-foot maximum on the $30 isocost (or production possibility) curve, output of recreation also increases—from $r$ to $r'$. And the situation is repeated at the other end of the curve. As output of recreation increases from $r_1$ to $r'_1$, sawtimber output is also increased from 275 to 350 board feet per acre. Under these conditions, $r'_1$ is the maximum output of recreation that can be produced at the $30-cost level, while (similarly) 875 board feet per acre is the maximum output of sawtimber that can be obtained at a $30-per-acre expenditure. Between these two points—between the respective maximum outputs of the two products at cost levels of $30—the output of either product can be increased only if output of the associated product is decreased.

What if an area such as $c$ were to be designated as a "timber key" area? Under the specified conditions the expansion path for management would become that of $T$–$T$ in the diagram. At nearly all expenditure levels the area would not only produce timber, but some recreation as well. As expenditures were increased, the output level of recreation would increase with every increase of timber output, even though timber production were maximized at each cost level. Using the expansion path of *T–T* one could proceed to determine the output that would maximize net revenue—a maximization under the constraint that timber yields were also maximized. People interested in timber production would have little to complain about as long as the combinations on *T–T* were being produced. Certainly there should be no complaints about the production of the indicated amount of recreation by timber advocates, for any reduction in recreation output under the conditions shown would either raise the production cost of timber or reduce its output. But note that the combinations along *T–T* can be termed rational from an economic point of view only if recreation is completely without value, for in this very unlikely case the isorevenue curve would be horizontal.

According to the production possibility curves of diagram (c) it would also be possible to declare recreation the primary product, in which case the optimum production path would be indicated by the curve *R–R*. Statements paralleling those made above for timber could also be made directly applicable to recreation if *R–R* were selected as the appropriate expansion path.

If both products have a positive value, the area between these two curves (between *T–T* and *R–R*) indicates the policy conflict area, for within this area the two products are competitive. Net revenue can be maximized only by producing a combination of the two products that does not include the maximum output of either.

One other point can be illustrated by this type of example. Suppose the situation were similar to that pictured in Fig. 18.5, a condition roughly comparable to that used earlier, but using specific figures for the two products (here, timber and wildlife). Assume the area is being managed primarily for timber, and that the cost budget is $10 per acre. The timber output will, therefore, be 250 board feet per acre, and that of wildlife will be two units (a unit of wildlife being composed of combinations of birds and animals). If the isorevenue curve is in reality that of *R–R*, the decision to make timber the primary product implies that 75 board feet of timber is worth the same as three units of wildlife (five units could be produced if revenues were maximized at a $10-cost outlay, but only two are produced because timber is the primary product). The op-

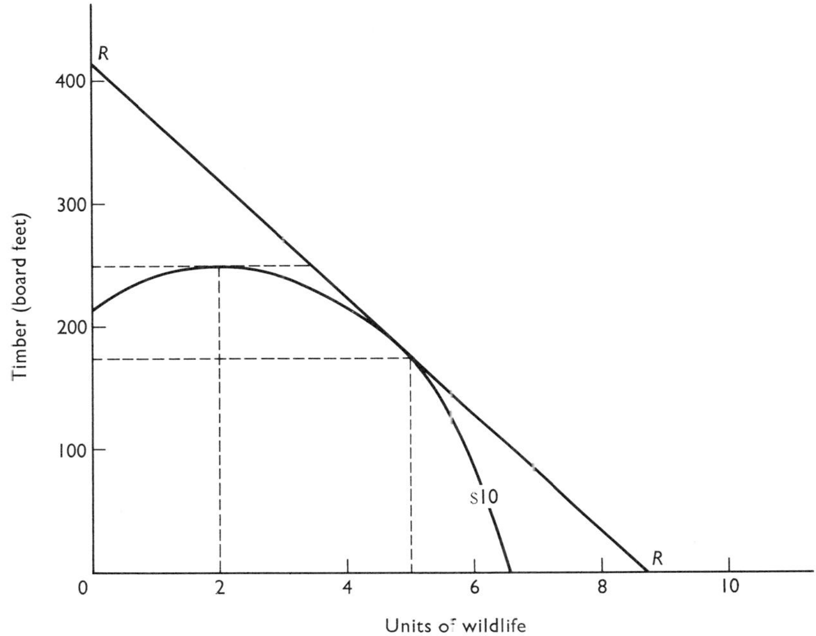


**Fig. 18.5** Single product management can be expensive. Here, wildlife production could be increased 150 per cent at a cost of 75 board feet of timber.

portunity cost of three wildlife units is, therefore, 75 board feet of timber; the "trade-off" being made under the conditions specified is 75 board feet of timber for three wildlife units. One might also observe that under present management policies (timber the primary product) wildlife output could be increased by 150 per cent at a cost of only 75 board feet of timber.

### Application Problems

The joint product approach to multiple use offers few conceptual problems. If one accepts economic efficiency as an adequate guide to management decisions then maximization of the difference between costs and revenues makes sense whether measured in dollars, in units of psychological satisfaction, or in terms of social values. The objection to the joint production model as a practical means for multiple use decisions does not lie in the validity of the results achieved, but in the difficulties of application.

**Measurement.** On the purely physical side there are several major, unsolved problems. One is the output unit; units for measuring timber and forage are well known and in everyday use, although one might argue the relative merits of cubic feet, board feet, cords, International or Scribner log scale, tons of dry weight, etc. Water is more difficult to quantify since quality and timing must also be specified. Quantities, however, can be measured in thousands of acre feet or in gallons per minute, hour, or day. But there is far less agreement on how forest recreation should be measured, and apparently little thought has been given to measuring wildlife production other than by making simple head counts. Both of these problems will be discussed briefly in the chapters on recreation and wildlife. Units for measuring some of the other products of forests are even less well-defined. Once the units have been identified and agreed upon the job of determining the physical relationships between inputs and output can be tackled seriously, and there is much to be done before production functions for wildlife and recreation—or water—can be approximated.

**Non-market values.** There is also the sticky problem posed by the non-market nature of many non-timber products and services. That recreation has value few would deny, but how much value? Lumber, logs, or even standing timber is bought and sold hundreds of times every day, but one does not often "buy" outdoor recreation, or songbirds, or hunting opportunities, or trout streams. There are, of course, some places where some of these products can be purchased, but most are not sold like groceries, or automobiles, or hardware.[4] Should products such as these be given a dollar value? If so, how may that value be established? In terms of the joint production model, how does one plot isocost and isorevenue curves for an analysis for timber and forest recreation, or for timber and trout streams?

Because of these problems forest managers and administrators tend to favor the primary product approach to multiple-use management. A particular product is designated the key item (i.e., a "timber key" area) and the forest then managed to maximize production of that product within the stated budget constraints. Other products can be produced, but the central attention of the manager

[4] This point is made in a most unusual fashion in "Trout Fishing in America" (Brautigan, 1969).

is concentrated on the key product. The decision to make one or the other output the key product is generally easier to make and to justify, is much easier to implement, and in many instances the product combination may even be reasonably close to the optimum. But there is no necessity for this latter to be so, and in many instances (as, for example, that of Fig. 18.5) the error can be very large.

### Value Systems and Value Judgments

One approach to decision making avoids the necessity to compare values of different products by recognizing that different kinds of values might exist. In effect the decision maker says "Recreation, timber, wildlife, water, beauty—all have value. But while all these values are important, they should not be directly compared. In making decisions involving alternative uses of forests, only like values should be compared." For example, suppose three alternative management plans for a public forest have the probable production possibilities shown in Table 18.1. Costs under all three management

**TABLE 18.1 PROBABLE PRODUCTION PATTERN, SHOWING OUTPUT OF FOUR PRODUCTS UNDER ALTERNATIVE MANAGEMENT PLANS.**

| Annual yields | Plan 1 | Plan 2 | Plan 3 |
|---|---|---|---|
| Timber (dollars) | 150,000 | 160,000 | 160,000 |
| Water (acre feet) | 50,000 | 50,000 | 40,000 |
| Summer recreation (visitor days) | 10,000 | 20,000 | 30,000 |
| Wildlife (deer harvested) | 5,000 | 7,000 | 7,000 |

plans are expected to be identical. Plan 1 is clearly inferior to Plan 2, for it produces less timber, less summer recreation, fewer deer, and the same amount of water. Plan 1 can therefore be discarded without valuing *any* of the four products (presumably timber could have been shown in board feet or cubic feet without changing the result). But things will not always be this easy. Between Plan 2 and Plan 3, for example, the choice is not clear, although one can say that the decision problem has been clarified. Plan 3 offers a "trade-off"; 10,000 man days of recreation for 10,000 acre feet of water. The decision maker must make up his mind (Hughes, 1968)!

The example of Table 18.1 serves to emphasize that different kinds of values may be important and that value judgments cannot always be avoided. In some cases such a guide leads directly to objective decisions; in others it only points out the areas in which value judgments must be made. The approach does not necessarily aid in identifying or measuring the values involved.

On the other hand, suppose the situation were as given in Table 18.2. Here again is a public forest, with three alternative plans for

**TABLE 18.2 SECOND SET OF MANAGEMENT PLANS, GIVING OUTPUT OF FOUR PRODUCTS UNDER ALTERNATIVE PROGRAMS. (TO BE COMPARED WITH TABLE 18.1.)**

| Annual yields | Plan 1 | Plan 2 | Plan 3 |
|---|---|---|---|
| Timber (dollars) | 150,000 | 140,000 | 128,000 |
| Water (acre feet) | 50,000 | 50,000 | 50,000 |
| Summer recreation (visitor days) | 10,000 | 10,000 | 10,000 |
| Wildlife (deer harvested) | 5,000 | 5,500 | 6,000 |

producing varying combinations of typical forest products. Only Plans 2 and 3 have been changed. Now suppose the decision maker, when offered his choice, decides on Plan 1. In this case his selection indicates that according to his value scale 500 deer are worth less than $20 apiece, for at that value or more he would have selected either Plan 2 or Plan 3. If he selects Plan 2 his choice indicates that according to his value system deer are worth at least $20 each, but *not* worth as much as $25, for in the latter case he would have taken Plan 3.

Whenever a choice between alternatives is made, something is indicated or implied about the value system of the decision maker. If dollars are involved, and if the dollar amounts differ between alternatives, one may be able to infer something as to comparative dollar values. But one must be very careful about imputing values by this or some similar method, and especially when the choice is made by political policy makers. One can never be certain that all the factors influencing the decision are available for study. Policy makers have not demonstrated any high degree of consistency in choices such as outlined here.

### Shadow Prices

Perhaps the most useful technique for improving multiple use decisions involves "shadow prices," a concept best described by an example. Suppose it was decided to manage the Voyageur National Forest *as if* timber could be sold at going market prices, canoeing could be sold at a price of five dollars per canoeist day, and deer hunting at a price of five dollars per day plus $25 per harvested animal. Management plans, including access roads, campgrounds, check points, patrol plans, concessions, etc., could then be designed to achieve the combination of timber production, canoeing, and deer hunting that would maximize the present net worth of the forest. All three products would be priced in dollar terms—timber at its market price, and canoeing and deer hunting at the "shadow prices" already indicated.

The term "shadow" price is appropriate to canoeing and deer hunting because, although obviously not "real," these prices can affect management just as if they were market determined—as if the "shadow" of a price were enough to influence the management decisions of the Forest Supervisor and District Ranger. Shadow prices need not be uniform throughout the nation, or even throughout a state or region; they could be set to reflect directly the needs and wants of the public using the forest. Moreover, a shadow price has a large advantage over one determined in a market, for it can be changed if it seems to bring on an undesired condition. Given a period for trial and error adjustment, it should be possible to establish a workable set of shadow prices, and these could be a useful device for guiding multiple use management.

## THE SQUEAKING WHEEL PRINCIPLE

Before leaving this topic of multiple use it might be observed that *maximization* of satisfaction, of social benefit, of present net worth, or of anything at all, is not the only possible goal for public land managers. One might venture that the apparent goal of at least some public land management is the *minimization of dissatisfaction*, and there is even considerable informal evidence to support the suggestion. With only a slight tongue-in-cheek feeling, it is proposed that the "principle of the squeaking wheel" may be the guide followed in managing some public properties. Consider the following:

1. In a nation as large and as complex as ours the single individual has difficulty being heard. Consequently, group activity tends to supplant individual action, primarily because it is much more effective.

2. Contented groups seldom make strong representation. It is the discontented groups—the *dis*satisfied users of forest land—that are most likely to express themselves to their governmental representatives.
3. Government agencies, of necessity, are very susceptible to pressure-group activity. The adage claiming that "The wheel that squeaks the loudest gets the oil" had its origin long before general acceptance of government by pressure groups, but it seems apropos.

These mutually reinforcing tendencies have the result that large organizations, particularly public agencies, grow ever more sensitive to criticism, and tend to direct their activities accordingly. The fact that we expect this to be so and applaud such sensitivity to public opinion is a major contributor to the danger—not just to public forest land management, but to other facets of our democratic life as well. Most of us drift along with decisions which minimize complaints.

The squeaking wheel principle is the essence of the art of compromise. A dark horse candidate secures nomination by being less *dis*liked by the several party factions than are any of the alternative candidates. The principle applies widely in the marketing field as well. Our supermarkets offer a wide choice of breads—all equally soft, equally uniform, and equally tasteless—so as to avoid *dis*satisfaction. At the cheese counter one has difficulty in finding a really tasty cheese—a good Roquefort or Gorgonzola. Why? If one sought to produce a cheese few could object to—aimed for the *least complaints*—he would probably "design" a bland, yellow spread completely lacking in distinctive flavor. Check in your local market and you'll find this kind of product readily available; one can't dislike its flavor—it has none!

Other examples could easily be cited, but the point should be clear; the principle of the squeaking wheel leads inexorably to acceptance of mediocrity, and before long mediocrity may become the standard. If multiple use management on public forests is aimed to minimize the complaints from timber users, campers, hunters, hikers, stockmen, birdwatchers, water users, etc., the end result for public forest management will be mediocrity—scarcely a suitable goal for forestry in this country!

# 19

# Public Production and Welfare

Throughout the discussion of timber and timber products it was usually assumed that wants were expressed by individuals (consumer sovereignty) and that most production would be under private control. These assumptions were satisfactory, for most wood products are produced to satisfy individual wants, the wants are expressed through a market, and the resulting price informs both private and public producers how much to produce. When nontimber products were first considered, however, it was pointed out that there are no organized markets for some products. Does it necessarily follow that such products should be produced in the public sector? And if the product is not priced through market transactions, what can be said about the quantity of the good that should be produced? In this chapter we will take a closer look at these related problems, first by examining briefly some of the developments of *Welfare Economics*, then by looking directly at *public production*.

## WELFARE ECONOMICS

Schumpeter has commented that the purpose of positive economics is to explain, and the purpose of welfare economics is to prescribe (Schumpeter, 1954). Welfare economics deals with problems concerning the total well-being, or welfare, of individuals and of society. Few would claim that the analytic framework devised is completely satisfactory, and in fact some critics believe the major contribution of welfare economics has been that of proving there is no way of dealing quantitatively with these problems. But in some cases welfare economics does provide at least partial answers, and, as a modern authority has commented, ". . . I feel that to provide partial answers to vital problems is at least as important as it is to provide complete answers to lesser questions" (Scitovsky, 1951).

One of the principal difficulties of welfare economics is that of measurement. Almost everyone rejects crass materialism as a guide to a full life (Man does not live by bread alone!), yet we usually accept some aggregate of material items as indicating wealth, then take wealth as synonymous with welfare. This gives rise to serious problems when comparing the "well-offness" of people having different tastes, cultures, or living standards, or those living in different time periods—problems well treated elsewhere (Boulding, 1952; Scitovsky, 1951). As long as our analysis is confined to problems of a single individual, however, we can be reasonably certain that people will make decisions in a manner they believe will result in the most benefit or least harm *to them.*

## Individual Preference Systems

This assumption—that people choose so as to maximize their own satisfaction—permits us to construct a "preference system" such as that of Fig. 19.1. Each of the curves (such as *aa* or *bb*) is an *indifference curve;* the individual is *indifferent* as to whether he has the quantities of $X$ and $Y$ indicated by points $P$, $P_1$, or $P_2$ on *aa.* At the same time this individual will always *prefer* a combination on *bb* to any on *aa,* because the combinations on *bb* always represent more of either (or both) $X$ or $Y$. Note that the *quantity* of satisfaction *does not have to be measured or known,* for the system is *ordinal* rather than *cardinal.* It is a "more than" and "less than" system, but not a "this much more than" or "this much less than" system. Any movement (on Fig. 19.1) toward the upper right-hand corner represents an increase in satisfaction, but we cannot say *how much* satisfaction is increased.

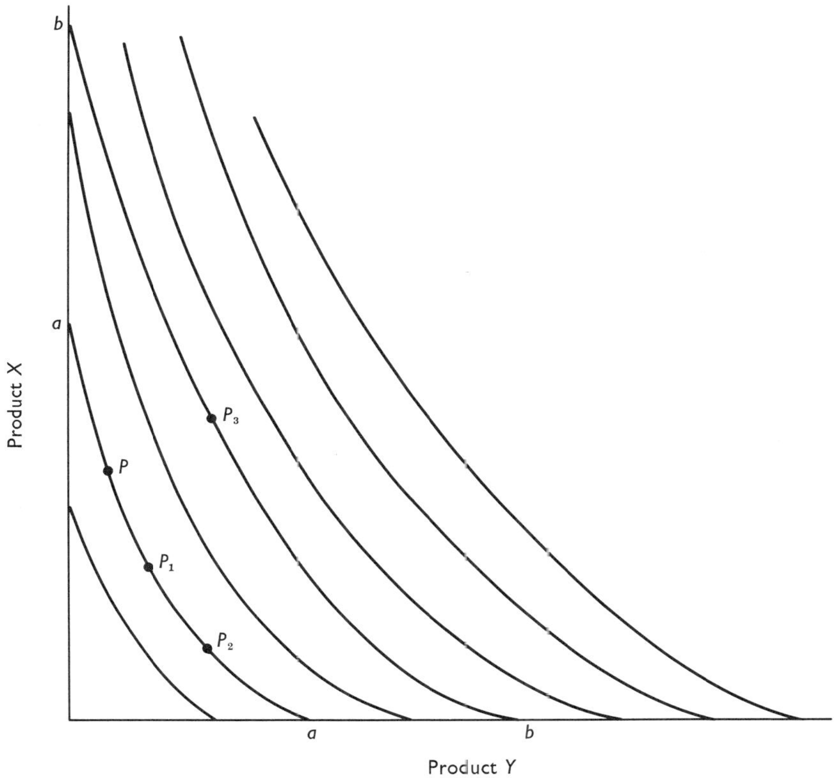


**Fig. 19.1** A set of indifference curves.

## The Pareto Optimum

In welfare economics, however, as in most aspects of public forestry, the concern is much less with the individual than with society as a whole, and additional problems arise when we try to measure the welfare of a group. Suppose we begin by dealing with just two individuals, Dave and Tom, and two commodities, timber and deer. Preference maps for each can be constructed, and might appear as those in Fig. 19.2(a) and (b).

If we assume further that the two commodities are limited in amount, and that the two axes in Fig. 19.2 show the total quantity of deer and timber available, then the diagrams can be combined as in Fig. 19.3. Here Dave's preference map is unchanged, while Tom's has been rotated 180 degrees, then superimposed directly

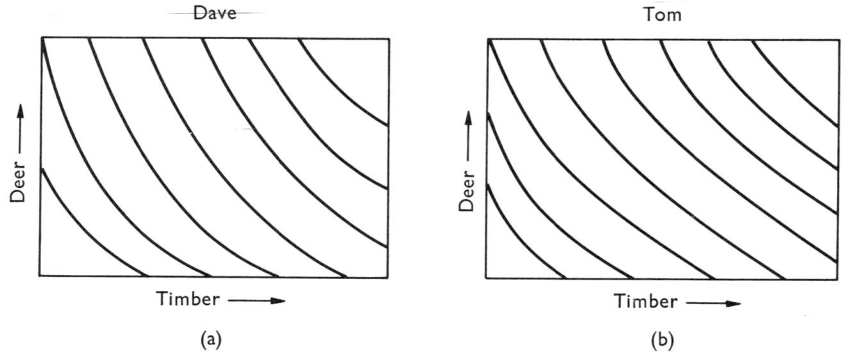


**Fig. 19.2** Preference maps for two individuals, where the two commodities are limited in amount.

over Dave's. Any point within the box now represents some specific allocation of deer and timber between Dave and Tom; at *P*, for example, Dave has $T_d$ of timber and $D_d$ of deer, while Tom has $T_t$ of timber and $D_t$ of deer.

But the particular division of the two commodities indicated by *P* is clearly not the best one possible! By moving along Dave's indifference curve (reallocating the two resources) Dave's state of satisfaction would be maintained while Tom's was being increased, and this improvement can be continued *until the point of tangency of the two sets of indifference curves is reached.* Alternatively, an indifference curve for Tom could be drawn through *P* and the two goods reallocated so Tom's satisfaction is held constant while Dave's welfare is increased—again until the point of tangency is reached. In fact, the allocation indicated by *P* could be changed so that *both* Tom and Dave are made better off. Note that we can recommend the indicated reallocation without knowing *how much* the satisfaction of either one is changed, *as long as the reallocation improves the position of one without harming the second.*

By connecting the points of tangency between the two sets of indifference curves a limited sort of "best" position is identified; if the allocation of the two goods falls anyplace other than on this line it will be possible to reallocate the two resources and make one person better off while keeping the other's welfare unchanged. Any point on this line, therefore, represents a kind of "optimum," called a *Pareto optimum* after the originator of the concept. It is a very

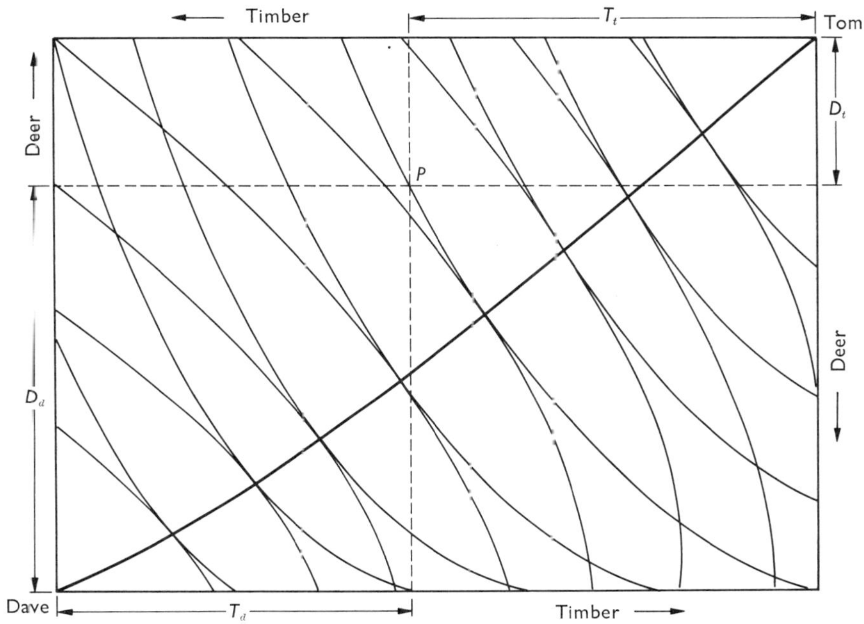


**Fig. 19.3** The two preference maps of Fig. 19.2 can be combined into a single system.

limited sort of optimum, but one against which policies can be judged (Pareto, 1927).

In spite of its optimal quality this line is generally called a "conflict line," and for good reason. Any movement along this line will be resented and resisted by one or the other of the two individuals, for while such a movement clearly benefits either Dave or Tom, it equally clearly harms the other.

## Social Welfare Functions

Now suppose we *could* measure welfare, and that we took the "conflict line" and stretched it out along the horizontal axis of a new graph, as in Fig. 19.4. When plotted, Tom's welfare would be increasing as we moved to the right, and Dave's would increase as we moved in the opposite direction.

If we could add the two together we would have a societal welfare function, with the highest point on this aggregate function identifying the point on the conflict line (and therefore the division

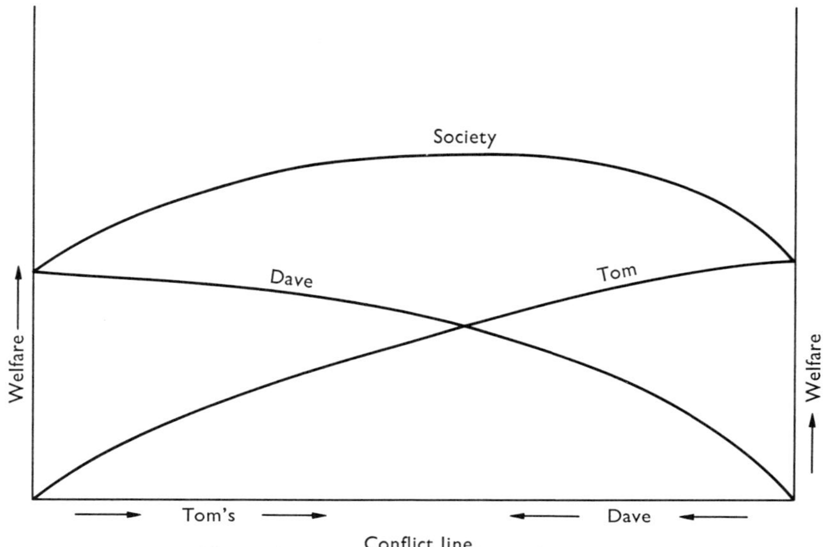


**Fig. 19.4** A "social welfare" function.

of the two products between the two men) that would maximize the social welfare of our two-man community. Unfortunately, the indifference maps were constructed on ordinal, not cardinal, bases. We know Tom's welfare goes up as we move from left to right, but we have no idea as to its slope at any point. We cannot even assume that Dave's welfare function could be graphed on the same scale as Tom's; Dave might be far more sensitive to timber and deer than is Tom. So, we can't draw either curve, much less add them together.

**Principle of compensation.** Under very restricted conditions a little more might be said about movements along the conflict line. If one of the two persons might gain more by a movement along the conflict line than the second would lose, the gainer might be willing to pay whatever the loser demands and still be better off than before. For example, *if* we could determine a welfare function for Tom and Dave and relate them to the same scale, the two functions might be very different (Fig. 19.5). Now if we moved along the conflict line from A to B (altering the division of the two products) Tom's welfare would increase far more than Dave's would decrease. Out of his gain, Tom could afford to compensate Dave's loss, and still be better off. True, we can't construct the graphs, but it might

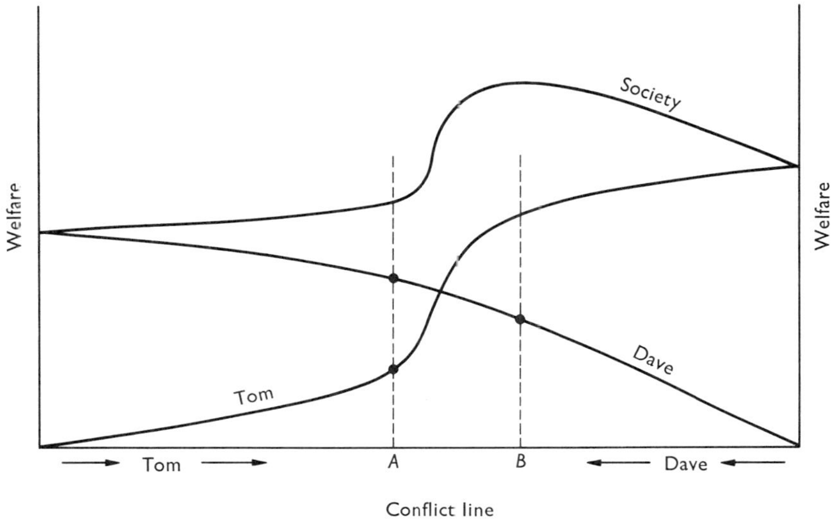


**Fig. 19.5** The principle of compensation.

be possible to increase one person's welfare enough to make him willing to compensate the loser.

This illustrates the so-called *Principle of Compensation* (Hicks, 1939; Kaldor, 1940). Once this principle was developed it was suggested that from a societal viewpoint it would only be necessary to show that compensation *could* be made in order to warrant the redistribution. But it was quickly pointed out that this latter statement necessarily involves an *interpersonal comparison;* it forces one to compare, directly, the welfare of one individual with that of another. Unless welfare has been measured in cardinal terms, this kind of comparison is meaningless; Tom's contour lines may have greater values than Dave's or vice versa. The comparison between individuals has no logical basis.

It is now generally agreed that under very rigid specifications the principle of compensation is valid as long as the compensation is actually paid, but *only* if it is paid. What is even more limiting is the fact that one cannot logically make a recommendation that the compensation *should* be paid, for this requires an implicit judgment about income distribution (Baumol, 1965). So we are left in a very uncomfortable position. Under some conditions we can say it will be OK to shift along the conflict line so long as the loser is compensated, but we cannot say the compensation *should* be paid, and we may even feel very strongly that it should *not* be paid (Boulding, 1952).

If one could "add up" all the welfare produced by, say, timber production on some specific area, then, similarly, total the welfare that would be produced if this same area were used for wilderness purposes, a direct comparison would reveal which use would serve society best—at least under the conditions existing when the comparison was made. However, since utility has yet to be measured acceptably and interpersonal comparisons cannot be made (unless we are willing to accept someone's judgment as to how important *our* welfare is), social choice functions cannot be derived.[1] We are left with the conclusion that the only social choices about which we can be positive are those in which no one is made worse off, and at least one person is made better off—the Pareto criterion. Fortunately for mankind, society goes ahead and makes social decisions anyway.

It would be unfair to leave welfare economics without stating that many refinements have been added to the basic approach outlined here. Under very restrictive assumptions (such as assuming cardinal measurement of utility) the analysis can be carried further (Bator, 1957). The principle most often utilized, however, is the Pareto criterion already met in the discussion of decision making under multiple use; any decision that makes some person better off without lowering the welfare of anyone else is a step ahead. As a matter of fact, it was the opposite side of this coin that was used in the very first chapter when pointing out the "normative" quality of the purely competitive model; given the existing distribution of income there is no shift in the use of resources that will permit any consumer or producer to be made better off without lowering the welfare of some other person.

## PUBLIC BUDGET POLICY OBJECTIVES

Economists concerned with governmental revenue and expenditure problems (with *public finance*) view a government as shaping

[1] Arrow has stated this as the "General Impossibility Theorem": "If we exclude the possibility of interpersonal comparisons of utility, then the only methods of passing from individual tastes to social preferences which will be satisfactory and which will be defined for a wide range of sets of individual orderings are either imposed or dictatorial" (Arrow, 1963).

its budget policies toward three kinds of objectives: (1) to adjust, where desirable, the allocation of resources; (2) to secure adjustments in the distribution of income and wealth, and (3) to promote the desired level of stability while maintaining growth (Musgrave, 1959). These same objectives were necessarily included as objectives of tax policy (Chapter 10). Treatment of these three objectives will necessarily be incomplete, and while each has an important bearing on the production of forest products, attention will be concentrated primarily on the allocation objective.

### Allocation

Resources are used in producing goods and services for the satisfaction of man's wants, and if the most wants are to be satisfied, then resources must be used most efficiently. These wants are of two very broad types—*private wants* and *public wants*—and the goods and services which satisfy these wants are appropriately called *private goods* and *public goods*. The private goods bundle is made up of all those goods that are *rival in consumption*. For example, the wood I burn in my fireplace is not available to others for heating; if my family lives in a house, that house is not available to other families. In contrast, public goods are those which are *non-rival in consumption*. I can enjoy a forest landscape whether I pay for it or not, and my enjoyment does not necessarily interfere with the enjoyment this same landscape provides others. Similarly, the benefits of a successful anti-air pollution ordinance are enjoyed by everyone living in the area; we all enjoy benefits from national defense, from police protection, and from the judiciary process (at least in theory).

The important difference from a production standpoint is that the market, operating through a system of consumer sovereignty, effectively allocates most *private* goods. But the market system fails to allocate *public* goods effectively. If the benefits I receive are not reduced by the fact that you share the same good, and if either (or both) of us can enjoy the benefits whether we contribute to their production or not, then the market system cannot signal to producers the correct quantity that is needed.

**Private goods.** In the United States, major reliance is placed on the market system to allocate resources efficiently. Private corporations or firms produce most of the products, which are mostly purchased by private consumers or investors as private goods. There are, however, some situations when the government must intervene to secure a better resource allocation by the market system, most of which have already been mentioned in other contexts. Anti-trust laws protect against monopoly power; public utilities are regulated (the guide here is to equate marginal cost to average revenue, not marginal revenue). If externalities are sufficiently important, public intervention may be required (anti-smoke ordinances, water pollution controls, etc.).

**Public goods.** Public goods are of two kinds: (1) those satisfying social wants are called *social goods*, and (2) those meeting merit wants are called *merit goods* (Musgrave, 1959). This particular distinction, while not universally accepted by students of public finance, is useful in dealing with public production of forest products.

A *social want* is one that, although determined in accordance with the individual's preference system, cannot be revealed through an existing market. The present wants of millions for cleaner air, lakes, and streams provides one example, while the desire for aesthetic enjoyment is another. For social wants, therefore, it is necessary to devise a governmental revenue–expenditure system to provide the money needed to produce these social goods, and simultaneously to collect their production cost from individuals in accordance with their desire for them. Note that this in no way means that the government should *produce* the goods; the goods could as well be produced by private means, but paid for and distributed by the government.

*Merit wants* are not determined by an individual's preference system, but by the preference system of some "higher" authority. In socialistic systems the higher authority is often the Central Planning Committee; in the United States it is usually a legislative body such as Congress. Merit goods may also be produced as private goods, but are not produced in quite the same way or same amount. Public housing, for example, is certainly produced within the private sector as a private good, determined according to individual preferences revealed through markets. But housing is also provided in response to governmental decision, and furnished to particular individuals either through subsidy or through direct action. Merit goods include things that "are good for you and ought to be available to United States citizens." Wildlife, water, and forest recreation are regarded by some as *merit goods*, even though they are also produced as private goods. As a merit good, forest recreation might be regarded as a responsibility of the Federal government, in which case the budgeting of expenditures for its production and the collection of taxes for its support become a concern of foresters as well as of those studying public finance.

### Distribution

What constitutes an ideal distribution of wealth or of income must, in the last analysis, be a matter of personal judgment. Equal monetary wealth would not yield equal welfare unless everyone placed the same value on possession of money, and this is certainly not true. There are many who feel equality should be interpreted in terms of opportunity, which others point out is closely associated with wealth and income. Yet there are some widely held notions about distribution in our nation. People should not be allowed to starve; some level of medical care should be provided to all; children should be fed, clothed, and educated even though their parents cannot bear the associated expense. And in general, people agree that any redistribution of income should be directed from the rich to the poor. The significant aspect of distribution here is that the allocation of public goods, either social or merit, together with the collection of taxes for their production, can influence the distribution of income and of wealth in the nation. Hence, the distribution of public forest goods should be in general accord with the ideas currently held by society of a proper income distribution.

### Stability and Growth

In this country the responsibility for maintaining high employment and reasonably stable prices has been assumed by the Federal government. In general, the fiscal means of attaining this objective involves compensatory action. If unemployment threatens, action is taken to increase the level of aggregate demand; if inflation threatens, then demand should be lowered. There are, of course, tools for attaining this objective in addition to those of a fiscal nature. In other countries, particularly in the so-called developing nations, the goal of growth may take precedence over all others.

## PUBLIC PRODUCTION [2]

In the preceding pages a three-way breakdown of private goods, social goods, and merit goods has been presented. The distinction between them came from the nature of the want which the product is intended to satisfy, and *not* from the production process or whether it was located in the private or public sector. We need now to examine the factors that influence whether production is to be under public or private control.

There are occasions when there is simply no market for the product being considered. If no one can be excluded from enjoying the product once it is produced (a condition sometimes described as meeting the "exclusion principle") it is unlikely that private enterprise will undertake its production, for there would be no means of establishing or collecting a price. The beauty enjoyed while traveling a mountain highway provides an example. Sometimes exclusion is possible but the collection of a user charge would be prohibitively costly; small forest campgrounds in wilderness areas are a case in point. Occasionally, private production and marketing would be possible but would be considerably more costly than provision by public means; it is unlikely that wilderness areas would be provided by private industry.

Public production can also arise because of externalities (Ayres & Kneese, 1969). If the externalities can be identified and marketed there may be no *necessity* for public production, but if the manner of provision of a primary product affects the kind or the timing of some accompanying effect there may be sufficient basis for public rather than private production. Forestry itself may be an example under certain instances, and the many externalities of forest management—watershed protection, flood control, air purification, provision of scenic beauty, wildlife habitat, etc.—are sufficient in the eyes of some to warrant far greater public ownership of forest land than that presently existing.

A need for public production may also arise from monopoly power or from the threat of such power, and from other market

[2] While preparing this section I was most fortunate in being able to read in manuscript form early drafts of two works later published, the first by Peter Steiner (1969) and the second by Harvey Brazer (1969). The earlier work of Musgrave (1959), Samuelson (1954), and Bator (1958) should also be reviewed by anyone interested in the topic of public production.

imperfections as well. The possibility of private owners monopolizing parts of Yellowstone National Park or the valley of Yosemite National Park loomed large in the minds of legislators when they delegated these areas to public control. Occasionally time lags produce market imperfections that lead to public production; the time required to produce large sawtimber, for example, might make it difficult for private suppliers to react sufficiently rapidly to market changes. Sometimes the transaction cost is simply too high—collection of user fees in small forest campgrounds has already been mentioned.

There are some goods which could be produced in the private sector but have been turned over to public production because the pattern of consumption under private markets would be undesirable. Urban playgrounds, hunting and fishing privileges, and many forms of outdoor recreation, for example, could be provided privately. But most people feel that these things should be available to all citizens regardless of their ability to pay; to at least some extent they are merit goods. Here one may observe that different societies may make very different decisions. In Scotland, for example, hunting and fishing rights are privately marketed while golf courses are heavily subsidized and provided (at a low cost) as public goods, and the pattern of consumption of these goods contrasts vividly with that observed in the United States. Some goods, then, are publicly produced simply because of an unwillingness to accept private market solutions. Irrigation water provides a somewhat different kind of example.

Some goods are necessarily "jointly supplied." If a good that provides utility for one person can also supply utility to others at no additional cost, it is by definition a social good and very frequently such goods are produced by the public sector. Highways and bridges provide classic examples.

Listing examples of how public production comes about does not of itself provide justification for public production. Public forest ownership, however, is one of the dominating institutional factors in the total United States forestry picture today. If we seek an understanding of forestry's role in our present economic society it is necessary to look a little more closely at how social decisions might come about, for social decisions are required in some forest production.

### How Much and What?

The *private* competitive market provides an efficient mechanism for aggregating individual wants into effective demand. Each consumer decides which and how much of the thousands of available products he will buy, thereby maximizing his own satisfaction, within his income. Each dollar spent represents a "vote" for the production of that particular good or service. The market "adds up" these votes so that, at least ideally, everyone gets as much as he is willing and able to pay for—what he "votes" for. Efficiency in production and in resource allocation is assured by competitive forces, which is the reason for looking askance at market imperfections. Since utility isn't measurable, and since interpersonal comparisons are ruled out, no one can *prove* that this particular pattern of production and consumption will maximize social welfare. *But no one can prove that any other pattern is better.*

On the other hand, effective markets do not exist for *public* goods, either social or merit. Governments, through exercising their police power and their ability to tax, can provide a kind of collective action to substitute for the market, assembling the necessary resources and organizing them to meet the aggregated individual wants for these public goods. But how does society decide how much of what should be produced by the public sector? How does society decide to spend dollars on building more camp sites instead of more timber access roads, or to take an area out of timber production and declare it to be a wilderness area?

Arrow concluded that if utility is not cardinally measured, and if interpersonal comparisons have been ruled out, there is no way of aggregating the individual wants for public goods; there is no means of weighting the individual wants (Arrow, 1963). The private market weights by the dollars spent, but this weighting system is unavailable when the product is a public good. But even though believing that individual wants cannot be aggregated, some problems can be at least partially resolved with the approaches developed previously.

**An example.** Take, for example, the problem of determining the proper mix of outdoor recreation and timber to be produced from some given public forest area.[3] The production possibilities under

[3] The problem is similar to the deer–timber illustration in the preceding chapter. Here, however, we are concerned with the production plan—to what use should

two different expenditure levels are as shown in Fig. 19.6, with curve *P–P* representing the present situation, and *P′P′* the possibility under a considerably increased cost. Timber output is measured in cubic feet and recreation in visitor days.

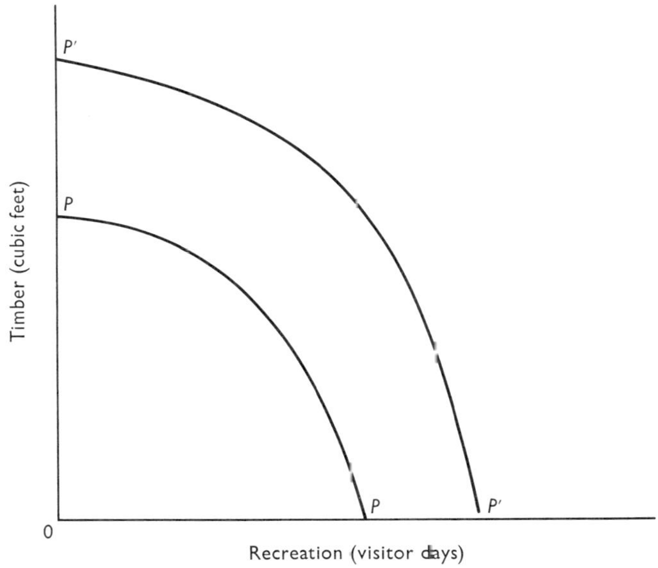


**Fig. 19.6** Production possibility curves for timber and recreation on a public forest.

Now suppose one could draw the preference maps representing the attitudes of two different groups of people. One group might be called timber advocates, who fervently believe God and Congress meant forests to "produce sawlogs, not play." Their preference map is shown in Fig. 19.7(a), with the timber and recreation scales kept the same as those of the production possibility curves in the previous figure. Directly beside this, Fig. 19.7(b) is the preference map—the set of indifference curves—that might represent some hiking, hunting, and camping group whose members believe forests should have a very different purpose, as the slopes of the indifference curves clearly show.

---

public money be put—whereas the previous problem dealt with the optimum division of two goods that had already been produced.

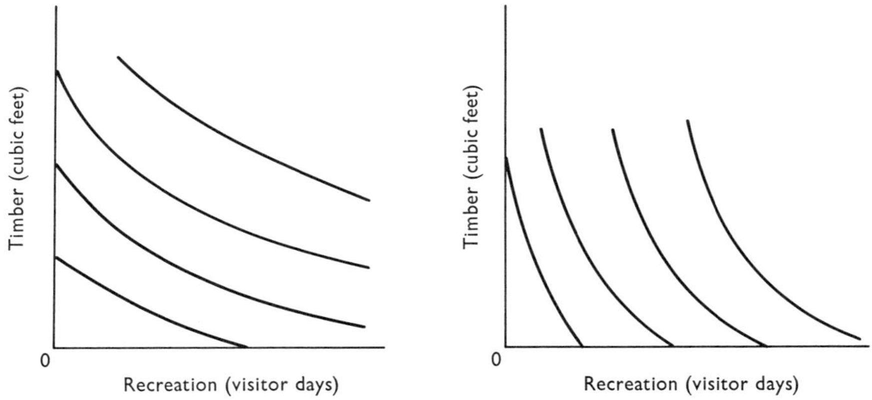


(a) Timber advocates (b) Hiking and hunting group

**Fig. 19.7** Preference maps for two groups of people: (a) timber advocates, and (b) "hiking and hunting group."

All three sets of curves have been combined in Fig. 19.8. Clearly, if the combination presently being produced is $P_t$ then the timber interests will be happy with the way available funds are being spent (although they may want larger appropriations). But the recreationists will not be content. They would only be satisfied if the combination were switched to $P_r$. According to the Pareto criterion, however, no change from the present spending pattern can be recommended, for any shift along the production possibility curve that would improve the lot of the recreationists will definitely harm those interested in timber.

There are two possibilities of obtaining cooperation from both groups. If the complete pattern of possibility curves gave the two optimum expansion paths shown by the dashed lines (Fig. 19.8), one optimum for timber, the other for recreation, then any combination that fell outside the enclosed (shaded) area could be improved; both the timber interests and the recreation interests could be made happier with no increase in cost. Alternatively, either group could be made better off while doing no harm to the interests of the second group.

The second possibility for cooperation is to advocate an increase in expenditure, for, to take the extreme, even if the increase were used to move from $P_t$ to $P_t'$ (thus keeping an optimum output from

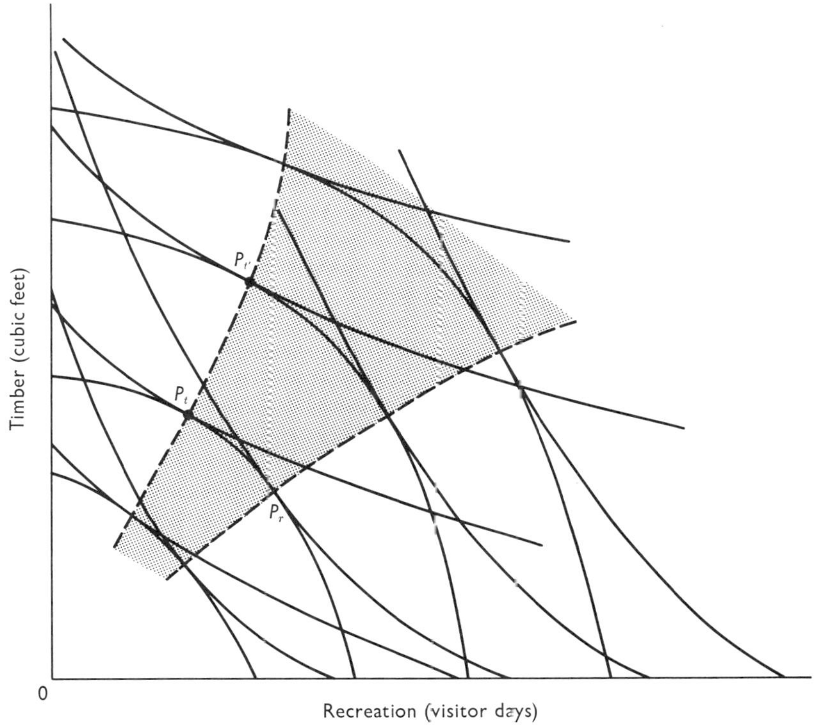


**Fig. 19.8** The differing optimum production paths for the two groups. Any combination lying outside the shaded area can be improved, with both groups benefiting. The shaded area is an area of policy conflict, unless costs are increased.

the standpoint of the timber interests), it nevertheless results in providing more outdoor recreation. Moreover, one would normally expect the agency involved to cooperate with the two groups in obtaining larger appropriations.

From an analytic viewpoint, however, no increase in the cost level can be recommended, even though both groups would benefit. Such a recommendation would imply knowledge of the preference map for *all* taxpayers (the basis for a true social welfare function), as well as that for all timber and recreation interests. This knowledge clearly implies both cardinal measurement of utility (or welfare) and interpersonal comparisons, both of which have been ruled out. It is worth noting, however, that there has been no reference to price or to value of either recreation or timber in the analysis. But how much easier and more definite the conclusions could have been if prices for the two products were available and comparable!

Acceptance of the Pareto criterion as a guide to public decisions provides an excellent argument for the maintenance of the status quo, for almost any change can be shown to reduce someone's satisfaction. As a consequence, any change in production of public goods tends to be limited to amounts and kinds that would make at least some persons better off in a clearly identifiable manner, but which reduce no one's satisfaction. Under this criterion, new production of public goods would be limited to amounts satisfying the taxpayer desiring the smallest quantity.[4]

The belief that individual wants are the only basis for determining social action is both logical and defensible, and leads to conclusions such as those illustrated above. But the belief is clearly not in accord with present public decisions. Many public decisions obviously make some people better off and others worse off. Timber areas have been designated as wilderness areas in spite of widespread opposition by timber interests, and there seems little doubt that in future years far more attention will be given outdoor recreation, air purification, watershed management, and perhaps other non-timber products, which may reduce public timber production far below present levels.

## Societal Choice Systems

Perhaps the approach of aggregating individual wants to derive social wants is overly narrow, based too much on our concept of private markets and consumer sovereignty.[5] Without pretense of analyzing such alternative approaches or even reviewing those currently being considered, two possibilities will be briefly presented.

One possibility is to regard individuals as possessing *social* as well as *individual* interests. This leads to the "pluralistic" viewpoint of sociologists, anthropologists, political scientists, and even of many economists; people have a "plural" set of values and loyalties—to family, to club, to university, to city, to state, to church, etc. If state decisions are being considered then the preferences

[4] Production of any more than this particular taxpayer wants takes more tax money than he is willing to pay.

[5] Here recall also the points made in Chapter 14, page 298, about consumer sovereignty and the objectives of large corporations, etc.

revealed by people's actions with respect to state decisions are appropriate rather than those typifying their choice pattern as individuals, and these latter may be very different. Thus people may be willing, for example, to accept being taxed for things they believe are socially desirable even though the tax paid exceeds the benefits (the welfare) they themselves obtain.

A second possibility carries this approach even further, regarding society as a distinct and separate entity. Under this view society possesses its own set of values and preferences and often approves (through political processes) actions to further develop these values. Society might, for example, decide that big trees are beautiful and clear lumber is important and desirable, sufficiently important and desirable to warrant production by long rotations. Such a decision would not rule out economic criteria; it would only mean that big timber should be produced—that timber size had become a production constraint. In effect one says, "The decision has been made to provide the public good called big timber. It is up to the agency responsible for timber production to grow these trees at as low a cost as possible." Similarly, if one accepts that society has its own values and preferences then one can argue that the public can decide to convert an area from commercial timber production to use as a wilderness area, even though this means a reduction in timber supply, an increase in timber costs, and may force some timber businesses to close!

Such approaches, and there are several variants, are sometimes criticized on the ground that aggregation is still necessary even though social, not individual, values are made the basis for action, that this aggregation could still involve interpersonal comparisons, etc. There is no really satisfactory reply to such criticism except to observe that it is not necessary to *understand* a social phenomenon completely before making use of its existence, any more than it is necessary to understand a physical phenomenon before using it to satisfy man's needs. Fire was of inestimable value to man for thousands of years before he began to understand the nature of combustion. There are many things only partially understood, but in everyday use. The fact that *social* choice methods are not completely understood should not restrict us to systems that clearly fail to explain what goes on in everyday life. If the view that society has its own set of values and creates its own preference pattern aids in making realistic decisions, then there seems little reason to reject the view unless a more usable one is available.

There is no single, universally accepted view of the manner in which society decides either how much or what kinds of public goods should be produced, or where they should be produced. That a tremendous variety of goods is now being produced in the public sector, however, is accepted as part of our economic system. The full extent of this production is seldom realized, but its significance is highlighted by the factual observation that in only one other country of the world, the USSR, does government control as much production as in the United States. If one makes public production the sole criterion of whether a country is socialistic or capitalistic, then the United States would necessarily be called the second largest socialistic power in the world today. It would also, of course, be the largest private enterprise power. With public goods looming this large in our economy it is only to be expected that rather special approaches have been devised for analyzing the need for some of this public expenditure. One of the most important of these is the procedure called *benefit/cost analysis*.

# 20

# Benefit/Cost Analysis

Prior to World War II, students of public finance were concerned primarily with the revenue side of government—with questions of tax incidence, tax effects on incentive, tax equity, and similar problems. As government expenditures grew larger their impact on the economy became much more significant, and in the early postwar years the expenditure side of government began to attract greater attention. Efficiency in government expenditures, as well as in obtaining revenues, became a matter of concern to an increasing number of economists.

## THE PPB SYSTEM

Their concern was matched, and to some extent preceded, by that of legislators and government executives. Successful wartime research in establishing military objectives, largely that in *systems analysis* by the Rand Corporation, was directed toward obtaining efficiency in government expenditures. In 1965 President Johnson issued an Executive Order establishing a comprehensive Planning-Programming-Budgeting (PPB) System throughout the Federal government, giving the Budget Bureau responsibility for its implementation (Haveman, 1969). According to the President, such a system would enable public decision makers to:

1. Identify our national goals with precision and on a continuing basis.
2. Choose among those goals the ones that are most urgent.
3. Search for alternative means of reaching those goals most effectively at the least cost.
4. Inform ourselves not merely on next year's costs, but on the second, and third, and subsequent year's costs of our programs.
5. Measure the performance of our programs to insure a dollar's worth of service for each dollar spent (Haveman, 1969).

Two kinds of analyses have characterized the PPB system. One, called *cost effectiveness analysis*, has been applied to investments in national defense and space programs. Once produced, everyone automatically benefits from these public goods, although the benefits cannot be evaluated quantitatively. This kind of analysis has not been applied extensively to forest investments and will not be treated further. The other, called *benefit/cost analysis*, has been applied to natural resource developments and, to a more limited extent, to transportation and human resource investments.

## BENEFIT/COST ANALYSIS

Probably the easiest way to gain an understanding of benefit/cost analysis is to follow through a simple example. Suppose one were called upon to evaluate a proposal for establishing a new recreation area, the "Forestshore Recreation Area" in the Susquehanna Valley of Pennsylvania. The project plan, we will assume, calls for acquiring the needed millions of acres, building a dam to impound a lake at least 60 miles long, reestablishing forests on the area, building campsites, etc. Three questions need answering: (1) Is the project worth undertaking? (2) How much should be spent if it is decided to be a worthy project? (3) How can this project be compared with others that might have equal claim to public funds?

Several products could be produced by Forestshore—electric power; perhaps water for supplemental irrigation; sawtimber and pulpwood of several species; recreation of many types, including hiking, camping, swimming, sailing, canoeing, powerboating, fishing; possibly a freshwater fishery; perhaps, after some time, navigation benefits; an improved water supply for municipalities downstream; flood control; wildlife for study and for hunting, and so on. Ignoring problems of aggregation and measurement for the moment, suppose we simply lump all these products together and call them "benefits." This, then, becomes our product—our output is public benefits. Then, assuming further that these benefits can be translated into dollars, the revenue part of the production framework will be under control.

It would seem that costs should be less complicated than benefits. Estimates will be needed of how much land, labor, equipment, seedlings, planting machines, concrete, steel, generators, picnic tables, and other items will be needed, but these can be worked up for almost any project. And since markets exist for all the needed inputs it should be no great problem to price the required materials. Then, with prices and quantities at hand, costs can be determined.

The benefits from a project such as this would continue for many years, so costs will be incurred and revenues received at different time periods. To make them comparable it will be necessary to discount all costs and revenues to some common point in time, a process already discussed in Chapters 14 and 15. Choice of the proper interest rate creates a problem, but for the moment this can be skipped over until we become familiar with the general approach.

## A Production Approach

One of the questions posed at the outset was "How much should be invested in the project?" Suppose we take a simple production approach to this question, which is just another way of asking what the scale of the undertaking should be. We have a kind of fixed factor here, the area of Pennsylvania that will be involved. To this area successive increments of all sorts of inputs are to be added, and from the resulting combination certain *public benefits* are expected. The general scheme can be visualized as a sort of aggregate production function, shown in Fig. 20.1(a), where benefits are indicated in millions of "bennys." A certain minimum level of inputs will be needed before any benefits can be realized, and once this minimum level has been reached the benefits will increase. Because of the fixed factor, sooner or later the benefits produced per unit of input will decline.

The transition to total costs and revenues poses no particular conceptual problem, as seen in Fig. 20.1(b). The inputs required for various output levels can be valued at market prices to yield the required discounted total cost curve, and the discounted curve of

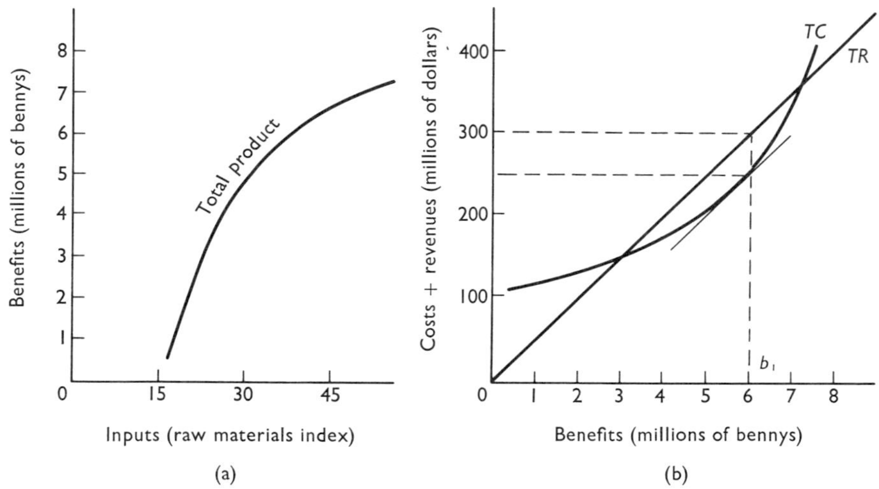


**Fig. 20.1** Production function for Forestshore is shown in (a) with resulting total cost and total revenue curves in (b).

public benefits, now expressed in dollars, provides the necessary total revenue curve. With these two curves at hand no further explanation is needed to see that the optimum output level should be $b_1$, the optimum expenditure level 250 million dollars, and that net discounted benefits (PNW, or net discounted cash flow) will be maximized when the slopes of the two curves are equal—where marginal costs equal marginal benefits. In dollar terms, net discounted benefit (PNW) is maximized at 50 million dollars.

## The Usual Benefit/Cost Approach

If one abstracts from all difficulties of identifying and measuring costs and benefits, this is about all there is to benefit/cost analysis. Usually, however, the approach is presented in a somewhat different guise. Because a project the size and scope of many government developmental schemes requires a great variety of inputs, it becomes impractical to devise a physical input scale such as used in the production function of Fig. 20.1(a). The size of the project, therefore, is usually measured by its total cost; each additional million dollars is assumed to be spent on the needed extra tons of concrete, thousands of seedlings, man-hours of labor, etc. And having decided to express inputs directly in cost terms, the same approach seems perfectly applicable to benefits. Ultimately all the different benefits have to be translated into dollars if they are to show up in a revenue curve, so why not express them directly?

With costs plotted on one axis and benefits on the other, and with both expressed in the same dollar units, the resulting figure becomes that of Fig. 20.2—the usual presentation of the benefit/cost model.

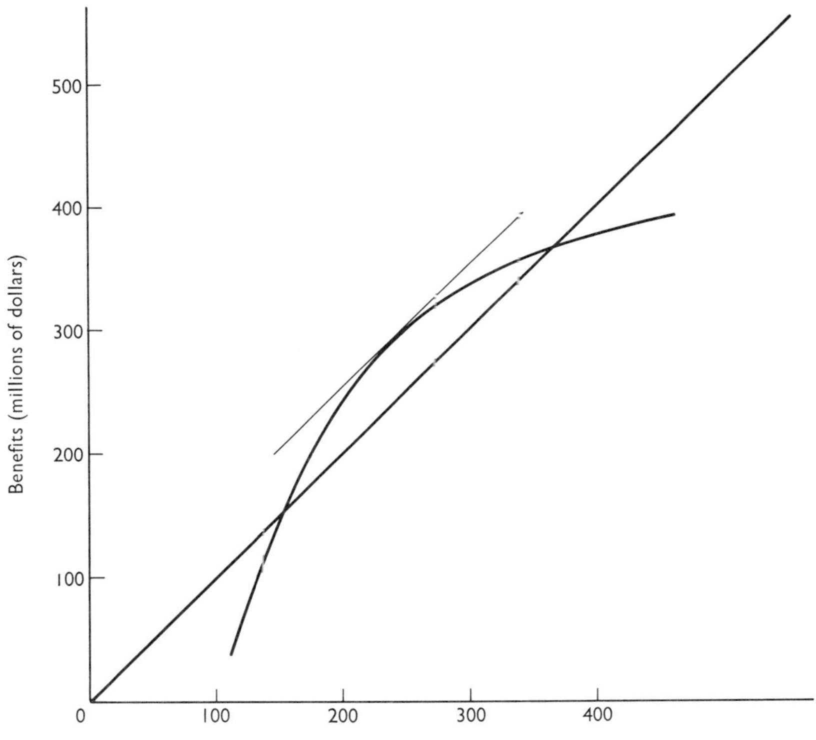


**Fig. 20.2** The usual benefit/cost model, here of the hypothetical Forestshore recreation area.

Any benefit level involves a specific cost level. Since the two axes are normally scaled in the same intervals, a line drawn at a 45-degree angle and passing through the origin identifies all points on the graph where benefits and costs are equal. With this diagram as the model we can return to the three questions posed at the outset of this discussion.

## Justification, Scale, and Ranking

Rather than asking "Is this project worthwhile?" legislators ask "Is the project justified?" Fig. 20.2 shows that benefits will exceed costs at any cost level between 150 million and 370 million dollars. Within this expenditure range, therefore, the ratio of benefits to costs will be one or more; that is, $B/C \geqq 1$. This has become the standard "justification" for public expenditures in the resource development area; if a project shows a benefit/cost ratio of unity or above, the project is said to be justified, for the benefits to society exceed the costs.

The second question is only slightly more difficult. What is the proper scale for the project? In the early days of benefit/cost development this question provoked considerable discussion. Eventually, however, it was agreed that the optimum scale for any project is the one maximizing *net benefits*. This is the marginal solution—marginal benefits must equal marginal costs. In our Forestshore example (Fig. 20.2) this involves an expenditure of 250 million dollars, with expected total benefits of 300 million dollars. At this point the slope of the curve is 45 degrees, for if both benefits and costs are changing at exactly the same rate, as long as the two scales are the same the slope of the curve will be 45 degrees. Here the B/C ratio is 300/250, or 1.2. Note that this is not the point where the B/C ratio is a maximum. Net benefits are maximized, not the benefit/cost ratio.

The third question—how to compare projects that compete for public funds—also provoked considerable discussion. Assuming all projects were developed until net benefits were maximized, there are several possible bases for comparison. Projects might be ranked according to their net benefits (benefits–costs), by the size of the benefit/cost ratio (calculated at the optimum scale), or according to their rate of return on the required investment (their internal rate of return). The first approach was rejected because cost levels would not be taken into consideration; a large project would always have a much better chance of being selected than a small one under this criterion. The difference between the last two criteria clearly lies in what is included in "costs"; if "cost" and "investment" are defined as equal, then the two approaches must yield the same order of ranking.[1] There seems little point in pursuing this question, particularly since the problem dwindles in practical importance when one realizes that actual priorities tend to be settled far more by political than by economic factors.

With this exception of determining project priorities there are no particular conceptual problems in the benefit/cost model. The difficulties are primarily the practical ones of identifying and measuring the benefits and costs.

## TYPES OF BENEFITS AND COSTS

Benefits are generally divided into two kinds; *project benefits* (sometimes called direct benefits), and *secondary benefits* (indirect benefits). Project benefits consist of all benefits coming directly from the project; the value of the electricity and timber produced, the fees paid by recreationists, etc. Secondary benefits are those additional benefits accruing to society because of the project—gas station operators' incomes generated by tourists or campers or fishermen attracted by the project, the incomes created by a pulp and paper plant built to use the pulpwood from Forestshore, incomes of motel operators, etc.

Costs are divided into two categories; *project costs* and *associated costs*. *Project costs* consist of the value of all goods and services used for establishing, maintaining, and operating the project—purchases of land, payment for labor and equipment, etc. Also included are costs such as those required for relocating transportation routes, businesses, homes, etc. *Associated costs* are those incurred (over and above project costs) to make the outputs of the project available—the costs of planting and harvesting the timber, of transporting the electricity if required, of distributing irrigation water, etc.

With both benefits and costs thus specified the benefit/cost ratio can be restated as follows:

[1] This is only saying that if one ranks by $\frac{B-C}{C}$ or by $\frac{B}{C}$ it makes no difference, since $\frac{B-C}{C} = B/C - C/C = B/C - 1$. The ranking is unchanged by subtracting a constant. It is necessary, however, that the benefit/cost relationship be that existing where net benefit is maximized rather than where the B/C ratio is maximized.

$$\frac{B}{C} = \frac{\begin{matrix}\text{total primary benefits} \\ + \ (?) \\ \text{secondary benefits}\end{matrix} \quad - \text{ associated costs}}{\text{project costs}}$$

Note the questionable inclusion of secondary benefits. This is a matter of continuing debate. Economists generally agree that in a nation such as the United States where reasonably full employment is a matter of public policy, secondary benefits should almost never be included. Such benefits seldom represent an addition to national income, usually being at best only a "transfer" of income from one region to another (Eckstein, 1958; McKean, 1958). This possibility of transfer, however, means that secondary benefits can be important from a *regional* standpoint.

## MAJOR APPLICATION PROBLEMS

The most intractable problem plaguing those who undertake benefit/cost analyses is almost surely that of measurement. Empirical analysts tend to divide all benefits into two classes determined by their measurement characteristics.

### Tangible Benefits

These are benefits measurable in commonly accepted units—not always units that are traded in established markets, but units that can be translated into dollar values at some specified price. Irrigation water, for example, can be measured in acre feet. The price by means of which irrigation water is given a dollar value may be assigned by tradition, by political expediency, or, in a few cases, by comparison with irrigation water purchased from private corporations in the same vicinity. Sawtimber or pulpwood that is to be produced as a result of the project's establishment offers no problem on the measurement side; these are definitely *tangible* benefits.

Even with tangible benefits major difficulties may arise. Suppose, for example, a benefit/cost analysis is to be made of some forest insect or disease control project. One might establish that certain control measures would, over a three-year period, probably save 20 billion board feet of, say, lodgepole pine presently in sawtimber size classes. If the current market price of lodgepole sawtimber stumpage is $20 per thousand board feet does the project then produce a saving of 400 million dollars? Surely not, for it is doubtful that 20 billion board feet of lodgepole pine could be *given* away over a three-year period, or even one of ten years. In fact, if the anticipated losses would not reduce the economic supply of lodgepole pine sawlogs (if the expected loss would not raise prices of lodgepole sawlogs) it is difficult to see how there could be any benefit whatever attributable to the project from a timber standpoint. The uncertainty attached to the possible loss and the possible savings acts further to make benefit measurement in this case a difficult empirical task. (The illustration also serves to show why economic analysis of projects is not always welcomed by agency administrators.)

### Intangible Benefits

These are benefits not presently measurable by accepted methods, which is usually equivalent to saying benefits that are not valued in the market. Scenic beauty, most types of recreation, wildlife values, values of sport-fishing (but not commercial fishing), etc.—these are considered intangible benefits. It is generally agreed that while these are very important, they should not be included in the benefit/cost ratio. To assure that decision makers consider such outputs, a verbal discussion of these benefits is usually included in the project analysis, together with the estimated costs of achieving the benefits. Frequently, estimates of physical use (visitor days, hunter days, etc.) are used to give a more vivid picture of the possible scale of benefits. The difficulty posed by including these intangibles is not that of introducing a serious error in the benefit/cost ratio (although this has certainly been the case in some past instances), but rather that the inclusion of non-market determined values tends to throw a cloud of suspicion on otherwise careful analyses. In spite of this suspicion there have been recent proposals for attaching standard value units to certain kinds of recreation, ranging from $1.50 to $6.00 per day. More will be said of the problems met in evaluating recreation in Chapter 22.

### Demand, Price, and Employment Levels

The investment program in some timber improvement and planting schemes extends over several years, while the benefits are often expected to continue for a much longer period. According to present instructions (Sen. Doc. 97, 1962) projections of product demand and consumption should be based on high-employment levels. Price levels, however, are more difficult. Prices used ". . . should reflect the exchange values expected to prevail at the time costs are incurred and benefits accrued" (Sen. Doc. 97, 1962). Unfortunately, satisfactory (mutually acceptable) long term price projections have yet to be developed, and current prices are usually used. Since it is the real costs and real benefits that are important, benefits and costs should be expressed in constant dollars to avoid effects of inflation or deflation.

### Interest Rates

With both costs and benefits occurring at different time periods and continuing for long time spans, the problem of making the two comparable is similar to that treated in Chapter 14. The question of the proper discount rate to apply to Federal projects has been debated vigorously for the past two decades.

The early committee of 1949–52 recommended that the coupon rate on Federal notes of at least 15 years duration be used, and when the guidelines of 1962 repeated the recommendation (Senate Document No. 97, 1962) the effective rate was about 3¼ per cent. In 1968 the Water Resources Council gave notice that it proposed to establish "a new formula for the determination each year of the discount rate for use in the formulation and evaluation of Federal plans for use and development of the Nation's water and related land resources" (United States Congress, 1968). The basic provision follows:

> The interest rate to be used in plan formulation and evaluation for discounting future benefits and computing costs, or otherwise converting benefits and costs to a common time basis, shall be based upon the average yield during the preceding fiscal year on interest-bearing marketable securities of the United States which, at the time the computation is made, have terms of 15 years or more remaining to maturity; provided, however, that in no event shall the rate be raised or lowered more than one-quarter per cent for any year.

The rate recommended by the Council for 1969 was 4⅝ per cent.

Testimony at the hearings brought out the wide disagreement on the suitability of any one interest rate. Most economists recommended rates of between 7 and 12 per cent as being appropriate for 1968, largely following the general approach outlined by Eckstein (Eckstein, 1958).

The significance of the interest rate used in calculating benefit/cost ratios is double-barreled. On the one hand this rate indicates the point at which funds should be taken out of the private sector and invested by the public sector, for the proper interest rate to use is the opportunity cost of Federal funds. Since these are taken, by taxes, from the private sector, a government project should be undertaken only if that project promises to yield social returns in excess of the returns that would have been achieved had the funds remained in private hands. That is, "the correct discount rate for the evaluation of a government project is the percentage rate of return that the resources utilized would otherwise provide in the private sector" (Baumol, 1969).

On the other hand, the interest rate used in calculating the benefit/cost ratio also determines the scale of the project and determines whether or not it will be "justified." An idea of the significance of this is given by calculations applying interest rates of 4, 6, and 8 per cent to the analysis of projects authorized by Congress in 1962 for construction by the Corps of Engineers (Fox & Herfindahl, 1964). If these rates had been used, 9, 64, and 80 per cent of the projects, respectively, would have had benefit/cost ratios of less than one; they would not have been justified. Moreover, since the design of the projects was predicated on lower interest rates, they had been "over-designed"—built to a scale that was much too large.

The going market rate in private markets may not always be appropriate to public projects, for the discount rate is meant to balance the productivity of an investment and the reluctance of society to sacrifice current for future consumption. If society has its own set of values there is no reason to suppose that individuals and society would have the same rate of discount (Hufschmidt et al., 1961). Individual time preference and social time preference are usually regarded as being quite different, with individual rates considerably higher than social rates.

## Period of Analysis

If it is known with reasonable precision, the actual life of the project is usually used. As a practical matter, anything in excess of 100 years is usually disregarded; benefits to be received 100 years or more from now have such a small present value there is little point in their calculation.

### Cost Allocation

Benefit/cost calculations are sometimes used for allocating cost of projects to different user groups. As long as this is nothing more than an accounting exercise no harm can be done, but if the cost allocation is made the basis for repayment by users, then troubles may ensue. Clearly, many and sometimes most of the costs are true *joint* costs, and any attempt to allocate such costs almost inevitably leads to difficulties.

## FUTURE OF BENEFIT/COST ANALYSIS

This description of benefit/cost procedures is in no way intended as a guide to benefit/cost calculations. It should suffice, however, to give any forester a knowledge of the general approach sufficient to enable appreciation of the problems faced by those making benefit/cost analyses. And perhaps more important, the discussion should give him an idea of the kinds of problems that make accurate analyses so very difficult.

There are some economists who would go so far as to say that the primary purpose served by benefit/cost procedures has been that of obscuring the true worth (or worthlessness) of public undertakings. Another group is merely skeptical, likening the problem

> . . . to appraising the quality of a horse-and-rabbit stew, the rabbit being cast as the consequences that can be measured and evaluated numerically, and the horse as the amalgam of external effects, social, emotional, and psychological impacts, and historical and aesthetic considerations that can be adjudged only roughly and subjectively. Since the horse was bound to dominate the flavor of the stew, meticulous evaluation of the rabbit would hardly seem worthwhile.[2]

Most economists would probably agree that benefit/cost analysis is a relatively undeveloped economic tool, and that there almost surely will always be some elements that cannot be appraised quantitatively and objectively. But they would also hold that the procedures in use today are far better than those used 20 years ago, and that improvements are constantly being introduced. Among the

[2] This was the example used to illustrate the range of attitudes expressed at a conference held to discuss measuring the benefits of government investments (Dorfman, 1965). Dorfman went on to say that the advocates of benefit/cost analysis termed this a "council of despair."

more interesting developments of recent years has been the recognition that analyses might be carried out at different levels (at national or regional levels, for example) and using different criteria (GNP and employment, for example). Such analyses could well result in different priority assignments, but might also provide better guides to allocating public funds

So far, relatively few timber production projects have been made the subject of benefit/cost analyses in this country. Planned expenditures for fire control and for insect and disease control have been so analyzed upon occasion, but this is not the usual procedure for forest managing agencies. Recreation expenditures, and investments for other non-timber products of forests, however, are sometimes made the subject matter for benefit/cost analysis. Many of these suffer from the usual problems of evaluating the so-called intangible benefits. It is interesting to conjecture what might result if investments in, say, sawtimber production on national forests were analyzed by benefit/cost procedures, especially if interest rates such as those suggested by the experts testifying before the Joint Economic Committee in 1968 and 1969 were used (United States Congress, 1968; 1969).

If current trends continue, particularly that implied by installing the PPB system in the Federal agencies, the end result may well be that of making benefit/cost analysis mandatory for most agency investments. There is no reason why Federal investment in timber production should be given special treatment as compared to Federal investment in water resource development, or education, or transportation. A uniform method of appraising alternative investment opportunities would be welcomed by almost everyone interested in efficient use of public funds and of other public resources, as long as adherence to procedure was not allowed to dominate common sense.

# 21

# Water and Watershed Management

It is almost axiomatic that little notice is taken of deteriorating natural resources until conditions become serious and widespread. In the later 1890's and early 1900's the situation in the Lake states combined with general excesses of industry to focus the nation's attention on forests. The 1930's saw the spotlight shift to soil conservation, brought on by the drought, the dustbowl, and the depression. The 1960's and 1970's may be marked by future conservationists as the decades of water consciousness in the United States, stimulated by the forced realization that the nation's rivers and lakes are rapidly changing from a national asset to a source of irritation and discontent for industry, government, and the private citizen. The rapid deterioration of Lake Erie and other major water bodies during the 1960's undoubtedly did much to arouse in the public a militant awareness of an imminent environmental crisis.

A great part of current water problems stems from the conflicting demands generated by an expanding industrial capacity and a population growing rapidly in both size and affluence. People everywhere want rivers and lakes to be sparkling and clear, for recreational purposes as well as for purely aesthetic reasons. At the same time, because these same people are stockholders, directors, and employees of the expanding industries and of mechanized agriculture, they want the "freedom" to dump vast quantities of industrial pollutants and municipal waste into the nearest stream, to use whatever herbicides, pesticides, or detergents they wish, and to have someone else bear the cost of cleaning up the rivers and lakes.

Forestry's role in pollution abatement will be a passive one. The primary part forests will play in the unfolding water drama is that of protector for the nation's watersheds. And because the forester, whether he recognizes his position or not, is a watershed manager, attention here will be concentrated on economic problems of watershed management and of determining water benefits and costs. Water pollution, navigation, flood control, power production, and even water recreation will not be treated directly, even though these are among the most challenging economic problems of our times.

Economic problems of water resource management, both qualitative and quantitative, have been a major area of research for Resources for the Future. The student seriously concerned with water economics will quickly encounter the pioneering works sponsored by this foundation, several of which will be cited.

## ECONOMIC ANALYSIS AND WATERSHED MANAGEMENT

Many of the problems faced by an economic analyst in reviewing watershed management projects are apparent as soon as one starts thinking in terms of product, costs, and revenues. Identification of the productive factors and of the product provide an excellent introduction to these problems (Fig. 21.1).

The fixed factor of watershed management should, from our particular standpoint, be that part of the watershed covered by the forest with which we are specifically concerned. It could be an entire watershed or parts of several smaller watersheds, but in either case the decision is not difficult and will be determined by the geography of the forest itself. For ease in presentation we will assume that all water (the product) flowing from the forest area that concerns us comes from a single watershed, and that the volume is measured as it flows from the forest.

To be meaningful, water must be regarded as a product with three dimensions, quantity, quality, and time. Product volume could be measured in any of several units: cubic meters, cubic feet, acre feet, etc. Regardless of the unit selected, two secondary but important problems must be solved—over what *period of time* is the

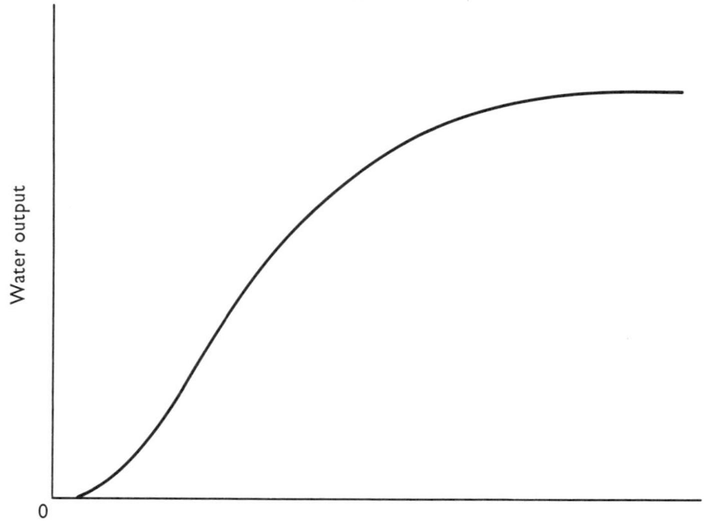


**Fig. 21.1** Production function for watershed management.

water flow to be considered, and what *quality* standards will be employed.

Quality of water is itself a many-dimensioned thing, and the measures of quality usually reflect the uses to which the water is to be put. Turbidity is one aspect of quality which can be importantly influenced by management of the watershed, while *purity* is a much more complicated matter, involving pollution. Biologists often classify pollution into five major types: (1) putrescible materials, (2) heated effluent, (3) toxic materials, (4) inert materials, and (5) radioactive elements or compounds.[1] All may be of concern to the forest watershed manager, although (1) and (3) are probably of most frequent interest—putrescible materials because of their effect on fish, wildlife, and recreational uses of forests; and toxic materials because forest use of herbicides and pesticides may be a major source of pollution. Concern with pollution is such a recent matter in this country that in many cases neither methods for detecting

[1] Putrescible materials are organic wastes that decompose, taking large quantities of oxygen. They include wastes of the ordinary household variety, water from papermills, etc. Toxic wastes do not settle out readily, are not easily broken down by biological agents, and tend to be poisonous to animals and plants. Phenols, herbicides, and pesticides are examples (Goldman & Shoop, 1967).

the pollutant, nor standards for limiting them, have yet been established.

The timing problem, so far as watershed management is concerned, stems largely from the fact that in most watersheds the quantity of water varies greatly from month to month, usually in a relatively predictable fashion. In most parts of the United States the spring months bring concern of too much water, while the late summer months often bring the opposite problem of too little. Hence some management practices may be designed to reduce peaks of flow (flood control) while others are instituted to increase flows at low-level periods. Fortunately, the two are not always competitive.

Once the difficulties of identification and measurement have been settled there are still large problems to be solved in the technical aspects of management. What happens, for example, when a specific cutting practice is instituted on the forest? How much will peak flows be retarded or low flows augmented? How much will water temperature and turbidity be affected? The science of watershed management is still very much in its infancy, yet unless one can state "If you do *this* to *these* acres it will usually *increase* (or *decrease*) the amount of water of the stated quality, at the specified time, by about *this* much," one cannot make any serious claim to be managing a watershed. And until this kind of information is at hand, the economics of watershed management will not be very meaningful.

## A Production Framework

Given the required data, however, the production analysis could be completed. Suppose, for example, one could measure the water output of a forest in units of hundreds of cubic meters of water per minute during the month of August, with the water being of a given minimum quality. Inputs might be measured in some unit combining hours of labor, specific materials, and the use of particular equipment. More than likely the inputs would be in terms of dollars spent. The resulting relationship, a production function for the forest watershed with inputs expressed in dollar units, could then be drawn as in Fig. 21.2(a), then easily transformed to the cost–output relationship shown in Fig. 21.2(b). (It is assumed that without management the low flow would be 200 cubic meters per minute; management of the watershed results in increasing this low flow according to the curve shown.)

The discussion of benefit/cost analysis has made clear that reve-

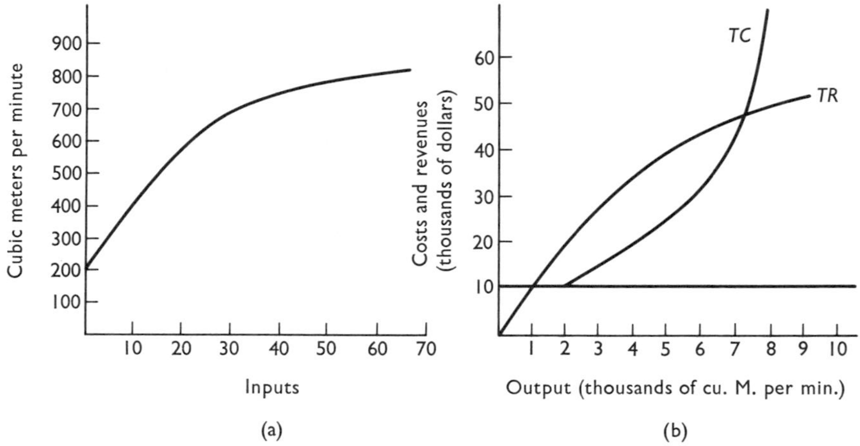


**Fig. 21.2** Usual production function for watershed, and resulting cost and revenue curves.

nue from water poses problems very different from those encountered in timber production. For the moment, however, we will assume that problems of identification and measurement have been solved, and that we have been given the data required to determine the dollar value of the water flowing from the forest watershed. These data have been used to construct the revenue curve given in Fig. 21.2(b). The gradually declining total revenue curve of the figure follows because unit values of water were assumed to decrease as output increased.

All costs and revenues will need to be discounted to some common point in time, with the social opportunity cost of capital probably being the most appropriate discount rate. The same financial objective of maximizing present net worth is as appropriate to watershed management as to timber management, although there will usually be constraints attached to the maximization. Problems will be met when estimating future costs of materials, of maintenance, of operation, and of product prices. The end result of the analysis would be achieved by obtaining the discounted total revenue and the discounted total cost curves, with the optimum level of output determined where the slopes of these discounted curves were equal —where marginal discounted cost equals marginal discounted revenue.

## The Benefit/Cost Framework

If the watershed is extensive, and if a variety of benefits are to be produced from the management activities, the framework of benefit/cost analysis would probably be more appropriate and is shown in Fig. 21.3(a). For comparison, the original total cost and revenue

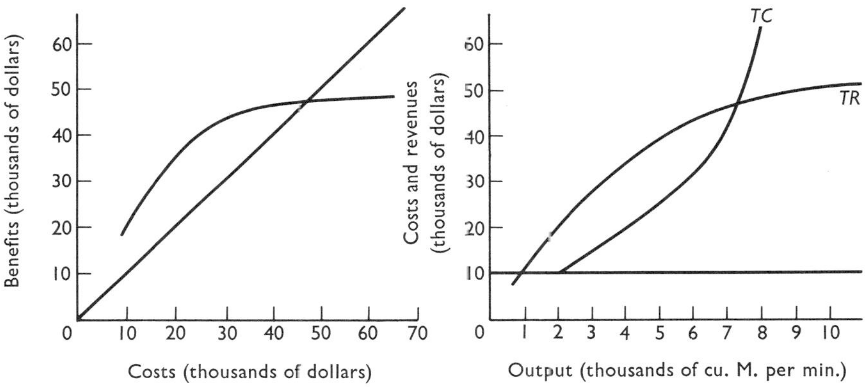


**Fig. 21.3** The benefit/cost approach.

curves are shown alongside in Fig. 21.3(b). The conversion to benefit/cost is easily made by recording, for each dollar level of costs, the corresponding dollar level of revenue. The optimum scale, of course, is located by determining the point at which the slope of the curve is 45 degrees, and indicates the same level of output as the more traditional form of analysis.

The cost side of watershed management is reasonably straightforward, with the major complications arising from our lack of knowledge of the physical relationships involved. Revenues, however, are a different matter!

### WATER BENEFITS

The questions here are the crucial ones of "What benefits should be included in the analysis?" and "How are these benefits to be measured?" If water had a single value that was determined in an ordinary market there would be no particular problems at this stage. But water values are not usually determined by market processes, and the same unit of water may have not only different values for different purposes, but because of the physical peculiarities of water it is possible for the same unit to take on any, or all, of these values. Water flowing in a forest watershed can be a source of high recreational value as a trout stream, then take on additional values as it is successively impounded in a reservoir (values for boating, swimming, etc.), released through a turbine to create electric power (power values), sent on downstream to become a municipal water supply and a source of cooling and cleansing water for industry, and, finally, used for irrigating a crop of cotton. And throughout much of its journey the water may be transporting barges, purifying both municipal and industrial sewage, providing boating and fishing to all who wish to participate, and even proving a source of inspiration to beauty lovers. It is this complexity of values that makes water economics such a fascinating, but frustrating, analytic subject matter for economists.

At the same time, this partial recitation of the different values that may be attributed to water gives one a better idea of why watershed management, and why water problems generally, are considered primarily within the context of public production. Many of these values are external to the production process of watershed management, and some cannot be captured by any private market mechanism. Some of the products are very definitely "social goods": once they have been made available they are available to all.

### Water Demand

Demand for water, as for most other natural resources, is a derived demand. In this case it is derived from the demand for other products or services that use water in their production process. These products may either be sold through the private market system or produced within the public sector, and either made subject to user charges or dispensed completely free of charge.

Water uses may be classified as being either *consumptive* (the water is consumed and hence no longer available as such), or *non-consumptive* (the water is used, but remains available for further use). The parallel between consumptive use and "rival in consumption," like that between non-consumptive use and "non-rival in consumption" is very close. Moreover, the classification of consumptive–

non-consumptive separates those uses in which water might be considered a *private* or a *public good.* Non-consumptive uses include the very important ones of navigation, power production (hydroelectric, but not steam), fish and wildlife production, most kinds of recreation, and many industrial uses as well. Consumptive uses include those industrial uses wherein water is evaporated and hence passes out of the water system. The major consumptive use is irrigation; any water used by plant crops is transpired, hence, no longer available. And of course, some municipal uses are consumptive, although modern recovery processes make this one of the lesser uses, in terms of volume.

Quantitative information on water consumption has been difficult to obtain and the estimates made by various sources show considerable disparity. In 1965 the Water Resources Council was established under the Water Resources Planning Act (P.L. 89–80) and their first assessment of the nation's water resources was released in 1968 (United States Water Resources Council, 1968). Table 21.1 gives the estimates of water use prepared for the council, together with the nation's water requirements projected ahead to A.D. 2000. In this report, consumptive use was compared to "withdrawals," representing somewhat of a departure from previous reporting procedures. No estimates were given for "instream" uses, the council reporting it had "been unable to adequately quantify requirements for instream uses of water."

**Domestic and municipal.** Water needs for these purposes are expected to be met with no problems over the foreseeable future. In 1965 total withdrawal for these uses amounted to 26 bgd (billion gallons per day), or about 2 per cent of the average annual discharge from the nation's watersheds. Per capita consumption of water averaged 157 gallons per day over the nation. Urban users consume considerably more than their rural counterparts; Chicago, for example, produced an average of 255 gallons per day for its citizens. Air conditioning, automatic washers, garbage disposers, etc., all tend to push consumption to higher levels. Costs of municipal water to the consumer varied widely, averaging slightly under 30¢ per 1,000 gallons.

**Industrial.** Industry uses water as a solvent (and for this reason industry is a major source of pollution); as a cooling agent (which adds heat as a pollutant—thermal pollution—and also increases the

**TABLE 21.1 ESTIMATED WATER USE AND PROJECTED REQUIREMENTS FOR THE UNITED STATES (MILLION GALLONS DAILY).***

| Type of use | Withdrawals | | Consumptive use | |
|---|---|---|---|---|
| | Used in 1965 | Projected requirements 2000 | Used in 1965 | Projected requirements 2000 |
| Rural domestic | 2,351 | 2,852 | 1,636 | 2,102 |
| Municipal (public-supplied) | 23,745 | 50,724 | 5,244 | 16,478 |
| Industrial (self-supplied) | 46,405 | 127,365 | 3,764 | 10,011 |
| Steam-electric | | | | |
| Fresh | 62,738 | 259,208 | 659 | 4,552 |
| Saline | 21,800 | 211,240 | 157 | 2,022 |
| Agriculture | | | | |
| Irrigation | 110,852 | 149,824 | 64,696 | 89,964 |
| Livestock | 1,726 | 3,397 | 1,626 | 3,077 |
| Total | 269,617 | 804,610 | 77,782 | 128,206 |

* From *The Nation's Water Resources,* Table 4.1, page 4.1 (United States Water Resources Council, 1968).

water's capacity as a solvent and reduces its capacity for degrading organic pollutants); and also as an integral part of certain industrial processes.

There is great variation in quantitative water use, even within the same industrial group. If water is abundant then the firm will use water freely, but if water is scarce then far less will be used. To illustrate: in 1959 the average water use per ton of paper manufactured was reported to be 57,000 gallons, yet pulp and paper production increased 26 per cent while water intake increased only 8½ per cent over the preceding five years (Bower, 1966). And in Wisconsin alone, one kraft mill reported using 100,000 gallons per ton of product, while another used only 28,000 (Mussey, 1961).

Manufacturing accounted for some 40 bgd (which, on the average, was recirculated some 2½ times). In the United States, industrial water use, like industrial output, is concentrated in a relatively few establishments. Ninety-eight per cent of the industrial water used was taken by just 8,925 establishments; a little over 5,000 plants in five major industry groups (food, pulp and paper, chemicals, petroleum and coal products, and primary metals) accounted for 88 per cent of the water withdrawals. If water is to be conserved it must be conserved by these particular industries.

The fact that less than 30 per cent of the water discharged by industry was treated in any way gives rise to serious misgivings about any hope of stopping industrial pollution. A better perspective on this problem, however, is gained from the fact that fewer than 9,000 establishments discharge about 10.5 bgd of process wastes, as compared to 0.7 bgd from the more than 300,000 other manufacturing establishments.

**Power production.** Hydroelectric or steam-electric plants produce almost all of our electric power, and water is a factor in all electric power produced in the United States. Consumption of water for power poses no problems for the future, but thermal pollution does. All states now regulate the amount of thermal pollution permitted, but procedures for alleviating the difficulty are expensive and will almost certainly become greater as urban sites demand more and more electric power.

**Agriculture.** More water is consumed by agriculture than by any other single industry in the nation. Moreover, irrigation consumes about 60 per cent of the total water withdrawn for such purposes. Most United States irrigation is, of course, in the 17 western states, but more and more irrigation water is used as a rainfall supplement, either to get over the occasional dry spell or to prevent crops from freezing or to bring about earlier marketing.

**Non-consumptive uses.** The above uses, though extensive, represent only the *consumptive* side of the picture. The non-consumptive uses, such as swimming, navigation, sailing, provision of a habitat for fish and wildlife, for scenic beauty, etc., all create a demand for water. And this is demand in its true economic sense in that there is, or would be, a relationship between the price a consumer would be willing to pay and the amount of water used.

## WATER COSTS

When treating demand for the usual consumptive uses, one of Marshall's points concerning the elasticity of derived demand is of prime importance. Derived demand (according to Marshall) will be more inelastic the smaller the fraction of total product cost that goes to the factor. Even in uses where water is metered and paid for by the thousand gallons, the actual cost of water per se comprises a very small proportion of total production costs. Irrigation is sometimes an exception, although irrigation water is usually very cheap. In most consumptive uses water is used as if the demand were highly inelastic. In fact, industries frequently have little idea how much water is required per unit of product. In a study made of the Muskingum River Valley during the late 1950's, most of the industrial water users interviewed were unable to give any figure at all on their water consumption. Unit costs for water were sufficiently low that there was simply no attempt to economize on its use.

Most supply systems calculate water cost as the cost of transporting and treating the water. Water per se—water at the intake pipe, or at the edge of the watershed—is not accorded a value at all. If one applies derived demand theory to determine the value of water at the watershed in an attempt to derive a value analogous to that for stumpage, the result is usually zero. In fact, if one begins with the demand curve for almost any product, then subtracts the costs of all productive factors other than water (the supply curve for all other productive factors), the remainder will usually be a very small item indeed. Yet this is the *value of water delivered to the user.* From this small value one must still subtract all the other costs that enter into the transportation and purification of the water. Most often the remainder from this last calculation is zero. It is not true that the water at the intake valve is valueless; it simply has no regular *market price* and is viewed, essentially, as a free good. This fact raises some problems for allocating funds to watershed management, for if the product being produced is not accorded a value, why should money and effort be devoted to its production?

In the United States almost every kind of governmental unit can be, has been, and is involved in water resource problems. In addition to the usual Federal, state, and municipal governmental units there are more than 10,000 special districts established under state law, authorized to plan, build, and operate local water projects of one or more types. There are also several interstate compacts, an international commission for treating water problems between Mexico and the United States, and an international corporation for operating the Saint Lawrence Seaway. Water management in the private sector is similarly diverse.

The Water Resources Council estimated that about 30 per cent of all funds spent annually for water and related land resource development were from Federal sources. An approximately equal amount came from non-Federal public funds, and the remaining 40 per cent were of private origin. Some 80 per cent of all irrigation waters, for example (and irrigation is the largest consumptive use), are financed *privately*.

## MEASURING BENEFITS

Since the Flood Control Act of 1936 specified that benefits of navigation and flood control must exceed estimated costs, the techniques and procedures for measuring these costs and benefits, and those of other water uses, have been undergoing intensive development. Some of the values of watershed management are not too difficult to determine and to specify in dollar terms. With flood control, for example, one would reason that a property owner would be willing to pay as flood protection fees an amount up to the damages averted by the protection. Flood protection is a good example of a social good; once the beneficial improvements have been installed the protection is available to all. Irrigation water could, theoretically, be valued in at least two ways. Most of the water so used is privately managed, so costs under the usual market forces should not be difficult to obtain. User charges for irrigation water supplied by Federal projects, however, are limited by law, so the actual price of Federal water would not necessarily be a proper guide. Since the use of irrigation water directly influences the income of those who use it, the change in personal income resulting from its use could be a measure of the benefits received. But current policies of Federal price support in the agricultural sector complicate the picture. How should one value water used to produce a crop if the crop price is supported by the same government that produces the water?

Electric power values, of course, can be directly estimated from the anticipated market sales and valued at well-established market rates. Yet here as well there have been severe problems. Should federally produced power be sold at the marginal cost of production, or at the maximum rate possible, or how? We know the usual answer would be in terms of marginal cost, but when power is produced jointly with many other products the answer is much more difficult.

Navigation improvements result in a savings in the total transportation bill of the nation. But because highways, railroads, and airlines are all subsidized by the Federal government, navigation poses real valuation problems. Moreover, experience with navigation on the Tennessee and other waterways indicates that there is a very decided time lag in developing water traffic. In some cases as much as 15 years is required before the changeover to water-borne freight is fully effective.

The usual approach to estimating benefits from recreation, fish, wildlife, and aesthetic values has already been mentioned. These benefits are not estimated in dollar terms but are described as carefully as possible in the project report along with the costs of achieving these benefits. The legislator or other public decision maker must then make his choice. Other approaches have been suggested for placing a dollar value on some of these uses, and a few of these will be discussed under the specific topics of wildlife and recreation.

It is plain that the evaluation of benefits from water resource developments is far from being either precise or foolproof. Most of the difficulties stem from the fact that not all benefits can be known, measured, or quantified equally. As stated by the then Secretary of the Interior:

> It is true that water and related land resources projects provide many benefits of a commodity or goods and services nature, more or less readily evaluated in private market terms. But in many instances the projects provide much more than goods and services. They provide also, in increasing instances, for the protection and preservation of the quality of the environment. Note the emphasis on water quality control for instance with its intangible as well as tangible benefits. Note also the enhancement of fish and wildlife, the contributions to aesthetics and the amenities, and the many recreational opportunities created by water and related land projects and programs. These benefits have generally been underestimated, partly because of the lack of explicit private market prices or values to represent them, and partly because the projects have been justified on other grounds, with these latter benefits given only incidental accounting. The significance of these kinds of benefits may well outweigh the more traditional benefits in the future (Udall, 1968).

# 22

# Forest Recreation

Outdoor recreation has been defined as "leisure time activity undertaken in a relatively non-urban environment characterized by a natural setting, for the primary purposes of enjoyment and physical or mental well-being" (Outdoor Recreation Resource Review Commission, 1962). *Forest recreation* is only part of this broad group of activities, and as a beginning might be defined as "leisure time activity voluntarily undertaken in a forest environment for the primary purpose of enjoyment." Even this encompasses a wide range of outdoor activities; hunting, fishing, hiking, horseback riding, birdwatching, camping, mountain climbing, snomobiling, skiing, swimming, canoeing, automobile riding, and many others.

While forest recreation is usually associated with some kind of activity, *recreation* and *activity* are not really identical. Many of the activities mentioned above can be, and are, undertaken by people for work as well as for fun. A timber cruiser hikes 15 miles through the woods, eats his sandwich by a brook, and calls it hard work. A New Yorker hikes 15 miles down the Appalachian Trail, eats a sandwich by a brook, and enthusiastically calls it recreation. The activity itself is not the entire story. Rather, it is the "experience" resulting from a very complex interaction between the individual, the forest environment, his objective, his usual environment, his cultural background, and perhaps other things as well.

The initial definition of forest recreation made *enjoyment* the primary purpose of the activity, a distinction often sufficient to determine when an activity is recreation and when it is not. The distinction becomes particularly significant to analyses of recreation demand, for if the primary purpose of the activity is enjoyment, then the activity must be a "discretionary" one; the individual or group must be free to choose whether or not to participate in the activity—the consumer is sovereign. Social pressures, the status gained through participation, family and other group pressures—all may play a role in the kind of recreational activity that individuals choose. These factors may limit, or constrain, the discretionary nature of recreation, but they do not eliminate it.

Yet enjoyment does not really answer the question of *why* people spend time in a forest environment. Most students of forestry, for example, like both fishing and camping, and both of these activities are typical forms of forest recreation. Yet the reward—the personal "enjoyment" one gains from camping—is very different from that gained from fishing. Furthermore, it is unlikely that two individuals would either gain the same kind, or the same amount, of reward from either activity. The interpersonal comparison problem of welfare economics is very real when dealing with values of forest recreation. Behavioral scientists, now beginning to explore this almost virgin territory, may provide valuable clues to the answers. There may be basic needs, or drives, or motivations, which forest recreation satisfies. Does aggression, for example, find a release in mountain climbing? Will man lose some of his territorial possessiveness by spending a month in a wilderness? Is the drive for material aggrandizement lessened if one "escapes" from the city with its traffic, its noise, its smells, and its hordes of competitive people, to a quiet glade where the only sound is that of the wind in the trees? Or will the genuine urbanite find the stillness of a forest unbearably oppressive? The answers to this kind of question might help us discover what kinds of forest recreation should be provided, and might even give leads on *who* should provide *what* for *whom*.

## PRODUCTION PECULIARITIES

Some significant peculiarities of forest recreation can be identified. From a production standpoint one of the most important is that, like timber, *forest recreation is produced on site*. It is "resource oriented," as contrasted to "user oriented," and the consumer must nearly always travel to the area.[1]

Like almost all outdoor recreation, *forest recreation is necessarily a kind of "package deal,"* involving anticipation, travel to the forest site, the activity itself, the return travel, and, finally, recollection (Clawson & Knetsch, 1966). The fact that it is convenient to refer to the activity by the place that the action occurs does not mean the other parts of the package can be ignored. In thinking of forest recreation, then, one must bear in mind that the forest part of the experience represents only part of this package of goods. In some ways the demand for forest recreation resembles the demand for houses; both are "packages," and our primary concern is with a component of that package. The demand for the forest site, like the demand for lumber or sawlogs, is derived from the demand for the total package.

*Forest recreation is often (but by no means always) produced jointly* with another product (or products). Private timberlands, the national forests, and the BLM lands, for example, are primarily managed for timber production, but these same areas supply a major part of forest recreation in the nation. On such areas production of forest recreation is often complementary to that of timber and other forest products, but examples of competitive production are common—wilderness versus timber; pure stands of conifers versus the aesthetically preferred mixed stands of conifers and hardwoods; campgrounds and sightseers versus loggers and logging roads; hunters versus birdwatchers; and so on for a long list. At the opposite end one can cite the national parks and many state parks, where management is primarily directed toward producing recreation. Here again other products are often produced, with cases of competitiveness and complementarity easily identified. No timber cutting or hunting permitted in national parks, yet excellent watershed protection and fishing, etc.

The *variety of the products* lumped together under the general name of "forest recreation" could scarcely be called a peculiarity, for certainly timber, wildlife, and other primary forest products share this feature, but it does complicate discussion of the topic. Some

[1] Three classes of outdoor recreation sites have been suggested (Clawson & Knetsch, 1966): (1) user oriented areas, (2) resource oriented areas, and (3) intermediate areas.

kinds of forest recreation are themselves complementary, others competitive; they are very often as distinct in form, as specific in their site requirements, and as variable in their production characteristics as are different tree species. The "spacing" of campsites, and the optimum "stocking" of hunters or of wilderness hikers, illustrate other parallels to timber production, although our knowledge of their effects on forest recreation is anything but precise.

## TRENDS IN FOREST RECREATION

At least two factors make forest recreation trend data of questionable accuracy. One deals with measurement; there is no general agreement upon how or in what units forest recreation should be measured. The other concerns the adequacy of the information itself; some kinds of activities have been reported only for the most recent years, and for forest recreation on private lands there is virtually no information whatever.

### National Parks

Gross recreational visitor numbers have been reported from several kinds of public areas. The National Park Service, for example, has reported the number of visitors to each of the 32 parks for many years. These parks, now comprising a total area of 13.6 million acres, were set aside in 1916 "to conserve the scenery and the natural and historic objects and the wildlife therein, and to provide for the enjoyment of the same in such manner and by such means as will leave them unimpaired for the enjoyment of future generations." The park system also includes a large number of other units such as national monuments, battlefields, seashores, lakeshores, and recreation areas.

While visitor statistics for national parks leave something to be desired from an analytical viewpoint, there is no question as to the general trend of use. Visitor numbers increased from slightly over 1 million in 1920 to around 20 million just prior to World War II; from a low of 7 million during the war to over 70 million in 1960; to 140 million in 1965, and in 1969 the number reported was 179.6 million. The pressure of increasing visitor numbers on these unique areas is creating a severe problem for long term maintenance; and it remains to be seen whether the organization can devise a management system that will permit "enjoyment" of these parks "in such manner and by such means as will leave them unimpaired for the enjoyment of future generations."

### National Forests

The statistics for national forests are less comparable and probably less accurate. Few national forests have entrance points that permit visitor counts to be maintained, and before World War II there were few restrictions on either entrance or on use of the established recreational facilities. But again, the available figures leave no doubt that the national forests have experienced the same sort of explosive increase as have the national parks. Numbers of visits for recreational purposes increased from 5 million in 1925 to 16 million in 1940, to 27 million in 1950, to 92 million in 1960, and to 147 million in 1965 (Clawson & Knetsch, 1966).

Table 22.1 presents the numbers of visitor days spent in various kinds of recreational activities on national forests during 1969 (as reported by the Bureau of Outdoor Recreation in a preliminary tabulation). Note that the unit of measurement has been changed—from *number of visits* to *visitor days*. The variety of activities included under forest recreation is almost as notable as the huge total of days spent in the reported activities. And Table 22.1 is a summary; the original information was reported under no less than 52 separate activities. Individual camping led all other recreation activities on national forests, accounting for fully 28 per cent of the total. Automobile driving was second with 21 per cent, and a combined "hunting and fishing" ranked third with 17.7 per cent. These three activities accounted for nearly two-thirds of all recreational visitor days spent on national forest lands.

## ECONOMIC ASPECTS OF FOREST RECREATION

Forest recreation became an economic problem when the resources involved in its production were scarce enough to command a price of some kind; at that time it began to affect the allocation of forest resources and, consequently, the production of other forest products. It became a problem of public finance when public funds were used for its production, either by political expediency or through social necessity. As an item of public finance it might affect either the allocation of public funds (efficiency), the distribution of income, or the growth and stability function of government. As will be seen, forest recreation could have an effect in all three of

**TABLE 22.1 SUMMARY OF ACTIVITIES ON NATIONAL FORESTS 1969.***

| Activity | Visitor day use (1,000's) | Per cent of total |
|---|---|---|
| Viewing | 55,427 | 3.3 |
| Auto driving | 342,962 | 21.0 |
| Hiking | 48,570 | 2.9 |
| Horseback riding | 21,705 | 1.3 |
| Other riding and looking | 65,471 | 4.1 |
| Canoeing, sailing, other water | 12,566 | .7 |
| Team sports and games | 5,855 | .3 |
| Swimming, skin diving, water-skiing | 35,367 | 2.1 |
| Fishing | 148,682 | 9.1 |
| Individual camping | 416,297 | 25.5 |
| Organization camping | 43,240 | 2.6 |
| Picnicking | 68,377 | 4.1 |
| Resorts | 40,332 | 2.4 |
| Recreation residence | 80,152 | 4.9 |
| Skiing | 47,774 | 2.9 |
| Other winter sports | 11,212 | .7 |
| Hunting | 141,479 | 8.6 |
| Nature study, programs, tours, etc. | 42,919 | 2.5 |
| Total | 1,628,381 | |

* Data supplied by Bureau of Outdoor Recreation (preliminary).

these public finance areas, but we shall first look at the private aspects.

## Private Production

As would be expected, private production of forest recreation is primarily concerned with goods that are either rival in consumption or those that, while non-rival in actual consumption, can still be made subject to the exclusion principle (an individual can be excluded from enjoying the particular commodity unless willing to pay the stipulated price to the operator). Resorts on privately owned forest areas probably offer the best example of the former, and there are literally thousands of such places. The variety of accommodations offered, the associated golf courses, tennis courts, swimming pools, stables, and other resort trappings are so well advertised that they require no further description. The total amount of capital invested in such facilities and the income and employment they generate is not known, nor is the part that can be attributed either directly or indirectly to forests. Certainly, people go to these kinds of places for status, for friendship, and for many reasons other than a desire to visit a forested area. In addition to those on privately used forests there are also many private hotels and resorts within public forest boundaries, either operated as leased facilities, as concessions, or on private "inholdings."

A ski resort offers an example of private enterprise providing a good that is not, strictly speaking, rival in consumption but is definitely subject to the exclusion principle. It closely resembles a bridge or highway; the good is "jointly supplied." Once a cable tow or chair lift has been installed, the cost of operation varies little with the number of people served by the lift or the number skiing on the slope. But it is possible to check those using the tow and to restrict use of either the tow or the slope to those who have paid for the privilege. Thus, even though the number of skiers on a slope can be varied widely without significantly changing costs (up to the point of crowding), exclusion is possible, and most organized skiing facilities are supplied by private operations.

Camping, especially that provided by enclosed and guarded camp grounds, is a special kind of forest recreation. In this activity one can find examples of nearly any kind of facility, ranging from those privately owned and operated strictly for profit, up (or down) to outstanding facilities operated at public expense but available without any specific charge. The most usual kind is probably the intermediate one of public operation, with a user charge that at least partially covers maintenance and operating costs.

In the private production of these kinds of forest recreation, the usual approach to optimum scale and optimum output patterns can be applied, and the usual financial objective of maximizing the present net worth of the total (privately owned portion of the) recreation facility is appropriate. Each specific application creates a special set of analytical problems, but as long as private operators are furnishing a service for which individuals are willingly paying dollars, and as long as the market for these services is reasonably free to entry, the production model should be appropriate.

One characteristic of some forest recreation operations does give the usual production model a somewhat different twist. Recreation facilities tend to involve relatively high fixed costs which are "lumpy" in nature. This tends to make interest charges relatively important, and can also make pricing policy of more than usual significance to the success or failure of the operation. Skiing again provides a good example. Some production questions can be answered by a straightforward application of marginal analysis—the original size of the resort to be constructed; the optimum width of slopes to be cleared (if this is not physically determined); the number and kind of tows or lifts to be installed, etc. But most of these decisions are "lumpy" rather than being truly continuous; one either buys a snow machine (or several) or relies on nature; one installs a rope tow or a chair lift. Such equipment often involves heavy capital investment, and once installed it becomes an element of fixed cost. Pricing policies in this situation tend to be more complex, for the marginal cost of supplying an additional skier is small while the revenue from the operation is importantly influenced by the price.

**Pricing problems.** Usually there is a significant degree of spatial monopoly in a ski operation, and there are other important differences between facilities that permit some degree of price control. The problem may be illustrated by Fig. 22.1. The total revenue

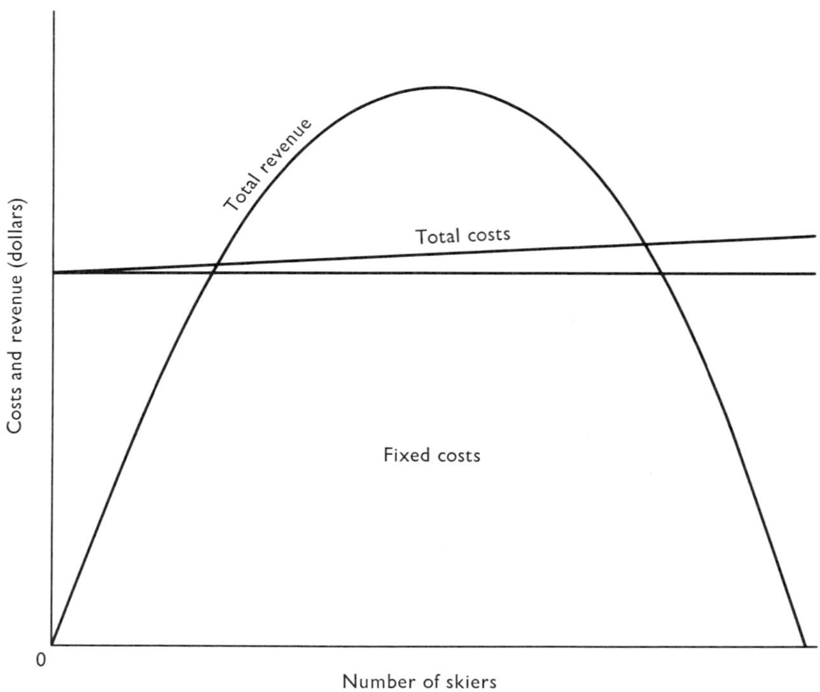


**Fig. 22.1** In some kinds of recreational activities the revenues may be changing more rapidly than the variable costs.

curve of the diagram is derived from a linear demand function and is shown together with the total and variable costs. Since operating costs vary little with the number of skiers using the facility, the marginal cost is low and virtually constant. Revenue, however, is very responsive to price, due in part to the demand assumption, and makes the major question facing the operators of this particular ski slope that of determining an optimum pricing policy. The degree of monopoly that any particular operator enjoys is surely limited, for skiers can usually choose among several alternatives, but the profit position for the operation—the rate of return that will be earned on the total investment—is clearly a function of the price that is charged.

The number of studies that have been directed toward determining demand schedules for such ventures is not believed large, and presumably a good many of those made would be for private use. Based on the slight evidence available, it appears that few operators have paid critical attention either to the problem of optimum pricing, or to that of optimum investment (Leuschner, 1969). It would seem that studies designed to identify the factors most important in establishing, and in shifting, the demand curve for such facilities would be very beneficial to the industry.

### Public Production

A large but unknown part of forest recreation is currently produced in the public sector, although not necessarily provided to the consumer free of charge. But the fact that forest recreation is often produced publicly does not necessarily mean it *should* be so produced.

There are some enthusiasts who believe outdoor recreation is essential for improving public morals, maintaining democratic principles, health, stability, the "American way of life," and perhaps even motherhood. These people clearly regard forest recreation as a *merit good*. Usually such advocates of this position also believe the needed facilities should be provided more abundantly, and at no cost (of which more will be said under distribution). The case for forest recreation's contribution to physical, social, and mental well-being, however, is not strong. There is little proof, for example, that a summer spent in forest surroundings or on a wilderness trip, produces better citizens, or has any positive social effects such as reducing the crime rate or juvenile delinquency, or produces even a mild lowering in the incidence of disease. Still, most people seem to feel there is something behind this general idea that forest recreation is "good" and, as stated earlier, behavioral scientists may yet establish a relationship between basic drives and man's desire to recreate in a forest environment. This could eventually establish forest recreation as a merit good. The strongest arguments for public production at the present, however, are believed the practical ones which follow.

**Arguments for public production.** *There are large areas of public forest land that can be used to supply forest recreation while simultaneously providing other goods such as timber.* This is certainly the case. Unfortunately, from a national standpoint, most public forest land is poorly located for recreation. Some 75 per cent of all public forests are in the western states, while most of the population is in the East. Purchases made over the last decade have reflected the newly available funds arising from the Land and Water Conservation Fund Act of 1965, specifically authorizing purchase of land for recreational purposes. Obviously, if lands are purchased to provide forest recreation the argument that there are already large areas of forest land in public ownership is scarcely relevant.

*Since there is no active private market for forest recreation, the government should exercise its powers, assemble the necessary resources, and organize them to meet the desire of its citizens.* Some people believe that once the demand for forest recreation becomes sufficiently strong, private accommodations will be made available. They further point out that one can scarcely expect a private market to be established if public facilities are being provided far below their cost, and often free. Some kinds of forest recreation, however, by the very nature of the activity, involve an extensive use of land. Because of their remote location or the relatively small number of participants at any one point, these forest activities might not begin to cover the cost of collecting fees or of enforcing user charges. With such activities, if forest recreation is to be provided at all it must be by the public. Legislation authorizing licenses similar to those presently sold for hunting and fishing, and enforced by the same kind of system (which relies largely on trust and good will) might provide an alternative means of financing such facilities.

*Even if forest recreation could be provided by private owners and organized for private gain, the resulting consumption pattern would be undesirable.* The major argument here is that forest recreation should be available to Americans without distinction as to color, creed, or income. The facts are, of course, that present consumption patterns are very heavily influenced by income, and probably by ethnic and racial groupings as well. This topic is better suited to treatment under the general topic of distribution; it is included here only because it is sometimes cited as a justification for public production.

*Once forest recreational facilities are constructed the marginal cost of supplying additional users is essentially zero.* This argument uses the concept of "jointness in supply"; that a recreation facility or area "is either provided or not in large indivisible lumps either as a pure public good or as a mixed good" (Cicchetti et al., 1969). Once provided, the marginal cost to additional users is zero (up to the point of crowding), much as is the cost of a bridge, etc. That forest recreational facilities are "lumpy" is hardly disputable; that the marginal cost of use is zero for such facilities might be accepted as a rough approximation. These facts, however, no more justify provision of forest recreation as a public good than would a parallel, and equally true, statement about transoceanic air transportation, or ocean shipping, or travel by rail, all of which are characterized by lumpy costs, and by marginal costs which approach zero.

How some elements of forest recreation came to be regarded as belonging within the public sector would make an interesting political and historical research subject. At any rate, it is clear that society has decided, so far, to produce a substantial part of forest recreation in the public sector. This raises the specific question: What determines how funds should be spent? In the absence of a functioning market for forest receation, how are public funds to be allocated? There should be some means of placing a *value* on forest recreation.

## Values for Forest Recreation

To many people the idea of placing a dollar value on an experience so "soul satisfying" as forest recreation seems almost sacrilegious. Yet these same persons are disturbed by none of these philosophical overtones when buying a ticket to a symphony concert, or helping on the annual every-member canvas of fellow church members. One objection frequently voiced to attempts at valuing forest recreation in dollar terms is that "every person reacts differently to a forest experience." This is undoubtedly true, but it is totally irrelevant to the determination of values. People do react differently to forest recreation experiences—and also to the music of Mozart, to paintings of Picasso, or to a bottle of beer. Yet these latter items all are valued in dollar terms. One can talk for hours about the "priceless beauty of a rugged shoreline" but if oil is discovered off that shore it will take more than talk to keep drilling rigs (and escaping crude oil) from destroying that beauty. If the purpose of the analysis is to aid the process of allocating funds efficiently, there seems little point in choosing any unit other than dollars.

Several methods of "valuing" outdoor recreation have been proposed within the past two decades, most of which have also been suggested as being applicable to determining the value of forest recreation. Among these methods are various attempts to measure "consumer surplus," and to arrive at "opportunity cost estimates" (Trice & Wood, 1958; Atkinson, 1956). Probably the oldest "method" was that of totaling the expenditures of users for travel, food, gear, etc., and regarding this total as the value of the recreation site ("after all, if it hadn't been worth this much to the individual to visit the area he wouldn't have come"). In fact, this same approach was put forth as recently as 1969 as a procedure for determining the "value" of the National Park System (Swanson, 1969). One of the more recent methods proposes use of "merit weighted user days" instead of dollars (Mack & Myers, 1965).

Most of these methods can be dismissed as being so economically naive they need be given no further consideration. The merit weighted user day, however, is somewhat different since it specifically recognizes that judgment decisions may be necessary in this particular area. Its proponents argue that social merit accorded a user day should vary:

> A day spent by a child at a day camp in the country has long-term value. . . . An adult's day of picnicking in a crowded park has, on the average, few of these same sorts of values. Thus, to assume that each day has about the same merit is incorrect, since the child's day will presumably produce benefits more nearly two or even three times those of the adult's day (Mack & Myers, 1965).

While granting that it may be necessary to exercise judgment as to relative importance between either uses or users, one may still question specific decisions. And when a quantitative comparison is ventured as in the statement above, serious questions should be asked.

Public expenditures for forest recreation should be in accord with the general objectives of all publicly financed operations; efficient resource allocation should be assured, the desired income distribution should be furthered, and economic growth and stability should be promoted. Proponents of publicly financed outdoor recreation have not always examined these aspects.

### Resource Allocation

If demand and supply curves for specific forest recreation sites could be established there would be little problem in establishing defensible admission prices for those areas. Provided all costs had been included in the marginal cost curve, and provided the indicated quantity was made available (so that the equilibrium number of visitors could be accommodated), the price indicated by the intersection of the demand curve and the marginal cost curve would result in an efficient resource allocation. Because user fees would almost surely be small relative to the total expense of participating in forest recreation, demand elasticity for desirable sites would probably be low (inelastic), which raises some interesting possibilities from the standpoint of forest recreation financing. Perhaps forest recreation should not only be self-supporting, but also be a means of financing construction of recreational facilities for those of lower income groups. This would mean a conscious use of forest recreation as a means of redistributing income in accordance with the usually desired pattern—a somewhat unusual role for this particular commodity to play!

At the present time, however, it would seem sufficient to use entrance fees as a device for financing the development of forest recreation facilities. The general lack of resistance to user fees since their quite recent introduction (principally since 1965) has done much to further their acceptance. Pricing could also be used to ration use of particular forest recreation areas (raise prices to the point where they restricted the number of visitors to the desired level). As long as consumers were allowed to choose, and as long as all costs including those of assuring future supplies were covered, the allocation criterion would still be met.

Provided they could be derived, would the intersection of demand and supply curves for the "total recreation package" be a proper indication of the "value" of forest recreation? Would this "price" cover all costs? Even more important, does it properly account for all benefits? Within the framework so far there is no allowance for the externalities that may accompany forest recreation, nor is there any recognition that there may be substantial *societal* revenues that do not enter into the private calculus of even this kind of price determination. Yet it would seem reasonably clear that, with minor modification to include some of these important additional revenues, some model of the total forest recreation package must be quantified if quantitative forecasts of future site needs are to be made. Recreation planners and policy makers in both Federal and state agencies, and perhaps even more, those in private agencies, need these estimates now, and need also to know how the quantities are shifting over time.

### Income Distribution

If forest recreation is to be produced within the public sector then collection of the needed funds and their expenditure should not only be efficient, but should also be in accord with the income distribution aims of the nation (Chapter 20). Without specifying an optimum income distribution, one can be reasonably certain that in the United States any redistribution should be from the richer to the poorer; the amounts may be questioned, but the direction of the flow does not seem a matter of dispute. This direction of flow is the only criterion needed to raise serious questions about our present approach to supplying forest recreation at public expense.

One of the peculiarities of forest recreation is the fact that it must be consumed *on site;* it is limited to individuals able to travel to the forest. Unfortunately, this means that (depending on the particular criterion of low income that one accepts) some 40 to 50 million Americans cannot participate in forest recreational activities. Of itself this does not necessarily imply that a problem exists, but it certainly points to the need for examining (1) the distribution effects of publicly providing forest recreation, and (2) the suitability of general taxes for raising the needed funds.

What income groups use publicly provided forest recreation facilities? There is little really solid information on income status of forest recreation users, but practically all existing evidence leads one to believe that the great majority of participants in forest recreation are "upper middle-class whites." Virtually all studies of use have shown that participation rates for forest recreation increase with increasing incomes and education. One author referred to the outdoor recreation areas of this country as "playgrounds for the affluent" (Hunt, 1969), and it would appear that this is essentially correct. A 1960 study (Reid, 1963) using a sample of some 26,000 visitors, showed that 51.4 per cent of those visiting national forests, and 57.4 per cent of those visiting national parks, had incomes in excess of $7,000—far above average incomes for that year. Families in lower income groups simply cannot afford the travel or the equipment needed to participate in forest recreation.

One might also raise a question about whether people in urban ghetto areas would choose to participate in forest recreation even if the choice was theirs to make. If forest recreation really has elements making it a merit good, then surely some expenditure should be devoted to making it available to those who need it most! This fact alone—that the group presently using publicly provided forest recreation facilities is a high-income group—raises a host of questions that badly need answering but have not been given serious study. For example, are universities performing their proper function in society by training recreation managers who are almost perfectly, even if unwittingly, "designed" to perpetuate past use patterns?

The other kind of question concerns the use of general tax funds to provide forest recreation. Clearly, facilities supplied *free of charge* represent a redistribution of income to those using the facility, and since only high-income groups use the facilities it is a regressive redistribution—from the poor to the rich. As seen above, much of the forest recreation presently provided at public expense is consumed by people well able to bear the full cost. To the extent that costs are not borne by the participants, forest recreationists are being subsidized by the general taxpayer, many of whom would prefer to have other uses made of the tax funds. Moreover, some taxes are paid by low-income groups presently unable, or unwilling, to participate in forest recreation, and to this extent also a redistribution of income takes place in which the poor are taxed to provide recreation for the rich. The extent to which this actually occurs, if at all, is not known.

## Growth and Stability Effects of Expenditures

Little can be said on this topic because so little is known. There have been some cases where outdoor recreational developments have resulted in substantial increases in local or regional income shares; reservoir developments probably offer the best examples. So far as growth and stability are concerned, the positive effects of recreation development would seem limited largely to the following:

1. Supplementary income is provided for local inhabitants. Because recreational activities tend to be highly seasonal in nature, local people who can adjust their time schedules satisfactorily are sometimes able to augment their incomes by part-time recreational employment. Such work is almost always part-time, however, because recreation is highly seasonal.
2. In low-income areas, recreational developments may add to family income by providing summer (or vacation) employment for children of resident families.
3. Motels, hotels, gas stations, restaurants, gift shops, and similar kinds of businesses do provide an increase in the tax base of communities. This increase, in turn, may aid development by making possible better schools, roads, and other facilities provided by public funds.
4. Some people believe that a recreational development, or the possibility of creating one, can act as a catalyst to community spirit. The process of economic growth is not well understood (see Chapter 25), but the "will to develop" is generally considered to be very important. *If* forest recreation does in fact act as a catalyst in this regard, its value in promoting regional economic growth would be much enhanced.

These positive factors are generally emphasized by outdoor recreation enthusiasts, and the situation has not been helped by economic growth studies made under the auspices of groups anxious to promote particular developments. There are several negative factors that must be given at least equal consideration:

1. The seasonal nature of forest recreation sharply limits the possibility of regional economic growth stemming from forest recreational development. An industry that is operative only two, three, or four months of the year is severely handicapped in the struggle for financial success. It is seldom wise to pick this kind of industry to promote economic growth and stability.
2. A varying, but usually high, percentage of the income generated by recreation does not remain in the region. Most of the required supplies and equipment come from external sources. Furthermore, major investments in recreational facilities are often made by large corporations. Motel and hotel "chains" provide one example, and such concerns frequently use centralized purchasing arrange-

ments to cut costs. Clearly, a business that buys most of its supplies from local sources tends to provide a greater increase in regional income.

3. Recreation industries, and the communities dependent upon them, are very sensitive to changes in the general economic activity of the nation and to seasonal factors such as weather. Because recreational expenditures are usually made from discretionary income, business reverses, or even a decline in business expectations, can have serious repercussions on recreation-oriented communities. Stability is not a characteristic of recreationally oriented economic activity.

Expenditures for recreation should usually be regarded as representing a transfer of income from one region to another, and not the creation of *new* income. These expenditures would probably have been made for other purposes if the particular recreational activity had not been available. This transfer nature of recreational expenditures is particularly important when considering the allocation of Federal funds, for normally these funds are directed toward achieving national, not local, objectives. Facilities that serve primarily local needs should either be supplied by local units of government or be fully paid for by the participants.

The concern of the Federal government for regional growth and development is a different matter. Because Federal recreational expenditures are of a transfer nature they may, very occasionally, offer a possibility for stimulating economic growth of depressed areas. The number of such possibilities, however, is not believed to be great, and before any large expenditure program is approved it should be subjected to a very careful analysis of benefits and costs in terms of the relevant goals.

# 23

# The Demand for Forest Recreation

If the relative importance citizens accord forest recreation as an opportunity for public investment is to be determined, some analysis of the demand for forest recreation will be needed. More specifically, aggregative demand functions would provide a basis for allocating public funds, while functions for specific sites and activities would make possible a comparison of demands for different areas, different kinds of facilities, and different activities, and would permit one to estimate the revenue that could be generated by user charges.

As seen in the previous chapter, there has been a tremendous increase in the number of people using public areas for forest recreation. Most presentations of such data refer to the "rapidly increasing demand" for recreation. The majority of "recreation demand" studies, however, are simply enumerations of how many people did something—visited a national park, camped in a forest campground, bought fishing licenses, etc. This in no way implies such studies are not useful; they are simply not studies of *demand*. The fact that far more people water ski in Michigan or Florida than in Nebraska, or hunt quail in Alabama than in Minnesota, signifies little with respect to demand for water skiing in Nebraska or quail hunting in Minnesota (although something may be implied with respect to conditions of supply). Such studies are much more concerned with *consumption* than with *demand.*

Demand is the relationship between *price* and *quantity,* and consequently, two peculiarities of forest recreation are of particular significance: (1) forest recreation must be consumed *on site*, and (2) forest recreation is a "package deal," the components of which are necessarily purchased together. The first peculiarity means that locational factors will be important to the demand for any specific site, and costs of getting to and from the site can be expected to be an important cost item. Its package nature implies that any analysis of demand for forest recreation should begin, at least conceptually, with the demand for the total package—the demand for the whole experience. The elementary model, then, is that of Fig. 23.1. It is

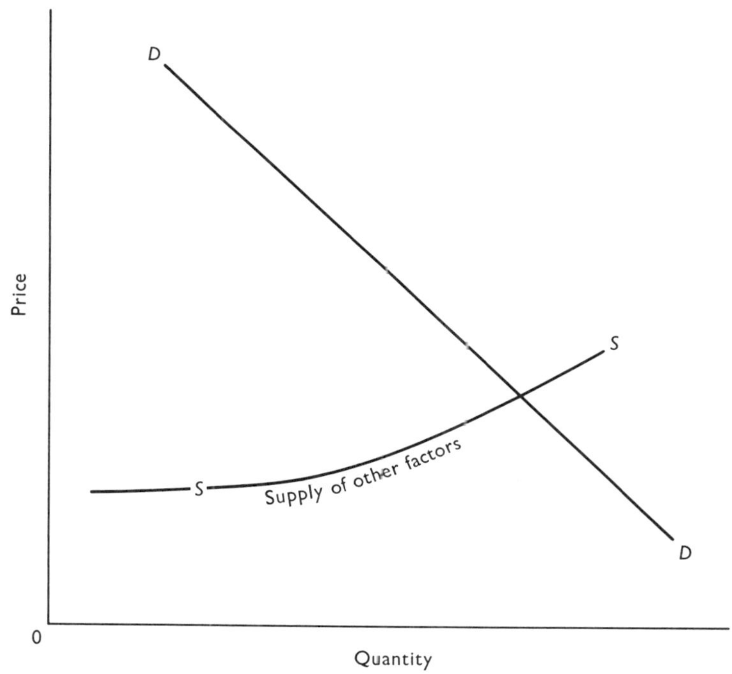


**Fig. 23.1** Demand for forest recreation, and the supply curve of all factors other than that for the site itself. The area between the two curves is the area under the demand curve for the site.

based on derived demand theory, for much the same reasons as made necessary derived demand when dealing with lumber, sawlogs, or stumpage.

## FACTORS AFFECTING DEMAND

As a starting point we should try to identify variables that might influence the position or shape of the demand curve for the total forest recreation experience over either a short or long term. And because we will want to explore possibilities of determining demand for the recreation site itself we will also be concerned with variables that could affect the supply of productive factors other than site. Clearly, since so many activities are included in forest recreation, many different variables are sure to be involved. Some of these are specific to particular activities; the availability of pack horses and guides, for example, probably has a heavy influence on the demand for wilderness pack trips but would have little effect on demand for weekend camping at a state park. On the other hand, the availability of leisure time would probably influence the demand for both wilderness pack trips and weekend camp space. Only the latter kind of factor will be discussed here, four of which seem particularly significant on the demand side; population, income, leisure time, and alternative opportunities. Each will be discussed briefly before turning to supply factors, first looking at their effects on demand within a short time period, then *over* time (long term).

### Population

Little need be said about population per se. If living habits remain unchanged, more people would mean more consumption of almost everything, and areas of high population density would be expected to have more people wanting forest recreation sites than those with less population. To eliminate this numbers effect in recreational studies, per capita demand is often expressed as the number of visits, or visitor days, per 1,000 population.

But living habits in this country are not the same everywhere. Urban people probably have a quite different demand for forest recreation than do those in rural areas. Would a New Yorker have the same "wants" as an Alabama farmer or a western rancher, or even a resident of Los Angeles? Part of the difference may be associated with population density. Some authorities claim that city living increases the desire of urban residents to renew contact with nature, that they feel a need to escape to more primitive, less complicated surroundings. The few indications there are do lead one to expect that the demand curve for the total forest recreation package rises with the degree of urbanization, but the basic reasons for this rise are at present only conjectural.

The age–class distribution of a population also significantly influences the recreational pursuits of those living in a particular area, for neither the young nor the very old are as active in forest recreation as are those of intermediate ages. Similarly, educational patterns probably influence the demand for forest recreation, as do ethnic characteristics. Any definitive study would need to test these variables for their possible effect on recreation demand, regardless of the time period involved, for these factors change from place to place and certainly shift over time.

## Income

In apportioning their income most individuals are by necessity forced to adopt some sort of priority system. Food, shelter, clothing, and transportation must be made available; doctors, dentists, and taxes must be paid. After allowing minimum amounts for these and other items, most United States families have some balance remaining, an amount available for *discretionary spending*. Much of this balance is put into upgrading the quality of food, shelter, clothing, transportation, doctors, dentists, schools, etc., and the distinction between "necessities" and "discretionary items" consequently becomes blurred. Recreation stands well toward the discretionary end of the "necessary items—discretionary items" scale. Money spent on recreation comes almost entirely from *discretionary income*, income available for spending as one likes. Decisions in spending this part of income may be more difficult to make simply because the choice is an individual one.

Families with higher incomes have an increased proportion of their income available for non-necessities. Higher incomes, therefore, make for increased spending on recreation of all kinds, including forest recreation. Families at the low end of the income scale participate little in forest recreation, partly because such families have definite travel restrictions. Urban families without cars have difficulty in getting to any forest area. There are also important relations between income and cultural patterns; an inhabitant of an urban ghetto or a rural slum is believed less likely to *want* forest recreation than is someone from an upper middle-class neighborhood. Very high-income groups are also believed to participate less than the average in forest recreation, probably because of the ready availability of alternative recreational pursuits.[1]

It seems unlikely that income elasticity of forest recreation demand would be constant as incomes changed over time. Most of the quantitative analyses made so far have used "cross-section" data to determine how the rate of participation in outdoor recreation activities varies from one income group to another at a given period in time. To generalize results of cross-section studies to long run influences requires one to assume that, for example, if average family income were to increase from \$5,000 to \$10,000, the average family would adopt the recreational pattern that presently characterizes a family with a \$10,000 income. This assumption is at least questionable.

Perhaps the major significance of changing income elasticities for forest recreation lies in its international aspects. Forest recreation in the developing nations is, to say the least, not highly developed, yet in these nations one might expect an increasing income elasticity of demand from their own populations. Added to these demands, however, may be those of the developed nations, including that of the United States, for as incomes in these latter countries increase, the possibility for international travel rises accordingly. It is very probable that forest recreation in Africa and South America could increase very rapidly with rising United States incomes. Over either the short term or long term, rising incomes mean more alternative opportunities.

### Leisure Time

Many writers have commented on the parallels between income and time, and especially on those leading to the use of a similar descriptive vocabulary; we *spend* time, *budget* time, *save* time, *earn* vacation time, etc. Probably the outstanding difference between the two concepts is that everyone has exactly 24 hours to *spend* every day, whereas incomes are far from identical. Leisure time is a special sort of time—time not used in fulfilling life's necessities such

[1] A recent study (Cicchetti et al., 1969) established by least squares regression a relationship between number of days camping in remote areas, family income, and (family income)$^2$ such as to imply that the number of days of camping in remote areas declines as incomes increase up to \$44,500. Above this income level, further income increases bring an increase in number of camping days. A decrease in the total numbers participating in remote camping was forecast for the end of the century. In contrast, "developed camping" was forecast to increase by as much as 460 per cent when A.D. 2000 was compared to 1965.

as working, preparing for work, eating, sleeping, etc. It is "discretionary time," and how it is spent is up to the individual. The relation between discretionary time and discretionary income is very close; how either one is spent is highly dependent upon the availability of the other.

Because forests are usually at least 2 or 3 travel hours from the average participant's residence, much forest recreation takes place during the consumer's vacation period or on weekends. Definite time limitations are thereby set on much of the participation in forest recreation; families take vacations during the summer months when children are out of school. The length and timing of the vacation period may also depend on the policy of the particular employer, hence both industrial management and labor union policy may affect the when and where of forest recreation activities.

Time budgets may be as important in recreation decisions as dollar budgets. If it takes one day to reach a particular skiing area, for example, and another day to return, that area is as out of reach to the wealthy weekend skier as to one in a relatively low-income group; neither skier can afford to spend the time. It seems likely that for some kinds of recreation "time cost" is at least as important as dollar cost.

Leisure time has been increasingly available to most people in the United States. The average work week of 100 years ago was about 70 hours; now it averages around 40 hours.[2] Further reductions are certainly possible and expected, and some say automation may increase leisure time by a substantial amount within the next two or three decades. Any large increase would have a major impact on United States living habits, including our recreational patterns.

It would seem almost a truism that the more leisure time people have the more they will participate in recreational activities. Its effect on forest recreation, however, depends very much on how,

[2] The author feels this kind of fact somehow tells only part of the story, for while this reduction in the work week cannot be disputed it is also true that there has been a decided increase in the degree to which our lives are ordered. I am not one who believes that the "old days were better," but it seems they were decidedly more "leisurely." The degree to which life in the United States is dominated by clocks is difficult to appreciate until one visits another culture (perhaps such as that of Latin America), where time accounting is considerably less precise. Freedom is a very important concept, and is certainly multidimensional. Freedom from a strict adherence to time schedules may not be important to some, but there can be little doubt that it is one of the freedoms most difficult to maintain in modern-day United States.

and to whom, this time is made available. The average work week could be reduced to about 35 hours in at least three different ways. A shortening of the work day from 8 to 7 hours would probably have little effect on the demand for forest recreation; few people spend their evenings driving to the forest. A reduction of the 5-day week to one of 4½ days would provide approximately the same increase in leisure time, but would have a greater effect on forest recreation. Some people presently believing a 2-day period is insufficient for the purpose would feel the extra half-day would enable them to spend weekends in forest areas. But if the same total amount of time were all put into a lengthened paid summer vacation period (it would accumulate to about four weeks per year), the effect on forest recreation might be overwhelming.

The kinds of people receiving the increased leisure time would also affect demand. Present trends indicate that professional, managerial, and supervisory personnel may share relatively little in the anticipated increased leisure time, yet it is these groups that presently generate much of the demand for forest recreation.

There are many interesting aspects and opinions regarding leisure time in this country. Some authorities doubt that there has been any real increase over the past century, citing the prevalence of moonlighting (holding two jobs), the increased commuting time, the extra burdens of social activity, and similar items to prove their point. Probably the major fact concerning quantitative relations between the beneficiaries of increased leisure time and forest recreation is the very simple one that we don't really know very much about it. There has been considerable speculation, but few rigorous studies.

### Alternative Opportunities

Almost by definition, an item of discretionary spending has many competitors. The short term demand curve for the forest recreation package is probably subject to decided shifts from locality to locality as other recreational opportunities are available. Suppose, for example, two campgrounds were almost identical, but one was closer to a large city than the other (in either time, distance, or cost). The campground near the population center would probably have a higher demand than the one farther away.

There have been some attempts to use so-called "gravity models" in analyzing this phenomenon, but the efforts have not been conspicuously successful.[3] Models capable of handling multiple forest recreational locations are decidedly more complex and require much more detailed information for their quantification than those dealing with a single area. Intuitively, most researchers feel that strong relationships should exist, for example, between attendance at forest campgrounds and the proximity of golf courses, movies, fishing streams, grocery stores, hiking trails, etc. These relationships have yet to be quantified in a definitive manner.

As new facilities are constructed—not only new forest recreational sites but anything that competes for the time and money of the possible participant—the demand curve for any given forest recreation area will shift. Yet not all new developments are competitive. New facilities, when added to older ones, may provide the variety necessary to encourage new participation. Families unwilling to purchase equipment for use at a single recreation area may buy camping equipment, boats, trailers, etc., when several alternatives are available. And the opportunity to visit several different areas on a single trip almost surely enhances the demand for any one of the sites.

## FACTORS AFFECTING SUPPLY

In discussing supply factors it is desirable to separate those effective over the short term from those operative over the long term. The short term is defined here as a single recreational season, since in most instances the capacity of the site (the campground, park, hiking trail, etc.) will not be changed during this period. The long term, then, becomes more than one season. It is also helpful to separate those factors that influence the whole recreation package (package factors), from those applicable only to the particular site (site factors).

### Short Term Package Factors

Of the several variables that might shift short term supply curves for the whole recreation experience, the most important is prob-

[3] Named because of their similarity to Newton's model of gravitational force (the attractive force between two bodies is directly proportional to their mass, and inversely proportional to the square of the distance between them). With such a model, population replaces "mass" and the distance between campgrounds, or between population centers and campgrounds, becomes the appropriate distance. Determining the "power" of the attraction (whether the square, cube, etc.) is usually the objective of the empirical study.

ably *travel cost.* The cost of traveling to and from any recreational area varies with distance and also with the mode of travel. If automobiles are used, then cost will also vary with type of car, type of road and road surface, and (at least after a certain traffic level has been reached) with the amount of traffic. Travel cost is frequently depicted as being directly related to visitor days, or visits, per 1,000 population, as in Fig. 23.2, a relationship often referred

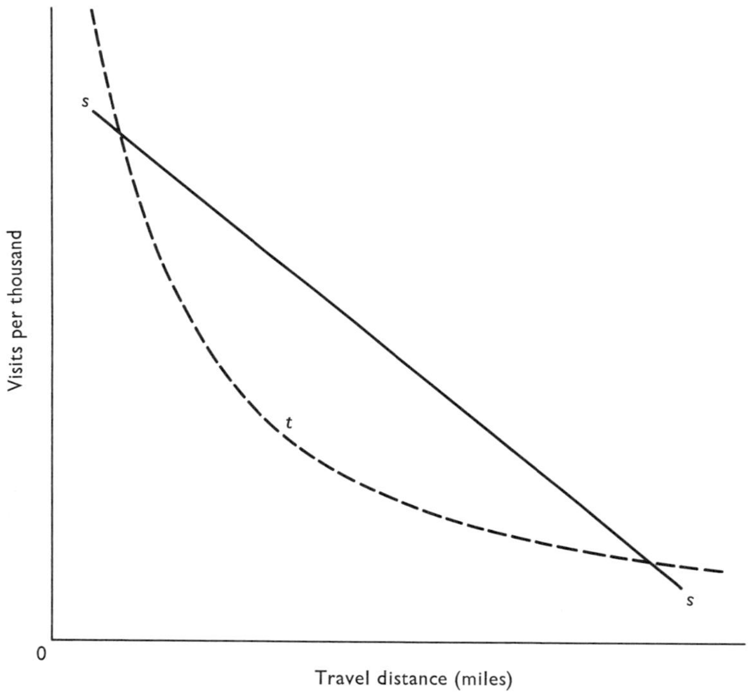


**Fig. 23.2** A typical "distance decay function."

to as a "distance decay function" (the number of visits per 1,000 population ""decays" as travel distance increases). Such relationships could be either linear (*SS*) or curvilinear (*t*). Distance might also be defined in terms of "cost distance," much as was done for stumpage, permitting travel zones to be established by the average cost of traveling to the activity site.

The significance of travel cost will often vary with the length of the visit, and because of this complication many analysts prefer to use a "visitor day" as the unit of consumption. With campgrounds this unit is usually defined in terms of overnight occupancy. Then, to the user, the cost of transportation to and from the campground declines with the length of stay.

### Short Term Site Factors

Attention here must be focused on marginal costs—the change in cost brought about by a change in the numbers using the site during a single season. Short term supply curves for forest sites would probably resemble the supply curve for public stumpage—very elastic up to the capacity of the facility, then highly inelastic. A forest campground, for example, normally has a given number of tent or trailer sites. Once these are taken the campground is full and no more campers are allowed. The curve would not be perfectly elastic since some costs (policing of the campground, repair work, supervision, etc.) depend on the number of visitors, but most operating costs would not vary over the short term. In other kinds of sites, however, the short term variables could be much more significant.

### Long Term Package Factors

These factors are particularly important to those holding policy and planning responsibilities. Two types of factors inducing long term shifts in the supply curve for a site will be used as examples: (1) equipment, and (2) mobility.

**Equipment.** As the time of the market period is increased beyond a single season the principal problems might appear to be those of cost allocation. What part of the cost of the family car should be charged to forest recreation? How is the cost of camping equipment to be calculated over time? Should part of the cost of a camera be included? The cost of a projector? Cost of film, etc.? The difficulty is easily exaggerated. Some items are clearly separable; cost of film, gasoline, oil, or of any other item consumed in a given season or trip would enter directly into the short term supply curve. In most other cases, such as that of allocating family automobile costs, these are true joint costs and simply cannot be allocated. Fortunately, for purposes of supply curve formulation there is no such need. Only the marginal costs are required—the additional expenses incurred in order for one to obtain the forest recreation package.

Yet there are still some sticky problems of identification and measurement. Supposedly one would depreciate the cost of camping equipment such as tents, stoves, sleeping bags, etc., over several seasons, and for these items there would seem to be no difficulty in establishing suitable depreciation rates. But the psychological influence of some major cost items could also be an important shifter of demand. For example, does ownership of camping equipment tend to shift outward an individual's demand for the forest recreation package? Would a decrease in cost of camping equipment shift the demand for forest campsites outward? Would the effect of such a decrease persist for more than one season?

**Mobility.** Any change in transportation technology or facilities would serve to shift the *supply* curve for the recreation package. There is little doubt that the increasing mobility of families in the United States is a major growth factor in the whole outdoor recreation picture. A new throughway that reduces travel time between a major city and a forest recreation site from, say, 3½ hours to 2 hours could completely alter the use pattern of that area, and also that of most other sites within a 3½-hour travel radius. One of the largest unknowns in some major forms of forest recreation, particularly skiing and big game hunting, is the effect of new, very large jet transports. Greatly reduced transocean fares, coupled with increased discretionary income in the United States, could change vacation habits of many thousands of people. For example, what would happen to the ski resort industry of the East and Midwest if the roundtrip air fare between New York and Switzerland drops to under one hundred dollars? The impact of such changes on remote areas, particularly on Alaska, South America, and Africa, could be devastating.

For the most part changes such as these are beyond the control of the recreation land manager, and usually beyond that of the recreation planner as well. Yet these changes have a pronounced effect on the use of the forest, and unless changes such as these are anticipated by those who plan for the expanding recreation demand, serious mistakes in resource allocation will be made.

### Long Term Site Factors

A change in the capacity of a forest recreational area will almost always shift the supply (cost) curve of that area. Hence the techniques of producing the facilities, as well as the technological nature of the facilities, their quality, etc., are important; in fact, they comprise the production function for the recreation site per se. Physical production of new forest recreation sites will normally be a long term phenomenon. Nothing will be ventured on the technical aspects of constructing forest recreational facilities other than to stress the need for coming to grips with the economic and physical problems of identifying the product and of relating the output of that product to the managerial inputs required for its production. Two observations might help to clarify this aspect.

1. The long time periods characteristic of timber production are less likely to cause problems in producing most forms of forest recreation. The required physical facilities, other than the forest itself, can be constructed over a relatively short time once the decision has been made as to what kind and how many are needed.
2. The pay-out period is likely to be considerably shorter for recreation than for timber. In fact, in most instances it probably has to be, since the life of recreational facilities requiring capital investment will be relatively short.

**Production costs.** As with most production endeavors, many kinds of costs will be involved in the construction of a forest recreation facility. Any investment required should be expected to earn a rate of return commensurate with other investments of equal risk; there is an opportunity cost of capital that cannot be avoided. Over time, then, any factor tending to change interest rates would tend to vary the initial cost of establishing a forest recreation facility or of adding to its capacity. Once built, of course, these costs become "sunk" and, like other fixed costs, should not influence the position of the supply curve.

The marginal cost of a campsite, for example, will always be the cost of (1) establishing the additional camping space, and (2) increasing the capacity of whatever is required in maintenance and supervisory costs of the campground due to the larger number of sites. Roads, parking facilities, beach construction, and protection, and the many amenities that are more and more expected by participants could all be factors that would serve to shift upward the supply curve of the forest recreation site as the tastes of consumers and techniques of forest recreation managers change over time.

**Problems of scale.** One other factor that should be mentioned is that of scale. There is no general agreement on the units best suited to identify the "scale" of a forest recreation area, but certain relations are clear. In forest campgrounds (to use again this most common example), the two factors determining capacity of the campground (assuming the other required amenities can be constructed) are the size of the individual camping site and the total area of the campground. Per acre costs and total cost would probably increase with total area alone, since maintenance, protection, supervision, etc., would tend to be more expensive. But per acre costs are important only as they affect the cost of the product —the recreation consumed by the participant. Just what the optimum size or scale of a forest recreation area would be in terms of, say, visitor days provided is at present simply another item requiring research.

## A DEMAND MODEL FOR FOREST CAMPGROUNDS

To illustrate the problems and opportunities of demand analysis in this field, suppose we try to derive the demand curve for forest recreation on some hypothetical public forest where the only type of recreation provided is that of forest camping. The market area will be determined by the location of the forest, its road system, the transportation network of the state, etc. We will assume campers are confined to organized and clearly identified campgrounds, and that we sample the attendance at the campground being studied to see how the number of visits per 1,000 population varies with distance. To keep the experiment as simple as possible, each visitor will be given a card that states:

> Please circle the figure that best represents the distance in miles from your home to this campground.
>
> 50 100 150 200 250

At the end of the season the results of this exercise are tabulated, providing the data of Table 23.1.

Two more bits of information are needed: (1) the number of people within each of the travel zones, and (2) the average cost of making a round trip to the campground from each zone. To determine the first, a series of concentric rings could be drawn about the campground, one ring at each of the distances already decided upon. Then, using population density maps, county census reports, city population figures, etc., the total population within each of

**TABLE 23.1 NUMBER OF VISITORS FROM EACH MILEAGE ZONE**

| Number | Miles |
|---|---|
| 250 | 50 |
| 400 | 100 |
| 4,500 | 150 |
| 1,600 | 200 |
| 3,000 | 250 |

the five distance zones could be determined. Average cost data could be obtained from automobile clubs, from national studies, or from any other suitable source. These data could finally be recorded in a table such as that of Table 23.2, and the visitation rates calculated for each zone.

**TABLE 23.2 SUMMARY TABLE, FOREST CAMPGROUND STUDY.**

| Zone | Cost (dollars) | Number of visitors | Total population | Visits per 1,000 population |
|---|---|---|---|---|
| 1 | 2 | 250 | 5,000 | 50 |
| 2 | 3 | 400 | 10,000 | 40 |
| 3 | 4 | 4,500 | 150,000 | 30 |
| 4 | 5 | 1,600 | 80,000 | 20 |
| 5 | 6 | 3,000 | 300,000 | 10 |

The usual problems of demand and supply shifts over time have been eliminated since the data pertain to a single camping season. Two key assumptions, however, must now be made, one with respect to demand, the other to supply. On the demand side *it is assumed that the demand curve for the entire recreation experience is constant throughout the market area.* People respond to travel cost changes in the same manner regardless of the travel zone in which they reside. Note that there is no recognition of urban concentrations, of education, of age levels, of income differences, etc. On the supply side it is assumed that all variation in cost of the total forest recreation experience is included in the travel expenditure. Here there is no recognition of highway patterns, road types, public transportation facilities, cost of equipment, etc.

Now since all costs are assumed to be included in the travel expenditure, it follows that within each of the five zones the supply curve will be perfectly horizontal. Each zone will have a different (horizontal) supply curve, and it will be located at the level of average travel costs from that zone. The five supply curves derived from the data of Table 23.2 are shown in Fig. 23.3. The significant

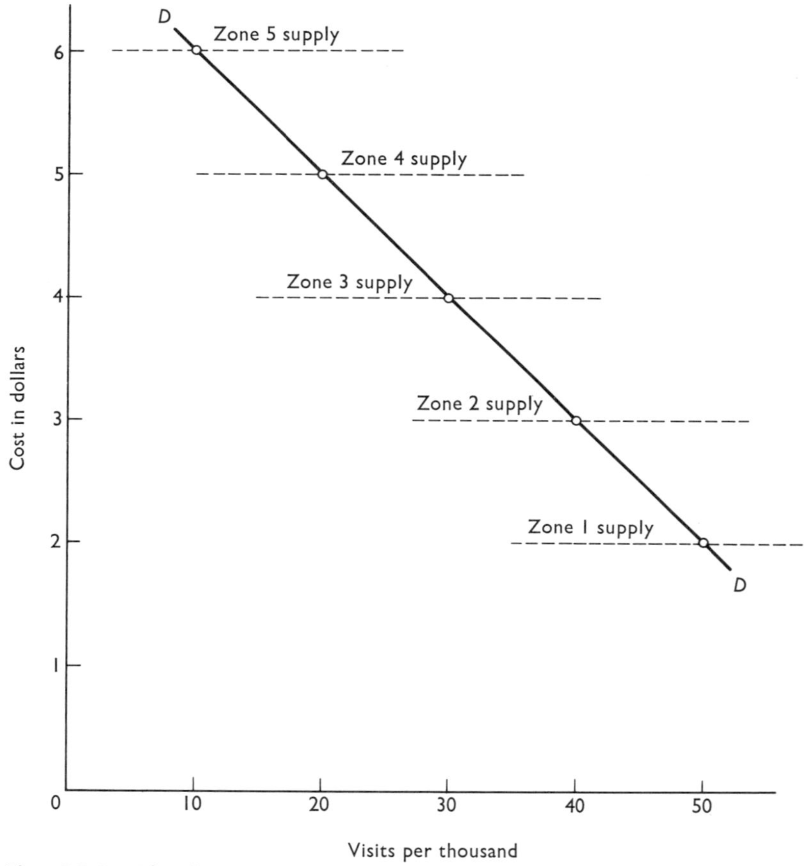


**Fig. 23.3** The five supply curves serve to identify the demand curve.

fact is that the *consumption* points on the five supply curves necessarily trace out the demand curve we have been seeking, for *demand was assumed constant* throughout the entire market area. By connecting these points the demand curve for the total forest recreation experience may be drawn, as shown in Fig. 23.3. The demand curve for the total recreation experience for each one of the five zones can now be drawn by multiplying the per 1,000 population demand value (i.e., the visits per 1,000 population values from the horizontal axis of Fig. 23.3) by the respective zonal populations in 1,000's. To illustrate, the demand curves for zones 1, 3, and 5 are reproduced in Fig. 23.4.

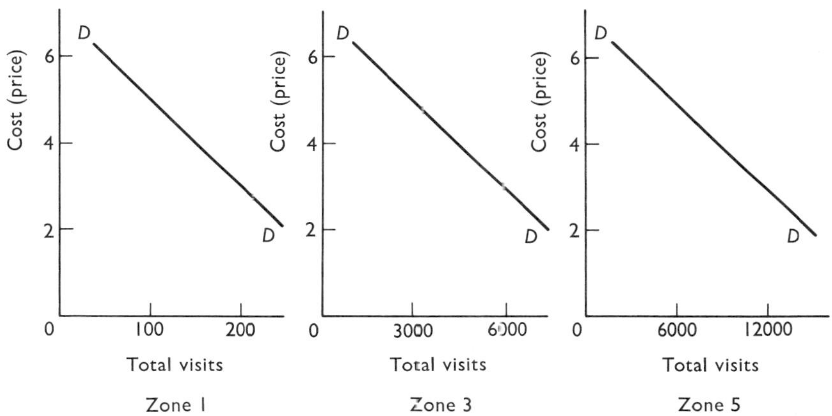


**Fig. 23.4** The demand curves for zones 1, 3, and 5. (Note change in horizontal scale.)

Having the demand curve for each of the five zones it should now be possible to derive the demand curve for the site itself. Derived demand, we know, requires knowledge of the supply conditions of all factors of production other than the one being considered, and *by assumption there are no production factors other than travel.* For each zone, then, the travel cost must be subtracted from the appropriate supply curve, leaving the demand curve for the forest recreation site in each of the five zones. The five demand curves, shown in Fig. 23.5, differ in two respects. The visitor scale has been changed by multiplying the per 1,000 scale by the population of the zone in 1,000's, and in each case the demand curve has been shifted down by the travel cost (the supply curve of other productive factors).

The final step can now be taken. To construct the *total* demand curve for the recreation site we need only aggregate the five demand curves—add up the number of visitors that will come from

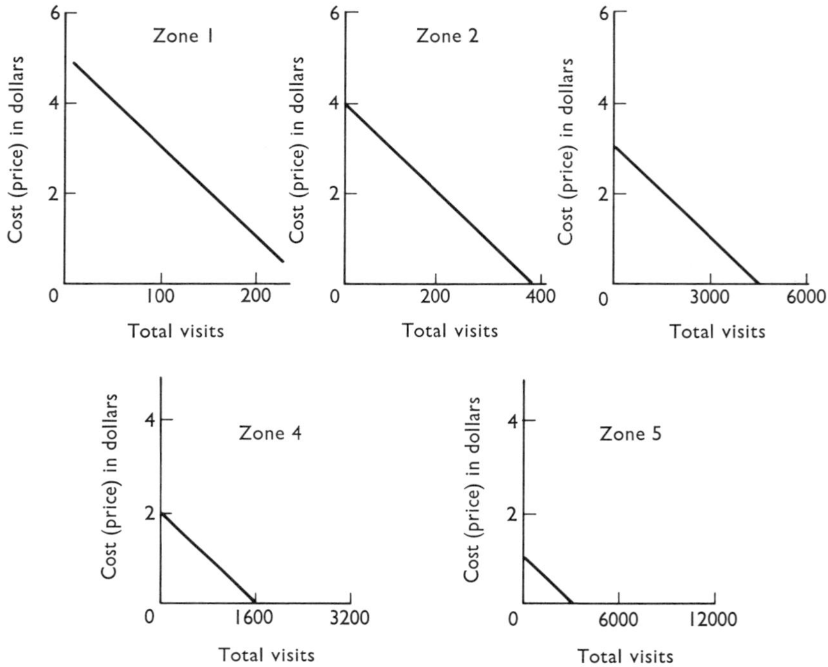


**Fig. 23.5** The demand for the forest recreation site (the campground) in each of the five zones (given the assumptions stated in the text).

each travel zone at any specified price level. The result is shown in Fig. 23.6, from which can be read the effect on number of visitors of any change in price, such as might result from imposing an entrance fee on the campground. Provided the set of assumptions is valid, this curve is the "demand for the forest campground."

Clawson and Knetsch used a very similar technique to derive estimates of the demand curve for the recreation resource around Lewis and Clark Lake, and for four national parks (Yosemite, Glacier, Grand Canyon, and Shenandoah.[4] After smoothing these curves, the results were used to illustrate how estimates of the num-

[4] The approach of Clawson and Knetsch, quite similar to the one presented here, was first proposed (for recreational studies, at least) by Hotelling (Hotelling, 1949). No explanation was given by Clawson and Knetsch as to how costs were allocated to individual parks when more than one park was visited, other than to emphasize that the cost sharing was "wholly an intuitive matter." Whatever the method, it could have a pronounced effect on the shape and location of the resulting demand curve (Clawson & Knetsch, 1966).

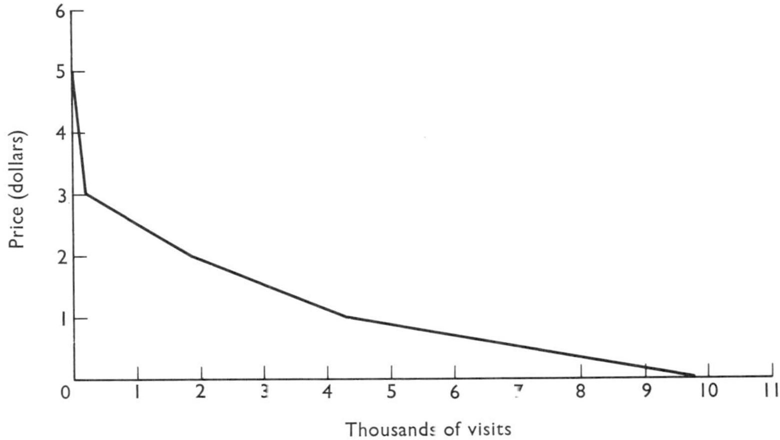


**Fig. 23.6** The demand for the forest campground (given the assumptions stated in the text).

ber of visits to the parks under various fee schedules might be derived.

## A REVIEW OF THE ASSUMPTIONS

Now return to the assumptions that were required. By assuming (1) that the demand for the total recreation package is constant throughout the market area; (2) that supply costs are constant within each travel zone; and (3) that the only variable shifting the supply curve is travel distance (cost); the rather complex derived demand approach has been reduced to the very simple (or naive?) model of Fig. 23.7. Since per capita demand (*dd*) has been assumed constant throughout the entire market area, as the demand curve is moved along the travel distance axis (from zone to zone) it creates a plane (*ddd′d′*). Similarly, the supply curve (*ss*) was assumed to be a linear function of travel distance, so rises from *ss* to *s′s′* as it moves along the travel axis, generating a supply plane (*sss′s′*). If the intersection of the two planes is projected on the base of the figure, visits per 1,000 population will be shown as a function of travel distance, the "distance decay function" of Fig. 23.2.

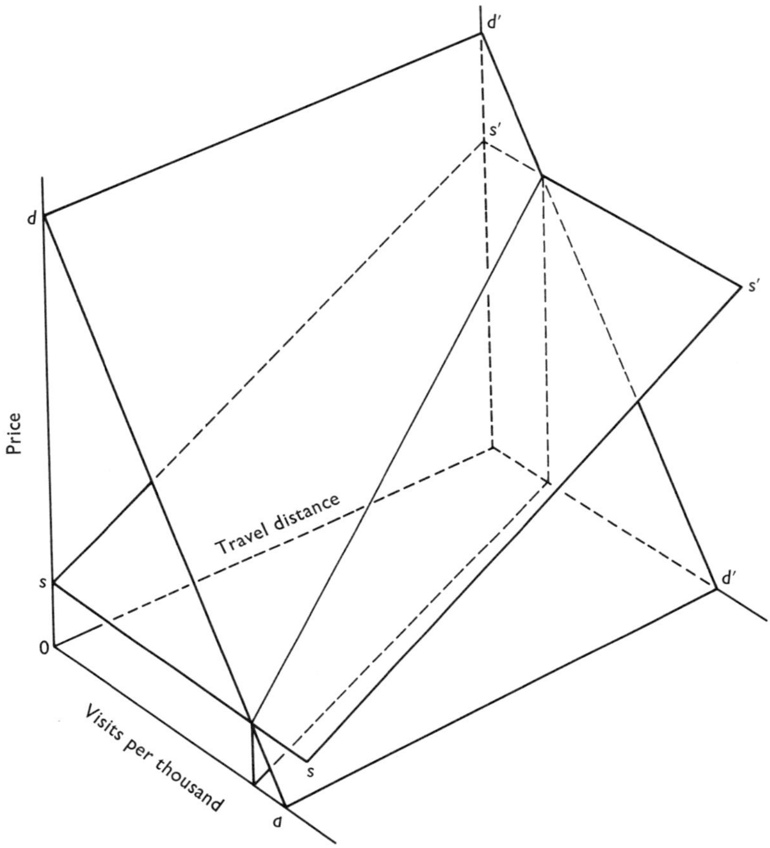


**Fig. 23.7** Three-dimensional representation of demand model used in preceding analysis.

## Effects of Omitted Variables

With the three-dimensional model of Fig. 23.7 one can begin to see the possible effect of some of the significant variables omitted from the model. An area of increased income, for example, would tend to shift the demand curve upward, producing a distortion or bulge in the demand surface at whatever distance the area was located from the recreation site. On the other hand, if an equally desirable campground was located near a population center, say at 100 miles from the area being investigated, the demand curve would be shifted downward. Similar effects would be produced by changes in the supply variables.

If travel zones were very large, and if the particular area being studied were unique, then perhaps one might ignore the influence of changing incomes, leisure time, education, and other normally important variables and still find a significant relationship to travel cost. Such might well be the case with national parks such as Yosemite or Yellowstone. When the same approach is used to estimate the "demand" curve for forest recreation sites, however, the result seldom bears much resemblance to a demand function. And in any case it would appear farfetched to call curves derived from such overly simplified models "demand curves" for forest recreation.

## Other Problems of the Model

There are several features of forest recreation that are difficult to overcome, and which this model does not explicitly consider:

1. An important segment of forest recreation in this country is supplied at either a zero price or at prices that are administratively determined. Direct taxes and indirect taxes often elicit quite different responses, and people would probably react differently to direct user charges than they do to indirect costs such as those incurred by travel, hotel bills, equipment costs, etc.
2. Whether one considers travel to be a cost or part of the pleasure, the recreation package that includes a day-long trip is quite different from one including travel of an hour or so.
3. The trip to a specific forest recreation site could make the total package appear very different to residents of different geographic areas. Compare, for example, the trip to Glacier National Park from Los Angeles with one to the same park from Chicago. Here, although the distance is roughly the same, the two "packages" are clearly different. These two geographic areas should not be combined in a single analysis. The same package must be used as the starting point.
4. Finally, conditions within the travel zones could be so different that the inhabitants would have almost totally different demands for a given forest site. Residents of Iowa, for example, go to northern Wisconsin or Michigan to escape the summer heat, but residents of Wisconsin and Michigan do not, nor do residents of Canada!

## RESEARCH OPPORTUNITIES

In spite of these deficiencies it seems possible that in many circumstances sensible demand curves might be derived. Travel zones do not need to be defined as concentric rings, and zones along highways could be identified where the "package" was essentially the same. Nor is it necessary to assume that incomes, or education, or leisure time, or other population characteristics are the same; so long as they can be identified and measured they can be taken into account in a multivariable (and possibly a multiequation) model. It is necessary, of course, first to discover, then to measure, the effect of the important variables on both shifting demand and supply curves, just as has been the case with demand analysis for other commodities in our economic system.

Hopefully, research in this area will develop new and better models. At the very least one would trust that demand and supply factors will be recognized for what they are, and that research will be directed toward something more useful than simple projections of past trends into the distant future. And in any event, it can be hoped that those trained in the rudiments of forest economics will question any suggestion that simple forecasts of future consumption should be taken as representing a societal "requirement." Before using such estimates as national, regional, or local production targets, one should at least examine the costs of their attainment, and match this against the benefits that are anticipated.

Recreation is a growing use of the forest, and study of its economic implications and problems is just beginning. Many foresters being trained today will find themselves devoting at least as much time to managing forest land for recreation as for producing timber. And as a source of revenue, recreation may become more important than timber to many forest land owners. The same economic principles will be used in its production and marketing as are used for producing and marketing timber, but at present the units of measurement, the methods of production, the factors affecting demand and supply, and the system of pricing are all in their early stages of development.

# 24

# Wildlife

The preceding chapters provide much of the basis for applying economic analysis to wildlife problems.[1] The last two chapters are particularly important in this regard, since in the United States the uses of wildlife are primarily recreational. Public finance principles concerning allocation, distribution, and growth apply to public funds for wildlife as much as to funds for producing other goods, and the principles of production economics are no less applicable. Nor should any further discussion be needed of variables such as leisure time, education, income, population, travel cost, or equipment costs as they relate to the demand or supply of hunting, fishing, birdwatching, or other wildlife-oriented activities. This chapter will be confined primarily to pointing out specific similarities between timber, water, and recreation on the one hand, and wildlife on the other, similarities that are useful in analyzing economic problems of wildlife, or in identifying those aspects of wildlife that might require distinctive treatment.

## GENERAL ASPECTS OF WILDLIFE ECONOMICS

From the standpoint of economic production much the same thing could be said for wildlife as was said of watershed and rec-

[1] The terms "hunting" and "wildlife" will be used throughout the chapter to mean "hunting, fishing, nature study, etc."

reation management. Scientific wildlife management has only recently reached a stage where one could ask a group of wildlife experts for data needed in production function analysis and expect anything but incredulity. Yet it is most difficult to imagine seriously claiming proficiency in management without knowing the general relations between managerial inputs and the outputs of the product. And this is true whether the product is wild ducks, white-tailed deer, quail, timber, or wine.

Admittedly, measuring wild populations can be difficult; a detailed census is costly and its accuracy seldom assured. New techniques may help, but for some time even measuring wildlife population changes will pose problems that must be solved before causal relationships between such changes and managerial inputs can be identified. The fugitive character of the resource and its extreme mobility raise other difficulties. When ducks that nest in northern Canada provide the shooting for duck hunters from Manitoba all down through the Mississippi flyway and finally winter on Mexican lakes, one must grant that measurement problems do exist, though this is probably an extreme example. Deer, rabbits, and many upland game birds pose less difficult problems, at least as far as determining numbers is concerned.

Problems of determining wildlife values have also been a stumbling block to economic analysis. Past efforts have often been directed toward determining a dollar value per animal or bird, or occasionally toward determining a value per pound. These efforts now appear to have been largely wasted. The approach discussed under forest recreation demand is believed more appropriate and will be partially developed a little later. The starting point for analyzing demand should surely be the total wildlife-oriented experience. From this one might proceed to derive a value for the output produced from forest land or for the land itself—the animals, birds, or fish per se or the hunting area. Few could argue sensibly that the modern deer hunter from Philadelphia, Detroit, or Los Angeles is only, or even primarily, concerned with obtaining venison, although meat could be the principal objective for some, especially those in remote areas or in the developing countries of Africa.

## Peculiarities of Wildlife Production

Since wildlife uses are primarily recreational, the peculiarities of wildlife production that influence economic analysis necessarily include those discussed in the previous two chapters; the wildlife-oriented recreational experience represents a "package" of which the wildlife itself is only a part; the activity must be produced and enjoyed "on site"; and wildlife is usually produced "jointly" with other forest products. There are also some peculiarities of wildlife-oriented activities that need particular emphasis.

**Ownership.** In this country wildlife is public property; birds, fish, and game animals in the wild state are not owned by individuals, and the capture, possession, and sale of such creatures is subject to special legal restrictions. Wildlife is for the most part regarded as a social good and is produced almost entirely within the public sector. But wildlife is found and is hunted upon private property, and access to hunting and fishing areas can be and is made subject to the exclusion principle.

**Migratory nature.** The migratory nature of much wildlife also represents a peculiarity of some significance and is the primary reason for its production being placed in the public sector. One can control access to the land upon which game animals or birds are found, but it is often difficult or impossible to control the wildlife. Game birds have no regard for property lines, and fences seldom effectively control wild animals. Sport fishing has its own peculiarity in that access to streams is usually open to all, even though the area through which the stream flows may be privately owned.

**Externalities.** The production of most forest products is accompanied by externalities, but wildlife could perhaps be regarded as peculiar in that so many of its values are external. For the most part these externalities are not subject to the exclusion principle. The startling explosion of a ruffed grouse, or the lift one feels when a beautiful buck bounds away over the brush, or the thrill of seeing a flock of wild geese making their wedge-shaped way south in the fall—or of suddenly, almost violently, being lifted out of city surroundings by their honking and being forced to relive for a fleeting moment experiences of much earlier and less complex ways of life—these are externalities that grow more valuable as they become more scarce. Yet their value rests heavily upon a background that is becoming almost foreign to an urbanized culture. The simple knowledge that someplace in a mechanized world it is still possible to see wild waterfowl, or to hunt bear and elk, may well be such a valuable externality of wildlife production that forestry and wildlife management will to at least some extent always remain a social good.[2]

## PRIVATE WILDLIFE PRODUCTION

Since World War II there has been a rapid increase in private production of wildlife-associated activities, particularly those of hunting and fishing. Hunting clubs and hunting cooperatives have become common in many parts of the country, as has the leasing of land for hunting privileges. In many sections there has also been a rapid increase in "commercial hunting," wherein privately owned areas are stocked and managed, with hunting privileges sold by the hour, day, or some similar unit. Sometimes a charge per bird or animal bagged provides the revenue; more generally such charges are an assessment made in addition to the per hour or per day fee. Occasionally charges are levied on the basis of number of shots fired.

On such strictly commercial operations the economic aspects of wildlife production and sale are being developed rapidly. The financial objective of management for such an area would normally be the maximization of present net worth, and the usual production framework would be fully applicable. *If* one could derive a function for, say, production of white-tailed deer, and *if* one could identify related production costs, and *if* one could attach a unique value (price) to the end product, *then* one could proceed to determine the inputs and outputs, convert to costs and revenues, and identify the number of deer or hours of deer hunting that should be produced each year to maximize present net worth of the management area. The first two "ifs" should be positively resolved as the science of game management is advanced, and as long as the product is sold in a private market the revenue side of the problem poses no particular difficulties.

Hunting clubs and hunting resorts offer a very close parallel to skiing resorts, although hunting clubs are more likely to own and operate hunting facilities for the exclusive use of their members than are skiing clubs. With such clubs and resorts the amenities

[2] The demand for this sort of knowledge—that it is possible to enjoy such things—is sometimes referred to as "option demand." This concept has been discussed by both Krutilla and Weisbrod (Krutilla, 1967; Weisbrod, 1964).

provided may be the major part of the package, and the variety offered is very large. Analyses of demand for such clubs or resorts must necessarily take into account such things as the quality of the lodge facilities, the degree of crowding permitted in the hunting area, the expertness of the guides (and the bartenders), and perhaps the probability of success in the hunting venture. There seems little doubt that social status might often be an important revenue gained from membership in such private operations, and status might be difficult to measure.

As was developed in the illustration of ski resorts (Chapter 22), the problem of pricing is probably as important as that of production itself. Because of the exclusive nature of hunting clubs and the element of status-seeking that could be significant, each enterprise tends to be of a monopolistic nature. In those that are member operated, objectives other than maximization of present net worth would probably be paramount, but whether returns are measured in dollars, in status units, or in satisfaction, the marginal principles developed earlier would be applicable to the production phase.

## PUBLIC WILDLIFE PRODUCTION

In 1968 some 15 million people in the United States paid nearly 90 million dollars for hunting licenses, tags or other permits, and 23 million people paid almost 80 million dollars for fishing permits of one kind or another. License fees constitute a special kind of tax. In a very rough way one can say that such taxes are levied in accordance with benefits received, for only hunters are taxed. Questions can be raised on grounds of income distribution, since rich resident hunters pay the same fee as those from rural or city slums.[3] Because license fees are usually small, however, no gross inequity is involved.

### Federal Financing

For the past decade and more, approximately 30 to 50 per cent of all game and fish management and research funds have come from Federal sources, even though most wildlife management ac-

[3] Hunting licenses are now required in all 50 states, although in some states licenses are not required of farmers to hunt on their own property, of veterans, or of other identified groups.

tivities are under state control. The Pittman-Robertson Act of 1937 provided for an 11 per cent excise tax on retail sales of sporting arms and ammunition, while the Dingell-Johnson Act imposed a similar (10 per cent) tax on certain items of fishing equipment. During 1968 these taxes yielded 40.6 million dollars for disbursement during 1969. At the close of this latter fiscal year the states were operating over 49.5 million acres as wildlife management areas open to public hunting, with some 653,000 additional acres in wildlife refuges. Pittman-Robertson funds are used to acquire new refuges and hunting areas, for research, for development and management (planting food and cover, making brush piles, etc.), and for coordination, but are *not* used for law enforcement activities. Although collected by the Federal government, these funds are returned to the states according to the state's proportion of total hunting license sales in the nation and the proportion of the total United States land area. States must first obtain Federal approval of a project, then build it with state funds. Pittman-Robertson funds can then be used to reimburse the state for 75 per cent of the project's total cost, resulting in an eventual three-to-one cost-sharing system. Similar regulations govern disbursement of Dingell-Johnson funds for fishing projects. In terms of Federal funds obligated in 1969, wildlife activities ranked in the following order: development, research, land acquisition, and coordination. Fishing activities were similar, but research ranked first, with development, acquisition, and coordination following (United States Department of the Interior, 1970).

### The Public Wildlife Package

Wildlife-associated activities on public lands provide a very different kind of package to the consumer than those privately controlled, although the wildlife production, per se, would not differ in an economic sense from that on private areas except for determination of values. The relationships between inputs and outputs must be determined, and costs attached to the inputs. It is the revenue side of public production that is essentially different from that in the private sector.

The five components of outdoor recreation identified by Clawson and Knetsch (1966) can provide a starting point for many wildlife problems. As a specific example, consider the recreation package we call deer hunting. *Anticipation* might be subdivided into two parts: (1) planning the hunt, and (2) acquiring, testing, and owning the necessary gear. Many hunters obviously gain great satisfaction from owning and occasionally testing a high-powered rifle. *Travel* to the hunting area is undoubtedly an important cost (supply) item, while travel *time* would be an important limiting factor in determining the demand for any particular area. Travel to a hunting area would probably be less likely to vary with geographic origin than would that for a summer vacation trip, since the purpose of the latter is less narrowly directed. *Participation* could also be divided into several parts: the hunting itself, the evenings spent at the hunting lodge or camp (with the usual poker games and other forms of entertainment), the renewal of old friendships with the hunting gang, the attractiveness of the hunting area, the degree of crowding, the probability of success and the problem of packing out the carcass—all would influence the demand for the total experience. The *return travel* would be a cost element (and might require weighting by whether the hunter had been successful or not). The *recollection* phase (a few dinners, the gifts of venison to protesting neighbors, swapping stories at the shop, etc.) would complete the major features of the deer hunting experience. The total package of deer hunting is indeed a large one; valuing a deer herd by multiplying the pounds of meat hunters take home by the current price of medium quality beef is little short of ludicrous.

### Demand Aspects

The approach to demand analysis outlined in the preceding recreation chapter offers some leads for analyzing the demand for specific kinds of hunting experiences such as deer hunting, duck hunting, upland bird shooting, or other types of wildlife-based recreation. The demand for various hunting packages includes the demands for hunting in a given *area*, and many of the same factors discussed as short and long term recreation demand shifters would be appropriate to wildlife demand studies as well. Income, alternative opportunities, and population variables such as age and level of education would surely influence the demand for a given hunting area, as would the attractiveness of the forest surroundings and the amenities afforded in the vicinity. Additionally, past exposure to hunting (whether the individual hunted in his youth) and other variables specific to hunting would probably prove significant.

One variable not discussed previously is the probability of success, which is expressed in probability terms. A most interesting experiment could be hypothesized in which areas with differing success probabilities (based on number of deer per unit area and weighted by cover, weather, etc.) were offered to hunters at definite but differing prices (license fees could be varied).[4] Average size of animal, number of other hunters allowed on the area, and the location with respect to major population centers could all be made variables if the experiment were sufficiently comprehensive or were replicated often enough in time. Areas such as those needed for an experiment of this kind already exist, people hunt in them, and they travel varying distances to achieve their goals and consequently spend different amounts. In this connection it is worth noting that in some parts of Europe the license fee is varied with the objective of the hunt; hunting for trophy heads requires special licenses, and very often special guides and special training.

## Supply Aspects

In analyses of the demands for a specific area or kind of hunting, a simultaneous analysis of supply would be necessary and, in many cases, is very possible. Travel cost and the (fixed) cost of hunting equipment would be needed to isolate demand for a specific opportunity or experience and many of the drawbacks previously cited would apply here. But the possibility of determining production costs for supplying various hunting levels (hunting success) should not be overlooked. The end result could well be one of determining not only the demand curve for, say, duck hunting, but also that of ascertaining how much duck hunters would be willing to pay for their duck stamp if such fees (revenues) were used for waterfowl production. A marginal cost price for duck stamps—to take an extreme example—might seem at present an allocative fantasy, yet such a price should not be impossible to determine. Supply and demand analysis for wildlife is believed to offer unusual opportunities for original research.

[4] I have been impressed by the fact that the number of deer hunters reported by the various states bears little relationship to the reported success ratio. Some minimum success ratio is surely required, yet a ratio of one could eliminate most of the attraction. In some hunting (?) clubs deer may be "staked out" for certain members, but few sportsmen want success to be guaranteed by such measures.

## WILDLIFE AS A "COMMON PROPERTY" RESOURCE

One aspect of wildlife poses an interesting, and quite different, possibility for economic analysis. Although it requires a level of sophistication not usually called for in applied courses it is believed worth presenting for those who may be interested in developing the embryonic field of wildlife economics.

In the United States, as in many, but by no means all, other countries, wildlife is a "common property" resource. Game animals can be hunted by anyone and everyone willing to obtain the necessary equipment and spend the time to "go hunting." In this respect wild game resembles marine fisheries, and the work of certain fishery economists (Gordon, 1954; Scott, 1955 & 1962) becomes both appropriate and enlightening. Consider, for example, the situation in which a given forest area is to be managed for, say, ruffed grouse.

### Production and Revenue

We will assume that the area is large enough that no one hunter will be able to deplete the stock sufficiently to have any noticeable effect on hunter success, and that inputs of hunting can be defined in terms of hours spent in the field. If a production function were to be drawn, relating the outputs (measured in birds bagged by hunters) to the inputs (measured in hunter hours spent in the field) it might appear as in Fig. 24.1. Bear in mind that this is the production function for the whole area, not that for a single hunter. All this production function tells us, so far, is that as the number of hunting hours is increased the number of birds bagged per hour of hunting will decrease. This seems a realistic relationship in most hunting areas, where initial success is high but drops off as the number of birds is depleted and the remaining birds "get wise."

Hunter satisfaction would probably be a function of birds seen, birds bagged, number of hunters, weather, and many other items. For present purposes, however, we will concentrate on the number of birds bagged and assume that hunter satisfaction is a linear function of birds taken. This implies that the value per bird remains the same throughout the season regardless of the number shot, so the production function of Fig. 24.1 can also be taken as a total revenue curve. To express the revenue in dollar terms requires

**Fig. 24.1** Production function for bird hunting.

that we know the value per bird, knowledge that at present is probably unavailable. For present purposes, however, the absolute figure is unimportant; we can assume some arbitrary figure as the average value and continue.

## Costs

Costs should be somewhat less of a problem. Ignoring possible relations between hunting cost per hour and the number of birds bagged (for example, an increase in birds bagged means more shells, etc.), we can assume the cost per hour of hunting is constant, and while the variation from hunter to hunter might be substantial, there is little reason to suppose the cost per hour could not be determined by using established research procedures. Having assumed that hunting costs *per hour* are constant, we see that the total variable cost curve would be a straight line when expressed as a function of total hours hunted. This total cost can now be plotted together with the total revenue curve (from Fig. 24.1), with the latter now shown in both dollar terms and in number of birds bagged (Fig. 24.2). Fixed costs of management, land value, and other similar costs that can be included are shown in the figure.

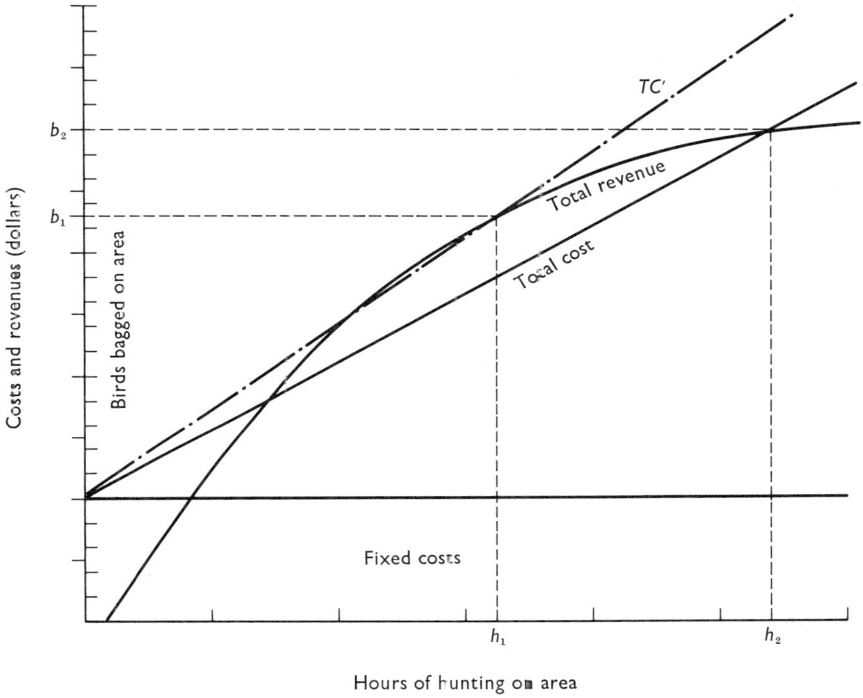


**Fig. 24.2** Total cost and revenue for hunting area (revenue function from production function of Fig. 24.1).

If this were a fully commercial operation—if the whole area were being handled so as to maximize the net hunter satisfaction—and providing the costs and revenues have been correctly estimated, then the optimum number of birds to be taken is that indicated by $b_1$, which would follow from an input of $h_1$ hunting hours. Here marginal cost and marginal revenues are equal, and *net* revenue is at a maximum.

## The Common Property Effect

But game is "common property," and the individual hunter will have no regard for maximizing net revenue from the area, for the net revenue does not accrue to him. The hunter's interest will be focused on *his own marginal costs and returns, not those appropriate to the game management area.* His marginal and average cost per hour of hunting is constant, and as long as the anticipated value per bird is above its cost to him, the hunter will continue to seek out birds and keep banging away. To operate at the point of maximum economic efficiency, taking $b_1$ birds, requires that some single person or agency control the whole show, a situation that almost never occurs in game areas in this country.

If the area were completely open, with no restriction other than a ban on hunting during the mating and nesting seasons, hunting will continue until $b_2$ birds have been taken (Fig. 24.2). This will have two possible effects. In the first place, the *net* gain will disappear entirely; total revenue and total cost are identical when $b_2$ birds are taken (when $h_2$ hunting hours are expended). In the second place, it is quite possible that the stock of birds will be depleted (through overshooting) to such a degree that the bird population cannot be maintained, and the number of hunting hours will be cut substantially below $h_2$. This last point is not a necessary conclusion, for the size of brood stock required is very much a technical (biological) matter, but as long as total revenue exceeds total cost, hunting will continue.

In any case it is clear that the danger of overshooting is much greater under these conditions than would be the case if a single agent had control, or if the entire operation were being managed for private gain. And this relationship is true whether the value per unit is correctly known or not. Resources under "common ownership" are always in much greater danger of depletion than are those privately owned. "Everybody's property is nobody's care."

The model also explains to some extent the "poaching" phenomenon, with the Florida alligator providing a good case in point at the present time. The possibility of apprehension must be included as one of the costs, and since this cost must be weighted by probabilities (the chances of being penalized vary with location, the skill of the poacher, the likelihood of a conviction, the severity of the judge, etc.) the model becomes complicated; both costs and revenues are uncertain. But as long as the probable revenue exceeds the probable cost—in the eyes of the poacher—one can confidently predict that poaching will continue. Law enforcement simply pushes up the costs to the poacher, and if pushed high enough will provide control. Alternatively, one could observe that the same result would be achieved by reducing the probability of revenue (by controlling sale or use of alligator hides). Whether the revenue or cost approach (or both) are taken should depend on the cost of instituting the proposed control measures.

### Policy Implications

The diagram of Fig. 24.2 reveals an interesting policy action that might prevent overshooting on the area studied. Suppose a fee were charged per bird, possibly paid at a checkout station. Then the hunting cost curve could be shifted up ($TC'$) so it would just touch the total revenue curve at $b_1$. Now, since any birds beyond the number indicated by $b_1$ would have a cost in excess of their value, hunters would no longer deplete the stock. The money obtained from fees might even be used to improve management of the area, perhaps increasing the number of birds, raising the production function, and hence shifting the total revenue curve upward. Conceivably, optimum production of birds might be shifted outward, perhaps even beyond $b_2$! Unfortunately, to achieve this balance requires that we know the actual revenue curve—the value of birds—and would require a certain amount of experimentation with fees to establish their proper level.

The important result of the foregoing analysis, however, is the fact that individual hunters continue to follow their (constant) marginal cost curve as long as total revenues exceed total cost. This fact, plus the often precarious biological balance between breeding stock, population, and hunting pressures has led to individual daily and seasonal bag limits for most kinds of game. In passing, it may be pointed out that shorter seasons will not necessarily accomplish the same objective of limiting the harvest, but may only compress the hunting, intensifying competition for the birds by crowding more hunting hours into the shortened season. And a flat license fee may also fail to limit the kill, for once paid it becomes a fixed cost to the hunter, who will still follow his own (constant) marginal cost. The flat license fee may, however, reduce the number of hunters if it is sufficiently high.

### Areas of Differing Productivity

The same general analytical approach is also helpful in comparing two hunting areas of either different productivity or differing costs (Gordon, 1954). Here, instead of examining a single hunting area where game is common property, the purpose is to examine probable results when hunters are faced with choosing between two areas. Fig. 24.3 illustrates the case by means of average and marginal cost curves. Hunting area (a) could be an area with more birds than area (b) or one with easier hunting conditions, or both.

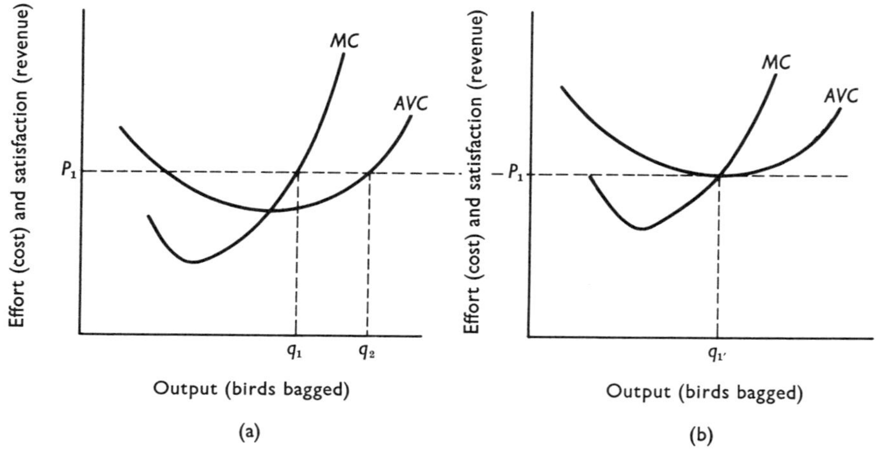


**Fig. 24.3** Average and marginal cost curves for two different hunting areas; (a) illustrates the richer area, (b) the poorer.

If average revenue (desirability or satisfaction) is at $P_1$, the optimum production level from area (a) would be $q_1$, and for area (b) would be $q_{1'}$. But hunters would not go to area (b) until an output of $q_2$ had been achieved from area (a) for only then would the average cost on (a) rise to the level of that for area (b). The good (or easy) hunting area will be "overshot" before area (b) is hunted at all unless some sort of control is instituted, and the control could consist of raising the fee (license) for hunting on area (a). The phenomenon of overshooting on good hunting areas is, of course, no new finding.

## WILDLIFE ECONOMICS—A NEW FIELD

Wildlife management economics is a virtually untouched area. As our knowledge of environmental control with respect to wildlife increases, game managers will turn to economics for managerial guides as have producers of other commodities. After all, the aim of wildlife managers *should* be that of providing satisfaction for people, not for (other) animals. The rapid spread of shooting preserves in this country could be an important influence in hastening application of economics to wildlife problems.

Quite possibly the major factor serving to push game management into economics will be the realization that "big game" is a prime tourist attraction. Tourism based on wildlife values has already developed into a highly important industry in several of the developing nations, particularly in Africa. An added spur in this latter case is the apparent likelihood that some combinations of game animals can produce far more protein from native ranges than can domesticated herds of cattle or other herbivores. Game management economics should prove to be a most interesting area for research.

# 25

# Forestry and International Affairs

There are many applications of economics to forestry not touched upon in this text, and not all were omitted because the author lacked the necessary interest. Forestry's role in the changing international picture, however, cannot be neglected. At least two of the international aspects are particularly important to foresters. One concerns the direct relationship of American forest industries to those of other nations—international trade in forest products. The second is concerned with forestry as an aid to economic development in the less advanced regions of the world.

International trade deals in part with questions of import and export, tariffs and other trade barriers, etc. It also treats problems of relative wage rates and raw material prices, and possibilities for export promotion and import substitution. Forestry's role in economic development is a quite different topic, though not unrelated to international trade. It is significant because of forestry's possible contribution to developing the economies of many nations. Development economics is a highly involved, quite specialized branch of economics, and its theoretical base is in the early stages of construction. Neither this latter topic, nor that of international trade, will be fully developed, but it is essential that both study areas be introduced as basic parts of the field of forest economics.

## INTERNATIONAL TRADE

Any kind of trade that is not forced requires both parties to feel their lot will be bettered by the proposed exchange. This is as true of trade between nations as between two small boys swapping jackknives or stamps. To see how this can be true, and to understand the basis for interregional or international trade, requires an awareness of two basic principles—specialization and comparative advantage.

### Principle of Specialization

The increased ability and superior performance an individual gains through specialization is so commonplace as to need no elaboration. This individual gain is multiplied by modern industry, perhaps reaching its highest point in the conveyer-belt assembly-line techniques so characteristic of Detroit automobile factories. And there are other important gains that accrue to industrial specialization—bringing together a team of research people who specialize in a particular field, developing a group of management executives who deal with the same kinds of problems, or creating a pool of skilled and semiskilled laborers having the required set of techniques. These and other external economies of scale are largely obtained only if a corporation, an area, or sometimes a nation or group of nations, specialize in certain productive endeavors. And there are occasions when even internal economies of scale can be obtained only if nations or groups of nations "specialize."

### Principle of Comparative Advantage

The second principle, that of comparative advantage, is less obvious but equally important. Probably the most used example is that of the leading lawyer who is also the town's best typist. It clearly pays this man to concentrate on the activity in which he enjoys the greatest comparative advantage and to hire a secretary to do his typing for him. Besides, there can be other advantages to having an attractive secretary—who may possess a comparative advantage in things other than typing!

To illustrate the significance of comparative advantage to trade, consider the following simple example. Suppose two individuals, Walton and Crockett, have retired to some remote wilderness and become hermits. One day they happen to meet. After comparing notes they discover each has certain talents. In terms of food production over a given unit of time (a day) Walton can catch four units of fish but only two units of game, while Crockett can get either three units of fish or four of game over the same time span. Then Walton gets a brilliant idea. "Suppose I do the fishing while you do the hunting. We'll meet at this same place after two days and swap."

Crockett agrees, and when they meet after the agreed time, Walton has produced eight units of fish and Crockett has eight units of game. Together they now have 16 units of food instead of the 13 they would have had if each had spent one day producing each product. Both men can now be better off, thanks to their specialization.

What about the terms of trade? Walton will go no higher than a "two-fish-for-one-game" exchange since he can do this well on his own. For the same reason, Crockett will say to himself "I've got to get at least a three-fish to four-game exchange rate." This means that the exchange rate of fish for meat will lie someplace between 2/1 and 3/4 (or between 2 and 0.75), but we can't tell exactly where it will be. The final result depends on the relative bargaining strength of the two individuals, and in this case it may be wise to remember that Crockett has a squirrel rifle and Walton a fly rod, which could, under some circumstances, influence the final terms of trading. Nevertheless, trade *could* be mutually advantageous to both Crockett and Walton.

Now suppose the production possibilities are slightly different—that the possible production of the two men (over the same time period) is changed to that of Table 25.1. Crockett is now clearly

**TABLE 25.1 PRODUCTION POSSIBILITIES OF WALTON AND CROCKETT.**

| Man | Fish | Game |
|---|---|---|
| Walton | 4 | 2 |
| Crockett | 5 | 4 |

the better provider, for he can out-fish and out-hunt Walton. He has an *absolute* advantage in producing both products. But suppose they decide (in spite of Crockett's superiority) to trade after each has concentrated on what he can do best, *comparatively*. Walton still has a comparative advantage in fish production (his fish-to-game ratio is 2.0 as compared to Crockett's ratio of 1.25) while Crockett's game-to-fish ratio is .80 compared to Walton's .50. With each man working for two days, their total production will be 16 units of food (compared to only 15 if each spends one day on each kind of production). The exchange limits within which the final exchange will be made can easily be determined, but not the exact ratio.

Note that two things had to exist before Walton and Crockett could benefit. Each had to specialize in the thing he did best *comparatively*, and the two men had to be willing to exchange the two commodities.

**Extension to international trade.** International trade is usually more complex, though basically similar. Very often trade is not a simple two-party exchange but involves several nations. Liberia may sell lavoa log to France, who sells hardwood plywood to Denmark, who sells furniture to the United States, who ships automobiles to Liberia. And there are, of course, important qualifications. Especially when dealing with large industrially developed countries one must bear in mind that costs tend to increase as output expands nationally beyond loosely defined limits. Japan, for example, might be able in 1972 to produce a million automobiles that compete very favorably in terms of cost with those produced in the United States—but not 10 million. Another example—the United States buys nearly 75 per cent of its newsprint from Canada but the United States still makes a lot of newsprint. The multinational example above, however, should show the fallacy of every nation trying to establish purely bilateral trade agreements. Everyone would suffer. There are also a good many qualifications of other kinds to be made, especially in the case of developing nations.

## Tariffs

In nearly any elementary text of economics one can find a discussion refuting the usual arguments advanced for protective tariffs—to earn revenue, to protect wage levels and special interest groups, for retaliation, etc. They need no review here except to comment briefly on the only one that makes a modest amount of sense—the argument that tariffs are needed to protect "infant industries." In large industrially developed nations infant industries almost never have enough political appeal (or pull) to gain the needed protection, while the older entrenched industries succeed in establishing and keeping protection (they "never shed their diapers"). There are some cases in developing countries, however, where protection of infant industries may be logical. It takes, for example, five to ten years after construction for a pulp and paper mill in a developing nation to get its crew properly trained, the machines working smoothly, the maintenance people made capable to handle the usual emergencies, markets established, etc. The odds of a new paper company being able to compete on a world-market basis with the older and very aggressive corporations of Scandinavia and North America are very slim; but a new mill may be able to compete successfully in its home market if given a chance to get established. The problem is that protection given industries in developing countries is most often designed to protect grossly inefficient operations. And, as observed above, although begun as a temporary measure, infant industry protection tends to be long continued and thereby encourages inefficiency.

The only real argument *for* free trade is the simple, but powerful one, that free trade promotes sensible and profitable international specialization, raises the real national product of all participating nations, and makes for rising living standards everywhere. Yet disputes over trade policy continue in a world where economic nationalism takes precedence over the ideal of increased world welfare.

## ECONOMIC DEVELOPMENT

A good many forestry problems of the United States today can be traced, directly or indirectly, to the growing affluence of our nation. The increasing relative demand for recreation as compared with that for timber products is an important illustration. A major segment of the world, however, has a quite different set of problems, a set involving the entire problem complex of getting an economy to move forward fast enough to provide populations with the bare necessities of life. Since World War II much has been said and written about the struggle of African, Asian, and Latin American nations to "catch up" to the standard of living enjoyed by the industrially advanced countries. The gap is widening; yet "catching up" is not really the problem, nor even a rational goal for much of the "other world." It has become crucial that *all* people be able to enjoy at least the minimum of health and leisure necessary for physical and mental peace and comfort—crucial to them *and to us.* This goal, the provision of minimum health, peace, and leisure, *is* realizable within the framework established by modern technology. While serving as the ambassador to India, John Kenneth Galbraith stated (Gill, 1963) that the attainment of this goal is ". . . without question . . . the most important and humane task in which men are now engaged."

## The Basic Problem

Economic development has become a crucial problem, and not just because of the inequalities between nations. A new element has been added to the mixture that gives it an explosive character. People in less developed areas today *know* the differences between western living standards and those of their own countries, and can see that technologies already exist for giving most people an adequate livelihood.

**Population.** Two major facts of life add to the severity of the problem: the population explosion, and the explosion of knowledge in the scientific world. The first has been described and emphasized in many publications. Dudley Stamp estimates it took from the time of Christ to about 1750 for the world's population to double, while at our present pace it will double in less than 40 years. Based on United Nations projections of 1966, the population of countries usually described as "developed" will rise by some 260 million by 1985. By the same date, populations in the developing countries (the rest of the world) will *increase by 1,370 million.* The rate of population growth alone poses a real strain on the possibility of economic growth for many nations. The economy is forced to grow faster and faster just to maintain per capita incomes. In El Salvador, for example, the population growth rate is estimated at 3.8 per cent annually. If this nation were able to achieve and maintain the United Nation's target of a 5 per cent annual growth in GNP, per capita income would *still* only be growing at 1.2 per cent, a rate scarcely fast enough to satisfy the long-suffering citizens.

**Growth of knowledge.** The explosion of scientific knowledge has also been well advertised. Someone recently estimated that 90 per cent of all scientists who have ever lived in the world *are alive today.* Beginning in Western Europe the scientific revolution spread quickly to North America where, in the United States, it achieved its most dramatic successes. The end result has been the creation of a society more abundantly endowed with material goods than any imagined even three or four decades ago. One may question whether this society is better, or happier, than those of earlier years, or even whether it is more "civilized," but one cannot question the fact that it is richer—far richer—than any society ever before dreamed of by man. And there can be no denying that the *potential* now exists for a truly "better world." Dr. A. Boerma, Director of the Food and Agriculture Organization of the United Nations, has suggested that we now have the technological knowledge to free the world from hunger, but unfortunately we lack the social, economic, and organizational ability to apply this knowledge.

**Revolution of "this-worldliness".** To these two potent forces for change Barbara Ward adds two "revolutions" (Ward, 1962). One of these she calls the "revolution of this-worldliness." For centuries, she says, mankind struggled along, firm in the belief that if one worked hard, endured his lot, did his best, and trusted his God, then eventually he would obtain his just reward of a happy life. Since for most people the reward was obviously not going to be granted in this world, these beliefs represented a philosophy of "other-worldliness." The revolution of "*this*-worldliness" follows from the conviction, now widespread, that it is not necessary to wait until the next world to enjoy a good life. *This* world can be a better one, if not for the present generation, then most certainly for their children.

**Revolution of equality.** The second of Barbara Ward's revolutions is the "revolution of equality." The United Nation's assembly epitomizes this revolution with its system of "one nation, one vote." The colonial era was dying before World War II and was thoroughly laid to rest by that struggle. Most students of international affairs agree that attempts at neocolonialism have failed regardless of which great power made the attempt. For the most part the drive by former colonies for independence is over. The leaders who have emerged have been successful in uniting their nations during the drive to independent statehood, but they now face an even harder problem. They are being asked "What now?" by their people. They must improve the way of life for their nation, and they look for help to those countries that have developed, the nations that are already rich. If our nation fails to rise to the challenge, others will respond. Barbara Ward put it succinctly by saying ". . . we prepare for our defeat simply by default." (Ward, 1962).

Because the probability of our children's survival in an unchanged world is small, this problem of economic development is crucial—crucial to the United States, to Western Europe, and to the Soviet Union as well as to nations such as India and China (who, incidentally, will together have very close to half of the world's population by 1980), and Africa and Latin America. By far the best hope these nations have is the one offered through the possibility of economic development. And it is in this field that the United States and other developed nations have a great opportunity.

### Foreign Aid Programs

To aid economic development the western world has established and administered large foreign aid programs. Assistance is given through international organizations, principally the various action agencies of the United Nations, as well as more directly through bilateral arrangements and special regional programs (such as the Alliance for Progress in Latin America). Private foundations have been of major assistance on the research front, and private industry has played an important role in supplying investment capital and technical knowhow. The Organization for Economic Cooperation and Development (OECD) reported that in 1968 the outflow of public and private assistance to developing nations from its 16 member countries increased to 12.9 billion dollars, fully 95 per cent of all such aid.[1] Of this total, 5.9 billion was from private sources, while official assistance amounted to 6.95 billion (Survey of International Development, 1969).

### Characteristics of Developing Nations

There is no precise or well-accepted definition of less developed (or "developing") countries. If compared to its potential, even the United States is presently underdeveloped. The nations almost always accepted as being less developed differ tremendously in their characteristics, yet possess some common features. The description of a nation in terms of these factors generally identifies it as being either developed or developing.

[1] $5.68 billion (44 per cent) came from the United States, placing it twelfth among the 16 OECD nations when assistance was ranked according to percentage of national income.

Developing nations have low incomes, almost incredibly low by our standards. The one billion people in the high income countries enjoy per capita incomes ranging from $600 to over $4,000, while the two-and-a-quarter billion people in the developing nations have incomes that average below $300, and the majority of the latter group have incomes of less than $100 (Westoby, 1966). Even though these income figures were converted to dollars through official exchange rates which underestimate purchasing power for necessities, and even though money incomes are not directly translatable into standards of living, it is painfully clear that much of the world is pitifully poor.

Many developing nations have very low rates of capital accumulation (low incomes usually mean low savings and savings are the main source of investment funds). Almost all these nations have high population growth rates, considerably higher than those of the developed countries. Most of their people—almost always over 50 per cent and sometimes as much as 80 per cent—gain their livelihood through agriculture. Characteristically there is a high rate of *un*employment, mostly in the urban areas, and an even higher rate of *under*employment in both rural and urban areas.[2]

Developing nations are also characterized by high illiteracy rates and low levels of technology. Many, but not all, poor nations are located in the tropics.

A lack of natural resource endowments is *not* a characteristic of developing nations. In fact, some of these nations are undoubtedly among the most richly endowed of all countries in terms of natural resources, with Indonesia and Brazil providing outstanding examples. This richness is usually contrasted with Switzerland, a highly industrialized nation that has virtually no stock of natural resources.

### Requirements for Development

While it is fairly easy to describe a condition of underdevelopment there is far less agreement on the basic requirements for development. Adam Smith's recipe was as follows: "Little else is requisite to carry a state to the highest degree of opulence from the

[2] Underemployment results when people work with inadequate equipment, or at tasks that do not fully occupy them, etc. In some rural areas of Asia it has been estimated that over 25 per cent of the workers could cease work entirely and national income would remain unchanged; the remaining 75 per cent could easily perform all the needed work. Underemployment is a sort of "share the work" kind of thing. It spreads the limited employment around.

lowest barbarism, but peace, easy taxes, and a tolerable administration of justice."[3] Present day development economists are inclined to believe Smith overlooked a few essential ingredients, although most would agree to include both the first and the last items.

*Capital* is almost always accorded a high place in the needs for development. Without capital accumulation there is no hope for building a modern industrial society. *Managerial ability* and *entrepreneurship* are also recognized as important factors. Partly because these abilities are enhanced by training, and even more because *technical knowledge* is always needed, *education* is usually ranked high on the list of developmental necessities. Some reasonable degree of *political stability* is believed a prerequisite, for without this there will be little investment from either internal or external sources. Inhabitants of unstable nations usually do little more than satisfy their immediate needs within their own borders while savings are placed in safer external holdings or in gold, jewelry, or some similarly hoardable commodity. A *will to develop* is being given more recognition as a necessary condition. People, not all, but enough to serve as a catalyst, must want to see their country develop and want it badly enough to work at it so hard that they inspire others by their example.

**Malthus, Ricardo, and Marx.** The theory of economic development is still very much in its evolutionary phase. There is, however, no lack of suggested theories. Two of these are of classic importance. Malthus and Ricardo drew a gloomy picture of increasing populations pressing ever harder on a limited natural resource base, with stagnation and poverty the inevitable outcome, a theory sometimes used to explain why poor nations are poor, and often referred to by those who regard population control as the one and only necessity to assure development.

The doctrines of Karl Marx have had a major impact on the thinking of most intellectuals in developing countries, and there is no denying that his philosophy is still an important force in most developing areas. The emphasis placed on centralized planning, the obvious developmental advantages gained through national ownership of natural resources and control of investment funds, the ease with which a central planning agency can establish priorities among developmental projects, and the ability to act that accom-

[3] Quoted by Hirschman (1959).

panies centralization of power—all these have an undoubtedly strong appeal to nations struggling to improve their economic machinery and keep it running. The fact that centralized planning does not assure *good* planning, that ease of establishing priorities does not assure any optimum selection of projects, and that ability to act quickly is accompanied by an ability to make drastic mistakes, is not always so readily recognized or acknowledged. Nor is it always recognized that the association between planning and economic development is not well established. One can make a fairly strong case for exactly the opposite. Japan and Mexico provide perhaps the best examples of successful development during the post-World War II era, yet they "enjoyed" far less intensive planning than have most of the less successful of the developing nations. It is a matter of fact, however, that some degree of centralized planning and some program of public saving and investment are almost always part of national development schemes.

**Post-war development theory.** In the 1950's and early 1960's development theory concentrated on three discussion areas, one of which concerned methods of breaking the so-called "vicious circles of poverty." At its worst the vicious circle approach reduces to saying "Poor nations are poor because they are poor," which is hardly helpful! But knowledge of the interrelations of these "vicious circles" can be very helpful, for they must somehow be broken. These circles arise in both demand and supply. On the demand side, low incomes lead to small markets for products, which lead to low inducement for investment, which leads to low production, which leads to low incomes, etc. And on the supply side, low incomes stem from the low productivity of workers, which is due to low capital investment in production facilities, which is traced to the lack of investment funds which results from the low incomes, etc. One way of breaking these particular vicious circles (and there are others) might be to inject capital into the stream, capital obtained by borrowing, by interesting foreign investors, or by obtaining direct foreign aid. In some instances new capital might be generated by increasing the exports of raw, or primary, materials. Exports have been called "the engine of the first stage of economic growth" (Lewis, 1966), but it must be noted that trade can work both ways. If exports increase faster than imports they can and will aid the growth process, but if imports increase faster than exports, growth may be slowed. Forests and forest-based industries can be important in both aspects.

Another debate centered on whether growth should or should not be "balanced" between the different sectors of the economy.[4] Some authors have observed that nations have pushed investment in industry almost to the exclusion of investment in agriculture, with the result that agricultural productivity has remained low, leading to low agricultural incomes, and consequently (since agriculture employs most of the population) to small markets for the newly available industrial materials. Others point to the need for transportation, power, communications, water, and other elements of "infrastructure" in amounts sufficient to support industry, and to the fact that investment in such "social overhead" tends to be very lumpy. The logical end of such reasoning is the "theory of the big push," the idea that many projects in different sectors should be undertaken simultaneously, an approach that usually calls for massive injections of new capital.

The alternative approach is generally called "unbalanced growth." This theory proceeds from recognition that growth is nearly always sparked by some leading sector or some leading industry. Practically speaking, growth is almost always a result of disequilibrium, not equilibrium. Growth occurs in spurts; it takes place at "growing points." Governments, under this approach, should look for unusual industrial opportunities, then push these to their limit. Other segments will be pulled along (backward linkages) or pushed ahead (forward linkages).[5]

Whatever the process of growth may be, and whatever means may prove best to initiate and sustain this process, the basic objective of developmental effort is one of inducing structural change in the economy of the concerned nation. This change must include releasing labor from agriculture and employing it in other, usually industrial, activities. Such shifts in employment are necessarily accompanied by shifts in populations, and the latter requires major investment in infrastructure. The rapid growth of urban areas in developing nations is posing an even more serious problem than urbanization in the industrialized countries of the West. In the

[4] The leading proponent of the "balanced growth theory" was probably Ragnar Nurske, while Hirschman espoused the "unbalanced growth" idea.

[5] Backward and forward linkages will be defined later in this chapter.

80-year period from 1920 to 2000 it is estimated that urban populations in developing nations will increase from 100 million to 2,000 million, while developed countries will experience a four-fold increase (Survey of International Development, 1969).

The third discussion area stems from the work of Rostow, who likened growth in developing economies more to the take-off of an airplane than to that of a biological organism—the latter has problems of immaturity, old age, etc. (Rostow, 1958). The first steps toward development are always uncertain and loaded with difficulty: development is very sluggish and hard to get underway. Then, almost suddenly, things begin to straighten out. A solid foundation has been established. And about this time the economy "takes off" into a period of fully self-sustained growth. There is, of course, always the possibility of engine failure and a crash, but it is the starting that poses the major developmental problems. It is observably easier to start an industry in developing countries than to keep it going!

**Theory topics of the 1960's.** Since the early 1960's discussion has tended to focus more heavily on the role of trade and of foreign exchange balances in economic development, and on the advantages and disadvantages of economic integration. "Import substitution" seemed to earlier workers an obvious way to reduce foreign exchange requirements; if a nation could manufacture its own paper, for example, it could save the major part of its foreign exchange presently spent on paper products. And this import substitution could be beneficial even if the home-produced product turned out to be more costly than the imported alternative and therefore required a substantial protective tariff. Today it is generally agreed that this *can* be so, but that the case for import substitution is easily exaggerated. Proponents ignored the loss of income to governments from tariffs on incoming goods, and there has also been an almost universal tendency to underestimate the most of home-produced manufactures. The principles of comparative advantage and of specialization are not repealed by the existence of a possibility for import substitution!

Success of the European Common Market has done much to promote economic integration schemes among developing nations. Whether such groupings can surmount the newly kindled nationalistic spirit so evident in developing nations remains to be seen. Economists have observed that almost anything that can be accomplished by a formalized economic integration mechanism, such as, for example, that represented by the Latin American Free Trade Association (LAFTA), or the Central American Common Market (CACM) can be done at least as easily without the formal arrangements. One can accept this as logically correct, but in fact such combinations do seem to promote development, at least in their earlier stages.

## FORESTRY'S ROLE IN ECONOMIC DEVELOPMENT

These introductory comments on development economics must inevitably leave most forestry students decidedly unclear as to the nature of the economic problems faced by over half the world's population. Hopefully, however, this sketchy background will prove sufficient to proceed with the major reason for introducing the topic in the first place—a discussion of where forestry fits into the scheme of economic development. The literature of forestry's role in economic development is not extensive. The most authoritative writings are those of Westoby (1962, 1966, 1968), whose work has been used freely in this chapter.[6]

### World Markets for Wood

Because the part forests and forest products can play in development will depend to no small extent on markets, an assessment of world markets for wood-based products is much in order. The Forestry Division of the Food and Agriculture Organization of the United Nations (FAO) published its first timber trend study in 1953. A series of similar studies followed—in the Far East, in Latin America, in Africa, and, in 1964, a second study of European timber trends (FAO, 1964). More specifically oriented studies of pulp and paper and of wood-based panels were completed during this same period. This work was brought together in the summary publication "Wood: World Trends and Prospects" which provided timber consumption projections by major product categories for 1975 (FAO, 1967).

In general these analyses show that world consumption of timber products will increase substantially if timber supplies are available.

[6] Jack C. Westoby has provided continuing inspiration and guidance to FAO's Forestry and Forest Industries Department since the early 1950's. The author gratefully acknowledges his own particular debt to this leader in international forestry.

Objections can be made to methods used in projecting consumption, and particularly to the lack of detailed demand analysis and to the level of assumed elasticities of consumption with respect to both income and price (Gregory, 1967). Still, the overall conclusions are indicative, and the changing relative importance of products is of particular interest. By 1975, based on data available in 1961, lumber consumption (sawnwood) is expected to rise by 23 per cent. The anticipated increase in consumption of pulp products is 110 per cent, while consumption of wood-based panels (plywood, particle board, fiberboard, etc.) is projected at 150 per cent above then-current levels. It is significant that 70 per cent of this extra industrial wood consumption is expected to take place in the already developed, highly industrialized nations (Westoby, 1966).

The elasticity of consumption with respect to both income and the general availability of wood has been studied by using international cross-sectional data (Gregory, 1965). The results, while far from definitive, indicate that income elasticity for sawnwood consumption on a per capita basis, and for all industrial wood, declines as incomes increase and as wood availability is held constant, and that it might even become negative at high per capita incomes (around $2,000 or more). Wood availability is almost as important as income; countries having little wood of their own have a relatively low-income elasticity, while high wood availability is accompanied by high-income elasticity—scarcely a surprising result!

From the standpoint of developmental possibilities two things stand out. First, it seems reasonably clear that markets for forest products, particularly for pulp and paper and for wood-based panels, should be very strong in the coming decade or beyond, and that a good part of the demand for these products will come from the developed countries.

Second, it follows that if a developing country has a prime forest resource it has a good chance of using its forests as a source of capital with which to promote economic development. Moreover, if it has resources which lend themselves to pulp and paper manufacture there may be an opportunity to make this one of the prime growth points for industrial development. Whether raw forest materials such as logs, or products as complicated as finished paper products are exported, such exports could be of substantial significance in increasing much needed foreign exchange.

## Timber Supplies in Developing Countries

The supply conditions in developing countries are extremely varied. Some 54 per cent of South America is covered with forest growth, most of which is composed of tropical hardwoods of virtually unknown quality (the Amazon forests were referred to by a tropical forestry expert as "that great green nothing"). Central and West Africa, Central America, and Southeast Asia have some magnificent stands of tropical hardwoods. Central America, and particularly Nicaragua and Honduras, have substantial areas of native conifers, as does Mexico, although Mexico is scarcely a developing nation in the sense used here. The forests of Southeast Asia are particularly rich, with as much as 75 per cent of the volume reported to be of commercial species, a sharp contrast to most of the world's tropical hardwood stands. Altogether, some 5½ billion acres (2¼ billion hectares) of forested land is found in developing countries.

Fast-growing plantations offer a very different kind of opportunity to developing nations, providing the conditions are right. The growth of *Pinus radiata* in Chile (see Chapters 14 and 15) furnishes an illustration of the potential some of these areas possess. Growth rates of up to five times the average obtained in most developed countries, including that of the southern pineries, are frequently reported. East Africa and New Zealand also have substantial areas of unusually rapid growth. (New Zealand is most assuredly not an underdeveloped country in the usual sense of the word, but it has based a major part of its industrial development on forest-based industries dependent on plantations for raw material supplies.) Because land is often low-priced and readily available in developing nations, outstanding opportunities for investment in pulp and paper manufacture may exist. At the same time it should be observed that when large-scale forest plantation programs are instituted there is typically a displacement of an impoverished population largely dependent on grazing. The support of such local populations during early development of the plantation (when labor requirements are low) can be a major cost in afforestation schemes.

It was noted earlier that much of the expanding market for timber products would probably be in the already developed nations. These are important export markets to developing areas, both quantitatively, and for earning foreign exchange. In many countries there is also a present need to reduce the drain on capital that results from the current net import situation, for in spite of their presently low levels of consumption many developing nations show a net deficit in value terms for their forest product trade balance.

### Potential Wood Markets

Far more important from a trade standpoint is the large potential market for forest products in the developing countries themselves. Earlier it was pointed out that the elasticity of consumption with respect to income tended to decrease as incomes increased. The corollary of this is that where incomes are low, an increase can and often does bring about a rapid rise in consumption. FAO estimated, for example, that the income elasticity of paper consumption in several of the developing nations was at least 2.0 (FAO, 1960). At the present time, paper consumption in these areas is small, varying from as little as 4 or 5 pounds per capita per year up to as much as 45 pounds per capita annually. Here one must remember that these nations contain over two-thirds of the world's population, and that these populations are increasing at rapid rates. If these countries succeed in raising per capita income levels even slightly, very large increases in paper product consumption could take place. The same general conclusions can be drawn with respect to wood-based panels, while consumption of sawnwood is much less likely to rise dramatically.

### Availability of Other Productive Factors

So far it would appear that (1) markets for forest products look very bright in both the industrialized and the underdeveloped nations; (2) some of the developing areas have excellent natural timber stands; (3) some can create fast-growing man-made forests; and (4) a few already have substantial tracts of these plantations. The review of developmental economics, however, indicated that markets and natural resource supplies were not enough to assure economic growth. There must also be a source of investment capital, a willingness on the part of a nation's people to develop, a range of technologies capable of utilizing the resource under consideration, sufficient political stability to warrant the investment, etc.

**Capital requirements for forestry.** Forestry, per se, requires a relatively small cash outlay. But plantation establishment calls for investment over very considerable time periods, and this is one of its major drawbacks that must be recognized. Wherever capital is in short supply, returns on the capital invested tend to be correspondingly high. The effect of high rates of return on probability of investment in man-made forests is well illustrated in a British Honduras policy analysis made for the United Kingdom by the then Senior Advisor to Her Majesty's Treasury. Because it covers several important points it is quoted at length (Downie, 1959).

> I am keenly aware that my views on the forestry programme conflict with those of others who have examined the economy of British Honduras and also with very deeply ingrained local sentiment. I am compelled to record nevertheless that I have the gravest doubts about the justification for a forestry programme of the present size. I appreciate the strong temptation to invest in something for which it is quite certain that the country is suitable, and how this is reinforced by the knowledge that present difficulties are so acute because of the heavy disinvestment in natural forest capital in the past. I accept further that the *ultimate* return on forestry is very large. But there is a strong *prima facie* presumption against a poor country investing a big proportion of its capital in a project which yields a return only after so long a period. When a peasant is hungry enough he eats even his seed corn. I do not suggest that British Honduras is in so desperate a position as that. Nevertheless, I do not believe that it can afford to wait anything from 40 to 100 years to get a return on its money.
>
> The Conservator of Forests was good enough to provide me with the information necessary to calculate a very rough rate of return on capital investment in forestry. The results I obtained were 8 percent on the relatively quickly yielding pine plantations and a rather higher rate on bringing pine under protection for natural regeneration. It may be suggested that, since money can be borrowed at 6 percent, the investment is justified. But this is an irrelevant comparison. For it is not possible for British Honduras to borrow unlimited amounts at 6 percent; nor indeed at any rate of interest, however high. Since forestry is competing for capital with other urgent needs—many of which can offer a much quicker return—the 6 percent rate at which a limited amount can be borrowed is not relevant; the effective rate of interest which capital should earn is far higher; and by this standard forestry does not pay. This conclusion is all the more true for investment in hardwoods, where the period of the investment is much longer and the true financial return correspondingly lower.
>
> I accept the need for a forestry service, to perform scientific and technical functions, manage the Crown estate, provide protection and supervise commercial operations. But I deprecate the investment of large sums of public money in the creation of forest capital.

Whether the analysis itself was sufficiently inclusive to warrant the negative conclusion is not known; it does drive home the need to use the best data available and to point out the many indirect benefits that often accrue from forestry projects. One wonders, for example, what prices were used for the timber, and on what products they were predicated. Did the analysis assume the end product to be export logs, or export sawnwood, or plywood manufactured locally; was pulpwood, or pulp, or manufactured paper to be produced? Low cost wood is a factor of great competitive advantage in almost any forest-based industry, but low-priced wood makes for low returns from forestry per se. Forestry and the forest product industries should be treated together in most developmental analyses.

**Capital needs for forest industries.** From an investment standpoint the forest-based industries (with the exception of pulp and paper) require relatively little capital. They are *labor* intensive, not *capital* intensive. Moreover, the return to the nation when measured in terms of dollars added to the gross domestic product is relatively high. The incremental capital-output ratio of lumber and plywood production is low, and even that for pulp and paper production will generally work out to be less than that for heavy industries.[7]

Because of the very wide range in capital requirements (sawmills range upward from $20,000, and at the other extreme are large paper complexes requiring as much as $100 million) the rural, wood-based industries offer a rare opportunity for initiating managerial and entrepreneurial interest. Practical experience in making investments and in managing a going concern is badly needed in developing nations. Moreover, because the range in required labor skills is large, the forest industries can provide a good starting point for industrialization. Transition from a subsistence agriculture to a modern industrial system is difficult at best. The transition is more easily made if mechanical skills can be acquired gradually, and forest-based industries have unusually flexible labor requirements.

**Linkage among forest industries.** When mentioning the unbalanced growth theory it was suggested that governments should look for "growing points" and encourage these as much as possible, thereby creating both a pulling and a pushing effect on other in-

[7] The usual capital-output ratio of developmental economics is a ratio of aggregates and in mature economies is usually regarded as being quite stable. As used here it is an "incremental capital output ratio" for a particular industry, and is the ratio between new capital invested, and the addition (increment) to the gross national product.

Work in FAO developed incremental capital output ratios for pulp and paper ranging from 4.5 to 6.0 or above. Those for lumber were frequently as low as 2.0.

dustries. Hirschman (1967) defined "backward linkage effects" in these terms:

> . . . every nonprimary economic activity will induce attempts to supply through domestic production the inputs needed in that activity.

Forward linkage effects were:

> . . . Every activity that does not by its nature cater exclusively to final demands will induce attempts to utilize its outputs as inputs in some new activity.

Hirschman ranked industries according to their interdependence through purchases of input factors (backward linkage). According to these data, paper and paper products were exceeded only by iron and steel, and by the non-ferrous industries in the "intermediate manufacturing group," while lumber and wood products ranked third in the final manufacture group.

The significance of this high ranking is simply that establishment of these industries creates strong inducements for the companion industries which supply the needed materials, and for those industries which use the manufactured products.

One other point should be mentioned—the relation of forest product industries to construction. In most countries the construction industry takes at least 50 per cent of all lumber produced, and construction is always a major—often *the* major—investment sector of developing economies. But construction is also of great political significance. Perhaps no single item is so important in the eyes of a developing nation's people as is housing. To relieve the drastic shortages of urban and rural housing that almost characterize the developing nations will take large volumes of low-cost construction materials.[8] Forests can often provide this material if the necessary investments are made.

## Problems of Forest-Based Development

A logical question to pose at this point might be "If forestry and forest products industries have so many advantages, how come these underdeveloped nations aren't going overboard to get their

[8] Adequate housing in developing nations will also probably require new methods of financing. Houses costing as little as $1,000 or $1,500 are beyond the means of those earning under $200 per year.

forests in shape and organized to take advantage of their golden opportunities?"

There are many reasons, for forestry and its related industries do have many of the problems common to other kinds of industrial activity—and it has a few of its own!

**Information shortages.** First, reliable information on timber resources is neither easy nor cheap to obtain. The Amazon River basin, for example, has long been regarded as a vast forest of magnificent tropical hardwoods. What kind of hardwoods? How big? How much will have to be invested to bring them to market? Are the markets already established? How long will it take to open up a logging area? And so on for a discouragingly long list of questions. To answer these takes reasonably detailed inventories, and developing nations have neither the money nor the technical knowledge to carry out these inventories.

**Species heterogeneity.** Second, a major problem of tropical forest development stems from the tremendous variety of species usually making up the stand. Some species are high-valued, with Central American mahogany providing an excellent example. These trees are sufficiently valuable that scouts (spotters) are employed to find the individual trees. Once located, each tree is examined by an inspector, and if found suitable a road is mapped out. Large caterpillar-type tractors are then used to build and bulldoze a trail into the tree, after which it is felled and the logs taken to the nearest landing or loaded directly on a truck and hauled to the mill, sometimes over 100 miles of very poor road. One tree to six or seven acres (2½ hectares) may be enough to warrant a commercial operation. But what of the other species? Removing the single tree leaves a virtually untouched forest with as many as 100 species to the acre, many of which produce a beautiful wood of high quality but which happens to be completely unknown in existing markets. Technological research in tropical hardwood timbers—means of drying them, their machining qualities, gluing qualities and characteristics, finishing techniques—is badly needed. If there were some process that permitted indiscriminate mixing of tropical hardwoods for pulping, at least part of the problem would be solved, but so far the pulping processes applied have produced relatively high-cost pulp.

**Small local markets.** Third, in many instances local markets are too small to warrant an operation on an efficient scale. This factor has been particularly important in pulp and paper manufacture, where economies of scale are pronounced. But even when producing logs for export the local market may be the decisive factor simply because of the above-mentioned species problem. In Liberia, for example, the successful log exporters were those able to market locally the lower grades and some non-export species. Logging operations everywhere tend to incur high fixed costs, and if production volumes are restricted to the very high-quality logs for export markets the opportunity for profit, or even that of breaking even, is slight. In Liberia this had the somewhat peculiar result that the log *export* market was restricted by the volume of lumber the local markets could absorb (Gregory, 1965).

These kinds of problems, along with those such as restrictions or high import duties on necessary machinery, problems of housing, of language, etc., are common to almost any industry establishing a foreign operation. But forest development has some of its own problems as well.

**Transportation.** To harvest wood efficiently requires a good transportation system, not just logging roads, but mainline highways and some semblance of an all-weather road system, or an established and operating rail line. In many of the developing countries the forest areas are virtually roadless, which forces any timber operation to undertake a major road-building effort and investment before logs can begin to flow into the newly established mill. Peru provides an extreme example; here the forests are found on the east side of the Andes, while the principal population centers are west of the mountains near the coast. Because there was no road across the Andes, until recently, the sawmill industry in Peru shipped its output some 2,500 miles down the Amazon, while the major part of Peruvian wood consumption was being imported from the United States and being used within 300 miles of the center of the Peruvian sawmill industry.

Liberia, Chile, and Nicaragua offer examples of another all-too-commonly missing part of the transportation network. All three nations have unusually fine sites for important forest industries, and could easily supply these industries with abundant raw materials, good water, etc. But port facilities for the specific sites are completely inadequate. Even with adequate ports, shipping is often a problem. Export volumes may be much too small and the shipments too irregular to warrant scheduled service by freighters properly equipped for transporting lumber or veneer. Much of the tropical hardwood from developing countries is carried at very high freight rates, often justified by the necessity for accepting small quantities, and for stopping at small ports that lack adequate loading or docking facilities.

**Other problems.** There are also peculiar problems arising from the extensive nature of forestry. Developing nations typically have few and badly kept records of land ownership, and even poorer systems of boundary surveying. The results may sometimes appear ludicrous, but they can also be disastrous to a newly formed company. The story is told of a Central American sawmill demonstration project which twice located a suitable forest area on government-approved land, built a road to the mill site, then purchased and installed modern equipment—only to have an "owner" appear and legally claim the sawmill because it was on his land. True or not, the story is illustrative of the "little" problems that make development work in forestry such a frustrating, yet fascinating, experience. Many of the problems have no parallel in an industrialized economy. And this is precisely why the job is so challenging.

### Planning Economic Development

The word "planning" evokes a variety of responses in this country when used in an economic context. The very idea of economic planning is anathema to some, while to others the general lack of acceptance of economic planning in this country seems ridiculous. Actually, our economy is planned to a very considerable extent, but a major segment is largely unstructured.

Most governments in developing countries have decided to plan their development insofar as possible. Usually this means they establish a ministry of planning and give this ministry at least authority to approve or disapprove new projects, either public or private. Developing nations have also often established developmental "corporations"—public agencies or corporations with varying degrees of autonomy that may perform services ranging from a sort of national "Chamber of Commerce" type of activity all the way to financing, organizing, and building major industrial projects. In Chile, for example, the steel plant, a tire corporation, a pulp and paper factory, and many other industries have been organized, built, and operated by the national development agency, then later sold to private investors. Similar examples could be found in other nations.

The major purpose of the planning agency, however, is to set up a "developmental plan." These vary tremendously in their complexity, depending on the availability of data, the people doing the planning, their particular objectives, etc. Present plans nearly always cover not only the public sector (the tax structure, the expected governmental income, the proposed budgets of each agency, and detailed expenditure plans for projects supposedly leading to development), but also the private sector as well. This last is inevitable, because the private and public sectors are so interrelated.

But if both public and private sectors are brought into the plan, with major industries reviewed, production targets established, quantitative estimates of internal and external demand made, etc., the plan necessarily involves a tremendous amount of work. Often the end result is the preparation of a macroeconomic model with interrelated inputs, outputs, imports, exports, etc. Such plans are beautiful exercises in logic and can be very helpful, but they are often quite divorced from reality.

> The principal danger of a macroeconomic exercise lies in its propensity to dazzle. . . . Attention shifts from policy to arithmetic. Consistency can be mistaken for truth. Revision is resisted. Yet the Plan is not necessarily right merely because its figures are mutually consistent. (Lewis, 1966).

Yet the planning exercise is a most useful and necessary one. The principal reason for emphasizing it here is to bring home the fact that unless foresters are prepared to enter into the planning process, forestry and the forest-based industries are likely to be ignored by the planning office. If forest-based operations are left out of the development plan it is most unlikely that forestry will attract more than passing attention when the budget is being drawn. For foresters, two aspects of planning are essential; sector planning, and project planning.

**Sector planning.** The basis for most forest-based development investment should be found in the developmental plan for the two sectors—forestry, and the wood-using industries. Approached as a problem of sector analysis, the national consumption of each product is first estimated, then forecasts of future requirements (future consumption under specified rates of increase in population and income) are prepared. These national requirements in terms of products are then translated into acres of new plantations, access roads to indigenous forests, capital, new industrial establishments, manpower, etc. From these estimates one can also derive the possible addition to GNP by each of the sector components, the need for foreign exchange, the contribution to balance of payments, the employment to be furnished, etc.

**Project planning.** More and more the form of analysis for planning developmental projects is that of benefit/cost analysis, but with several important differences. In the United States, where benefit/cost analysis originated with water resource development planning, the concept of secondary benefits has been dismissed. As long as full employment and reasonable factor mobility can be assumed as a usual condition, and where normal economic growth is self-sustaining, any secondary benefits of either the stemming or induced variety are offset by reductions in other parts of the economy (Chapter 20, page 434). But developing countries seldom enjoy full employment of either labor or resources, and the principal objective of projects is to initiate the process of development and growth.

> Development projects in these countries have precisely the mission to reduce prevailing unemployment and underemployment, to absorb immobile factors of production, and, in general, to *ignite growth sequences* rather than merely to step, as it were, on an already fast-moving escalator. (Hirschman, 1967.)

In developmental projects the idea of establishing a single ranking device that would assure selection of the best projects under all conditions is little short of absurd. Even if interest rates could be objectively selected, and if all backward and forward linkages could be established (the secondary benefits) there are so many "spin-offs" of a good project in terms of training, inspiration, political change, and changes in human conduct, that no single criterion could or should be adopted.

Usually, several separate analyses will be needed to provide the necessary information for deciding among alternative projects. Perhaps the most fundamental criterion is the project's contribution to growth of the GNP. Hence one analysis will usually be in terms of discounted cash flow, from which value added can be directly estimated. The corresponding basis for establishing project priorities is that of the internal rate of return. In most cases, however, additional analyses will be needed. These may be in terms of the project's effect on the national foreign exchange balance, on employment, on the expenditure of restricted funds from foreign aid, or on its contribution to either GNP or employment from a regional rather than a national viewpoint.

Shadow pricing of factors used by the project is often used as a device to bring some of these elements into a single analysis. Labor, for example, may be priced at its social opportunity cost rather than at the going wage rate, or the product may be priced at world market levels rather than at the local level. Shadow pricing must be used with caution, however, if suspicion is not to be thrown on the results of the analysis.

The process of economic development is not simple, and it is at present incompletely understood. But the task of initiating growth in the underdeveloped nations of this world is one which must not be shirked and in which we cannot afford to fail. In some nations forestry has a unique role to play in this, the most important task of our century. In these nations no one is challenged more than those trained to understand the economic implications of professional forestry.

# Bibliography

AINSWORTH, J. H. 1958. *Paper and you.* Book I in series *Paper, the fifth wonder.* Kaukauna, Wisc.: Thomas Publishing Co.

ALLEN, R. G. D. 1933. *Mathematical analysis for economists.* London: Macmillan and Co.

AMIDON, E., and GOULD, E. M., JR. 1962. *The possible impact of recreation development on timber production in three California national forests,* Technical Paper no. 68, Forest Service, U.S.D.A. Berkeley, Calif. Pacific Southwest Forest and Range Experiment Station.

ARROW, K. J. 1963. *Social choice and individual values.* 2nd ed. Cowles Commission Monograph no. 15. New York: Wiley.

ATKINSON, W. A. 1956. *A method for the recreational evaluation of the forest.* Master's thesis, University of California.

AYRES, R. U., and KNEESE, A. V. 1969. Production, consumption, and externalities. *American Economic Review* (June 1969): 282–97.

BAIN, J. S. 1958. *Pricing, distribution, and employment.* New York: Holt Rinehart and Winston.

BATOR, F. M. 1957. The simple analytics of welfare maximization. *American Economic Review* (March 1957): 22–59.

———. 1958. The anatomy of market failure. *Quarterly Journal of Economics* 72:351–79.

BAUMOL, W. J. 1965. *Economic theory and operations analysis.* Englewood Cliffs, N. J.: Prentice-Hall.

———. 1969. On the discount rate for public projects. In *The analysis and evaluation of public expenditures: the PPB system.* A Compendium of Papers submitted to the Subcommittee on Economy in Government of the Joint Economic Committee, vol. 1, pp. 489–503. Washington, D. C.

BENTLEY, W. R. 1965. *An economic study of timber allocation policies of the U. S. Forest Service.* Ph. D. dissertation, University of California.

———. 1968. Forest service timber sales: a preliminary evaluation of policy alternatives. *Land Economics* 44:205–18.

BENTLEY, W. R., and FIGHT, R. D. 1966. A zero rent comparison of forest rent and soil rent. *Forest Science* 12:460.

BENTLEY, W. R., and TEEGUARDEN, D. E. 1965 Financial maturity: a theoretical review. *Forest Science* 11:76–87.

BOULDING, K. E. 1948. *Economic analysis.* 2nd ed. New York: Harper and Row.

———. 1952. Welfare economics. In *A survey of contemporary economics,* vol. 2, pp. 1–36, ed. B. F. Haley. Homewood, Ill.: Richard D. Irwin.

BOWER, B. T. 1966. *Water research.* Edited by Kneese and Smith. Baltimore: Johns Hopkins Press, Resources for the Future.

BRAUTIGAN, R. 1969. *Trout fishing in America.* New York: Dell Publishing Co.

BRAZER, H. 1969. "Outdoor recreation as a public good and some problems of financing." In *Elements of outdoor recreation planning.* Proceedings of a National

Short Course held in Ann Arbor, Michigan, May 6–16, 1968. Edited by B. L. Driver. pp. 121–30. Ann Arbor: University Microfilms.

Briggs, C. W. 1962. *Tax treatment of timber.* 4th ed. Washington, D. C.: Forest Industries Committee on Timber Valuation and Taxation.

Carlson, S. 1937. *A study in the pure theory of production.* Westminster, England: P. S. King and Sons, Ltd.

Carson, R., 1962. *Silent spring.* Boston: Houghton Mifflin Co.

Chamberlin, E. 1933. *Theory of monopolistic competition.* Cambridge: Harvard University Press.

Cicchetti, C. F.; Seneca, J. J.; and Davidson, P. 1969. *The demand and supply of outdoor recreation; an econometric analysis.* Bureau of Economic Research. New Brunswick, N. J.: Rutgers University Press.

Ciriacy-Wantrup, S. von. 1938. Multiple and optimum use of wildland under different economic conditions. *Journal of Forestry* 36:665–74.

———. 1952. *Resource conservation.* Berkeley: University of California Press.

Clawson, M., and Knetsch, J. L. 1966. *Economics of Outdoor Recreation.* Baltimore: Johns Hopkins Press, Resources for the Future.

Dana, S. T. 1956. *Forest and range policy.* New York: McGraw-Hill.

Davis, K. P. 1966. *Forest management.* New York: McGraw-Hill.

Davis, K. P. et al. 1962. Determination of allowable cut on forty-two western national forests. U. S. Forest Service Report.

Dernberg, T. F. and McDougall, D. M. 1968. *Macroeconomics.* New York: McGraw-Hill.

Dorfman, R. 1953. Mathematical or linear programming. *American Economic Review* 43 (1953): 797–825.

———. 1965. Introduction to *Measuring benefits of government investments.* Edited by R. Dorfman. Washington, D. C.: The Brookings Institution, pp. 1–10.

Downie, Jack. 1959. *An economic policy for British Honduras.* Report submitted to Ministry of Trade and Industry, Belize City, British Honduras, from Her Majesty's Treasury, Treasury Department, London, England.

Duerr, W. A.; Fedkiw, J.; and Guttenburg, S. 1956. *Financial maturity: a guide to profitable timber growing.* U. S. Department of Agriculture Technical Bulletin. 1146.

Ecklund, R. 1967. *Economic strategy in the integration of the forest industries.* Paper given at the Symposium on Integration in the Forest Industries, Geneva, Switzerland (Economic Commission for Europe, Timber Committee).

Eckstein, O. 1958. *Water resource development: the economics of project evaluation.* Cambridge: Harvard University Press, 1958.

Fairchild, F. R. et al. 1935. *Forest taxation in the United States.* Washington, D. C.: U. S. Department of Agriculture Miscellaneous Publication 218.

Fedkiw, J. 1965. *Forest industry capacity, production, and available log supplies in the Douglas fir subregion, Portland, Oregon.* Forest Service Research Paper P.N.W. 11.

Fisher, J. L. 1969. The Viewpoint of the Public. Paper read at Virginia Symposium: Forest Resources Undergraduate Education and the Forestry Profession. Roanoke, Virginia.

Food and Agriculture Organization of the United Nations. 1960. *World demand for paper—1975.* Rome: FAO.

———. 1963. *Latin American timber trends and prospects.* New York: United Nations.

———. 1964. *European timber trends and prospects, a new appraisal: 1950–1975.* New York: United Nations.

———. 1967. *Timber trends and prospects in Africa.* Rome: FAO.

———. 1967. *Wood: world trends and prospects.* Basic Study no. 16. Rome: FAO.

Forbes, R. B. ed. 1955. *Forestry handbook.* New York: The Ronald Press.

Fortune Directory. 1967. *The 500 largest U. S. industrial corporations.* Chicago: Time, Inc.

FUSFELD, D. R. 1963. Economic education—or indoctrination? *Challenge,* pp. 15–17.
FOX, I. K., and HERFINDAHL, O. E. 1964. Attainment of Efficiency in Satisfying Demand for Water Resources. *American Economic Review* 54 (May 1964): 198–218.
GAFFNEY, M. 1960. *Concepts of financial maturity of timber and other assets.* Raleigh, N. C.: Department of Agricultural Economics, State College.
GALBRAITH, J. K. 1967. *The new industrial state.* Boston: Houghton Mifflin Co.
GILL, R. T. 1963. *Economic development—past and present.* Modern Economic Series. Englewood Cliffs, N. J.: Prentice-Hall.
Goldman, M. I., and Shoop, R. 1967. What is pollution? In *Controlling pollution,* ed. M. I. Goldman, pp. 59–70. Englewood Cliffs, N. J.: Prentice-Hall.
GORDON, J. S. 1954. The economic theory of a common property resource. *Journal of Political Economy* 62:124–42.
GREELEY, W. B. 1951. *Forests and men.* New York: Doubleday and Co.
GREGORY, G. R. 1955. An economic approach to multiple use. *Forest Science* 1:6–13.
———. 1965. *Report to the Liberian government on industrial possibilities of forestry.* Monrovia, Liberia: Liberian Planning Commission Report no. 30.
———. 1965. Estimating wood consumption with particular reference to the effects of income and wood availability. *Forest Science* 12:104–17.
———. 1967. Forecasting consumption and price. *Report of international union of forest research organizations, section 31. Working Group 4:* pp. 83–98.
GUTHRIE, J. A. 1950. *The economics of pulp and paper.* Pullman, Wash.: State College of Washington Press.
HAIR, D. 1963. *Use of regression equations for projecting trends in demand for paper and board.* Forest Resource Report no. 18, U. S. Department of Agriculture. Washington, D. C.: United States Forest Service.
———. 1963. *The economic importance of timber in the United States.* U. S. Department of Agriculture Miscellaneous Publication 941. Washington, D. C.: United States Forest Service.
HALL, C. R. 1935. *The forest tax problem and its solution summarized.* U. S. Department of Agriculture Circular no. 358. Washingotn, D. C.: United States Forest Service.
HALL, O. F. 1962. *Evaluating complex investments in forestry and other long-term enterprises using a digital computer.* Purdue University Research Bulletin 752. Lafayette, Ind.: Purdue University Press.
HAVEMAN, R. H. 1969. The analysis and evaluation of public expenditures: an overview. In *The analysis and evaluation of public expenditures: the PPB system,* vol. 1, pp. 1–10. Submitted to the Subcommittee on Economy in Government of the Joint Economic Committee, 91st Congress, 1st Session. Washington, D. C.
HICKS, J. R. 1939. The Foundations of Welfare Economics. *Economic Journal* 49:696–712.
———. 1946. *Value and capital.* Oxford: At the Clarendon Press.
HIRSCHMAN, A. C. 1959. *The strategy of economic development.* New Haven: Yale University Press.
———. 1967. *Development projects observed.* Washington, D. C.: The Brookings Institution.
HOLLAND, I. I. 1955. *Some factors affecting the consumption of lumber in the United States, with emphasis on demand.* Ph.D. dissertation, Berkeley: University of California Press.
HOTELLING, H. 1949. Letter to Director, National Park Service, June 18, 1947. In *The economics of public recreation—an economic survey of the monetary evaluation of recreation in the national parks,* ed. Roy Prewitt. Mimeographed. Washington, D. C.: National Park Service.
HUFSCHMIDT, M. M.; KRUTILLA, J. V.; and MARGOLIS, J. 1961. *Standards and criteria for formulating and evaluating federal water resources development: a report of a panel of consultants to the Bureau of the Budget.* Washington, D. C.: U. S. Government Printing Office.
HUGHES, J. M. 1968. Wilderness and economics. *Journal of Forestry* 66:855–58.

HUNT, J. D. 1969. America's outdoor recreation areas—playgrounds for the affluent. Paper read at Annual Meeting of the Rural Sociological Society, San Francisco, California, August 28–31. Mimeographed.

HUSCH, E. 1963. *Forest mensuration and statistics.* New York: The Ronald Press.

KALDOR, N. 1940. A note of tariffs and the terms of trade. *Economica*: 377–80.

KAPP, K. W. 1950. *The social costs of private enterprise.* Cambridge: Harvard University Press.

KENDRICK, J. W. 1961. *Productivity trends in the United States.* Princeton: University Press.

KNEESE, A. V., and D'ARGE, R. C. 1969. Pervasive external costs and the response of society. In *The analysis and evaluation of public expenditures: The PPB system,* vol. I, pp. 87–115. Submitted to the Subcommittee on Economy in Government of the Joint Economic Committee, 91st Congress, 1st Session. Washington, D. C.

KREPS, C. H., JR., and PUGH, O. S. 1967. *Money, banking, and monetary policy.* New York: The Ronald Press.

KRUTILLA, J. V., and ECKSTEIN, O. 1958. *Multiple purpose river development: studies in applied economic analysis.* Baltimore: Johns Hopkins Press.

———. 1967. Conservation reconsidered. *American Economic Review* 57:777–86.

LEUSCHNER, W. A. 1969. *Factors associated with financial success in the midwestern skiing industry.* Ph.D. dissertation, University of Michigan.

LEWIS, W. A. 1966. *Development planning.* London: George Allen and Unwin.

LUTZ, F. A., and LUTZ, V. C. 1951. *The theory of investment of the firm.* Princeton: University Press.

MACK, R. P., and MYERS, S. 1965. Outdoor recreation. In *Measuring benefits of government investments,* ed. R. Dorfman, pp. 71–101. Washington, D. C.: The Brookings Institution.

MARKUS, R. 1967. *Ostwald's relative forest rent theory.* Munich: Satz und Druck.

MARQUIS, R. W. 1952. *Forest yield taxes.* Circular no. 899. Washington, D. C.: U. S. Department of Agriculture.

MARSHALL, ALFRED. 1947. *Principles of economics.* 8th ed. London: Macmillan and Co.

McKEAN, R. 1958. *Efficiency in government through systems analysis.* New York: Wiley.

McKILLOP, W. M. 1967. Supply and demand for forest products—an econometric study. *Hilgardia* vol. 38. Berkeley; University of California Press.

MEAD, J. W. 1966. *Competition and oligopsony in the Douglas fir lumber industry.* Berkeley: University of California Press.

MUSGRAVE, R. A. 1959. *The theory of public finance.* New York: McGraw-Hill.

MUSSEY, O. D. 1961. *Water requirements of the pulp and paper industry.* Geological Survey—Water-Supply Paper 1330-A. Washington, D. C.: U. S. Government Printing Office.

Outdoor Recreation Resources Review Commission. 1962. *The future of outdoor recreation in metropolitan regions of the United States.* Report to the Outdoor Recreation Resources Review Commission no. 21, vol. 2. Washington.

PARETO, V. 1927. *Manuel d'economic politique.* 2nd ed. Paris: Girard.

PERRY, T. 1948. *Modern plywood.* 2nd ed. New York: Pitman Publishing Corp.

PINCHOT, G. 1947. *Breaking New Ground.* New York: Harcourt Brace Jovanovich.

POTTER, N., and CHRISTY, F. T. 1962. *Trends in natural resource commodities.* Baltimore: Johns Hopkins Press, Resources for the Future.

Public Land Law Review Commission. 1969. *Study of public land timber policy.* Vol. 2. Prepared for Public Land Law Review Commission, April, 1969, by George Banzhaf & Co. Milwaukee: George Banzhaf & Co., 1969.

REID, L. M. 1963. *Outdoor recreation preferences—a nationwide study of user desires.* East Lansing: Michigan State University.

Resources for the Future. 1958. *Forest credit in the United States.* Report of a Committee Appointed by Resources for the Future. Washington, D. C.

Ricardo, D. 1911. The principles of political economy and taxation. London: J. M. Dent and Sons.

———. 1951. *Works and correspondence of David Ricardo.* Edited by Piero Sraffa, Vol. 1. Cambridge: At the University Press.

Robinson, J. 1934. *The economics of imperfect competition.* London: Macmillan and Co.

Rostow, W. W. 1958. The take-off into self-sustained growth. In *The Economic Journal* (March, 1956). Reprinted in *Economics of Underdevelopment* ed. A. N. Agarwala and S. P. Singh, pp. 154–86. New York: Oxford University Press.

Row, C. 1963. *Determining forest investment rates-of-return by electronic computer.* U. S. Forest Service Research Paper SO-6. New Orleans: Southern Forest Experiment Station.

Samuelson, P. Q. 1954. The pure theory of public expenditures. *Review of Economics and Statistics* 36:387–89.

Schumpeter, J. A. 1954. *History of economic analysis.* New York: Oxford University Press.

Schweitzer, D. L.; Lundgren, A. L.; and Wambach, R. F. 1967. *A computer program for evaluating long-term forestry investments.* St. Paul, Minn.: North Central Forest Experiment Station.

Scitovsky, J. 1951. *Welfare and competition.* Chicago: Richard D. Irwin.

Scott, A. 1955. The fishery: the objectives of sole ownership. *Journal of Political Economy* 63:116–24.

———. 1962. The economics of regulating fisheries. In *Economic Effects of Fishery Regulation.* Fisheries Report no. 5, pp. 25–64. Rome: Food and Agriculture Organization.

Shirley, F. C. 1969. *Forest plantation investment, financial maturity, and the Federal income tax.* Ph.D. dissertation. University of Michigan.

Stanford Research Institute. 1954. *America's demand for wood: 1929–1975.* Tacoma, Wash.: Weyerhaueser Timber Corp.

Steiner, P. O. 1969. *Public expenditure budgeting.* Washington, D. C.: The Brookings Institution.

Stevenson, L. E. 1940. *The background and economics of American paper making.* New York: Harper and Row.

Subcommittee on Benefits and Costs. 1950. *Proposed practices for economic analysis of river projects.* Report to the Interagency River Basin Committee. Washington, D. C.

Survey of International Development. 1969. Vol. VI, no. 7 (Sept. 15). Washington, D. C.: Society for International Development.

Swanson, E. W. 1969. *Travel and the National Parks.* National Park Service. Washington, D. C.: U. S. Department of the Interior.

Sweezy, P. M. 1939. Demand Under Conditions of Oligopoly. *Journal of Political Economy* 47:568–73.

Trice, A. H., and Wood, S. E. 1958. Measurement of Recreational Benefits. *Land Economics* 34:195–207.

Udall, S. L. 1968. Statement read at hearings of the Subcommittee on Economy in Government of the Joint Economic Committee, U. S. Congress, on July 30, 1968.

U. S. Bureau of Labor Statistics. 1953. *The case study data on productivity and factory performance: veneer and plywood.* Report no. 37. Washington, D. C.: Bureau of Labor Statistics.

U. S. Code 16. 1960. Sections 528–31. Multiple Use-Sustained Yield Act.

U. S. Congress. 1943. *Senate Report* no. 627, 78th Cong., 1st sess., C. B. 1944, 973 at 993; Revenue Act of 1943.

———. 1960. House of Representatives. Hearings on the Multiple Use-Sustained Yield Bill.

———. 1962. *Senate Document* 97, 87th Cong., 2nd sess. *Policies, standards, and procedures in the formulation, evaluation, and review of plans for use and development of water and related land resources.* Washington, D. C.

———. 1963. House of Representatives, 88th Cong., 1st sess. Hearings before the Committee on Ways and Means on the President's 1963 Tax Message. Part 6 (March 21, 1963).

———. 1968. Hearings of July 30–31, and August 1, 1968, before the Subcommittee on Economy in Government of the Joint Economic Committee, U. S. Congress. *Economic analysis of public investment decisions, interest rate policy, and discounting analysis.*

———. 1969. *The analysis and evaluation of public expenditures: the PPB system.* Vol. 1. Submitted to the Subcommittee on Economy in Government of the Joint Economic Committee, 91st Cong., 1st sess.

U. S. Department of Agriculture. 1940. Forest credit to facilitate sustained yield management of privately owned forests. Inter-Bureau Coordinating Committee on Forest Credit. Mimeographed. Washington, D. C.: United States Forest Service.

———. 1946. *Forests and national prosperity.* Miscellaneous Publication 668. Washington, D. C.: United States Forest Service.

———. 1958. Historical forestry statistics of the United States. Statistical Bulletin no. 228. Washington, D. C.: United States Forest Service.

———. 1958. *Timber resources for America's future.* Forest Resource Report no. 14. Washington, D. C.: United States Forest Service.

———. 1965. *Timber trends in the United States.* Forest Resource Report no. 17. Washington, D. C.: United States Forest Service.

———. 1965. *Wood pulp mills in the United States and Canada: 1965.* Washington, D. C.: United States Forest Service.

———. 1965. *The demand and price situation for forest products, 1964.* Miscellaneous Publication No. 1009. Washington, D. C.: United States Forest Service.

———. 1971. *The demand and price situation for forest products 1970–71.* Miscellaneous Publication no. 1195. Washington, D. C.: United States Forest Service.

U. S. Department of Commerce. 1969. *Business statistics.* (Published biannually.) Washington, D. C.: Office of Business Economics.

———. 1970. *Statistical abstract of the United States 1970.* Bureau of the Census, Washington, D. C.: United States Government Printing Office.

———. 1971. *Survey of current business.* Vol. 51, no. 4. Washington, D. C.: Office of Business Economics.

U. S. Department of the Interior. 1970. *Federal aid in fish and wildlife restoration, 1969.* Annual Report on Dingell-Johnson and Pittman-Robertson Programs for the Fiscal Year Ending June 30, 1969. Washington, D. C.: Wildlife Management Institute.

U. S. Treasury Department. 1969. *Your Federal income tax.* 1969 ed. Publication no. 17. Washington, D. C.: Internal Revenue Service.

U. S. Tariff Commission. (n.d.) *The production, importation, and marketing of hardwood plywood in the United States.* Escape Clause Investigation no. 39.

U. S. Water Resources Council. 1968. *The nation's water resources.* Washington, D. C.: Water Resources Council.

Viner, J. 1932. Cost Curves and Supply Curves. *Zeitschrift Fur Nationalokonomie* 3: 1932. Reprinted in Clemence, R. V. *Readings in Economic Analysis.* Vol. 2, pp. 8–35. Cambridge: Addison-Wesley Press.

Wallis, J. R. 1965. *A factor analysis of soil erosion and stream sedimentation in northern California.* Ph.D. dissertation, University of California.

Ward, B. 1962. *The rich nations and the poor nations.* New York: W. W. Norton and Co.

Weisbrod, B. 1964. Collective consumption services of individual consumption goods. *Quarterly Journal of Economics* 78:471–77.

Westoby, J. C. 1962. *The role of forest industries in the attack on economic development.* Reprinted from *The State of Food and Agriculture, 1962,* pp. 8–128. Rome: FAO.

———. 1966. *World trends in wood resources and requirements.* The Introductory Statement to Plenary I. Madrid: World Forestry Congress.

———. 1968. *The forester as agent of change.* Paper read at the International Conference on the Role of the Professional as an Agent of Economic, Social, and Political Change in Low Income Countries, May 24–26, at University of California, Berkeley.

WICKSTEED, P. H. 1933. *The common sense of political economy.* Edited by Lionel Robbins, Vol. 1. London: George Routledge & Sons.

WILLIAMS, E. T. 1963. *The timber owner and his Federal income tax.* Agriculture Handbook no. 274. Washington, D. C.: United States Forest Service.

WORRELL, A. C.; LOCKWOOD, A. N.; and LAURIDSEN, M. L. 1963. *A general review of U. S. Forest Service timber appraisal policies and procedures.* Report of the Timber Appraisal Review Committee. Washington, D. C.: United States Forest Service.

WRIGHT, M. G., and PHELPS, R. B. 1967. *Particleboard, insulation board, and hardboard.*

YOHO, J. 1965. Economic impact of local taxes and death taxes in timber resources. *Timber Tax Journal* 1:50–65.

ZAREMBA, J. 1963. *Economics of the American lumber industry.* New York: Robert Speller.

ZIVNUSKA, J. A. 1955. Supply, demand, and the lumber market. *Journal of Forestry* 53:547–53.

———. 1963. The future for wood in a competitive market. Paper read at a joint meeting of the Puget Sound and Columbia River Sections, Society of American Foresters, Longview, Washington.

# Index